Electron Microscopy

SECOND EDITION

METHODS IN MOLECULAR BIOLOGY™

John M. Walker, SERIES EDITOR

METHODS IN MOLECULAR BIOLOGY™

Electron Microscopy

Methods and Protocols

SECOND EDITION

Edited by

John Kuo

Centre for Microscopy and Microanalysis
The University of Western Australia
Crawley, Western Australia, Australia

HUMANA PRESS ✳ TOTOWA, NEW JERSEY

© 2007 Humana Press Inc.
999 Riverview Drive, Suite 208
Totowa, New Jersey 07512

www.humanapress.com

All papers, comments, opinions, conclusions, or recommendations are those of the author(s), and do not necessarily reflect the views of the publisher.

This publication is printed on acid-free paper. ∞
ANSI Z39.48-1984 (American Standards Institute) Permanence of Paper for Printed Library Materials.

Cover illustration: *Foreground (left to right):* Fig. 3, Chapter 22, courtesy of Saara Mansouri; Fig. 2B, Chapter 22, courtesy of William Kournikakis; and Fig. 5, Chapter 22, courtesy of Steve Schmitt. *Background*: E. coli GroEL using the negative staining-carbon film procedure, courtesy of Robin Harris.

For additional copies, pricing for bulk purchases, and/or information about other Humana titles, contact Humana at the above address or at any of the following numbers: Tel.: 973-256-1699; Fax: 973-256-8341; E-mail: orders@humanapr.com; or visit our Website: www.humanapress.com

Photocopy Authorization Policy:

Printed in the United States of America. 10 9 8 7 6 5 4 3 2 1

eISBN 1-59745-294-7

ISSN 1064-3745

ISBN 13: 978-1-58829-573-6

Library of Congress Cataloging-in-Publication Data

Electron microscopy : methods and protocols. — 2nd ed. / edited by John Kuo.
    p. ; cm. — (Methods in molecular biology ; v. 369)
  Rev. ed. of: Electron microscopy methods and protocols / edited by
M.A. Nasser Hajibagheri. c1999.
  Includes bibliographical references and index.
  ISBN-10: 1-58829-573-7 (alk. paper)
  1. Electron microscopy—Laboratory manuals. 2.
Histology—Technique. I. Kuo, John. II. Electron microscopy methods
and protocols. III. Series: Methods in molecular biology (Clifton,
N.J.) ; v. 369.
  [DNLM: 1. Microscopy, Electron—Laboratory Manuals. 2.
Histocytological Preparation Techniques—Laboratory Manuals. W1
ME9616J v.369 2007 / QS 525 E38 2007]
  QH212.E4E39824 2007
  570.28'25—dc22

Preface

This second edition of *Electron Microscopy: Methods and Protocols* is written for established researchers as well as new students in the field of molecular biology. It is not only for biomedical but also for general biological science research and its application. The combination of microscopical and chemical analyses has provided the basis for our current understanding of cell biology. The high-resolution electron microscope with its associated equipment can serve as a powerful tool for analyzing molecular structure, interactions, and processes.

The last decade not only saw the ever-expanding development of electron microscopy and microanalytical techniques but also witnessed their widespread application in biomedical laboratories and the molecular biology and plant science fields. Excitingly, electron microscopy techniques such as transmission electron microscopy (TEM) crystallography, cryo-TEM, and cryo-TEM tomography have been used in the study of membrane proteins, macromolecules, and 3D virus imaging, together with electron energy-loss spectroscopy (EELS) for elemental analysis in biological specimens. Scanning electron microscopy and X-ray microanalytical attachments have also been applied in molecular biological research for many years. Furthermore, ion mass spectrometry for elemental analysis and mapping used in non-biological fields has also been utilized in biological tissues and cells. All these electron microscopical and microanalytical techniques are major additions in the second edition of *Electron Microscopy: Methods and Protocols*. In this edition, some of the established conventional procedures and techniques of electron microscopy have been modified and fine tuned.

The second edition now consists of two main areas: the first relates to TEM, and the second covers scanning electron microscopy (SEM) and mass spectrometry (MS). The TEM area comprises several sections:

1. Conventional and microwave-assisted specimen preparation methods for cultured cells and biomedical and plant tissues, for the benefit of those who only require routine TEM images through ultramicrotomy sectioning and then positive staining.
2. Cryo-specimen preparation by high-pressure freezing and cryoultramicrotomy.
3. Negative staining and immunogold labeling techniques for samples prepared through conventional, cryosectioning, or high-pressure freezing methods. Some of these chapters are presented in correlative approaches using TEM with fluorescent or confocal microscopy. Quantitative aspects of immunogold labeling in resin embedded samples are also included.

4. TEM crystallography and cryo-TEM tomography for the study of membrane proteins, macromolecules, organelles, and cells. One chapter on the application of EELS to biomedical tissue is included.

For the SEM area, conventional-, variable pressure-, environmental-, and cryoscanning microscopy techniques are covered. The application of X-ray microanalysis in SEM and mass spectrometry in elemental and molecular mapping concludes this volume.

The methods and protocols presented in the first edition were designed to be general in their application. The accompanying Notes provided the reader with invaluable assistance in adapting or troubleshooting the protocols for transmission electron microscopy. This strength remains in this revised and expanded second edition. Our aim is to provide the reader with self-explanatory, practical instructions on how to process biological specimens and, to some extent, offer a discussion on the principles underlying the various processes. Here, in the new edition of *Electron Microscopy: Methods and Protocols,* each chapter is again provided by experts with first-hand experience and in-depth knowledge of the techniques that they describe.

John Kuo

Acknowledgments

It was my great honor to have been introduced to the fascinating world of microscopy by Professor Margaret E. McCully, and to the field of structure-functional biology by Professor John S. Pate. It is my great pleasure to thank Dr. Martin Mueller and Dr. Ralf Stezler for their introduction to cryo-EM techniques.

I am grateful to John Bozzola, Kent McDonald, Jose Mascorro, Martin Saunders, Richard Stern, and Paul Webster for their recommendations on chapter topics and chapter authors for this second edition. Special thanks to John Murphy for his ever-reliable computer skills and support of the work done at the Centre for Microscopy and Microanalysis, as well as to my University of Western Australia colleagues for their encouragement. I am very grateful to Professor John M. Walker for his continuous guidance and help throughout the entire editing processes, as well as to the capable staff of Humana Press for their professional work on the publication of *Electron Microscopy: Methods and Protocols, Second Edition.*

John Kuo

Contents

Contributors

JOHN J. BOZZOLA • *IMAGE Center, Southern Illinois University, Carbondale, IL*

WILLIAM (BILL) BRIEHER • *Department of Systems Biology, Harvard Medical School, Boston, MA*

WAH CHIU • *National Center For Macromolecular Imaging, Verna and Marrs Mclean Department of Biochemistry and Molecular Biology, Baylor College of Medicine, Houston, TX*

HUI-TING CHOU • *Molecular & Cellular Biology, University of California, Davis, CA*

DUŠAN CMARKO • *Institute of Cellular Biology and Pathology, 1st Faculty of Medicine, Charles University, Prague, Czech Republic*

MARGARET (PEG) COUGHLIN • *Department of Systems Biology, Harvard Medical School, Boston, MA*

ANCA DRAGOMIR • *Department of Medical Cell Biology, University of Uppsala, Sweden*

E. ANN ELLIS • *Microscopy and Imaging Center, Texas A & M University, College Station, TX*

PETER ENGELHARDT • *Haartman Institute, Department of Pathology and Virology, University of Helsinki; Laboratory of Computational Engineering, Helsinki University of Technology, Helsinki, Finland*

JAMES E. EVANS • *Molecular and Cellular Biology, University of California, Davis, CA*

BEAT FREY • *Soil Ecology, Swiss Federal Research Institute, Birmensdorf, Switzerland*

KIICHI FUKUI • *Department of Biotechnology, Graduate School of Engineering, Osaka University, Osaka, Japan*

PETER GEHR • *Institute Anatomy, University of Bern, Bern, Switzerland*

MARIANNE GEISER • *Institute Anatomy, University of Bern, Bern, Switzerland*

BRENDAN J. GRIFFIN • *Centre For Microscopy and Microanalysis, The University of Western Australia, Crawley, Western Australia, Australia*

NICOLE GRIGNON • *Biochimie et Physiologie Moléculaire des Plantes, Centre National de la Recherche Scientifique, Institut National de la Recherche Agronomique, Ecole Nationale Supérieure Agronomique de Montpellier, Université Montpellier 2, Montpellier, France*

HERB HAGLER • *Department of Pathology, University of Texas, Southwestern Medical Center, Dallas, TX*

JEFF D. HARDIN • *Zoology Department, University of Wisconsin, Madison, WI*

J. ROBIN HARRIS • *Institute of Zoology, University of Mainz, Mainz, Germany*

CHYONG-ERE HSIEH • *Resource For Visualization of Biological Complexity, Wadsworth Center, Empire State Plaza, Albany, NY*

BRUNO M. HUMBEL • *Electron Microscopy and Structural Analysis, Cellular Architecture and Dynamics, Department of Biology, Faculty of Sciences, Utrecht University, Utrecht, The Netherlands*

WEN JIANG • *Department of Biological Sciences, Purdue University, West Lafayette, IN*

NADINE KAPP • *Department of Veterinary Anatomy, Veterinary School, University of Bern, Switzerland*

KAREL KOBERNA • *Department of Cell Biology, Institute of Experimental Medicine and Institute of Physiology, Academy of Sciences of the Czech Republic, Prague, Czech Republic*

JOHN KUO • *Centre For Microscopy and Microanalysis, The University of Western Australia, Crawley, Western Australia, Australia*

VELI-PEKKA LEHTO • *Haartman Institute, Department of Pathology, University of Helsinki, Finland*

NICHOLAS P. LOCKYER • *Surface Analysis Research Centre, School of Chemical Engineering and Analytical Science, University of Manchester, Manchester, UK*

MICHAEL MARKO • *Resource for Visualization of Biological Complexity, Wadsworth Center, Empire State Plaza, Albany, NY*

KENT L. McDONALD • *Electron Microscope Laboratory, University of California, Berkeley, CA*

TERRY M. MAYHEW • *Centre for Integrated System Biology and Medicine, School of Biomedical Sciences and Institute of Clinical Research, University of Nottingham, Nottingham, UK*

JOSÉ A. MASCORRO • *Department of Structural and Cellular Biology, Tulane University Health Science Center, New Orleans, LA*

JARI MERILÄINEN • *Haartman Institute, Department of Pathology, University of Helsinki, Helsinki, Finland*

MARY MORPHEW • *The Boulder Laboratory For 3-D Electron Microscopy of Cells, University of Colorado, Boulder, CO*

THOMAS MÜLLER-REICHERT • *Electron Microscope Facility, Max Planck Institute of Molecular Cell Biology and Genetics, Dresden, Germany*

RYOMA (PUCK) OHI • *Department of Systems Biology, Harvard Medical School, Boston, MA*

GODFRIED M. ROOMANS • *Department of Medical Cell Biology, University of Uppsala, Uppsala, Sweden*

YUZOU SANO • *Laboratory of Wood Biology, Graduate School of Agriculture, Hokkaido University, Sapporo, Japan*

HEINZ SCHWARZ • *Max-Planck-Institut für Entwicklungsbiologie, Tuebingen, Germany*

PAUL A. SIMS • *Zoology Department, University of Wisconsin, Madison, WI*

HENNING STAHLBERG • *Molecular and Cellular Biology, University of California, Davis, CA*

DANIEL STUDER • *Institut für Anatomie, University of Bern, Bern, Switzerland*

LUCA STUDER • *Institut für Anatomie, University of Bern, Bern, Switzerland*

SUSUMU UCHIYAMA • *Department of Biotechnology, Graduate School of Engineering, Osaka University, Osaka, Japan*

YASUHIRO UTSUMI • *Shiiba Research Forest, Graduate School of Agriculture, Kyushu University, Miyazaki, Japan*

DIMITRI VANHECKE • *Institut für Anatomie, University of Bern, Bern, Switzerland*

PAUL VERKADE • *Electron Microscope Facility, Max Planck Institute of Molecular Cell Biology and Genetics, Dresden, Germany*

ALEXANDRE WEBSTER • *Department of Molecular, Cellular, and Developmental Biology, University of California, Santa Barbara, CA*

PAUL WEBSTER • *Ahmanson Advanced Electron Microscopy and Imaging Center, House Ear Institute, Los Angeles, CA*

FANG ZHAO • *Haartman Institute, Department of Pathology, University of Helsinki, Finland*

1

Conventional Specimen Preparation Techniques for Transmission Electron Microscopy of Cultured Cells

John J. Bozzola

Summary

This chapter covers the conventional methods and considerations for preparing cultured cells for examination in the transmission electron microscope. Techniques for handling cells grown in liquid culture, as well as on substrates such as culture dishes or agar, are discussed. Directions are given on how to prepare the most commonly used buffers, fixatives, enrobement media, and embedding resins. These methods may be applied to most cultured organisms, from bacteria to mammalian cells.

Key Words: Transmission electron microscopy; cultured cells; preparative procedures; bacteria; eukaryotic cells.

1. Introduction

Because cultured cells for transmission electron microscopy (TEM) may be grown either in suspension in a liquid medium, or on a substrate such as agar or plastic, detailed methods for handling the cells will vary. Nonetheless, the basic steps are the same: fixation in buffered aldehyde, postfixation in osmium tetroxide, dehydration in ethanol, embedding in plastic, sectioning, staining, and examination using TEM. Some general references are listed at the end of this chapter (*1–4*).

A wide variety of cells, ranging from prokaryotic bacteria to mammalian cells, may be grown in culture. Consideration must be given, therefore, to the types of chemical fixatives and buffer combinations that will be used. An aldehyde fixative such as glutaraldehyde, perhaps augmented with formaldehyde, generally is used as the initial step. Aldehydes penetrate cultured cells rapidly, crosslink proteins, and halt the dynamics of the living cell. Although many enzymes will remain active and immunological reactivity may be preserved after treatment with aldehyde, motility and other processes will be stopped and

From: *Methods in Molecular Biology, vol. 369*
Electron Microscopy: Methods and Protocols, Second Edition
Edited by: J. Kuo © Humana Press Inc., Totowa, NJ

the cell is considered fixed or dead. The aldehydes preserve mainly proteins and other macromolecules associated with proteins, such as lipoproteins and histoproteins associated with DNA. Glycogen may be preserved, but the majority of carbohydrates will be extracted during the processing of the cells.

To augment the fixation, cells are subsequently treated, or postfixed, with a solution of osmium tetroxide. Although slower to penetrate the cell than aldehydes, osmium tetroxide is a strong oxidizer that reacts vigorously with the double bonds of unsaturated lipids. Osmium solutions also react slowly with proteins, including the histoproteins, hence helping to preserve associated DNA. Carbohydrates are not preserved. After the oxidative reaction, the heavy metal osmium is reduced onto the macromolecules, thereby enhancing contrast when viewed in the TEM. Enzymatic activity and most immunological reactivity are destroyed. Because the cells are considerably hardened after osmication, they are brittle and easily damaged by rough treatment during pipetting or centrifugation.

Several different buffers may be used to protect the cells during the fixation process. The most commonly used buffers are phosphate, cacodylate and organic buffers such as PIPES (i.e., 1,4 piperazine bis [2-ethanosulfonic acid] *[5]*). Buffers are not entirely innocuous, however, and some fine detail may be altered. For example, high concentrations of phosphates may damage mitochondria; consequently, one should keep the buffer concentration as low as possible while still maintaining the pH within the desired range. Phosphate and cacodylate buffers predominate in electron microscopy, but the organic buffers should be strongly considered as a replacement because they have fewer detrimental effects on fine structure and are nontoxic. Organelles such as microfilaments and microtubules are better preserved, and the increase in the overall density of the cell suggests that extraction is minimized with the organic buffers *(6)*.

After fixation and postfixation in buffered systems, the preserved cells are dehydrated, usually in ethanol, by gradually replacing the water with an increasing concentration of ethanol. After the cells reach 100% ethanol, they are infiltrated with a plastic monomer, most commonly an epoxy resin. Ultimately, the epoxy is polymerized, resulting in a chemically preserved cell completely filled and surrounded by a hard plastic. It is then possible to cut ultrathin sections of the plastic-embedded cells and stain them for contrast and examination in the TEM as described in other chapters in this book.

2. Materials

2.1. Equipment

1. Graduated cylinders (100 mL, 10 mL).
2. Disposable polypropylene tubes (50 mL, 10 mL).
3. Glass Erlenmeyer flask, 100 mL.

Table 1
Preparation of PIPES, 0.2 *M* Buffer

pH	6.4	6.6	6.8	7.0	7.2	7.4	7.6
N NaOH,mL	5.0	7.5	10.0	12.2	14.4	16.2	18.2

To 100 mL of 0.2 *M* PIPES solution (6.1 g PIPES/100 mL distilled water), add the appropriate amount of *N* NaOH (4.0 g/100 mL distilled water) to obtain the desired pH. The final working dilution of this buffer is 0.1 *M*.

4. Glass pipets (5 mL, 1 mL) and rubber bulb.
5. Calibrated pH meter.
6. Conventional centrifuge or, preferably, high speed microcentrifuge.
7. Centrifuge tubes (conical, 10 mL) or microcentrifuge tubes.
8. Top loading balance for measuring buffer salts and resin components.
9. Hot plate with magnetic stirrer and stirring bar.
10. Glass stirring rods, magnetic stir bars.
11. Wide-mouthed glass bottle with tight lid.
12. Polypropylene embedding capsules and holder.
13. Oven adjusted to 60°C.
14. Culture vessels (Petri dishes, plastic flasks, culture tubes).
15. Dissecting needle, wooden applicator sticks.
16. Hot water bath adjusted to 40°C or 60°C, if agarose or agar enrobement is used.

2.2. Buffers

The two most commonly used buffers are Sorenson's phosphate and sodium arsenate or cacodylate. PIPES buffer is slowly replacing these buffers and offers the advantages of being nontoxic and easier to prepare. The buffers are used at a final, working dilution of 0.1 *M*.

1. PIPES buffer: 0.2 *M* PIPES, pH adjusted with 1 *N* NaOH. PIPES buffer is prepared using 1,4 piperazine bis (2-ethanosulfonic acid), with a formula weight of 302.4 g/mol. To prepare a 0.2 *M* stock solution, dissolve 6.1 g of PIPES in 100 mL of distilled water. Adjust the 100 mL of PIPES solution to the desired pH using N NaOH (4.0 g/100 mL) as shown in **Table 1** (*see* **Note 1**).
2. Sorenson's phosphate buffer: 0.2 *M* NaH_2PO_4 and 0.2 *M* Na_2HPO_4 combined in the appropriate amounts to obtain the desired pH. Sorenson phosphate buffer consists of two parts: 0.2 *M* monobasic sodium phosphate, NaH_2PO_4, and 0.2 *M* dibasic sodium phosphate, Na_2HPO_4. The weight of each component will vary depending on the waters of hydration. The pH is adjusted by mixing the two components as shown in **Table 2**.
3. Cacodylate buffer: 0.2 *M* $(CH_3)_2AsO_2Na$ adjusted to the proper pH using 0.2 *N* HCl. Cacodylate buffer consists of a 0.2 *M* stock solution of sodium cacodylate in distilled water (4.28 g/100 mL) and the pH is adjusted by adding the appropriate

Table 2
Preparation of Sorenson's, 0.2 *M* Phosphate Buffer

pH	A, mL	B, mL	pH	A, mL	B, mL
6.0	87.7	12.3	7.0	39.0	61.0
6.2	81.5	18.5	7.2	28.0	72.0
6.4	73.5	26.5	7.4	19.0	81.0
6.6	62.5	37.5	7.6	13.0	87.0
6.8	51.0	49.0	7.8	8.5	91.5

Solution A: monobasic sodium phosphate (2.76 g of $NaH_2PO_4 \cdot H_2O$ or 3.12 g of $NaH_2PO_4 \cdot 2H_2O$ in 100 mL of distilled water).

Solution B: dibasic sodium phosphate (2.84 g of Na_2HPO_4 or 5.36 g of $Na_2HPO_4 \cdot 7H_2O$ or 7.17 g of $Na_2HPO_4 \cdot 12H_2O$ in 100 mL of distilled water).

Table 3
Preparation of Cacodylate, 0.2 *M* Buffer

pH	6.0	6.2	6.4	6.6	6.8	7.0	7.2	7.4
0.2 *M* HCl, mL	59.2	47.6	36.6	26.6	18.6	12.6	8.4	5.5

To 100 mL of 0.2 *M* cacodylate solution (4.28 g/100 mL distilled water), add the appropriate amount of 0.2 *N* HCl (1.7 mL concentrated HCl/100 mL distilled water) to obtain the desired pH.

volume of 0.2 *M* HCl (1.7 mL concentrated HCl/100 mL distilled water) to the 100 mL stock as shown in **Table 3**. (**CAUTION:** cacodylate buffer contains arsenic, poisonous, carcinogenic, substances that can be absorbed through the skin; wear gloves.)

2.3. Miscellaneous Reagents

1. Loose specimens may be enrobed in an agar enrobement medium. This is prepared by boiling 4 g agar in 100 mL of distilled water or buffer until the agar is completely dissolved and the solution is no longer turbid. Once dissolved, store in a 60°C oven in a capped container until needed. It will be useable for several days. Preferably, 4% agarose should be used to enrobe the cells since it will dissolve in water or buffer at 45°C. This much lower temperature will have fewer damaging effects on cells.
2. Specimens can also be enrobed in a solution of 10% gelatin. This is freshly prepared by dissolving plain gelatin (purchased at any food store) in boiling water or buffer. This may be stored at room temperature until needed. Because gelatin solutions are used at room temperature, they are less likely to damage cells.
3. 1.0 *M* $CaCl_2$ (1.1 g/10 mL distilled water).
4. 0.2 *M* HCl (1.7 mL concentrated HCl/100 mL distilled water).

2.4. Fixatives (Use Electron Microscopy or Analytical-Grade Reagents)

(**CAUTION:** Fixatives are poisonous irritants; work in a fume hood and wear gloves.)

1. Electron microscopy grade glutaraldehyde, 25%, in sealed ampoules. Glutaraldehyde fixatives are easily prepared from 25% solutions of electron microscopy grade glutaraldehyde in sealed glass ampoules by making a 1:10 dilution in the buffer of choice.

    50 mL   0.2 *M* buffer stock solution at proper pH
    10 mL   25% glutaraldehyde (electron microscopy grade)
    2 mL   0.1 *M* $CaCl_2$ (**IMPORTANT:** do not use $CaCl_2$ with phosphate buffer as a precipitate will form)
    40 mL   distilled water

2. Glutaraldehyde–formaldehyde fixative. Glutaraldehyde–formaldehyde mixtures often are used on hard to fix cells or in specimens for which a fixation protocol is not available. Formaldehyde fixatives are used in combination with glutaraldehyde since formaldehyde alone does not adequately preserve ultrastructural detail. A typical fixative consists of 4% formaldehyde–2.5% glutaraldehyde in the buffer of choice. Formaldehyde fixatives should be freshly prepared by depolymerizing paraformaldehyde powder using N NaOH (*see* **Note 2**). (**CAUTION:** irritating fumes can be produced; work under a fume hood). Formaldehyde Stock Solution: Prepare an 8% aqueous stock solution of formaldehyde by placing 50 mL of distilled water in a 100-mL flask with a stirring bar. Add 4.0 g of electron microscopy-grade paraformaldehyde powder and stir the mixture while slowly increasing the temperature to 70 to 85°C. Most of the powder will depolymerize, but it is usually necessary to add several drops of N NaOH. After stirring for several minutes, the solution should be clear.

    Glutaraldehyde/formaldehyde buffered fixative:

    40 mL   0.2 *M* buffer stock solution at proper pH
    10 mL   25% glutaraldehyde (electron microscopy grade)
    2 mL   0.1 *M* $CaCl_2$ (**IMPORTANT:** do not use $CaCl_2$ with phosphate buffer as a precipitate will form)
    50 mL   8% aqueous formaldehyde stock solution

3. Osmium tetroxide fixative. Osmium tetroxide is used as a secondary fixative after aldehyde fixation because its rate of penetration is too slow to prevent artifacts if used initially. (**CAUTION:** osmium tetroxide is toxic, and the volatile fumes are very corrosive, especially to mucous membranes. It is essential that osmium solutions are handled in a fume hood and used osmium solutions be disposed of properly; *see* **Note 3**).

    Prepare a 2% aqueous solution (1 g of osmium tetroxide in 50 mL of distilled water). The sealed glass ampoule containing the electron microscopy grade osmium tetroxide must be scrupulously clean, as should all glass items contacting osmium (*see* **Note 4**).

A working fixative is prepared just before use by mixing equal parts of 2% aqueous stock osmium tetroxide solution with an equal part of 0.2 *M* buffer.

2.5. Dehydration Series

CAUTION: flammable; propylene oxide is carcinogenic.

1. 25, 50, 75, 95% ethanol in distilled water.
2. 100% (absolute) ethanol in sealed pint containers (*see* **Note 5**).
3. Propylene oxide or acetonitrile may be used as an intermediate solvent with cell pellets before embedding in plastic (*see* **Note 6**).

2.6. Embedding Resins

CAUTION: both are hyper-allergenic and VCHD is carcinogenic.

1. Epoxy resin 812. Epoxy resin 812 consists of epoxy resin (originally designated Epon 812), the hardeners dodecenylsuccinic anhydride and methyl Nadic anhydride, and an accelerant such as benzyldimethylamine (BDMA) or 2,4,6-Tris(dimethylaminomethyl)phenol. Prepare epoxy 812 resin embedding medium by pouring measured amounts (usually volumes) into a graduated, disposable polypropylene tube such as a 50-mL centrifuge tube with a tight sealing cap. A mixture of medium hardness consists of the following components:

Epoxy resin 812	20 mL	(24.0 g)
Dodecenylsuccinic anhydride	16 mL	(16.0 g)
NMA	8 mL	(10.0 g)
BDMA	1.3 mL	(1.5 g)

   Mix the resins thoroughly to obtain satisfactory results. The best way to accomplish this is to place the resin components, uncapped, in a 60°C oven for 15 to 20 min, to reduce the viscosity, as recommended by Glauert and Lewis *(4)*. After measuring the warmed ingredients into a 50-mL polypropylene tube, tighten the cap and invert the tube end over end continuously for 5 min (*see* **Note 7**).
2. Spurr's resin. The classical formulation of Spurr's resin consists of vinylcyclohexene dioxide (VCHD), a diglycidyl ether of polypropylene glycol (DER 736, Dow Epoxy Resin 736), nonenylsuccinic anhydride, and an accelerant such as BDMA or dimethylaminoethanol. Several years ago, VCHD, a known carcinogen, stopped being produced. The substitute for VCHD is ERL-4221, a cycloaliphatic epoxide, used in exactly the same way as VCHD.

   Spurr's embedding medium is prepared by weighing components in a 50-mL graduated centrifuge tube on a top-loading balance. Generally, a firm mix is prepared by warming the components as described for the epoxy 812 resin. Prepare Spurr's resin of firm hardness resin as follows:

VCHD (or ERL-4221) resin	10.0 g
DER 736	6.0 g
Nonenylsuccinic anhydride	26.0 g
Dimethylaminoethanol	0.4 g

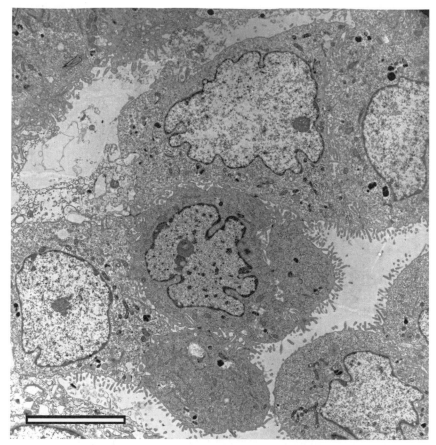

Fig. 1. Mammalian tissue culture cells after centrifugation to form a compact pellet before processing for TEM. Cells were prepared as described in **Subheading 3.1.1.** Bar = 7.5 μm.

Spurr's resin is very sensitive to traces of moisture; therefore, anhydrous conditions must be maintained (keep containers capped, including embedding capsules) or an unusable, brittle plastic will be produced.

3. Methods

3.1. Preparation of Cultured Eukaryotic Cell Suspensions

Eukaryotic cells grown as suspensions are consolidated (rather than handled as suspensions) during processing for TEM. This is accomplished either by centrifugation to form a pellet (*see* **Fig. 1**) or by enrobing the cells in agar or gelatin.

3.1.1. Centrifugation Procedure

For the centrifugation procedure, prepare an initial fixative by mixing one part of either glutaraldehyde or glutaraldehyde/formaldehyde fixative (warmed to the same temperature as the culture) with an equal part of the cell culture. A recommended fixative is 4% formaldehyde/2.5% glutaraldehyde in PIPES buffer.

1. Centrifuge the culture/fixative mix at room temperature to generate a cell pellet. Excessive centrifugation must be avoided, but the pellet should be compact enough to remain a firm mass. Typically, mammalian cells are centrifuged at 1,000–2,000 rcf (relative centrifugal force (*see* **Note 8**) for 5 to 10 min.
2. Carefully remove the supernatant and gently flow several milliliters of pure buffered fixative, at room temperature, on top of the pellet taking care not to disturb the pellet.
3. After 15 min, gently dislodge the pellet with either a pointed spatula or sharpened wooden applicator stick so that the opposing side of the pellet is exposed to the fixative for an additional 15 min at room temperature.
4. Remove the fixative with a pipet and replace with 0.1 *M* buffer.
5. Transfer the pellet into a Petri dish by flowing buffer into the tube and allow the intact pellet to move into a pool of buffer in the dish.
6. With a clean razor blade or scalpel, cut the pellet into small cubes no larger than 1 to 2 mm.
7. Transfer the cubes into a small vial containing several milliliters of buffer for 10 min. This is easily accomplished by tilting the Petri dish so that the cubes move into a common location. Lift and transfer the cubes with a sharpened applicator stick or piece of filter paper cut into a sharp point.
8. After two additional changes of buffer, each for 10 min, postfix the cubes in 1 to 2 mL of 1% buffered osmium tetroxide for 1 h at room temperature.
9. Rinse in distilled water and dehydrate the cubes in an ethanol series (25, 50, 75, 95% ethanol, each for 10 min).
10. As the specimens are undergoing dehydration, prepare two mixtures of propylene oxide:epoxy 812 embedding medium consisting of 1:1 and 1:3 parts, each. Use 10 mL of graduated, disposable polypropylene tubes with tight fitting lids. The propylene oxide:epoxy 812 embedding medium is best mixed by gentle aspiration and stirring with a pipet rather than inversion of the tubes because the caps may be dislodged as the result of pressures inside the sealed tube. Acetonitrile is a safer alternative than propylene oxide *(7)*. After the two mixtures are prepared, store capped until needed in **step 12.**
11. After three changes of absolute ethanol, 10 min each, use three changes of propylene oxide (or acetonitrile), each for 15 min.
12. Remove most of the propylene oxide from the specimen cubes but leave a trace to prevent the specimens from drying out. Place 1 to 2 mL of 1:1 propylene oxide: epoxy 812 embedding medium over the specimen cubes and gently rotate the capped container five to six times during a period of 20 min.

13. Remove the 1:1 mixture and replace with 1:3 mixture of propylene oxide:epoxy medium and rotate the capped container five to six times during a period of 45 to 60 min.
14. Remove the 1:3 mixture and pour on several milliliters of pure epoxy 812 embedding medium and rotate the uncapped container five to six times during a period of 90 min. The specimens should gradually sink to the bottom of the tube. If not, leave them overnight, uncapped in a dry environment.
15. Fill several polypropylene embedding capsules half-way with epoxy 812 embedding medium. Hold the capsules upright in commercially supplied holders or produce a capsule holder by puncturing holes in a small cardboard box. Dislodge any air bubbles that form at the tip of the capsule using the sharp tip of a broken, wooden applicator stick or a dissecting needle. Use the wooden applicator stick to lift and transfer individual specimen cubes into the capsules. Allow the specimens to sink to the bottom tip of the capsules. This may take an hour or so. We often leave the capsules overnight at room temperature.
16. Place an identifying label in each capsule. Use pencil (ink will dissolve in resin) and small pieces of paper approximately 5×20 mm. Place the labels inside the capsules so that the numbers are legible. The easiest way is to roll the label and insert it so it is pressed against the inside wall of the capsule, the numbers facing out.
17. Transfer the capsules into a 60°C oven for 24 to 48 h until the epoxy resin is completely hardened and ready to section.

3.1.2. The Agar-Enrobement Procedure

This procedure should be used with suspension cell cultures that do not form a firm pellet (*see* **Fig. 2**).

1. Mix one part of 4% formaldehyde/2.5% glutaraldehyde in PIPES buffer (warmed to the same temperature as the culture) with an equal part of the cell culture.
2. Lightly centrifuge the culture/fixative mix at room temperature to generate a very loose cell pellet. Typically, mammalian cells would be spun at 400 to 500 rcf for several minutes.
3. Carefully remove the supernatant and gently flow several milliliters of the buffered glutaraldehyde/formaldehyde fixative, at room temperature onto the loose pellet of cells. Do not be concerned if the cells are re-suspended.
4. After 15 min, lightly centrifuge the cells to form a loose pellet and remove the fixative using a pipet, taking care not to aspirate the cells.
5. Using a warmed plastic pipet, quickly transfer approx 50 μL of warm agarose (or agar) onto the loose cells and gently stir the cells with the tip of the plastic pipet (*see* **Note 9**) to suspend the cells in the warm agarose. Do not use too much agarose or too few cells will be enrobed for viewing in the TEM. It is also possible to centrifuge the cells, while in molten agarose, to form a concentrated pellet (*see* **Note 10**).
6. Transfer the tube containing the cells suspended in agarose to an ice bath, or place in a refrigerator until the agarose solidifies (15 min). Do not touch or disturb the tube during the solidification process, or it will not harden properly.

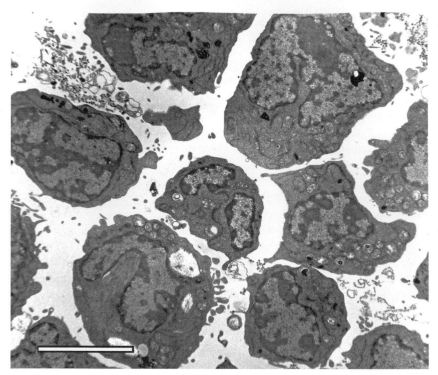

Fig. 2. Mammalian tissue culture cells after enrobement in agarose as described in **Subheading 3.1.2.** The cells were concentrated by centrifugation in agarose before gel formation. Note that cells are closely placed but not touching. Bar = 15 μm.

7. Place several milliliters of buffer into the tube containing the solidified agarose and use a spatula to dislodge the agarose plug containing the cells. The agarose may break into several pieces at this point but it should not crumble into many tiny pieces. If the latter occurs, the agarose was either diluted by excessive liquid in the cell culture or the cells were not adequately mixed into the agarose. Salvage the larger pieces for further processing.
8. Transfer the agarose pieces containing the enrobed cells into a Petri dish containing buffer and trim the agarose into 1- to 2-mm cubes using a sharp razor blade.
9. Follow **steps 7–17** as described in **Subheading 3.1.1.**

3.1.3. The Gelatin Enrobement Procedure

Cells can also be enrobed in 10% gelatin solutions instead of agarose. Although more steps are involved, gelatin-enrobement is advantageous because cells are handled at much lower temperatures. The procedure is identical to the procedure in the previous section with a few exceptions:

1. Follow **steps 1–5** in **Subheading 3.1.2.** with the exception that 10% gelatin, at room temperature, is used instead of agar.

2. After suspending the cells in 50 μL of gelatin at room temp, centrifuge the cells to form a compact pellet and transfer the tube into a refrigerator or ice bath until the gelatin becomes firm.
3. Overlay the gelatin with several milliliters of aldehyde fixative for 20 min at room temperature. This will permanently cross link the gelatin so that it can be treated like a cell pellet as described in **Subheading 3.1.1.**
4. Follow **steps 3** through **17** in **Subheading 3.1.1.**

3.2. Preparation of Cultured Bacterial Cell Suspensions

3.2.1. Culture Suspensions

Bacterial cells grown as suspensions are handled and enrobed in agarose (*see* **Fig. 3**) in the same way as eukaryotic cells described in **Subheading 3.1.2.** with the exception that longer fixation times and higher centrifugation speeds are needed to pellet the cells, as follows.

1. Centrifugation forces of 7,000 to 10,000 rcf for 5 to 10 min are needed to sediment bacterial cells. Although bacteria are sturdier than eukaryotic cells and can be taken up by pipets, they should be resuspended gently using pipets with wide bores (enlarged by heating or cutting with a blade).
2. Longer fixation times are needed for the 4% formaldehyde/2.5% glutaraldehyde in PIPES buffer. After mixing equal parts of the fixative with cell culture and centrifugation, re-suspend the cells in fixative and keep at room temperature from 4 hr to overnight, depending on the organism.
3. After rinsing the bacterial cells three times in 0.1 *M* buffer, each for 15 min by repeated centrifugations, fix in 1% buffered osmium from 4 hr to overnight at room temperature.
4. After rinsing several times in distilled water, each for 10 min, enrobe the cells in agarose and process as described in **Subheading 3.1.2., steps 5–9**.
5. Spurr's embedding medium is recommended since it infiltrates the bacterial cells better than epoxy 812 resin. Extend the times in the propylene oxide: Spurr's resin mixtures to 2 h each and the pure Spurr's resin mixture to overnight. Keep the capsules **capped** since this resin will absorb moisture and give an improper polymerization.

3.2.2. Others

Hard-to-fix bacterial cells grown in suspensions will benefit from the lengthy fixation protocol described by Ryter and Kellenberger *(8)*.

3.3. Preparation of Cells Grown on Substrates

3.3.1. Hard Substrates

Some cultured cells grow on substrates such as plastic flasks, Petri dishes (*see* **Fig. 4**) or Permanox slides surrounded by a chamber (*see* **Fig. 5**). Glass surfaces

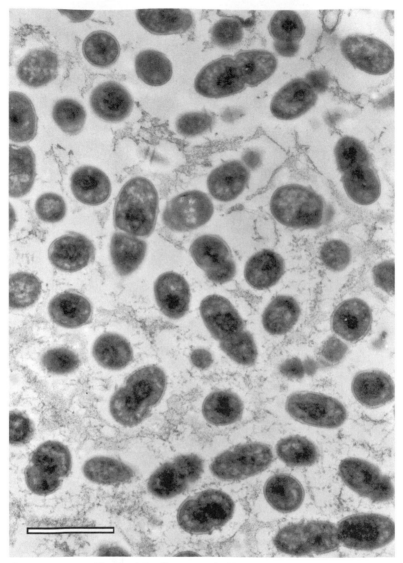

Fig. 3. Streptococcal bacterial cells grown in liquid culture, centrifuged and enrobed in agarose prior to sectioning for TEM. Prepared as described in **Subheading 3.2.1.** Bar = 1 µm.

are unsuitable in this procedure as it is impossible to safely separate the mono-layer of cells from the glass. Attached cells are processed as follows.

1. Decant (or gently aspirate) the culture medium and replace with 4% formaldehyde/ 2.5% glutaraldehyde in PIPES buffer (warmed to the same temperature as the culture). Fix at room temperature for 30 min.

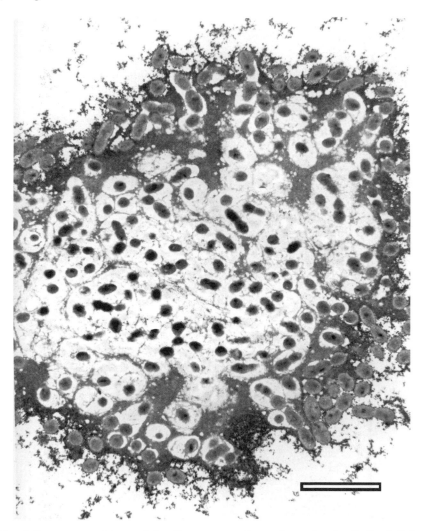

Fig. 4. Streptococcal bacterial cells grown on the surface of a plastic Petri dish after culturing in liquid medium. The specimen was sectioned through a growing micro-colony of bacteria. The bacteria exhibit different growth characteristics compared to growth in suspension culture as seen in **Fig. 3**. Prepared as described in **Subheading 3.3.1.** Bar = 1 µm.

2. Carefully remove the fixative solution and rinse three times with 0.1 *M* PIPES buffer, each for 5 min.
3. Postfix for 30 min at room temperature in 1% osmium tetroxide freshly prepared by mixing equal volumes of 2% aqueous osmium tetroxide and 0.2 *M* PIPES buffer stock.
4. Dehydrate in graded ethanol series, 10 min each.

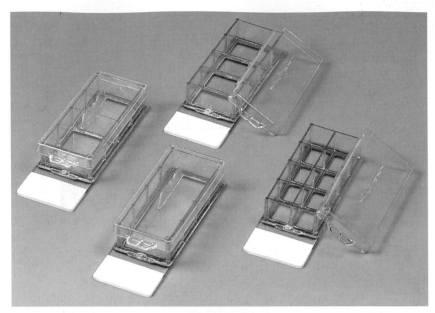

Fig. 5. Lab-Tek II plastic culture vessels consisting of Permanox slide surrounded by plastic chamber. A variety of cells, both eukaryotic and prokaryotic, can be cultured in these chambers. The fixation and embedding procedure is conducted *in situ* with minimal disturbance of the growing cells. (Courtesy of Electron Microscopy Sciences.)

5. As the specimens are undergoing dehydration, prepare three mixtures of absolute ethanol:epoxy 812 embedding medium consisting of 3:1, 1:1, and 1:3 parts, each. Use 10 mL of graduated, disposable polypropylene tubes with tight fitting lids. The ethanol:epoxy 812 embedding medium is best mixed by stirring with a pipet followed by inversion of the capped tubes. After the three mixtures are prepared, store them capped until needed in **step 6**.

6. Pour off the third change of absolute ethanol and gently pour on the 3:1 mixture of absolute ethanol:epoxy 812 embedding medium. Gently swirl the culture vessel five to six times over a period of 60 min to assist infiltration of the mixture into the cells. Similarly, repeat this procedure with the 1:1 and 1:3 mixtures.

7. Decant the 1:3 ethanol:epoxy mix and replace with pure epoxy 812 embedding medium. Swirl gently, as perviously, during a period of 60 min and repeat the embedding medium exchange an additional time.

8. Replace the second epoxy 812 embedding medium with one final change and leave overnight, uncovered to facilitate evaporation of any residual ethanol (*see* **Note 11**). The layer of epoxy should not exceed 2 to 3 mm as it may be difficult to separate from the plastic container.

9. Polymerize the epoxy 812 embedding medium for 24 h at 45°C and 48 h at 60°C. If two ovens are not available, then use 60°C for 48 h.

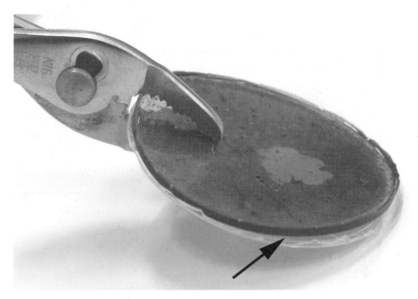

Fig. 6. One method to separate polymerized epoxy containing embedded cells from the plastic culture dish it to use a pliers to press firmly against a hard surface. The epoxy bends whereas the polystyrene either shatters off or separates as a complete sheet. Arrow shows separation of the upper epoxy layer from underlying polystyrene dish.

10. After the epoxy has polymerized, strip it from the underlying polystyrene (*see* **Fig. 6**) or Permanox plastic (*see* **Note 12**), glue selected areas of the sheet of epoxy-embedded cells onto stubs, or clamp in flat holder (*see* **Fig. 7**), and section in an ultramicrotome.

3.3.2. Agar Substrate

If the cells are growing on an agar surface, excise 1×1-cm pieces of agar containing the cells and place in a glass Petri dish or other vessel containing warmed 4% formaldehyde/2.5% glutaraldehyde in PIPES buffer to submerge the cells. Be sure to maintain the orientation of the agar with the cells uppermost. Process the specimens as follows.

1. After 60 min fixation at room temperature, rinse the agar blocks in PIPES buffer for 10 min.
2. Treat the blocks as solid tissues (keeping the cell layer uppermost) and process as described in **Subheading 3.1.1.** following **steps 8–14**.
3. Pour epoxy 812 embedding medium into a Petri dish to a depth of several mm. If a flat embedding mold (of polypropylene or silicone) is available, use this instead of a Petri dish.

Fig. 7. Portions of the separated epoxy layer containing embedded cells are broken into small pieces, or excised, and clamped into holder for ultramicrotomy. Arrow shows epoxy clamped into holder whereas arrowheads show areas of the epoxy that were excised using a cork borer. A portion of the polystyrene dish is shown still attached to the underlying epoxy layer containing specimens.

4. Transfer the agar blocks into the epoxy embedding medium with the cell side down, against the mold surface.
5. Follow **steps 16** through **17** in **Subheading 3.1.1.**
6. Separate the polymerized epoxy resin from the underlying mold, locate the cells of interest in a light microscope, trim the blocks using a hacksaw or cork borer.
7. Mount the epoxy blocks on an appropriate stub using epoxy glue and section in an ultramicrotome.

4. Notes

1. The PIPES solution will have a milky appearance initially but will clear as the pH is raised by adding NaOH and the PIPES goes into solution. Be aware that the pH will change as the PIPES goes into solution and one should verify the pH using a pH meter once the solution has cleared.
2. It is important to completely depolymerize the paraformaldehyde powder into formaldehyde gas that is highly soluble in water. If the powder only partially depolymerizes, short chain formaldehyde molecules will be produced. These molecules will be trapped inside cells as they start repolymerizing and react with the osmium solution to generate a fine precipitate sometimes referred to as "pepper" when sections are viewed in the TEM.
3. A given volume of osmium tetroxide solution may be disposed of by combining with two to three times the volume of corn oil. Spilled osmium tetroxide solution

should be covered with dried milk powder and then transferred to a container for disposal in accordance with local restrictions.

4. Completely remove any attached labels and clean the ampoule using soap and water. The ampoule, and all containers coming in contact with osmium solutions, must be clean otherwise the osmium will be reduced to metallic osmium and the fixative will be worthless. New bottles should be washed with soap and water, rinsed four or five times in double distilled water and used only for storing osmium solutions. Working in a fume hood, carefully score the clean ampoule (some ampoules are pre-scored), snap open and quickly drop both pieces into distilled water in a glass bottle. Electron microscopy supply houses sell ampoule openers to safely break the glass ampoule in half. Immediately cap and tighten the lid on the bottle. Osmium solutions are slow to dissolve and should be prepared 1 to 2 d in advance. Place the bottle containing the osmium solution inside a second container (as an old mayonnaise jar) and store in a fume hood away from direct sunlight. Withdraw needed amounts from the stock bottle using a pipet, taking care not to disturb the glass debris from the ampoule. Small shards of glass may be deposited in the specimen vial and damage the knives used in ultramicrotomy.

5. The term, absolute ethanol, refers to 100%, pure ethanol with no adulterants, including traces of water. Although it is possible to produce absolute ethanol by dehydrating 95% ethanol with molecular sieves or other desiccants, we purchase pints of pharmaceutical-grade, absolute ethanol and use fresh containers when absolute ethanol is needed. The pint containers of ethanol remain absolute for several openings, especially if one recaps them immediately after use. Eventually, they are diluted with water to produce the graded ethanol dehydration series.

6. Acetonitrile has been recommended as a substitute for propylene oxide since it is not carcinogenic or flammable yet is miscible with ethanol and the epoxy resins.

7. Do not shake the tube containing the resin components too vigorously or bubbles will be introduced that will interfere with the embedding. The few bubbles that form during the inversion process will rise to the surface and not pose a problem. It is best to prepare the resin several hours in advance and freeze any unused quantities for use the following day. Be sure to allow the frozen tubes to come to room temperature prior to opening since moisture will condense in the resin, possibly ruining the embedding.

8. The formula for relative centrifugal force is $RCF = 11.2r \, (RPM/1000)^2$, where r = radius in centimeters, RPM = revolutions per minute.

9. Glass pipets can be used but they should be fire-polished to prevent the release of small chips of glass in the specimen. Such glass chips will damage the knives used in ultramicrotomy.

10. To concentrate the cells in the pellet, transfer them to a micro-centrifuge tube after **step 3**. After a quick centrifugation (using a high speed burst for several seconds), suspend the lightly sedimented cells in 100 µL of warm agar and very quickly recentrifuge using the high speed burst feature of the micro-centrifuge. The goal is to form a compact pellet surrounded by agarose at the bottom of the tube. Avoid excessive centrifugation, however, as cells will be damaged and incompletely sur-

rounded by agarose. After centrifugation, process the cells as described in **steps 6–10**.

11. To assist the removal of traces of ethanol, remove the lids of the Petri dishes and Lab-Tek Permanox chamber slides. To open up plastic culture flasks, melt a series of holes in the upper side of the flask using either a soldering iron or a heated glass rod. Do this in a fume hood as noxious fumes are generated. Place the open culture vessels on top of a warm spot, such as an incubator, as this will speed the evaporation of ethanol.

12. Polymerized epoxy plastics can be stripped from conventional plastic Petri dishes relatively easily with a pair of pliers. Immediately after removing the plastic Petri dish from the oven, use the pliers to remove all polystyrene making up the wall of the plastic dish, leaving only the bottom of the plate with the overlying layer of epoxy. Grab the Petri dish with the pliers so that the arms maximally extend over the plastic. With the epoxy layer uppermost, press the Petri dish against a hard surface as if one were trying to bend the plastic. The epoxy, being more flexible than the polystyrene, will bend whereas the polystyrene will remain stiff and separate from the epoxy in large sections. Move the pliers around the plate to free additional areas until the entire epoxy later has separated from the polystyrene. It is possible that the epoxy may shatter, however, the large epoxy pieces will strip from the polystyrene if one continues the bending procedure. Permanox can be separated using either the bending procedure or by slipping a single edge razor blade between the epoxy and Permanox plastic and prying the two apart.

References

1. Bozzola, J. J. and Russell, L. D. (1999) *Electron Microscopy Principles and Techniques for Biologists.* Jones and Bartlett Publishers, Sudbury, MA.
2. Dykstra, M. J. (1992) *Biological Electron Microscopy Theory, Techniques, and Troubleshooting.* Plenum Press, New York and London.
3. Hayat, M. A. (1989) *Principles and Techniques of Electron Microscopy Biological Applications.* CRC Press, Inc., Boca Raton, FL.
4. Glauert, A. M. and Lewis, P. R. (1998) *Biological Specimen Preparation for Transmission Electron Microscopy*, vol. 17, *Practical Methods in Electron Microscopy* (Glauert, A. M., ed.), Portland Press, London and Miami.
5. Good, N. E., Winget, G. D., Winter, W., Connolly, T. N., Izawa, S., and Singh, R. M. M. (1966) Hydrogen ion buffers for biological research. *J. Biochem.* **5,** 467–483.
6. Baur, P. S. and Stacey, T. R. (1977) The use of PIPES buffer in the fixation of mammalian and marine tissues for electron microscopy. *J. Microsc. (Oxford)* **109,** 315–321.
7. Edwards, H. H., Yeh, Y. Y., Tarnowsky, B. I., and Schonbaum, G. R. (1992) Acetonitrile as a substitute for ethanol/propylene oxide in tissue processing for transmission electron microscopy: Comparison of fine structure and lipid solubility in mouse liver, kidney, and intestine. *Microsc. Res. Tech.* **21,** 39–50.
8. Ryter, A. and Kellenberger, E. (1958) Etude au microscope electronique de plasma contenant de l'acide desoxyribonucleique. I. Les nucleotides des bacteries en croissance active. *Z. Naturforsch.* **13b,** 597–609.

2

Processing Biological Tissues for Ultrastructural Study

José A. Mascorro and John J. Bozzola

Summary

Biological tissues are passed through numerous procedures before they can be studied at the ultrastructural level with the electron microscope. Chemical fixation is widely used as a method for preserving structural detail and can be performed by simple immersion or total body vascular perfusion. A 2 to 4% solution of glutaraldehyde buffered with 0.1 M sodium phosphate, or a combination of similarly buffered glutaraldehyde and paraformaldehyde, can be used successfully to preserve the fine structure of biological tissues. The material next is washed briefly in the buffer vehicle and then secondarily fixed in 1% osmium tetroxide (osmic acid), which also is buffered with sodium phosphate. The tissue then is thoroughly dehydrated in solutions of ethanol at increasing concentrations of 50%, 70%, 95%, and 100%. After dehydration, tissues are infiltrated for a prescribed time interval with an epoxy embedding medium. After infiltration, specimens are transferred into fresh epoxy resin and polymerized at 60 to 70°C for several hours. This orderly process ultimately yields fixed tissues that are encased in hardened blocks that can be thin-sectioned with an ultramicrotome. The thin sections are counterstained with solutions of heavy metals to add contrast. The material then can be subjected to the electron beam in an electron microscope to produce useful images for ultrastructural study. This overall procedure has been used successfully since the advent of biological electron microscopy to define the minute details of cells and tissues.

Key Words: Biological tissue; chemical processing; TBVP fixation; excision/immersion fixation; dehydration; infiltration; embedding.

1. Introduction

Biological tissue is delicate and must be protected in the harsh internal environment of the electron microscope, which includes a high vacuum and bombardment by a stream of electrons. To accomplish this, the tissue must pass through a series of well-defined preparatory steps. The material initially must be stabilized by treatment with specialized chemicals called fixatives (*1–4*). After fixation, specimens must be dehydrated (*5,6*), usually with ethanol, and

From: *Methods in Molecular Biology, vol. 369*
Electron Microscopy: Methods and Protocols, Second Edition
Edited by: J. Kuo © Humana Press Inc., Totowa, NJ

infiltrated with a liquid resin, which is hardened by applying gentle heat *(7,8)*. Hardened resin blocks containing the tissue specimens are thinly sectioned, and the sections treated with solutions of heavy metals in order to impart contrast *(9,10)*. It is only at this point that the thin sections can be placed in the electron microscope column for observation and study.

The purpose of chemical fixation is to stabilize the tissue so that it will present a final appearance not substantially unlike its original form. This process should preserve the structure of the component cells with minimal alteration of volume, morphology, and spatial relationships *(11,12)*. Dehydration is necessary because the epoxy resin used to infiltrate and embed the specimen is not miscible with water. Dehydration should be accomplished without introducing swelling or shrinkage artifacts that would distort normal fine structure *(13)*. After fixation, dehydration, and embedding in epoxy resin, the tissues not only acquire the consistency essential to tolerate the physical forces incumbent in ultramicrotomy but also the stability to withstand the harsh environment inside the electron microscope.

Preparing biological tissues for ultrastructural study using the transmission electron microscope is a lengthy exercise that requires many hours, from tissue procurement to completion. All steps in this paradigm are completely interdependent, and all must be successfully accomplished to produce tissue worthy of examination. Although the previously discussed steps are constants for processing tissues in general, many different variations in the protocols can be used for different tissue types. For example, plant tissues with impervious cellulose walls may pose a particular problem during infiltration and therefore special vacuum methods may be required *(14,15)*. The overall process discussed here is designed to be as simple as possible and minimizes the total time tissues are in contact with chemicals. A very important reason for its success is that mature, fully differentiated organ tissues lend well to processing. Although organ tissues may not always produce perfect results, they do present fewer problems than tissue types that are of a delicate nature or highly hydrated.

This chapter describes both total body vascular perfusion (TBVP) as well as excision and immersion methods to prepare soft mammalian organ tissues for transmission electron microscopy. The TBVP procedure is a very simple and direct processing method that has served the first author (J.M.) well for more than 40 yr.

2. Materials

2.1. The Perfusion Apparatus for TBVP

The apparatus can be quite simple and consists of two containers suspended approx 90 cm (36 inches) above the properly anesthetized animal. Dropping the solutions from this elevation by normal gravity produces a perfusion pressure that does not damage the vascular bed. One container delivers a physiological saline solution (*see* **Subheading 2.4.**) for clearing the vascular bed, whereas the second delivers the fixative of choice. Flexible tygon tubing from each container is joined by a Y connector into a single tube that is inserted into the circulation of the animal.

2.2. Fixatives

2.2.1. Glutaraldehyde (GA)

25% GA, electron microscopy (EM) grade (distilled and purified) or 50% GA, aqueous solution, EM grade (Electron Microscopy Sciences, USA; TAAB, UK). The GA should be stored in a refrigerator. The fixative of choice for TBVP is GA (*see* **Note 1**). GA also can be prepared in various combinations with formaldehyde, depending upon the primary focus of the experiment. For general fixation purposes, TBVP can be accomplished with GA buffered with Sorensen's phosphate buffer. The perfusion fixative is prepared by diluting pure 25% GA to 6% with pure distilled H_2O. Six percent GA then is mixed with an equal volume of 0.2 M phosphate buffer to produce a final fixative of 3.125% GA in 0.1 M phosphate buffer at pH 7.2 to 7.3. Two hundred milliliters of perfusion fixative can be prepared as follows:

1. 25 mL of 25% GA + 75 mL distilled water gives 100 mL of 6.25% GA.
2. 100 mL 6.25% GA + 100 mL 0.2 M sodium phosphate buffer gives 200 mL of 3.125% GA in 0.1 M sodium phosphate buffer.

2.2.2. Paraformaldehyde

EM grade, purified, solid granular form; also available as 10%, 16%, 20%, or 32% solution. Work in hood and avoid vapors. The final, working strength of formaldehyde is 4% to 8%.

2.2.3. Osmium Tetroxide (Osmic Acid)

Either aqueous solution (2% or 4%) in 2-mL prescored amber ampoules or crystalline, 99.95% purity in 1, 2, 4, and 6 g of prescored amber ampoules. This chemical is highly toxic to mucous membranes and eyes. Work in hood and avoid vapors. Osmium tetroxide is used as a secondary fixative that is similarly buffered with sodium phosphate. Stock 2% osmium tetroxide is prepared by dissolving 1 gram of crystalline osmium tetroxide in 50 mL of distilled water. The fixative is prepared as follows:

1. 10 mL 2% osmium tetroxide + 10 mL 0.2 M phosphate buffer gives 20 mL 1% osmium tetroxide in 0.1 M phosphate buffer.

2.3. Other Reagents and Equipment

2.3.1. Sorensen's Phosphate Buffer (Mixed Sodium Salts)

Sorensen's 0.2 M dual sodium solution is prepared from a combination of anhydrous dibasic sodium phosphate (Na_2HPO_4) and monobasic sodium phosphate ($NaH_2PO_4 \cdot H_2O$; see **Note 2**).

1. Prepare 0.2 M solution A by combining 27.6 g of NaH_2PO_4 in 1 L of distilled H_2O; prepare 0.2 M solution B by combining 28.4 g of Na_2HPO_4 in 1 L of distilled H_2O.
2. A large amount of stock 0.2 M Sorensen's sodium phosphate buffer solution is prepared as follows: place 500 mL of di-sodium solution B in appropriate container and add, while stirring on a magnetic stirrer and monitoring with a pH meter, the necessary quantity of solution A to lower the pH of solution B to a range of 7.2 to 7.3. The resulting 0.2 M sodium phosphate solution is stable and can be stored in the refrigerator. When reusing, allow the buffer to reach room temperature to re-dissolve any sodium salts that may have crystallized while stored in the cold.

2.4. Ethyl Alcohol (Ethanol, Anhydrous)

1. Dehydrating solutions are prepared in 50%, 70%, and 95% concentration. Solutions are stored in the refrigerator and used cold during the dehydration process.
2. A new bottle of absolute ethanol should be opened for each dehydration procedure. Unused absolute ethanol is utilized for preparing the graded series of alcohols.

2.5. Propylene Oxide (PO)

PO (1,2-epoxypropane), EM grade. **Caution:** highly flammable, carcinogenic. Use in a fume hood. This solvent traditionally is used as the transitional solvent with epoxy resins and gradually displaces the dehydration agent during infiltration and final embedding. If preferred, this step may be eliminated.

2.6. Embedding Media Kits

Embedding media are marketed as kits by the various electron microscopy supply houses. Each prepackaged kit contains the particular resin marketed by that supplier (Embed 812, Medcast, LX112, Polybed 812, SciPoxy, Eponate 12, Epon 812). Each kit also comprises two anhydrides and a catalyst to accelerate the polymerization reaction. The anhydrides usually included are dodecynyl succinic anhydride and nadic methyl anhydride. The traditional catalyst in embedding kits is 2,4,6-(dimethylaminomethyl) phenol.

2.7. Specimen Vials

The vials have a 20-mL capacity with screw caps for tissue processing. Vials to be used for dehydration should be stored permanently in a 60°C drying oven to ensure that they are completely dry prior to use.

2.8. Physiological Saline

Prepare by dissolving 8.5 g of NaCl in 1000 mL of deionized water. It should be used within a week because microorganisms will contaminate the solution. It is possible to purchase pharmaceutical grade, sterile saline from hospital supply houses.

2.9. 18-Gage Surgical Needle

This needle can be purchased from Quest Medical, Inc.

2.10. Polyethylene Embedding Molds

Molds that hold approximately 1.5 mL of epoxy resin are available from electron microscopy supply houses (Electron Microscopy Sciences, Ernest Fullam, Structure Probe Inc., Ted Pella). The most favored mold produces a hardened epoxy block with a 1 × 1-mm face atop a truncated pyramid. The specimen should have settled just under the surface of the pyramid tip.

3. Methods

3.1. TBVP Preparation

3.1.1. TBVP Fixation

To prevent coagulation of the blood by the aldehyde fixative, blood is flushed from the vasculature by flowing 0.9% chilled (4°C) saline solution through the circulatory system. This is accomplished from the perfusion apparatus by inserting an 18-gage surgical needle directly into the right ventricle of the heart. The right atrium is snipped immediately after insertion to prevent recirculation of the solution. Proper insertion and flow of the solution can be judged by the appearance of the liver, which should clear immediately after commencing perfusion (*see* **Note 2**). After clearance of blood from the circulatory system, tissues are fixed using TBVP, by flowing a solution of chilled 3.125% GA in 0.1 *M* sodium phosphate buffer at pH 7.2 to 7.3. Total perfusion time depends entirely on animal size, with larger animals (e.g., cats, dogs, primates) requiring a longer period than smaller rodents (e.g., mice, rats, hamsters). A total perfusion time of 2 h ensures that the largest animals are adequately fixed, whereas 60 min or less seems adequate for the smaller animals (*see* **Note 3**).

3.1.2. Trimming TBVP Fixed Tissues

After TBVP, the whole organ (or selected portions) are removed using sharp blades (razor blade or scalpel), placed into the chilled fixing solution and quickly cut into slices no larger than 1 mm in one dimension. These slices are then re-trimmed to give pieces no larger than 1 mm^3 and placed in vials containing GA for the subsequent procedures.

3.1.3. Washing and Storing TBVP Tissues

Before tissues are passed through a secondary fixing solution, they should be very thoroughly washed to minimize chemical interaction between molecules of the primary and secondary fixation chemicals. Washing preferably is done with the same buffer system used in the primary perfusion mixture. The advantage of using the same vehicle for washing is that it avoids sudden changes in the tissue environment which could harm membranes or cell organelles. For small tissue blocks, washing may be accomplished in as little as 20 to 30 min with two or three changes. However, larger tissue blocks may require longer periods. If there is any doubt about washing time, simply store the tissues in the buffer solution overnight in the refrigerator. If necessary, tissues may be stored for short periods (1–2 wk) in the buffer solution in the refrigerator. If long-term storage is necessary, tissues may be left in the original GA perfusion fixative (*see* **Note 4**).

A total washing time of 60 min, with several changes, usually is sufficient and prepares tissues for secondary fixation without the danger of traces of fixative remaining within the material. A thorough removal of residual GA is very important because interaction between GA and osmium will produce a fine, dense precipitate of reduced osmium. This artifact will appear all over the thin sections and become clearly visible with the electron microscope. The desired organs are excised and cut into slices or small cubes for further processing. The tissue cubes are washed thoroughly in 0.1 M sodium phosphate buffer for a minimum of 60 min, with changes every 15 to 20 min.

3.1.4. Postfixation

After the buffer wash, tissues are postfixed in 1% osmic acid in 0.1 M sodium phosphate buffer for approx 45 min at room temperature (*see* **Note 5**). Excess osmic acid is removed by passing the tissues through two rinses in distilled water, 1 to 2 min in each.

3.1.5. Dehydration

Epoxy resins used for infiltrating and embedding are not miscible with water. Thus, all free water from the previously fixed and rinsed specimens must be replaced with a suitable organic solvent before proceeding with infiltration and embedding (*see* **Note 6**). Water can be removed stepwise by passing the specimens through a series of solutions of increasing concentrations of an organic solvent. The solvent of choice usually is absolute ethanol, although acetone and acetonitrile are other suitable choices.

The ethanol dehydration solutions are prepared in increasing concentrations of 50%, 70%, and 95%. Dehydration is competed in freshly opened 100%

(absolute) ethanol. Dehydration usually is performed with alcohols that have been stored in the refrigerator and remain cold during dehydration. A standard dehydration schedule can be performed as follows:

Dehydrate in ice-cold ethanol in increasing concentrations:

1. 50% ethanol at 4°C for 15 min
2. 75% ethanol at 4°C for15 min.
3. 95% ethanol at 4°C for15 min.
4. 100% ethanol 15 min in room temperature *(tissues acclimate to room temperature)*.
5. 100% ethanol: four to six changes 10 to 15 min each in room temperature.

3.1.6. Embedding and Polymerization

1. Most epoxy resins are miscible to some degree with ethanol, but they mix more readily with certain organic solvents such as PO. This intermediate solvent routinely is used following dehydration and prior to infiltration (*see* **Note 7**). Tissues can be passed through two changes of PO, 5 min in each change.
2. Tissues initially are infiltrated with a combination of PO and embedding medium for a prescribed time period. Initial infiltration is done in a 1:1 mixture of PO and embedding medium for 30 to 45 min with the tissues placed on a rotator.
3. After the initial infiltration in the combination of PO and resin, the tissues are fully impregnated with the final embedding medium. Final infiltration can be achieved according to the following schedule:
   Full embedding medium: 3 to 6 h, or
   Full embedding medium: overnight (15 h).
4. After final infiltration is completed using the desired schedule, tissues are placed in appropriate molds with freshly prepared embedding medium and the molds then placed in an oven in order to polymerize (harden) at a temperature of 60 to 70°C. One additional period, approx 15 h, will harden the blocks sufficiently to allow them to be initially sectioned. Blocks can remain in the oven for a longer period (an additional day) if the first blocks do not produce adequate sections because of insufficient hardness.

   Ultrathin sections from the hardened tissue blocks are produced by sectioning with ultramicrotomes and are collected on copper grids (*see* Chapter 5 in this volume). The sections are counter stained with salts of heavy metals (*see* Chapter 6 in this volume) and then examined with the transmission electron microscope.

3.2. Excision and Immersion Method

Although TBVP is certainly the preferred method for preparing tissues from the central nervous system, there may be instances in which a more conventional method may be used. This method involves excision of the tissue followed by immersion in the fixative. The excision method may be chosen by individuals who lack experience with TBVP or who are not working with tissues taken from the central nervous system. Similar reagents as were used in TBVP are also used in the excision-immersion method but in a different manner.

1. The animal may be anesthetized or humanely sacrificed. Tissues are surgically removed as rapidly as possible to prevent artifact formation, or degradation of ultrastructural detail. Generally, a portion of the tissue is removed and placed into a container of fixative, usually a Petri dish with buffered GA such as a 2% solution of GA in Sorenson's buffer.
2. While in the GA fixative, the tissue is cut into smaller pieces no larger than 1 mm in any one direction. For example, tissue initially may be cut into 1-mm thick slices that are then trimmed into 1-mm cubes as described in **Subheading 3.1.2.** The cubes are left in the fixative for approximately 1 h. This is most often carried out in a container of crushed ice or with the specimen vials placed into a refrigerator.
3. After the primary fixation in GA, the tissue cubes are processed exactly as the tissue from the TBVP procedure (*see* **Subheadings 3.1.3.** to **3.1.7.**).

A summary of TBVP and Emission protocols for processing biological tissues is presented in **Table 1**. Ultrastructural images of selective tissues prepared from both methods are also illustrated in **Figs. 1–8**.

4. Notes

1. Cell cultures, isolated cells, pellets, plant materials, or biopsy specimens easily can be fixed *in situ* or by simple immersion *(5)*. Organ systems that are easily accessible, such as liver or heart, can be removed rapidly and also processed by immersion *(3)*. However, this choice is not the best in this instance because fixation by immersion depends on diffusion of the fixative into the tissue, a rather slow process. TBVP, although it is lengthy and technically more difficult, is the method of choice for preparing animal organ tissues *(16)*. With this process, the fixative is gently "pushed" throughout the vascular system. The fixative reaches the cellular level quickly as it only needs to diffuse through the very thin capillary walls. The perfusion apparatus can be complex and expensive, or as simple as hanging two containers (flushing solution of 0.85% physiological saline and fixative solution of buffered 3.125% GA) above the supine animal and allowing the fixative to flow under normal gravity. Inserting the 18-gage needle (cannula) correctly into the left ventricle is absolutely essential for proper distribution of the fixative throughout the circulatory system. Snipping the right atrium to prevent recirculation of the flush and fixative also is paramount. The test of a successful perfusion technique can be judged by the appearance of the liver. This large organ will receive the flushing vehicle immediately and should quickly clear from a deep, dark red to a uniform light appearance, indicating that the flush and the fixative are flowing unimpeded through the vascular system. Failure of the liver to "clear" is a certain sign that perfusion is not progressing and will not be successful. In this case, the organ systems and tissues will not be properly fixed. If such is the case, the cannula should be removed, repositioned and perfusion started anew. Improper perfusion at this initial step will yield tissues not properly preserved and guarantee failure for the entire process. Because electron microscopy is overall a lengthy procedure, an unsuccessful perfusion will result in the loss of many hours of valuable time, effort, and chemicals.

Table 1
Comparison of TBVP and Excision Immersion Methods
of Biological Specimen Preparation for Transmission Electron Microscopy

	TBVP	Excision immersion
Specimen source	Anesthetized animal. Perfusion of tissue with primary fixative.	Anesthetized or humanely sacrificed animal. Surgical removal of tissue and trimming to 1 mm^3 in primary fixative.
Primary fixative	Perfuse with 3.125% GA in 0.1 M Sorenson's buffer, pH 7.2, 5–10°C, 1–2 h. Removal and trimming of tissues into 1 mm^3 in primary fixative.	Place excised tissue in 2% GA in 0.1 M Sorenson's buffer, pH 7.2, 5–10°C, 1–2 h.
Washing and storage	0.1 M Sorenson's buffer, pH 7.2, 5–10°C, 3 changes, 15–20 min each. Can store specimen here for 1–2 wk in refrigerator.	0.1 M Sorenson's buffer, pH 7.2, 5–10°C, 3 changes, 15–20 min each. Can store specimen here for 1–2 wk in refrigerator.
Postfixation	1% osmium tetroxide in 0.1 M Sorenson's buffer, pH 7.2, room temp, 45–60 min.	1% osmium tetroxide in 0.1 M Sorenson's buffer, pH 7.2, room temp, 45–60 min.
Rinse	Distilled water, 2–3 changes, 2 min each.	Distilled water, 2–3 changes, 2 min each.
Dehydration	50%, 75%, 95%, 100%, 100% ethanol, 10 min each, 5–10°C.	50%, 75%, 95%, 100%, 100% ethanol, 10 min each, 5–10°C.
Intermediate solvent, PO	PO, 2 changes, 5 min each, 5–10°C.	PO, 2 changes, 5 min each, 5–10°C.
Infiltration with epoxy resin at room temp (RT)	(a) 1:1 PO/epoxy resin, 30–45 min. (b) Epoxy resin 6–18 h (overnight).	(a) 1:1 PO/epoxy resin, 30-45 min. (b) Epoxy resin 6–18 h (overnight).
Transfer into mold	Put individual specimens into polyethylene molds, allow to sink, room temperature.	Put individual specimens into polyethylene molds, allow to sink, room temperature.
Polymerization	1–2 d at 60–70°C.	1–2 d at 60–70°C.
Section and stain sections	Sectioning (*see* Chapter 5) Staining (*see* Chapter 6)	Sectioning (*see* Chapter 5) Staining (*see* Chapter 6)

GA indicates glutaraldehyde; PO, propylene oxide; TBVP, total body vascular perfusion.

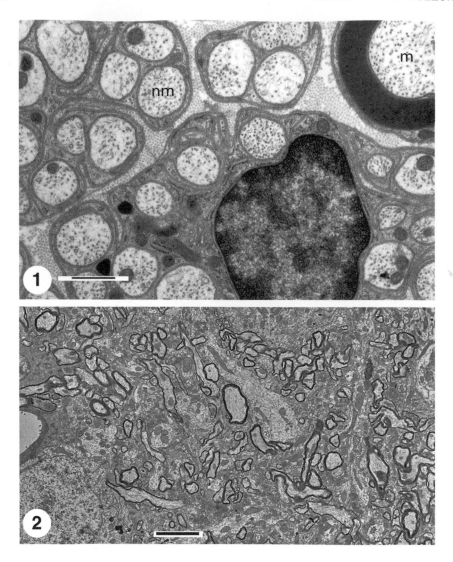

Fig. 1. Peripheral (sciatic) nerve from rat processed using the total body vascular perfusion (TBVP) method as described in **Subheading 3.1.** A portion of myelinated (m) nerve extends into the field from top right corner and numerous nonmyelinated (nm) nerves dominate the left half of this image. All nerves show microtubules and neurofilaments appearing as small spots and fibers inside the nerve. Bar = 2 μm.

Fig. 2. Low magnification view of central nervous tissue from rat brain. The dark, irregularly shaped profiles are myelinated nerve fibers. Bar = 5 μm.

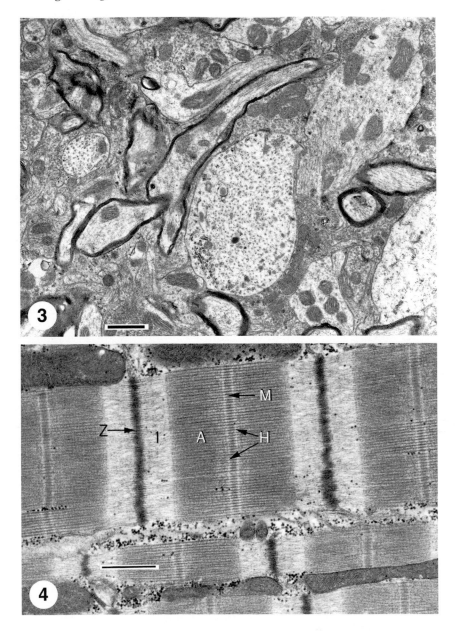

Fig. 3. Higher magnification of central nervous system tissue from same specimen shown in **Fig. 2.** The microtubules and microfilaments inside the neurons show good structural preservation (*see* large, central neuron). Bar = 1 μm.

Fig. 4. Cardiac muscle from rabbit prepared using the total body vascular perfusion (TBVP) method. Note the presence of well-preserved bands (Z, I, A, H and M, respectively). Bar = 0.25 μm.

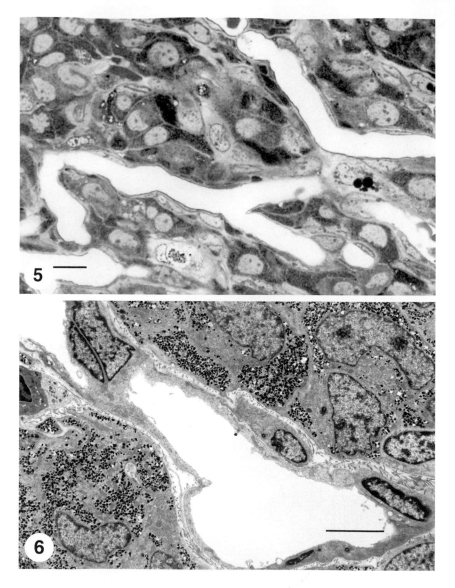

Fig. 5. Light micrograph of chromaffin tissue from celiac ganglion of rabbit. The large, clear tracts are capillaries that have been flushed of blood using physiological saline followed by glutaraldehyde and osmium fixatives (*see* **Subheading 3.1.**). The examined section is 1.0 μm thick and stained with 1% toluidine blue dissolved in a solution of 1% sodium borate. Bar = 10 μm.

Fig. 6. Electron micrograph of same tissue shown in **Fig. 5**. The electron dense bodies are chromaffin granules characteristic of this type of cell. Bar = 10 μm.

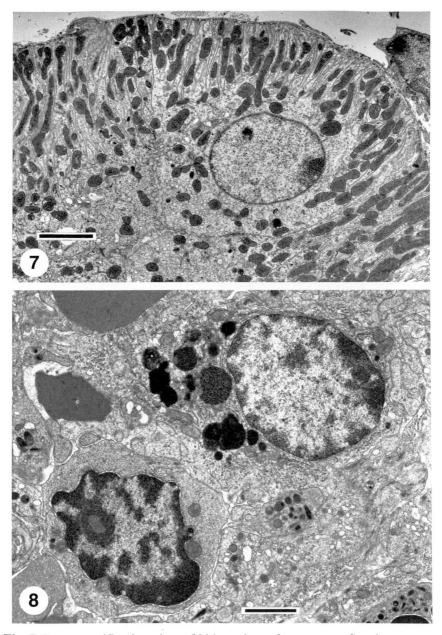

Fig. 7. Low-magnification view of kidney tissue from mouse. Specimen was processed using excision/immersion method as described in **Subheading 3.3.** Bar = 3 μm.

Fig. 8. Spleen tissue from mouse showing two cells in top-right and lower-left corners. The solid, dense structures in the upper-left corner are portions of red blood cells which were not flushed out of the tissue as in the total body vascular perfusion (TBVP) method. Prepared using excision/immersion procedure. Bar = 2 μm.

2. Several buffer systems are available for the double-fixation process (17–19). Sorensen's di-sodium phosphate buffer is preferred because it is simple to prepare, remains stable when stored, and is more physiological than others (5). When preparing the final buffer, solution A is added carefully to a given volume of solution B to achieve a range of pH 7.2 to 7.3. Once the fixative and sodium buffer are mixed, the final pH of the total mixture will remain very stable within the desired range. If necessary, a dilute solution of NaOH or HCL can be used to make minor acid-base adjustments in pH.

3. When used at room temperature, diffusion of GA into a tissue is a slow and complex process. Post-mortem changes leading to loss of cell materials within the tissue can result before the fixative reaches the cellular level. For this reason, perfusion fixation with GA, as well as secondary fixation with osmic acid, often is performed with refrigerator cold (4°C) solutions (5,6). In general, fixation with lower temperature fixatives by TBVP is preferred as this represents the best compromise for minimizing cell extractions (5). Also, it is likely that extraction of cellular materials would be more pronounced when applying immersion fixation with fixatives at room temperature.

4. For best results, tissues should be processed from beginning to end without long pauses, other than those prescribed within the overall process. Tissues should never be allowed to stand for long times in alcohol or PO to avoid cellular extractions, even from fixed material (13). However, if necessary or as a matter of convenience within the workday, tissues may remain in the 0.1 M sodium phosphate buffer wash solution for a few hours or overnight. If a particular tissue needs to be re-examined at some future time, samples may be stored on a semipermanent basis in the primary phosphate-buffered GA fixing solution, in a cold environment, with no apparent damage. Presumably, the crosslinking of tissue components by the GA ions proves protective for the tissue. Storing materials long-term, although not usually done by the first author (J.M.), would preclude the necessity for performing an entirely new perfusion procedure.

5. Primary fixation with GA, followed by a secondary fixation with osmic acid, provides very satisfactory cellular preservation and became the norm for processing animal tissues many years ago (2,20,21). GA, when delivered effectively by TBVP, penetrates quickly and cross-links with proteins, thereby stabilizing cellular structure (22). Osmic acid, on the other hand, serves not only to further fix and stabilize tissues, but also adds contrast by depositing metal ions upon cell structures, particularly membranous structures. Contact with osmic acid should be as limited as possible to prevent loss of cellular proteins (23). It has been determined, through many years of personal experience, that a double fixation process utilizing 3.125% GA followed by 1% osmic acid, both adjusted to pH 7.2 to 7.3 with sodium phosphate buffer, produces excellent results in terms of general preservation of most animal organ tissues. Other tissue types, however, require different processing protocol.

6. Dehydration usually is performed with refrigerator-cold ethyl alcohol, although acetone also can be used to remove water from the specimens (24). Acetone, however, rapidly absorbs water vapor and is not widely used for dehydrating tissues. Proteins, carbohydrates, and lipids reportedly are lost during the dehydration process,

even from thoroughly fixed tissues *(1)*. This loss may be minimized by using ethanol at the lower temperature. Ideally, tissues must remain in contact with the dehy-
drant for a period of sufficient length to achieve dehydration while minimizing
loss of cell components. Personal experience and the tissue type being processed
is the best guide for designing and implementing a dehydration schedule. Many
dehydration schedules exist, and some are designed for specific tissues. Generally, a total dehydration of 2 h or less will adequately remove water from biological tissues and prepare them well for subsequent infiltration and embedding.
Striated muscle and liver are particularly susceptible to dehydration damage, such
as extraction of cell components and swelling of cell organelles *(13)*. On the basis
of the appearance of these sensitive tissues, a schedule of sufficient length can be
designed that will achieve the purpose while minimizing the loss of cell material.

7. PO traditionally is used as a transitional solvent for facilitating infiltration following dehydration and prior to embedding *(1,5)*. PO is the best organic solvent for
decreasing the viscous character of the epoxy resins used for formulating embedding media. It is necessary to reduce the viscosity of the embedding media to increase
infiltration capability into the tissue. PO, however, can severely deplete lipids
from fixed tissues and is a known carcinogen. The loss of lipids, and perhaps other
cell components as well, is exacerbated by lengthy, stepwise infiltration schedules
that gradually reduce the concentration of ethanol while increasing that of PO. Such
schedules can keep tissues in contact with PO for 1 h or longer *(1,5–7)*. Ethanol
does not reduce the viscosity of embedding medium to the same degree. However,
past experience shows that ethanol is sufficiently effective for reducing viscosity
and that resin infiltration into the tissues is adequate *(25)*. It is normal practice for
the first author to eliminate completely the use of PO as a transitional solvent. By
eliminating PO from the process, the tissue, as well as the microscopist, is spared
contact with a volatile organic compound.

References

1. Glauert, A. M. (ed.) (1974) *Fixation, Dehydration and Embedding of Biological Specimens: Practical Methods in Electron Microscopy*. North-Holland, Amsterdam.
2. Sabatini, D. D., Bensch, K., and Barnett, R. J. (1963) Cytochemistry and electron microscopy. The preservation of cellular ultrastructure and enzymatic activity by aldehyde fixation. *J. Cell Biol.* **17**, 19–58.
3. Fahimi, H. D. (1967) Perfusion and immersion fixation of rat liver with glutaralde-hyde. *Lab. Invest.* **16**, 736–751.
4. Trump, B. F. and Bulger, R. E. (1966) New ultrastructural characteristics of cells fixed in glutaraldehyde-osmium tetroxide mixture. *Lab. Invest.* **15**, 368–379.
5. Hayat, M. A. (1989) *Principles and Techniques of Electron Microscopy: Biological Applications*. CRC Press, Boca Raton, FL.
6. Glauert, A. M. and Lewis, P. R. (1988) *Biological Specimen Preparation for Transmission Electron Microscopy*, Princeton University Press, Princeton.
7. Luft, J. H. (1961) Improvements in epoxy resin embedding methods. *J. Biophys. Biochem. Cytol.* **9**, 409–414.

8. Finck, H. (1960) Epoxy resins in electron microscopy. *J. Biophys. Biochem. Cytol.* **7,** 27–30.

9. Reynolds, E. S. (1963) The use of lead citrate at high pH as an electron opaque stain in electron microscopy. *J. Cell Biol.* **17,** 208–212.

10. Watson, M. L. (1958) Staining of tissue sections for electron microscopy with heavy metals. *J. Biophys. Biochem. Cytol.* **4,** 475–478.

11. Maunsbach, A. B. (1966) The influence of different fixative and fixation methods on the ultrastructure of rat kidney proximal tubule cells. I. Comparison of different perfusion fixation methods and of glutaraldehye, formalin and osmium tetroxide. *J. Ultrastruct. Res.* **15,** 242–282.

12. Maunsbach, A. B. (1966) The influence of different fixative and fixation methods on the ultrastructure of rat kidney proximal tubule cells. II. Effects of varying osmolality, ionic strength, buffer system and fixative concentration of glutaraldehyde solutions. *J. Ultrastruct. Res.* **15,** 283–309.

13. Mollenhauer, H. H. (1993) Artifacts caused by dehydration and epoxy embedding in transmission electron microscopy. *Microsc. Res. Tech.* **26,** 496–512.

14. Bain, J. M. and Gove, D. W. (1971) Rapid preparation of plant tissues for electron microscopy. *J. Microsc. (Oxford)* **93,** 159–162.

15. Gahan, P. B., Greenoak, G. C., and James, D. (1970) Preparation of ultrathin frozen sections of plant tissues for electron microscopy. *Histochemie* **24,** 230–235.

16. Bowes, D., Bullock, G. R., and Winsey, N. P. J. (1970) A method for fixing rabbit and hind limb skeletal muscle by perfusion. *Proc. 7th Annual Int. Cong. Electron Microsc.* **1,** 397a.

17. Hayat, M. A. (1972) *Basic Electron Microscopy Techniques.* Van Nostrand Reihnold, New York, NY.

18. Wood, R. L. and Luft, J. H. (1965) The influence of buffer systems on fixation with osmium tetroxide. *J. Ultrastruct. Res.* **12,** 22–45.

19. Bennett, H. S. and Luft, J. H. (1959) s-Collidine as a basis for buffering fixatives. *J. Biophysic. Biochem Cytol.* **6,** 113–114.

20. Karnovsky, M. J. (1965) A formaldehyde-glutaraldehyde fixative of high osmolarity for use in electron microscopy. *J. Cell Biol.* **27,** 137a.

21. Schultz, R. L. and Case, N. M. (1970) A modified aldehyde perfusion technique for preventing certain artifacts in electron microscopy. *J. Microsc. (Oxford)* **92,** 69–84.

22. Ericsson, J. L. E., Saladino, A. J., and Trump, B. F. (1965) Electron microscopic observations of the influence of different fixatives on the appearance of cellular ultrastructure. *Z. fur Zellforsch.* **66,** 161–181.

23. Luft, J. H. and Wood, R. L. (1963) The extraction of tissue protein during and after fixation with osmium tetroxide in various buffer systems. *J. Cell Biol.* **19,** 46a.

24. Lee, R. M. K. W. (1984) A critical appraisal of the effects of fixation, dehydration and embedding on cell volume, in *Science of Biological Specimen Preparation: SEM Proceedings* (Revel, J.-P., Barnard, T., and Haggis, G. H., eds.), Scanning Electron Microscopy, Inc. AMF O'Hare, IL, pp. 61–70.

25. Mascorro, J. A. (2004) Propylene oxide: to use or not to use in biological tissue processing. *Microscopy Today* **12,** 45–46.

3

Processing Plant Tissues for Ultrastructural Study

John Kuo

Summary

This chapter describes conventional chemical fixation methods and techniques for studying the cellular and organelle ultrastructure of plant tissues under transmission electron microscopy. The general methods and procedures for the plant specimen preparation (including fixation, dehydration, infiltration, and embedding) and the composition of fixatives, buffers, dehydration solvent, and embedding media are similar to those for animal tissues. However, certain special characteristic features of plant tissues, such as a thick cellulosic cell wall, waxy substance in the cuticle, large amount of gases in the intercellular spaces, the presence of vacuoles, have created fixation and resin filtration difficulties and, therefore, special modifications of the protocols used for animal tissues are required. The addition of chemicals such as caffeine in fixative can stabilize the phenol in the vacuole; however, the rupture of vacuole caused by the fixative still cannot be controlled, particularly for plants with highly vacuolated cells. The application of vacuum infiltration during the initial fixation stage to remove gases from the tissues is described. Additional vacuum infiltration during resin infiltration procedure to improve the efficiency of resin penetration is implemented.

Key Words: Plant tissue; ultrastructure; chemical fixation; dehydration; vacuum infiltration; resin embedding; cell wall; vacuole.

1. Introduction

In higher plants (flowering vascular plants), the surface of organs (leaves, stems, roots, flowers, fruits, and seeds) usually is covered by waxy substances in the cuticle that are hydrophobic. Unlike animal tissues, plant tissues contain large amount of gases in the intercellular spaces, particularly the spongy parenchyma in leaves. These features have posed a serious barrier for the penetration of fixative. In addition, the plasma membrane, which protects each plant cell, as in the case of animal cell, is always covered by cell walls with diverse thicknesses of

From: *Methods in Molecular Biology, vol. 369*
Electron Microscopy: Methods and Protocols, Second Edition
Edited by: J. Kuo © Humana Press Inc., Totowa, NJ

extracellular materials. The physical and chemical properties of cell walls vary with cell types and the location of plant cells. For example, the presence of suberin materials in hypodermal cells and endodermal cells in most plant root tissues and the appearance of lignin in xylem cells and sclerenchyma can create the fixation and resin infiltration difficulties.

Plant cells, unlike animal cells, normally contain large central vacuoles, which possess various substances, including phenolic compounds and hydrolytic enzymes such as nuclease and protease. Vacuolar membranes (tonoplast) are very sensitive to fixatives and are easy to rupture during fixation, releasing vacuolar contents into the cytoplasm. The presence of vacuoles and thick cell walls has caused plant cells to have a considerably higher osmotic pressure than the surrounding media and, consequently, the compartmentalization of acids will be lost as the result of the damage to the vacuolar membrane during the process of fixation. The detrimental outcome is the release of organic acids from the vacuoles to the cytoplasm before adequate stabilization of the cytoplasm has taken place. Therefore, preventing leakage of vacuolar contents at least during the initial fixation will improve the preservation of the cytoplasm. This process can be accomplished by adding caffeine to glutaraldehyde to precipitate phenols within the vacuole and prevent their release into the cytoplasm (*1*).

Although glutaraldehyde has been considered to be the best fixative for general use since it was introduced by Sabatini et al. (*2*) more than 40 yr ago, it should be realized that its application on the plants with highly vacuolated cells causes vacuolar rupture. The eruption of canalicular vacuoles and the transformations of cell membranes and tonoplasts in living cells of the petiolar hairs and onion epidermal cells caused by the application of buffered aldehydes (glutaraldehyde or formaldehyde) have been reported (*3,4*). Unfortunately, no effective fixative agent has been discovered to tackle the problem of vacuolar rupture for the highly vacuolated plant cells.

General specimen preparations of plant tissues for ultrastructural investigations basically are similar to those for animal cells that are described in various books (*1,5–12*), including an initial fixation with a buffered fixative (such as glutaraldehyde, perhaps augmented with formaldehyde), postfixation in osmium tetroxide, dehydration in ethanol, infiltration, and embedding with resins. However, specific details of specimen preparations for plant cells are yet to be thoroughly researched. This chapter provides a brief, condensed account of chemical specimen preparations for plant cells, with special emphases on adding caffeine in fixatives to stabilize the phenol in the vacuoles and using vacuum infiltration to facilitate the removal of gases in tissues during fixation as well as resin infiltration.

To minimize the artifact formation caused by the prolonged specimen preparation generally used in the conventional chemical preparation method, rapid specimen preparation (up to eight-hours) for ultrastructural studies of plant cells has been suggested *(13)*. Shortening the preparation time to only 4 h can be achieved with the advanced microwave-assisted method *([14]*; *also see* Chapter 4 in this volume).

2. Materials

1. Plastic Petri dishes (90-mm diameter).
2. Filter paper (no. 1, 90-mm diameter, Whatman cat. no. 1001 090).
3. Single-sided razor blade and double sided razor blade (Gillette stainless-steel blades).
4. 0.1 *M* Sorenson's phosphate buffer, pH 7.0 (*see* **Note 1**).
5. 2.5% to 5% Glutaraldehyde, in 0.1 *M* phosphate buffer.
6. 0.1% to 1.0% Caffeine (*see* **Note 2**; *[1]*).
7. Karnovsky's fixative (*see* **Note 3**; *[15]*).
8. Glass vial (8 mL, 12-mm width mouth, ACI, Victoria, Australia) with screw lid or its equivalent.
9. Vacuum oven with an attached vacuum gauge for specimen initial fixation and for specimen infiltration and polymerization.
10. Flint glass disposable pipettes (23-cm length, 2 mL, Chase Instrument, Glens Fall, NY, cal. no. 093) and rubber bulb.
11. 2% Osmium tetroxide in 0.1 *M* phosphate buffer, pH 7 (*see* **Note 4**).
12. 100% (absolute) Ethanol (AnalaR Ethanol 99.7-100% v/v, BDH) and 25%, 50%, 75%, 95% ethanol in distilled water.
13. Propylene oxide (BDH; cat. no. 28290).
14. Top loading scale for measuring resin components.
15. Spurr's resin (*see* **Note 5** *[16]*).
16. Home-made rotary shaker.
17. Sharpened tooth picks or sharpened bamboo sticks for transfer specimens from vials to embedding moulds.
18. Aluminum foil weighting dishes or Peel-away embedding moulds (ProSci. Tech., Queensland, Australia; *see* **Note 6**).

3. Methods

3.1. Primary Fixation and Vacuum Infiltration

1. Cut fresh tissues (not larger than 1 cm^3) initially with single sided razor blade in a Petri dish with a filter paper wetted with 2.5 to 5% glutaraldehyde fixative.
2. Cut the 1-cm^3 block into 1- to 2-mm^3 pieces with double sided razor blade in the same fixative in the Petri dish (*see* **Note 7**).
3. Transfer specimens into the glass vials containing the fixative using a pair of forceps (*see* **Note 8**).
4. Vacuum infiltration for 3 to 5 min or longer if necessary, until the specimens have sunk to the bottom of glass vials (*see* **Note 9**).
5. Leave in the fixative at ambient temperature or with crushed ice for 2 h or longer.
6. Replace the fixative with the buffer 3X within 30 min.

7. Postfix with 2% OsO_4 in phosphate buffer at room temperature for 1 to 2 h.
8. Replace OsO_4 with distilled water 3X within 30 min.

3.2. Dehydration (see Note 10)

1. Replace distilled water with 50%, 70%, 90%, and 95% aqueous ethanol sequentially, each for 30 min or longer.
2. Replace with new 100% ethanol, 2X, each for 2 h or longer.

3.3. Infiltration (see Notes 11 and 12)

1. Replace 100% ethanol with infiltration medium 100% propolyne oxide, 2X, each for 1hr or longer.
2. Replace 100% propolyne oxide with the infiltration medium at the ratio of 100% propolyne oxide: Spurr's resin = 2:1, and place vials in a home-made rotary shaker for 2 h or longer.
3. Replace the old infiltration medium with the new infiltration medium at the ratio of 100% propolyne oxide: Spurr's resin = 1:1, and return vials to the rotary shaker for 2 h or longer.
4. Replace the used resin with a new infiltration medium at the ratio of 100% propolyne oxide: Spurr's resin = 1:2, and return vials to the rotary shaker for a further 2 h or longer.
5. Replace the old infiltration medium with the new 100% Spurr's resin and return vials to the rotary shaker for a final 2 h or longer.

3.4. Embedding

1. Use sharpened wooden sticks to remove specimens from vials to flat embedded mould containing fresh Spurr's resin.
2. Place embedding moulds with specimens in the oven at 70°C for at least 8 h.

Ultrastructural images of selective plant tissues are illustrated in **Figs. 1–6**.

4. Notes

1. Phosphate buffer generally is chosen because it mimics certain components of extracellular fluid and is nontoxic to cells in culture. However, it should be aware that phosphate buffer is more likely to cause precipitates during fixation than other buffers and can slowly be contaminated with micro-organisms. Sorenson phosphate buffer consists of two components: a monobasic sodium phosphate (NaH_2PO_4 2.78g/100 mL H_2O) and a dibasic sodium phosphate (Na_2HPO_4 5.36g/100 mL H_2O). The pH will be adjusted according to the variation of the volume of each stock solution that is shown in **Table 1**. Osmolarity at pH 7.2 is 226 mosmols, and an addition of 0.18 M sucrose can raise it to 425 mosmols.
2. For the high plant, particularly highly hydrated specimens, the leakage of vacuolar contents into cytoplasm will cause a considerable dilution of the fixative concentration. To overcome this effect, it is desirable to use a relatively higher concentration of glutaraldehyde. If necessary, 0.1% to 1.0% caffeine can be added into

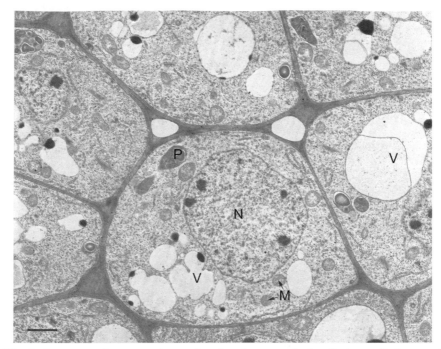

Fig. 1. A young wheat root tissue showing these cortical cells have distinct nuclei (N), plastids (P), small mitochondria (M), and various sizes of vacuoles (V), some of which contain electron opaque materials. Bar = 1 μm.

glutaraldehyde fixative to stabilize vacuolar saps (e.g., phenols). Caffeine precipitates phenols within vacuoles and prevents its release into the cytoplasm. If required, $CaCl_2$ (1–3 mM), sucrose or NaCl can be added in the final concentration of the fixative to adjust osmolarity.

3. The final concentration of Karnovsky's fixative is 2% paraformaldehyde and 2.5% glutaraldehyde in 0.1 M phosphate buffer. It should be noted that methanol-free formaldehyde must be freshly prepared from paraformaldehyde. The benefit of using both paraformaldehyde and glutaraldehyde in Karnovsky's fixative is based on the fact that formaldehyde can penetrate more rapidly than glutaraldehyde, so cellular structures can be more rapidly but only temporarily stabilized by formaldehyde and subsequently crosslinked and fixed permanently by glutaraldehyde. For immunolabelling works, a fixative of 2% paraformaldehyde in 0.1 M phosphate buffer is recommended in order to retain more antigenicity in tissue.

4. To prepare 2% OsO_4, 1 g of ampoule is used. After removing all attached labels on the surface of the ampoule, the cleaned container is placed in the liquid nitrogen for 10 to 20 s. During the freezing, osmium tetroxide will be recrystallized and can be removed easily and cleanly when the ampoule is snapped open. Pour crystallized osmium tetroxide carefully into a glass beaker containing a buffer solution and stirred until osmium tetroxide is dissolved completely. The leftover diluted osmium

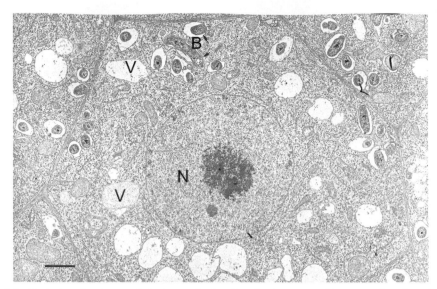

Fig. 2. A developing nodule in cortical cells of a *Lupinus albus* root showing these cells have prominent nucleus (N) and many small vacuoles (V), some of them have bacteria (B). Bar = 1 µm.

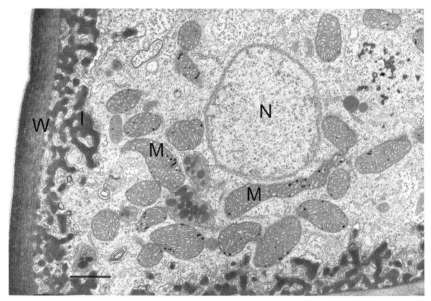

Fig. 3. A digestive gland cell of *Byblis* sp. showing a distinct nucleus (N), numerous mitochondria (M) with distinct cristae, and electron dense lipids as well as prominent wall ingrowths (I) extending from cell walls (W) to increase the plasma membrane surface area. All these unusual cellular features indicating this are a metabolic very active cell. Bar = 1 µm.

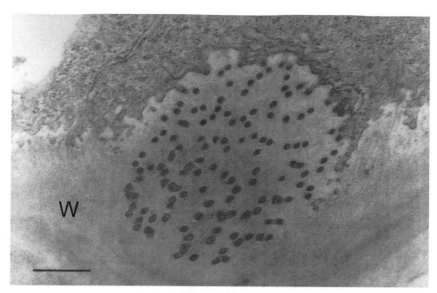

Fig. 4. Numerous plasmodesmata from a pit-field on cell walls (W) of bundle sheath cells of *Atriplex* sp. leaf. Plasmodesmata are important cellular communication channels between the adjacent plant cells. Bar = 3 μm.

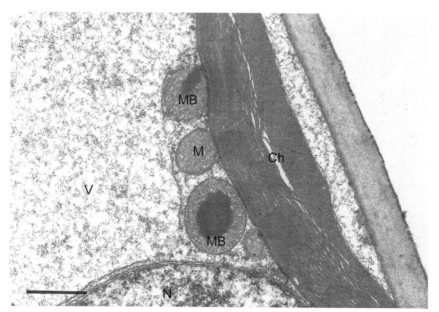

Fig. 5. A mesophyll cell of a Western Australian native plant showing nucleus (N), mitochondria (M), microbodies (MB) are situated near by. Note that the chloroplast (Ch) without distinct granna and a large vacuole contains amphorus materials. Bar = 3 μm.

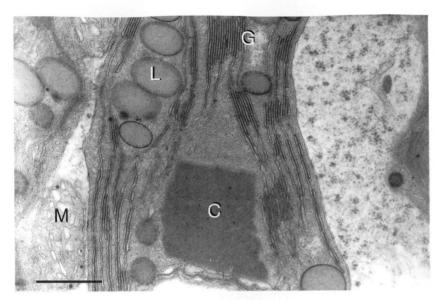

Fig. 6. An electron micrograph showing chloroplasts contain a large crystal (C), distinct grana (G) and many lipid bodies (L). Mitochondria (M) are also associated with chloroplasts. Bar = 5 μm.

tetroxide solution can be placed into 5- to 10-mL glass vials with the caps being sealed completely using Parafilm and stored in a freezer. For future usage of stored osmium tetroxide, vials should be defrosted in a fume hood at room temperature.

5. Spurr's resin is a mixture of two epoxides (ERL 4206, vinyl cyclohexene dioxide and DER 736, diglycidyl ether of polypropylene glycol) and one anhydride (NSA; nonenyl succinic anhydride) embedding medium. DER is an aliphatic and has the property of a flexibilizer whereas NSA is a hardener and with a variation of the compositions, different degrees of hardness can be achieved. ERL has a very low viscosity; therefore it can facilitate rapid penetration into tissues and is particularly suitable for plant tissues with hard cell walls. Dimethylalminoethanol (DMAE) has the function as an accelerator to increase the viscosity and the rate of polymerization. It should be noted that DMAE should never be mixed with NSA alone because it may cause an exothermic reaction. Weighting and mixing compositions for Spurr's resin of firm hardness are as follows *(16)*:

ERL 4206	10.0 g
DER 736	6.0 g
NSA	26.0 g
DMAE	0.4 g

Spurr's embedding medium is prepared by weighing components in a 50-mL disposable plastic bottle on a top-loading scale and stir for 30 min before use. The prepared Spurr's resin should be used within 3 d after it is mixed. Unused Spurr's resin should be stored in a tight-lid bottle and kept in a freezer. Because Spurr's

Table 1
Preparation of Sorenson's
0.1 *M* Phosphate Buffer with pH Value

A (mL)	B (mL)	pH of buffer
51	49	6.8
45	55	6.9
39	61	7.0
33	67	7.1
28	72	7.2
23	77	7.3

resin is very sensitive to traces of moisture, it is essential to make sure that stored resins are brought up to room temperature before uncapped.

6. For embedding, we prefer to use aluminum foil weighting dishes or peel-away embedding moulds. Multiple samples from each vial (under the same experimental condition) are embedded in the same mould. This arrangement allows for an easy control of cutting specimen in a corrected orientation or even gluing cut samples onto blank blocks. Other embedding methods include Bean capsules or gelatin capsules, which commonly are used for minute tissues when orientation is not the concern. Sometimes the whole process from fixation to polymerization can take place in Bean capsules using centrifuge tubes for tiny tissues. Slide embedding method is another choice in which tissues (normally algae or other small tissues) are placed into a drop of resin on a polished slide that is pretreated with dry film lubricant. A second treated slide is lapped (treated side down) over part of the first. After polymerization, the slides are separated and cut out a desirable area and glue onto a resin filled capsule.

7. We prefer using double sided razor blades for the specimen cutting because they are thinner and sharper than single sided razor blade or scalpel and consequently can avoid squeezing or tearing plant tissues.

8. The entire process from fixation to infiltration can take place in glass vials using Flint Glass disposable pipettes for removing old and adding fresh solutions/resins.

9. Vacuum fixation is an essential process in removing gases from plant tissues as these gases can obstruct the infiltration of fixatives, solvents and resins. Vacuum treatment is recommended at 25 to 30 psi and should be conducted intermittently with an interval of one minute or so during the primary fixation stage. The operation should continue until there is no string of tiny bubbles being released from the tissue and specimens all sink to the bottom of glass vials. Any specimens remain floating on the surface of fixative should be discarded from further specimen preparation procedure. In some cases, the air bubbles trapped inside the hair cells can be released using a small brush gently rubbing the specimen surface while the specimen is in the fixative solution or simply vigorously shaking the vial.

10. Most embedding media are not soluble in water so consequently fixed specimens have to be dehydrated by passing them through a sequence of dehydrating solution. The most widely used dehydrating agents are ethanol and acetone.

11. The resin infiltration is more difficult in plant tissues than animal tissues. Although Spurr's resin is miscible to some degree with ethanol, it can be mixed more readily with propylene oxide (1,2-epoxypropane). Therefore, this intermediate solvent has been used routinely after ethanol dehydration to assist resin infiltrating into plant tissues, even though the solvent has been considered as carcinogenic and abandoned in some animal tissue preparation *(17)*. **Caution**: Propylene oxide is carcinogenic and highly flammable. If concern exists, this step may be eliminated.

12. It has been reported that air bubbles can be formed in certain plant tissues during resin infiltration by solvent cavitations due to hydraulic pressure decreases within the cells because the thick lignified and suberized cell walls are freely permeable to the solvent but not to the resin molecules *(18)*. According to England et al. *(18)*, the bubble can form not only within the intercellular spaces but also inside the lumens of a particular type of cells such as parenchyma. The presence of these bubbles prevents complete homogeneous infiltration of the tissue with the resin, which can cause the collapse of cells and is particularly common in endodermal and hypodermal tissues containing suberin in their cell walls. Such tissue blocks are almost impossible to section. The primary cause of these problems is the differential penetration of various components of resin mixture. Vacuum infiltration of resin infiltrated specimens can assist in removing these bubbles. However, the anhydrous freeze-substitution prepared tissue has been demonstrated to be successful at preventing bubble formation and the collapse of cells *(18)*.

References

1. Hayat, M. A. (2000) *Principles and Techniques of Electron Microscopy—Biological Application*, 4th ed. Cambridge University Press, Cambridge, UK, pp. 439–471.

2. Sabatini, D. D., Bensch, K., and Barrentt, R. (1963) Cytochemistry and electron microscopy. The preservation of cellular structure and enzymatic activity by aldehyde fixation. *J. Cell Biol.* **17,** 19–59.

3. O'Brien, T. P., Kuo, J., McCully, M. E., and Zee, S.-Y. (1973) Coagulant and non-coagulant fixation of plant cells. *Aust. J. Biol. Sci.* **26,** 1231–1250.

4. Mersey, B. and McCully, M. E. (1978) Monitoring of the course of fixation of plant cells. *J. Microsc. (Oxford)* **114,** 49–76.

5. Dykstra, M. J. and Reuss, L. E. (2003) *Biological Electron Microscopy: Theory, Techniques, and Troubleshooting*, Kluwer Academic/Plenum Publishers, New York.

6. Dashek, W. V. (ed.) (2000) *Methods in Plant Electron Microscopy and Cytochemistry*, Humana Press, Totowa, NJ.

7. Harris, N. and Oparka, K. J. (eds.) (1994) *Plant Cell Biology: A Practical Approaches*. Oxford University Press, Oxford, UK.

8. Hall, J. L. and Hawes, C. (eds.) (1991) *Electron Microscopy of Plant Cells*. Academic Press, London, UK.

9. Vigil, E. L. and Hawes, C. (eds.) (1989) *Cytochemical and Immunological Approaches to Plant Cell Biology*, Academic Press, London, San Diego, CA.

10. Robards, W. A. (ed.) (1985) *Botanical Microscopy*, Oxford, London, UK.

11. O'Brien, T. P. and McCully, M. E. (1981) *The Study of Plant Structure—Principles and Selected Methods*. Thermacarphi Pty Ltd, Melbourne, Australia.

12. Hall, J. L. (1978) *Electron Microscopy and Cytochemistry*. Elsevier North Holland Biomedical Press, Amsterdam, The Netherlands.

13. Bain, J. M. and Gove, D. W. (1971) Rapid preparation of plant tissues for electron microscopy. *J. Microsc. (Oxford)* **93,** 159.

14. Giberson, R. T. and Demaree, R. S., Jr. (1999) Microwave processing techniques for electron microscopy: a four-hour protocol. *Methods Mol. Biol.* **117,** 145–158.

15. Karnovsky, M. J. (1965) A formaldehyde-glutaraldehyde fixative of high osmolarity for use in electron microscopy. *J. Cell Biol.* **27,** 137a.

16. Spurr, A. R. (1969) A low-viscosity epoxy resin embedding medium for electron microscopy. *J. Ultrastruct. Res.* **26,** 31–43.

17. Mascorro, J. A. (2004) Propylene oxide: To use or not to use in biological tissue processing. *Microscopy Today* **12,** 45–46.

18. England, W. E., McCully, M. E., and Huang, C. X. (1997) Solvent vapour lock: an extreme case of the problems caused by lignified and suberized cell walls during resin filtration. *J. Microsc. (Oxford)* **185,** 85–93.

4

Microwave-Assisted Processing and Embedding for Transmission Electron Microscopy

Paul Webster

Summary

Microwave processors can provide a means of rapid processing and resin embedding for biological specimens that are to be sectioned and examined by transmission electron microscopy. This chapter describes a microwave-assisted protocol for processing, dehydrating, and embedding biological material, from living specimens to blocks embedded in sectionable resin in 4 h or less.

Key Words: Transmission electron microscopy; resin embedding; rapid fixation; microwave processing.

1. Introduction

Microwave processors have become essential tools in the histology laboratory and have been incorporated into routine protocols. Microwaves are used to assist in specimen fixation (*1–3*), paraffin embedding (*4,5*), antigen retrieval (*6–8*), staining (*9–11*), immuno-labeling (*6,12–18*), and *in situ* hybridization (*19, 20*). Microwave processor are also being used for the rapid decalcification of bone (*13,14,21–23*). In all instances in which microwave-assisted processing has been compared with established protocols, the authors have reported significant reductions in processing times with no detectable differences between the two methods. Often, the investigators have reported improved specimen morphology or antigenicity in specimens processed using microwave-assisted protocols.

Similar reports of microwave-assisted processing for electron microscopy (EM) indicate rapid processing times, improved specimen morphology, and increased antigenicity of specimens being used for immunocytochemical experiments (*24–33*). Microwaves have been used to embed specimens for subsequent immuno-

From: *Methods in Molecular Biology, vol. 369*
Electron Microscopy: Methods and Protocols, Second Edition
Edited by: J. Kuo © Humana Press Inc., Totowa, NJ

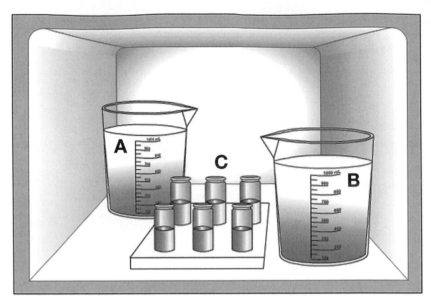

Fig. 1. Microwave processor chamber floor plan schematic. Two 500-mL glass beakers (**A** and **B**) filled with water are placed at opposing corners of the chamber. Beaker A, at the back left of the chamber, can be connected to a circulating, cooled water supply. The specimen vials are placed in the center of the chamber (**C**), between the two beakers of water (the cold spot). A temperature probe can be placed in a blank specimen vial that is treated in the same way as those containing specimens.

labeling and for reducing immunolabeling times *(34,35)*. Microwave-assisted resin polymerization has been used to embed thawed, thin cryosections *(36)*, and diluted antibodies have been irradiated by using microwave to increase their labeling efficiency for EM immunocytochemistry at the electron microscopic level *(34)*. However, laboratories using EM have not readily adopted the microwave processor into their routine protocols.

The protocol described in this chapter has been extensively tested in different laboratories using different microwave processors and has produced reproducible results. The only variable we have discovered is with the Epon substitutes available from different suppliers. The Spurr-Epon recipe we have provided will polymerize differently if Eponate 12 is not used. However, although all Epon substitutes can be used, the final result may require some adjustment of the resin hardness. We recommend testing all resin mixtures before using them to embed specimens.

The complete protocol can be applied as is, or, as is the case in my laboratory, parts of the protocol can be incorporated into existing longer protocols. For example, we occasionally perform dehydration and resin infiltration using the

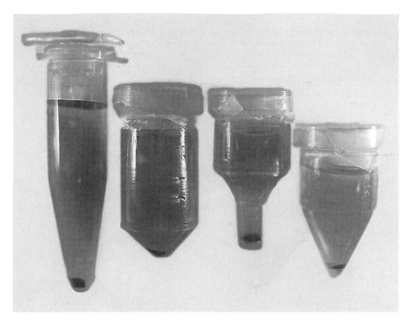

Fig. 2. Tubes for microwave-assisted resin polymerization. Epon resin (recipe 2 from **Table 2**) was used to embed small pieces of mouse liver. The specimens were embedded in resin using tubes of different plastic and shapes. The first tube is a 0.5-mL Eppendorf centrifuge tube. The other tubes are BEEM capsule molds obtained from microscope supply companies.

microwave processor but polymerize the resin overnight at 60°C in a regular oven. We also use rapid resin polymerization to re-embed polymerized specimens that need reorienting for sectioning.

2. Materials

2.1. Equipment

A microwave processor designed for laboratory use with the following features *(37,38)*: magnetron prewarming; variable wattage; forced gas extraction system; thermocouple temperature probe; load cooler; and programmable presets (*see* **Fig. 1**).

2.2. Consumables

1. Glass or plastic specimen vials. We use 60-mL scintillation vials. Of course, other glass or plastic vials can be used (*see* **Note 19**). For safety reasons, it is important that lids be removed before being irradiated in the microwave chamber. This will prevent closed vials from exploding.
2. Eppendorf tubes (1.5-mL and 0.5-mL size).

Table 1
Resin Components Used in Microwave-Assisted Polymerization

Spurr's resin	Epon 812	LR White
Comprising:	Comprising:	This resin is sold premixed and
ERL 4206	Epon 812 substitute	should be stored at −20°C following
DER 736	DDSA	the supplier's instructions
NSA	NSA	
DMAE	BDMA	

3. Two 500-mL glass beakers.
4. Rubbermaid sandwich boxes.
5. Plastic Pasteur pipets.
6. Ice buckets.
7. BEEM capsules (*see* **Fig. 2**).
8. Teflon capsule holders or Eppendorf tube racks.
9. Parafilm® (Parafilm).

2.3. Chemicals

1. 2.5% Glutaraldehyde in 100 mM sodium cacodylate buffer, pH 7.2.
2. 4% Formaldehyde in 100 mM N-Hydroxyethylpiperazine-N'-2-ethanesulfonate (HEPES) buffer, pH 7.0.
3. 1% Aqueous osmium tetroxide.
4. 100 mM Sodium cacodylate, pH 7.2.
5. 100 mM HEPES buffer, pH 7.0.
6. Increasing concentrations of acetone in water (50%, 70%, and 90%).
7. Dry acetone.
8. Embedding resins for electron microscopy. The resins are available in kit form, and the individual components can be purchased from any electron microscopy supply company (*see* **Tables 1** and **2**).
9. 1% Osmium tetroxide containing 0.3% potassium ferrocyanide.
10. 1% Aqueous uranyl acetate.
11. 1% Uranyl acetate dissolved in 50 mM sodium maleate, pH 5.2.
12. 0.12% Glycine in phosphate-buffered saline (PBS) buffer.
13. 10% Gelatin in water or PBS.

3. Methods

3.1. Setting Up the Microwave Processor

It is recommended that laboratory microwave processors, and not household microwave ovens, be used to perform the following protocols (*see* **Note 1**). The commercially available laboratory machines meet the necessary safety standards required when handling the hazardous materials encountered in biomedical

Table 2
Resin Formulations for Microwave-Assisted Polymerization

1. SPURR-EPON Resin (from Giberson, R. T. and Demaree *(28)*
   To make approx 20 mL, mix the following together in a tube:

ERL 4206	2.5 g
DER 736	1.0 g
NSA	6.5 g
Eponate 12	6.25 g
DDSA	3.25 g
NMA	3.0 g

   Store aliquots frozen in glass or polypropylene tubes
   Before use, thaw and add:

DMAE	6 drops
DMP-30	8 drops

2. EPON Resin (from Cavusoglu *[24]*)
   Prepare and mix
   Solution A: Epon 812 (1.86 g) + DDSA (2.4 g)
   Solution B: Epon 812 (2.48 g) + NMA (2.17 g)
   Mix equal amounts of solutions A and B and add BDMA, 4 drops per milliliter.

3. Epon Resin (simplified recipe for making large batches)
   Mix the following in a large glass jar:

NMA	110 mL
DDSA	130 mL
Eponate 12	230 mL

   When the ingredients have been well mixed, the resin is poured into small tubes in aliquots of 4 to 6 mL and stored frozen.
   For use, warm a tube and add four drops of BDMA per milliliter. Use immediately and discard unused resin.
   Preparing and storing resin this way will ensure that resin ingredients are not wasted and all the tubes will have resin of the same consistency. The mixture can be tested before use by polymerizing one tube of resin. All subsequent tubes of resin will have the same polymerization and sectioning qualities as this first tube.

4. LR White
   This resin is supplied premixed. Warm to room temperature before use.

It is possible to prepare the resin formulations listed in this table without adding the catalyst. The mixed resins can be stored in small aliquots at −20°C for long periods without risk of the resin polymerizing (*see* **Note 23**). When the resin is ready to use, an aliquot is warmed to room temperature and catalyst is added dropwise from glass Pasteur pipettes. The numbers of drops added has been calibrated by using glass Pasteur pipettes (1 drop = 0.01 g).

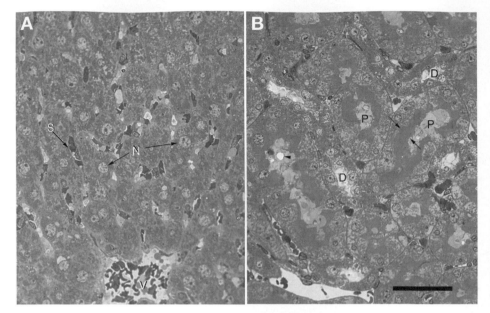

Fig. 3. Light micrographs of microwave processed epoxy resin-embedded mouse liver and kidney. The tissues were fixed overnight in 4% buffered formaldehyde at 4°C. The next day, the tissues were processed following the protocol described in the Methods section and embedded in Epon using recipe 3 from Table 2. **(A)** The liver section shows the portal vein at the bottom of the image (V) filled with erythrocytes. The nuclei (N) of the hepatocytes are visible and the sinusoids (S) between the hepatocytes are filled with erythrocytes. This tissue has not been perfusion-fixed so there is no expansion of the sinusoids. **(B)** The kidney section reveals sectioned proximal tubules (P), identified by the presence of the characteristic brush border (arrows). Profiles of distal tubules (D) also are present. Although the epoxy resin appears to have efficiently penetrated the cells in this section, there are holes in the amorphous material (arrowhead) present in the lumen of the proximal tubules, suggesting incomplete polymerization of resin. Bar = 300 µm.

laboratories. Follow the manufacturer's recommendations when installing and operating the microwave processor (*see* **Notes 2** and **3**) and pay special attention when connecting the gas extraction system. It must be attached to an outlet suitable for handling volatile, toxic gasses.

Once the processor has been connected, the hot spots in the oven chamber should be identified. It is then necessary to create a cold spot within the chamber where specimens can be irradiated (*see* **Fig. 1**) (*see* **Notes 4–6**). The cold spot is created by placing a water load within the microwave chamber that consists of two 500-mL beakers of water close to the hot spots. The water load will absorb the heat produced by the microwaves and create an area on the chamber floor where microwave irradiation is more evenly spread. Test the cold spot with either a neon bulb array or an array of small tubes filled with water (*see* **Notes 7–9**). Prebuilt devices consisting of a flat chamber through which cold water is pumped, thus creating a cold spot without the need for extensive testing, can be purchased (*see* **Note 9**).

3.2. Microwave-Assisted Primary Fixation Using Glutaraldehyde/Formaldeyde

Cells and tissues that are to be processed for examination by EM usually are crosslinked using chemical agents such as glutaraldehyde or formaldehyde. The biological material is either immersed in the chemical fixative or the fixative is perfused through the specimen. Successful crosslinking depends on the chemical fixative to be able to rapidly penetrate and react with cellular components. How chemical fixatives work is still not fully understood (for a review of chemical fixation *see* Chapter 2 in *[39]*). The mechanisms by which microwaves assist in chemical fixation are even less understood (*see* **Note 10**). However, many protocols are available that have been reported to work successfully. In this instance, a fixation schedule proposed by Giberson and Demaree *(28)* is suggested.

1. If the microwave processor has a temperature probe connected to the magnetron, set the temperature limit to 30°C.
2. Carefully dissect out the target tissue and immerse it immediately in a Petri dish containing 2.5% buffered glutaraldehyde. Cut the tissue into 2- to 3-mm^3 specimen blocks using two thin, flat razor blades and transfer these blocks to glass or plastic vials containing fresh 2.5% buffered glutaraldehyde.
3. Cool the vials (without lids) on ice until they have stabilized at 4°C and then transfer them to the cold spot in the microwave chamber. The cold spot is the area between the two beakers of water where microwaves are evenly spread.
4. Irradiate at 100% power for 40 s and then transfer them to ice outside the microwave chamber. Leave for 5 min.
5. Place the vials back in the microwave chamber and irradiate again at 100% power for 40 s.

3.3. Microwave-Assisted Postfixation Using Osmium Tetroxide

1. Keep the temperature limit on the microwave processor set to 30°C.
2. Remove the primary fixative from vials and add sodium cacodylate buffer immediately, and leave for 5 min on ice outside the microwave chamber.
3. Remove the buffer and add 1% aqueous osmium tetroxide that has been pre-cooled to 4°C.
4. Microwave at 100% power for 40 s.
5. Transfer the vials back to ice outside the microwave chamber and leave for 5 min.

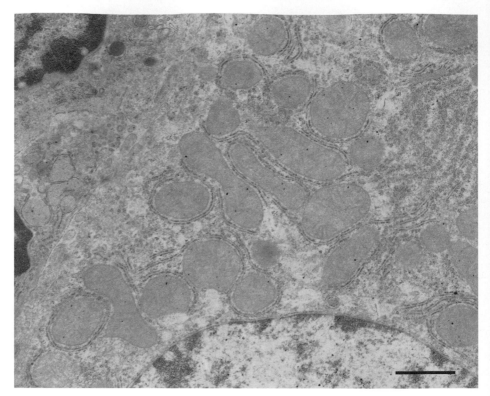

Fig. 4. EM of a section through the immersion-fixed liver specimen shown in **Fig. 3**. Cellular components are easily identified. However, membrane contrast in this section is low, probably due to the tissues being fixed first in 4% buffered formaldehyde alone. Bar = 1 μm.

6. Keeping the specimens immersed in the 1% osmium tetroxide, return the vials to the microwave chamber and microwave at 100% power for 40 s (*see* **Note 11**).

3.4. Microwave-Assisted Dehydration

1. Remove the osmium tetroxide and replace with distilled water. This is a quick change only (*see* **Note 12**).
2. Specimen dehydration consists of exposing the specimens to increasing concentrations of acetone. The dehydration steps are single changes in 50%, 70%, and 90% acetone followed by two changes in 100% acetone (*see* **Note 13**).
3. For each change in the required acetone solution, microwave the specimens at 100% power for 40 s (*see* **Note 14**).
4. Remove the specimens from the microwave processor after each step and replace the acetone with the next, increasing concentration of acetone. A protocol using LR White for immunocytochemical experiments will have slight changes. These include using 4% formaldehyde in 100 m*M* HEPES buffer as the primary fixative

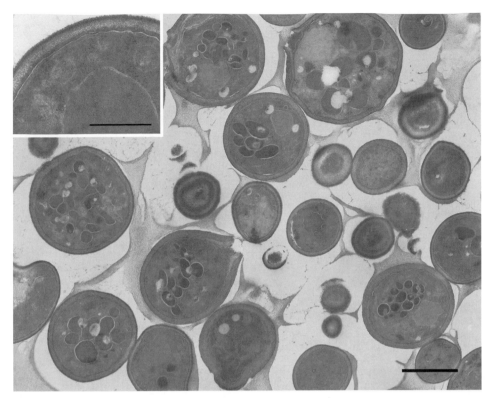

Fig. 5. Yeast cells, with intact cell wall, processed and embedded using the microwave protocol. The cells were fixed in suspension in 2.5% buffered glutaraldehyde washed in buffer containing 0.12% glycine and then embedded in 10% gelatin. Small blocks of the embedded cells were microwave processed and embedded in Spurr-Epon resin (recipe 1 in **Table 2,** *see* **Note 16**). The images were taken from a study where all specimens were prepared using a microwave-assisted protocol *(41)*. Bar large image = 2 μm. Bar insert = 200 nm.

(washing in 100 m*M* HEPES buffer), and using ethanol as the dehydration agent. Fixation using osmium tetroxide can be omitted (*see* **Note 15**).

3.5. Microwave-Assisted Resin Infiltration

1. Set the temperature limit on the microwave processor to 45°C.
2. Replace the 100% acetone in the vials with a mixture containing equal volumes of 100% acetone and epoxy resin (*see* **Table 2** for formulations). For LR White embedding, *see* **Note 15**).
3. Irradiate the vials at 100% power for 15 min.
4. Remove the acetone-resin mixture and replace with 100% resin.
5. Irradiate the vials at 100% power for 15 min.
6. Remove the old resin and replace with fresh 100% resin.

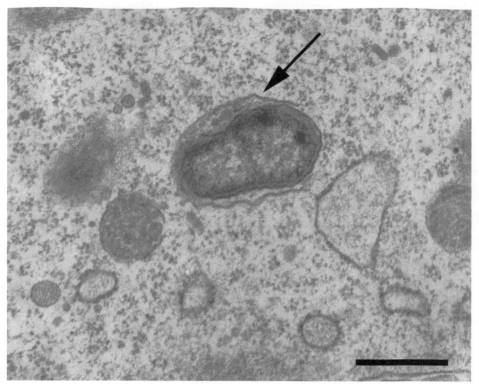

Fig. 6. A section through a human epithelial cell containing an internalized bacterium. The bacterium, the gram-negative non-typeable *Haemophilus influenzae* a common otitis media pathogen, is in the process of breaking out of the enclosing phagosome membrane (arrow). The cell pellets used in this study were fixed in 2.5% buffered glutaraldehyde and processed using the microwave protocol described in the Methods section. Bar = 500 nm.

7. Irradiate the vials at 100% power for 15 min (*see* **Note 16**).

3.6. Microwave-Assisted Resin Polymerization

1. Remove the specimens from vials and place each specimen block individually into a BEEM capsule. The BEEM capsules should have the lids cut off and be firmly held in a Teflon rack. Unique identifier labels can also be placed into the tubes at this stage (*see* **Note 17**).
2. Fill the capsules with fresh resin containing catalyst and cover the top with a small square of Parafilm.
3. Cover the Parafilm with the previously removed BEEM capsule lid and press the lid firmly in place. The aim of this step is to seal the BEEM capsule and keep moisture out (*see* **Note 18**).
4. Place the Teflon tray containing the BEEM capsules into a Rubbermaid tray and cover the capsules with tap water.

5. Place the tray in the microwave chamber, place the temperature probe into the water and set the temperature limit to 101°C (*see* **Note 19**).
6. Microwave at 100% power for 90 min (60 min plus 30 min; *see* **Note 20**). Shorter exposure to microwaves is required for LR White polymerization (*see* **Note 21**).
7. Remove the tray containing the BEEM capsules and hot water (*see* **Note 22**).
8. Remove the BEEM capsules from the water and leave to cool.
9. Using a razor blade or BEEM capsule press, remove the plastic mold from the block and prepare for sectioning. Schematic summaries of typical microwave-assisted specimen preparation protocols are presented in **Table 3**. Examples of microwave-processed biological samples are presented in **Figs. 3–7**.

4. Notes

1. The methods described in this chapter for rapid fixation, dehydration, and resin embedding have been tested using a regular high-power household microwave oven and a commercially available laboratory microwave processor. For safety reasons, we do not recommend the use of a regular high-power household microwave oven. However, a regular high-power microwave oven with no shielding or ventilation system is not recommended. Aldehydes, osmium tetroxide, and the resins used for fixing and embedding biological materials give off toxic vapors that are harmful to humans. Acetone and ethanol are flammable substances that may be ignited in an unmodified and unshielded microwave processors.
2. Microwave processors should not be operated if the chamber is empty. Doing so may cause damage to the magnetron.
3. All microwave processors have the ability to heat objects exposed to microwaves. The heating is unpredictable in that contents of containers can become much hotter than the container. Taking objects out of a microwave processor should be performed with extreme caution after exposure to the microwaves and all objects should be treated as burn hazards.
4. Calibration of laboratory microwaves as been extensively covered in a book by Login and Dvorak *(40)*.
5. Hot spots are areas within the microwave chamber that occur as a result of the resonating microwaves generated within the closed chamber. The microwaves produce an uneven distribution in the form of a standing wave.
6. Hot spots in the microwave chamber can be detected using a neon bulb array or liquid crystal temperature strips (both available from many EM supply companies). These indicators provide instant, visible result. The neon bulbs light up, and the temperature strips change color when located on hot spots. However, the hot spots also can be determined using an array of small tubes containing equal volumes of water. Place the array on the chamber floor and turn on the processor at full power for a short time (e.g., 5 s). The hot spots are revealed by an increase of water temperature in the tubes. Although not recommended, Styrofoam sheets have also been used to detect hot spots *(37)*. After irradiation, patches of melted plastic identified the hot spots!

Table 3
Annotated Microwave Specimen Preparation Protocols

Time	Solution	Microwave	Temp limit
1. Embedding in epoxy resin (for specimens already chemically fixed) Use resin formulations A or B from Table 2			
10 s	Fixative	Yes	30°C
20 s	Fixative	No	n/a
10 s	Fixative	Yes	30°C
5 min	Buffer	No	n/a
40 s	1% osmium tetroxide	Yes	37°C
30 s	Water	No	n/a
40 s	50% Acetone	Yes	37°C
40 s	70% Acetone	Yes	37°C
40 s	90% Acetone	Yes	37°C
40 s	100% Acetone	Yes	37°C
40 s	100% Acetone	Yes	37°C
15 min	1:1 Acetone:resin	Yes	45°C
15 min	100% resin	Yes	45°C
15 min	100% resin	Yes	45°C
10 min	Resin	Yes	101°C
80 min	Resin	Yes (50% Power)	101°C
2. Embedding in LR White resin (for specimens already chemically fixed) Use LR White Resin (Formulation C from Table 2)			
10 s	Fixative	Yes	30°C
20 s	Fixative	No	n/a
10 s	Fixative	Yes	30°C
5 min	Buffer	No	n/a
30 s	Water	No	n/a
40 s	50% Ethanol	Yes	37°C
40 s	70% Ethanol	Yes	37°C
40 s	90% Ethanol	Yes	37°C
40 s	100% Ethanol	Yes	37°C
40 s	100% Ethanol	Yes	37°C
15 min	1:1 ethanol:resin	Yes	45°C
15 min	100% resin	Yes	45°C
15 min	100% resin	Yes	45°C
45 min	Resin	Yes	95°C

Unless noted otherwise, all microwave irradiations are conducted with the processor operating at full power and containing a full water load (two full beakers of water with one placed at the front and one at the back of the microwave chamber) (*see* **Fig. 1**).

7. The location of the cold spot will remain constant if the water loads remain the same and in the same location.

8. Giberson and Demaree *(28)* recommend temperature changes of more than 10°C but less than 15°C in 600-μL volumes of water exposed to microwaves at 100% power for 40 s. Larger volumes of water can be used for the water load if required, either by increasing the size of the beakers used, or by adding extra beakers of water.

9. The water load can be recycled through a water chiller to help minimize the heating within the chamber. A flat chamber placed on the floor of the microwave through which cold water is circulated, can be used to create a cold spot (available from Ted Pella Inc, cat. no. 36115). Specimen vials are placed on top of this chamber when being irradiated.

10. Microwave-assisted chemical crosslinking using buffered glutaraldehyde or formaldehyde is still not well understood. Earlier studies of this process are difficult to assess because although short exposure times to microwaves were used, subsequent soaking of the irradiated biological material in dilute fixative for long periods (which continues the cross-linking process) invalidate conclusions that address only the microwave effects.

11. Increased contrast can be obtained by replacing the 1% osmium tetroxide with reduced osmium. An aqueous solution of 1% osmium tetroxide containing 0.3% potassium ferrocyanide is added to the specimen vials and irradiated at 100% power for 40 s. Cool the specimen vials on ice as described.

12. An *en bloc* staining step can be inserted after the osmium tetroxide after fixation. After the osmium solution is removed, the specimens are washed in distilled water and then irradiated at 100% power for 40 s in aqueous 1% uranyl acetate. Using 1% uranyl acetate dissolved in 50 mM sodium maleate (pH 5.2) instead of distilled water may help increase specimen contrast.

13. Although the protocol presented here uses a routine dehydration protocol with relatively large increases in the concentration of dehydrating solution, we have recently observed that whole rodent embryos processed for embedding in epoxy resin will shrink significantly when transferred directly from 70% to 90% acetone (or ethanol). To counter this shrinkage, we now dehydrate most of our specimens using more gradual increases in the concentration of the dehydration solvent (e.g., 70% to 75% to 80% to 85% to 90%).

14. Large specimen blocks or impermeable materials such as nerve biopsies, yeast, embryos or plant material may require longer fixation and dehydration times to ensure complete removal of water from the specimens. Such impermeable materials may also require extended exposure to osmium tetroxide and resins during infiltration to enable complete penetration of these substances into the specimen. The exact times to use will depend on the penetration rates into the specimens and the size of the specimen blocks. A good indication of incomplete dehydration, infiltration or polymerization is when sections from a polymerized specimen block do not hold together when sectioned. Usually the section pulls apart as it floats on the water surface during sectioning. If this occurs then new specimens can be prepared using longer processing times (*see* **Note 16**). Wrinkled sections also may

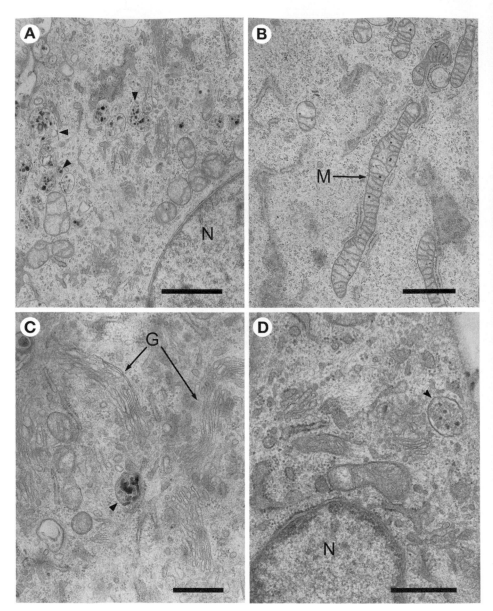

Fig. 7. Images of sections through cells taken from a study to examine the uptake and intracellular fate of synthetic particles used for DNA delivery *(45)*. All the specimens used in the transmission EM part of this study were prepared using the protocol described in **Subheading 2**. Selected images appeared on the front, cover page of the journal in which the work was published. The most remarkable thing about these images is that they appear as if they have been prepared using more traditional methods. (**A**) The DNA particles are taken into membrane-bounded vacuoles (arrowheads) that accumu-

indicate a less severe incomplete infiltration of resin. Specimens that are difficult to embed in resin can be left soaking in uncatalyzed unpolymerized resin for many days if necessary. Catalyst is added only when the resin is to be polymerized.

15. The suppliers of LR White do not recommend using acetone when dehydrating biological specimens because residual amounts of acetone can interfere with the polymerization process. When using LR White in the microwave assisted protocol, ethanol can be substituted for the acetone without problem.

16. Some tissues and cells are particularly difficult to process, even when not using microwave-assisted methods (*see* **Note 14**). We have encountered difficulties with human nerve biopsies that required much longer exposure to osmium tetroxide than most other tissues. Longer exposure can take place during exposure to microwaves or outside the microwave chamber. We also have encountered difficulties in embedding yeast cells and rodent embryos. In both instances, the problem originated from incomplete infiltration with resin. The most extreme case was with the yeast cells that were first exposed to the preparation protocol described in this chapter. To obtain complete resin infiltration, the cells were taken from the microwave chamber, placed on a rotating table and soaked for an additional 3 d in uncatalyzed resin, changing the resin each day. Embedding was performed in fresh resin containing catalyst *(41)*.

17. Early reports on microwave-assisted processing strongly advise against embedding specimens in soft, cylindrical containers such as BEEM capsules. The heat produced in the microwave chamber resulted in the capsules melting and deforming. However, if the resin is polymerized in the presence of water loads in the microwave, and if the BEEM capsules are immersed in water (it is not necessary to completely submerge the capsules), then successful resin polymerization can occur.

18. Earlier protocols are confusing in that they report poor sectioning properties of embedding resins if they are polymerized in the presence of water vapor (such as is produced when the water load is heated *(42–44)*. However, sealing BEEM capsules with Parafilm appears to prevent moisture from interfering with the polymerization process. In fact, blocks are often immersed in water to make the microwave exposure more uniform. If there is a problem with using Parafilm or submerging in water, then the resin can be polymerized in regular Eppendorf tubes (1.5-mL or 0.5-mL sizes). Eppendorf tubes can be used for LR White or epoxy resin embedding, special sealing precautions are not required and the tubes do not need to be filled. Even small amounts of resin can be polymerized. To ensure uniform exposure to microwaves the tubes are partially immersed in water and two water loads are included in the chamber (*see* **Fig. 1**).

Fig. 7. *(Continued)* late in the region around the nucleus (N). Bar = 1 μm. **(B)** The mitochondria (M) in these cells are well preserved. Bar = 1 μm. **(C)** The structures containing the phagocytosed DNA delivery particles (arrowhead) are transported to the region of the cell containing the Golgi complex (G). Bar = 500 nm. **(D)** A cell profile containing phagocytosed DNA particles (arrowhead). The nucleus (N) and other cellular structures appear normal. Bar = 500 nm.

19. If the microwave processor has a fitted temperature probe (usually a thermocouple), then it can be immersed in and used to monitor the liquid temperature around the specimen vials. However, it is recommended to place the probe in a blank vial containing the solutions the specimen is exposed to. It is possible to obtain specimen baskets with porous sides and/or base that can be placed in shallow dishes for bulk processing of multiple specimens. If specimen baskets are used, then the temperature probe is placed away from the specimens and immersed in the processing liquid. Be aware that there is still some controversy about the usefulness of temperature probes in the microwave processor. It is thought that the probe may produce localized heating by attracting microwaves and thus produce inaccurate readings.
20. LR White polymerization requires only a 45-min exposure to microwaves at 100% power.
21. Although the above protocol suggests using the microwave processor set at 100% power for polymerization, resins will polymerize using lower power irradiation. Cavusoglu et al. *(24)* suggest using a preliminary exposure at 100% power but then switching to reduced power for the remainder of the polymerization time. Polymerization experiments are very simple to perform so new users are encouraged to experiment with different conditions.
22. **SAFETY!** Irradiated materials in the microwave processor can get very hot. Polymerized resin blocks will be immersed in water close to boiling point. Remove the tray from the chamber carefully.
23. The advantages of storing mixed resins that do not contain catalyst are convenience and reproducibility. If large batches of resin are prepared and frozen down in small aliquots the formulation can be tested and then used when needed. The absence of catalyst will ensure that the first aliquot will be the same as the final aliquot used. There will be no rapid polymerization of resin occurring during storage, an event that occurs in prepared resins containing catalyst.

Acknowledgments

Thanks to Swaroop Mishra for providing the images used in **Fig. 7**. Siva Wu, Debbie Guerrero, and Alexandre Webster provided invaluable assistance in preparing the figures. This work was supported in part by the Hope for Hearing Foundation of Los Angeles and by grants from the National Institutes of Health (5 R03 DC005335-03) and the Deafness Research Foundation.

References

1. Marani, E., Boon, M. E., Adriolo, P. J., Rietveld, W. J., and Kok, L. P. (1987) Microwave-cryostat technique for neuroanatomical studies. *J. Neurosci. Meth.* **22,** 97–101.
2. Ruijter, E. T., Miller, G. J., Aalders, T. W., et al. (1997) Rapid microwave-stimulated fixation of entire prostatectomy specimens. Biomed-II MPC Study Group. *J. Pathol.* **183,** 369–375.

3. Laboux, O., Dion, N., Arana-Chavez, V., Ste-Marie, L. G., and Nanci, A. (2004) Microwave irradiation of ethanol-fixed bone improves preservation, reduces processing time, and allows both light and electron microscopy on the same sample. *J. Histochem. Cytochem.* **52,** 1267–1275.
4. Boon, M. E., Wals-Paap, C. H., Visinoni, F. A., and Kok, L. P. (1995) The two-step vacuum-microwave method for histoprocessing. *Eur. J. Morphol.* **33,** 349–358.
5. Kok, L. P. and Boon, M. E. (1995) Ultrarapid vacuum-microwave histoprocessing. *Histochem. J.* **27,** 411–419.
6. Kanai, K., Nunoya, T., Shibuya, K., Nakamura, T., and Tajima, M. (1998) Variations in effectiveness of antigen retrieval pretreatments for diagnostic immunohistochemistry. *Res. Vet. Sci.* **64,** 57–61.
7. Kahveci, Z., Minbay, F. Z., Noyan, S., and Cavusoglu, I. (2003) A comparison of microwave heating and proteolytic pretreatment antigen retrieval techniques in formalin fixed, paraffin embedded tissues. *Biotech. Histochem.* **78,** 119–128.
8. Shi, S. R., Key, M. E., and Kalra, K. L. (1991) Antigen retrieval in formalin-fixed, paraffin-embedded tissues: an enhancement method for immunohistochemical staining based on microwave oven heating of tissue sections. *J. Histochem. Cytochem.* **39,** 741–748.
9. Kahveci, Z., Minbay, F. Z., and Cavusoglu, L. (2000) Safranin O staining using a microwave oven. *Biotech. Histochem.* **75,** 264–268.
10. Minbay, F. Z., Kahveci, Z., and Cavusoglu, I. (2001) Rapid Bielschowsky silver impregnation method using microwave heating. *Biotech. Histochem.* **76,** 233–237.
11. Marani, E., Guldemond, J. M., Adriolo, P. J., Boon, M. E., and Kok, L. P. (1987) The microwave Rio-Hortega technique: a 24 hour method. *Histochem. J.* **19,** 658–664.
12. Munoz, T. E., Giberson, R. T., Demaree, R., and Day, J. R. (2004) Microwave-assisted immunostaining: a new approach yields fast and consistent results. *J. Neurosci. Meth.* **137,** 133–139.
13. Tinling, S. P., Giberson, R. T., and Kullar, R. S. (2004) Microwave exposure increases bone demineralization rate independent of temperature. *J. Microsc. (Oxford)* **215,** 230–235.
14. Cunningham, C. D., 3rd, Schulte, B. A., Bianchi, L. M., Weber, P. C., and Schmiedt, B. N. (2001) Microwave decalcification of human temporal bones. *Laryngoscope* **111,** 278–282.
15. Boon, M. E. and Kok, L. P. (1994) Microwaves for immunohistochemistry. *Micron* **25,** 151–170.
16. Boon, M. E., Hendrikse, F. C., Kok, P. G., Bolhuis, P., and Kok, L. P. (1990) A practical approach to routine immunostaining of paraffin sections in the microwave oven. *Histochem. J.* **22,** 347–352.
17. Suurmeijer, A. J., Boon, M. E., and Kok, L. P. (1990) Notes on the application of microwaves in histopathology. *Histochem. J.* **22,** 341–346.
18. Jackson, P., Lalani, E. N., and Boutsen, J. (1988) Microwave-stimulated immunogold silver staining. *Histochem. J.* **20,** 353–358.
19. Mabruk, M. J., Flint, S. R., Coleman, D. C., Shiels, O., Toner, M., and Atkins, G. J. (1996) A rapid microwave-in situ hybridization method for the definitive

diagnosis of oral hairy leukoplakia: comparison with immunohistochemistry. *J. Oral. Pathol. Med.* **25,** 170–176.

20. Van den Brink, W. J., Zijlmans, H. J., Kok, L. P., et al. (1990) Microwave irradiation in label-detection for diagnostic DNA-*in situ* hybridization. *Histochem. J.* **22,** 327–334.

21. Hellstrom, S. and Nilsson, M. (1992) The microwave oven in temporal bone research. *Acta Otolaryngol. Suppl.* **493,** 15–18.

22. Madden, V. J. and Henson, M. M. (1997) Rapid decalcification of temporal bones with preservation of ultrastructure. *Hear. Res.* **111,** 76–84.

23. Keithley, E. M., Truong, T., Chandronait, B., and Billings, P. B. (2000) Immunohistochemistry and microwave decalcification of human temporal bones. *Hear. Res.* **148,** 192–196.

24. Cavusoglu, I., Minbay, F. Z., Temel, S. G., and Noyan, S. (2001) Rapid polymerisation with microwave irradiation for transmission electron microscopy. *Eur. J. Morphol.* **39,** 313–317.

25. Arana-Chavez, V. E. and Nanci, A. (2001) High-resolution immunocytochemistry of noncollagenous matrix proteins in rat mandibles processed with microwave irradiation. *J. Histochem. Cytochem.* **49,** 1099–1109.

26. Giberson, R. T. and Demaree, R. S., Jr. (1995) Microwave fixation: understanding the variables to achieve rapid reproducible results. *Microsc. Res. Tech.* **32,** 246–254.

27. Giberson, R. T., Demaree, R. S., Jr., and Nordhausen, R. W. (1997) Four-hour processing of clinical/diagnostic specimens for electron microscopy using microwave technique. *J. Vet. Diagn. Invest.* **9,** 61–67.

28. Giberson, R. T. and Demaree, R. S., Jr. (1999) Microwave processing techniques for electron microscopy: a four-hour protocol. *Methods Mol. Biol.* **117,** 145–158.

29. Giberson, R. T., Austin, R. L., Charlesworth, J., Adamson, G., and Herrera, G. A. (2003) Microwave and digital imaging technology reduce turnaround times for diagnostic electron microscopy. *Ultrastruct. Pathol.* **27,** 187–196.

30. McLay, A. L., Anderson, J. D., and McMeekin, W. (1987) Microwave polymerisation of epoxy resin: rapid processing technique in ultrastructural pathology. *J. Clin. Pathol.* **40,** 350–352.

31. Wendt, K. D., Jensen, C. A., Tindall, R., and Katz, M. L. (2004) Comparison of conventional and microwave-assisted processing of mouse retinas for transmission electron microscopy. *J. Microsc. (Oxford)* **214,** 80–88.

32. Wouterlood, F. G., Boon, M. E., and Kok, L. P. (1990) Immunocytochemistry on free-floating sections of rat brain using microwave irradiation during the incubation in the primary antiserum: light and electron microscopy. *J. Neurosci. Meth.* **35,** 133–145.

33. Kok, L. P. and Boon, M. E. (1990) Microwaves for microscopy. *J. Microsc. (Oxford)* **158,** 291–322.

34. Chicoine, L. and Webster, P. (1998) Effect of microwave irradiation on antibody labeling efficiency when applied to ultrathin cryosections through fixed biological material. *Microsc. Res. Tech.* **42,** 24–32.

35. Rangell, L. K. and Keller, G. A. (2000) Application of microwave technology to the processing and immunolabeling of plastic-embedded and cryosections. *J. Histochem. Cytochem.* **48,** 1153–1159.
36. Keller, G. A., Tokuyasu, K. T., Dutton, A. H., and Singer, S. J. (1984) An improved procedure for immunoelectron microscopy: ultrathin plastic embedding of immunolabeled ultrathin frozen sections. *Proc. Natl. Acad. Sci. USA* **81,** 5744–5747.
37. Kok, L. P., Visser, P. E., and Boon, M. E. (1994) Programming the microwave oven. *J. Neurosci. Meth.* **55,** 119–124.
38. Marani, E. and Horobin, R. W. (1994) Overview of microwave applications in the Neurosciences. *J. Neurosci. Meth.* **55,** 111–117.
39. Griffiths, G. (1993) *Fine Structure Immunocytochemistry*, Springer-Verlag, Heidelberg.
40. Login, G. R. and Dvorak, A. M. (1994) *The Microwave Tool Book*, Beth Israel Hospital, Boston.
41. Zhong, Q., Gvozdenovic-Jeremic, J., Webster, P., Zhou, J., and Greenberg, M. L. (2005) Loss of function of KRE5 suppresses temperature sensitivity of mutants lacking mitochondrial anionic lipids. *Mol. Biol. Cell* **16,** 665–675.
42. Glauert, A. M. (1991) Embedding, in *Practical Methods in Electron Microscopy,* vol. 3 (Glauert, A., ed.), Elsevier North-Holland Biomedical Press, Amsterdam, pp. 123–176.
43. Hayat, M. A. (1989) Rinsing, dehydrating and embedding, in *Principles and Techniques of Electron Microscopy* (Hayat, M., ed.), MacMillan Press, Hong Kong, pp. 79–137.
44. Giammara, B. L. (1993) Microwave embedment for light and electron microscopy using epoxy resins, LR White, and other polymers. *Scanning* **15,** 82–87.
45. Mishra, S., Webster, P., and Davies, M. E. (2004) PEGylation significantly affects cellular uptake and intracellular trafficking of non-viral gene delivery particles. *Eur. J. Cell Biol.* **83,** 97–111.

5

Ultramicrotomy for Biological Electron Microscopy

Herbert K. Hagler

Summary

This chapter describes the practical details involved in the successful sectioning of plastic-embedded specimens for electron microscopy. The focus will be on those parts of the process that are really key to the successful sectioning of any biological material, i.e., the proper shape and size of the specimen, sharp knives and how to make them, and the orientation of specimen and production of ultrathin sections onto water-filled boats attached to either glass or diamond knives. This chapter is designed to assist the beginner in microtomy in understanding more fully the terms and basic principles involved in producing ultrathin sections of most plastic-embedded biological specimens.

Key Words: Plastic sectioning; thin sectioning; ultramicrotomy; glass knives; diamond knives; specimen mounting; ultramicrotome design.

1. Introduction

Ultramicrotomy of biological materials requires the transformation of specimens from the living, hydrated state to a dry state that faithfully maintains the structural and biological relationships that an investigator is interested in studying by light and electron microscopy. The changes introduced into the specimen during this transition to the electron microscope for observation and measurement can be quite significant. This chapter will focus on the steps involved with the ultramicrotomy of plastic-embedded biological material for observation in routine transmission electron microscopy (TEM). Many good books (*1*) have been written on the fixation and plastic embedding of biological material, and this chapter will not go into extensive detail other than to assume that the specimen of interest has been adequately fixed and infiltrated with a plastic material that is suitable for the cutting of 40- to 150-nm thin sections.

From: *Methods in Molecular Biology, vol. 369*
Electron Microscopy: Methods and Protocols, Second Edition
Edited by: J. Kuo © Humana Press Inc., Totowa, NJ

2. Materials

1. Glass knife production: glass strips 6- or 10-mm thick, 25-mm wide by 400-mm long, detergent for cleaning the glass strips, fingernail polish, aluminum tape, truf-style plastic boats, small boxes suitable for storage of glass knives.
2. Sectioning: distilled water, compressed air, hot plate, toluidine blue stain (2), dilute nitric acid (i.e., two drops concentrated nitric acid in 10 mL of distilled water), source of eye lashes or Dalmatian dog hairs, soft drinking straws, fingernail polish, microscopic slides for thick sections, EM grids, grid boxes, double action forceps, a good light microscope for evaluation of thick sections and a basic TEM.

3. Methods

3.1. Glass Knife Production: The Balanced Break Method

1. The use of the balanced break for producing excellent glass knives is quite simple in principle, i.e., to use an equal weight of glass plus breaking force on each side of the score (a scratch that is made in the glass surface using a diamond or tungsten-carbide wheel or scribe) used to initiate a fracture in the glass strip. This method was used when glass knives were first routinely broken using a pair of glazier's pliers. All scores were made in the middle of the glass sheet to be broken and an equal force was applied symmetrically to each side of the score. With the availability of the very first LKB 7800 glass knife maker, this fundamental principle was discarded and the glass knives that have been made in most EM laboratories for the past 30 yr have been made using a nonsymmetrical fracture from one end of a strip of glass. The development and distribution of this instrument has practically guaranteed that everyone has been using very poor-quality glass knives for routine plastic and cryosectioning ever since. In early work by Tokuyasu and Okamura (3), they described the importance of using the balanced break in the production of glass knives for cryoultramictomy. The importance of this observation and additional details were also described by Griffiths et al. (4). It was only during a cryo course 10 yr ago that this principle came to our attention (5), and we began to perfect the method using existing commercial knife makers for the making of excellent glass knives. Heinz Schwarz was instrumental in helping determine the optimum time for the break in experiments performed at an EMBO course in Dundee (H. Schwarz, personal communication, 1994). Because most laboratories still have access to an "old" LKB 7800 glass knife maker and there is such strong resistance to switching over to a better technique (5), I will describe how to "fix" this knife maker to make it possible to put into practice the principle of the balanced break. These details are described in **Note 1**. It is still very difficult to get a good balanced break with all of the currently manufactured glass knife makers. They all seem to misunderstand the fundamentally simple principle of the balanced break and produce knife makers that make the balanced break difficult to achieve.
2. A diagram of the glass knife and the terms associated with it are shown in **Fig. 1**. This figure illustrates that the smaller (closer to 45°) the actual knife angle, the

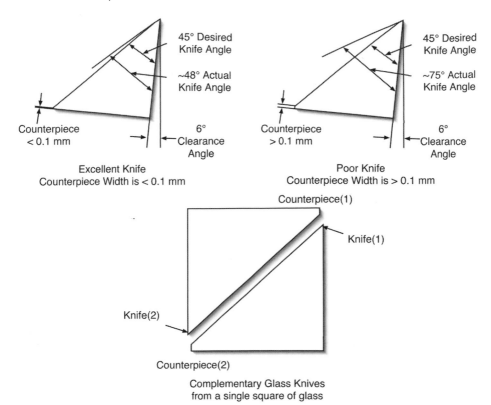

Fig. 1. The optimal 45° glass knife is shown in top left of the figure and has a counterpiece width of <0.1 mm. This knife has an actual knife angle of approx 48°. The 6° clearance angle is used most commonly with glass knives. The knife at the top right illustrates an unacceptable glass knife with an actual knife angle of approx 75° produced with a counterpiece width >0.1 mm, more on the order of 1 to 2 mm as was recommended in the instruction manual that came with the LKB 7800 glass knife maker. This glass knife will not be sharp, and the actual knife angle will vary along the knife edge. Shown at the bottom is a diagram of a glass square after the diagonal break is made. Illustrated are the two knives and their associated complementary counterpieces. The respective counterpiece that is associated with each knife is used for evaluation of the knife and is explained more fully in following figures and text.

sharper the knife. The making of excellent glass knives requires patience in the production process, and they must be prepared well before the day of intended use. Because ultramicrotomy demands the use of excellent glass knives, this production process cannot be left until the last minute when the specimen is mounted in the ultramicrotome demanding to be sectioned right now! The use of excellent glass knives is absolutely essential to the successful production of ultra-thin sections for electron microscopy. Fully 95% of the failures to obtain good sections on the ultramicrotome can be directly attributed to the use of an unsuitable glass knife or dull diamond knife.

It is important to remember that our goal is to produce very high-quality glass knives that can be reused many times, so that the time involved is not wasted by throwing away the knives after a single use. The old mythical tale that glass knives must be made fresh just is not true. If care is taken in the making of high-quality glass knives described here, they may be stored for months to years and used repeatedly until they become unusable after 5 to 15 or more uses for thin sectioning. We do get more uses when the knives are used for cryosectioning (*4*); however, we routinely use a good glass knife 5 to 15 times when cutting thin sections, for example, of Spurr's plastic-embedded specimens. The glass knives do not last as long when they are used for the cutting of thick sections because of the increased stress that is placed on the knife edge, which also is true if diamond knives are used routinely for thick sections, that there is a greater danger of damage to the diamond knife.

3. The process begins with the acquisition of glass strips typically measuring 6 or 10 mm thick × 25 mm wide × 400 mm long. The manufacturer's edges are inspected carefully to insure that they are flat surfaces. Many times the manufacturing equipment used to produce these strips is not correctly calibrated and thus produces strips that have a slightly concave edge on one side of the strip and a convex edge on the other side. These poor-quality strips are not suitable for the reproducible production of high-quality glass knives because they will result in distinctly curved edges in the final glass knife edges if the fracture is allowed to propagate into the curved manufacturer's edge.

4. The first step is to clean the glass strip with soap and water, rinse under hot running tap water, and dry the glass with a clean cloth or paper towel.

5. **Figure 2** illustrates the next part of the process. The long strip is first balanced on the balance pins of a "fixed" LBK 7800 (*see* **Note 1** *[5]*), gently clamped to hold the glass strip perpendicular to the desired fracture line (*see* **Fig. 3**, left side), a breaking force applied (*see* **Fig. 4**, items 6, 7, and 8), and then the glass is scored. When the glass is preloaded with a breaking force before scoring the glass, the fracture begins immediately and progresses very slowly through the glass strip. The force applied (*see* **Fig. 4**) with the glass knife maker is adjusted so that it takes about 3 min (H. Schwarz, personal communication, 1994) for the glass to separate under the score. The term separate is used to emphasis that we are not trying to achieve a violent break, but a gentle, controlled fracture perpendicular to the long axis of the glass strip.

6. It is critical for the person breaking the glass to understand that, for this process to be reproducible, the process must be followed as depicted in **Fig. 2**, completely finishing all of the same-sized glass strips from each row before progressing to the next row. This step insures that the balanced break of each strip of the same length is more reproducible. Because each length has half the weight of the previous row, the balance "feel" will change slightly and must be mastered by the practice of balancing all of the lengths that are the same weight at the same time. This

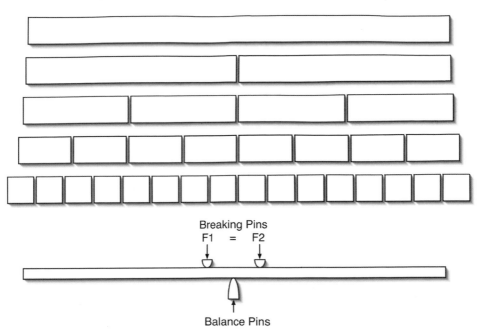

Fig. 2. The first step is to clean the glass strip with soap and water, rinse under hot running tap water, and dry the glass with a clean cloth or paper towel. The long strip is then balanced on the balance pins of a "fixed" LBK 7800 (*see* **Note 1**), and broken into two equal pieces. These two pieces, shown in the second row, are then balanced and broken. The resulting four pieces are then balanced and broken, and finally the resulting eight pieces are then balanced and broken, which should result in 16 square pieces that are all have identical dimensions. The two original end pieces may not yield usable knives because they contain manufacturer-produced edges. The edges produced in this fashion should be perpendicular to the surface of the glass strip and very flat. It is important to note that the new surfaces created by this process will be the ones that will contain the knife edge. The preparation of these surfaces requires as much care and patience as the final 45° fracture for the production of the knives. Details of the clamping and breaking of the glass are shown in **Fig. 3.**

process of balancing, preloading with the breaking force, and scoring is repeated with each equal length of glass until we have as a final product 16 squares of glass that are the same shape and size. Experience has shown that most people have mastered this process by the time they have processed at least three complete glass strips.

7. The breaking of the complete strip of glass should take approx 45 to 60 min to complete (approx 3 min/fracture), resulting in the production of 16 equal-sized squares. It is important to note that the new surfaces that are created will be super clean and will ultimately contain the knife edge, which is produced in the next steps. These

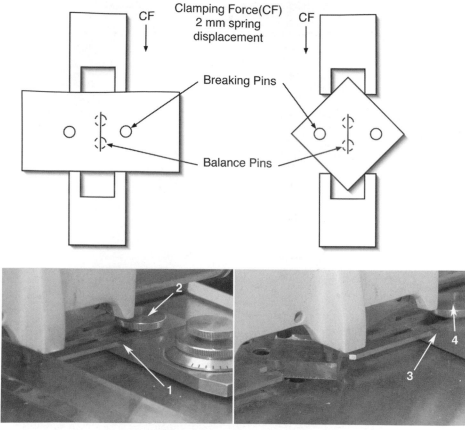

Fig. 3. This figure illustrates the light clamping force that is used to gently position the glass while making the balanced break to create the squares on the left and the final diagonal break shown on the right side. The millimeter scale (1) is used to set the 2-mm force against the glass strip using clamp screw (2). The millimeter scale (3) is also used to set the 2-mm force against the glass square using clamp screw (4). Typically, the force is set for the glass strip and then readjusted if necessary for fracturing the diagonal of the resulting squares.

fractures require as much care and patience as the final 45° fracture used for the production of the knives.

8. **Fig. 5** illustrates the production of the glass knives from the final square pieces of glass. The squares at the top show clean (**Fig. 5c**, freshly broken) and dirty (**Fig. 5d**, manufacturer produced) edges of the final squares. The cleanest edges are the freshly made surfaces. The squares are rotated counterclockwise 90° and positioned between the holding forks of the glass knife maker as illustrated in the right side of **Figs. 3** and **5**. The squares are positioned under the score wheel (*see* **Figs. 4** and **5**, which show the score setting that is used for all scores) such that the

Fig. 4. Details of the LKB 7800 glass knife maker showing (1) and (4) the picto-grams detailing the direction to turn the upper fork and lower fork adjustment knobs to achieve the alignment of the score such that the fracture propagates into the corners of the glass square. Numbers (2) and (5) show the setting of the score wheel cam which is used to score the strips and the final diagonal break of the squares. The score knob (3) is used to pull the score wheel along the top surface of the glass. The lower right panel shows a mark that is made on the breaking force knob. Number (6), which is aligned with a mark (7), indicates the zero or reset position of the breaking knob. Using trial and error with a stopwatch a mark, number (8) is drawn at the clockwise position, which gives, on average, an approx 3-min time to fracture for each break made with the glass knife maker.

fracture propagates just into the clean glass at each of the top and bottom corners as illustrated in **Fig. 5**. The distance between the knife edge and the corner of the glass square should be <0.1 mm so that a knife with an actual knife angle of approx

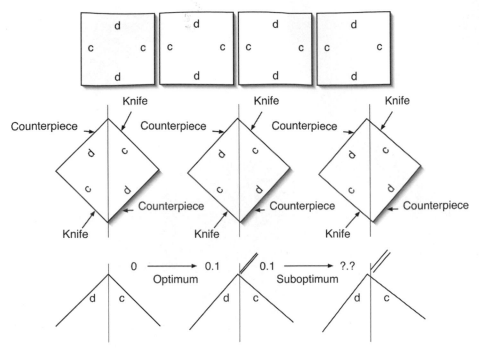

Fig. 5. Details of positioning the final glass squares is shown. In the top row the clean (c) and dirty (d) edges of the glass squares are shown. In the second row, the glass square has been rotated 90° counter-clockwise, which places the clean (c) glass surface in the upper right and lower left quadrant. There are three fractures depicted, left, directly into the apex, center, with a counterpiece width of <0.1 mm and, right, >0.1 mm. The optimum counterpiece width is shown on the left in the bottom row, whereas unacceptable knives are shown on the right in the bottom row. The direction of the fracture is controlled by moving the upper and lower apexes either left or right to place the fracture into the clean side of the apexes to produce the optimum thickness counterpiece.

48° will be produced (**Fig. 1**). Because the squares have all been made by balancing and are all the same size, it is relatively easy to adjust the upper holding fork and the lower holding fork (shown in right side of **Fig. 3**) by moving each of them either left or right until the desired <0.1 mm distance is obtained at both corners of the glass square. The best way to understand the direction to move the upper (**Fig. 4**, number 4) and lower (**Fig. 4**, number 1) holding forks is to ask the following question. Which direction do the upper and lower apexes of the glass square have to be moved to position the fracture <0.1 mm from the apexes into the clean side? Several glass squares may have to be fractured before the optimal setting is achieved. Once this final adjustment is complete most of the remaining knives should be perfect, consistently producing counterpieces that have <0.1-mm widths. This adjustment is usually only required for the first set of glass squares. Subsequent glass strips will require little or no additional adjustments and this is usually

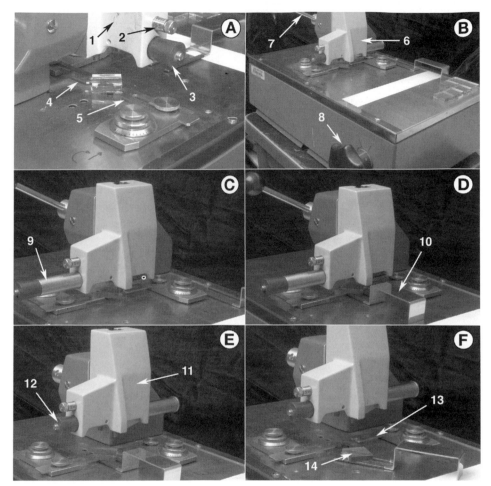

Fig. 6. Shown are the steps involved in breaking the glass squares. **(A)** the anvil (1), with score setting (2) and scoring knob (3), are lowered by gravity onto the diagonally aligned glass square. The upper fork (4) is released and holds the square against the lower fork (5). **(B)** the anvil (6) is locked into place using the locking lever (7), then the breaking force is applied using the breaking knob (8). **(C)** the scoring knob (9) is firmly pulled out. **(D)** the pickup tool (10) is placed under the glass square. **(E)** after the fracture, the anvil (11) is raised by releasing the locking lever (7) and the scoring knob (12) pushed in. **(F)** the upper fork (13) is retracted and the fractured glass square (14) removed using the pickup tool.

true for four to five complete strips of glass. By spending about a day making glass knives, there will be enough to last for several weeks to months.

9. The operational steps are always the same whether glass strips or final diagonal fractures are being produced. They are illustrated in **Fig. 6**: (a) balance the glass

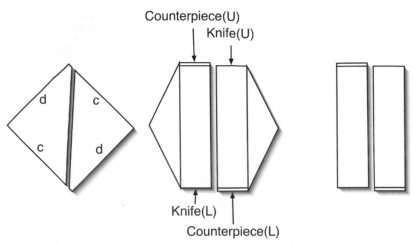

Fig. 7. The fractured glass square is removed from the knife maker, the left and right corners are then folded away from the microtomist, and the knives with their counter-pieces next to the knife that is complementary to it held as shown on the right. The upper (U) knife and counterpiece (U) are kept side by side as are the lower (L) knife and coun-terpiece (L). The counterpieces are then judged as described in the next figure.

   strips or squares on the Balance Pins; (b) apply the 2-mm clamping force to hold the strips or square in position (**Fig. 6,** numbers 4 and 5); (c) lower the anvil (**Fig. 6**), from position (1) to position (6), bringing the breaking pins into contact with the top of the glass and then use the locking lever (**Fig. 6,** number 7), to hold the glass in position; (d) apply the predetermined breaking force using the breaking knob, (**Fig. 6,** step 8; also *see* **Fig. 4**), the mark on knob (6) is rotated clockwise to alignment with position (8), this applies the amount of force that results in a approx 3-min fracture; (d) score the top surface of the glass by pulling out the scoring knob, **Fig. 6,** (9); (f) after the score is made, the preloaded glass begins to fracture over the course of the next approx 3 min until the fracture is complete, then the pickup fork is inserted under the glass square (**Fig. 6,** number 10); (g) the anvil, (**Fig. 6,** number 11) is raised by releasing the locking lever; (h) the scoring knob is pushed back in, (**Fig. 6,** number 12); and (i) the glass carefully removed using the pickup tool for the glass squares, (**Fig. 6,** number 14).

10. **Fig. 7** illustrates the handling of the glass knives for examination of each comple-mentary knife and counterpiece from the two corners of the diagonal fracture. The broken square is removed gently from the knife maker, the left and right corners are folded away, and the two knives with their complementary counterpieces held together. This step allows the examination described in the next step and keeps the knife edge adjacent to the complementary counterpiece that produced it.

11. Each final glass square has the potential of yielding two perfect knives. **Figure 8** illustrates the judgment criteria that are used to select which knives will be kept

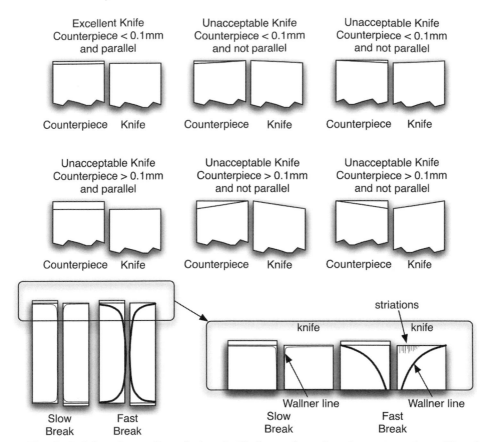

Fig. 8. Judging the quality of glass knife from observing the counterpiece. Six of the many possibilities are illustrated. Only one, the upper left illustrates a "perfect" glass knife that is suitable for use with ultramicrotomy. The other possibilities demonstrate imperfect knives and the reasons that they would not be chosen for cutting thin sections. In the case in which the counterpiece is not parallel, the knife will have an actual knife angle that changes all the way across the knife, which makes the sectioning characteristics difficult to reproduce from knife to knife. When the counterpiece is too thick, the resulting knife has a very large actual knife angle, thus making it difficult to impossible to cut thin sections. These other knives may be used to rough trim specimens but there is a risk of damaging the specimen because the knives are really dull. The importance of the slow break (>3 min) is illustrated in the bottom portion of the figure. With the slow break, the Wallner line (stress line) will be parallel and very close to the knife edge and the striations and variation in actual knife angle across the knife will be minimal. The illustrated striations associated with a fast break and the very strong Wallner line will be absent in the slow break condition. The two criteria, the counterpiece <0.1 mm and a slow (>3 min) break are required to produce the excellent knife. The excellent knives will cut more sections before they become damaged than will the poor quality knives.

for future ultramicrotomy. As illustrated in the **Fig. 8**, only knives that meet the criteria (produced using a slow break and a counterpiece <0.1 mm and parallel edges in the counterpiece) are chosen for ultramicrotomy. All of the other knives are tossed away. Some of the unacceptable knives may be used for rough trimming of specimen blocks, but the final block faces used for thin sectioning should never see a less than perfect glass knife.

A perfect glass knife produced with less than 0.1-mm counterprieces and using a slow break will result in the entire knife edge being suitable for thin sections. The slow break (approx 3 min or longer) will ensure that the striations normally associated with the classic method described in the literature will be absent from the knife edge.

The best knife, based on the aforementioned criteria, is chosen and the opposite knife with its counterpiece is placed against the bottom of a storage box. Only one of the knives from each pair is kept and stored with the counterpiece that produced it. It is bad luck if both are perfect, then one just has to choose one to place on the bottom of the box. Storing the knife with its counterpiece allows the later observation of the counterpiece to verify the quality of the knife that it produced. Thus, we have a way to look at a knife indirectly through observation of the counterpiece and judge its quality today, next week or next month. It is important to note that there is no reliable way to observe a knife edge without its complementary counterpiece to choose the best knife. There are methods described in the literature using a dissection microscope for this purpose, but the method of using the counterpiece is a more precise and reliable method in practice.

The literature on glass knife making describe vertical striations that run perpendicular to the knife edge and can extend more than 80% of the knife edge. These striations, which are caused by the fast breaking of the glass, will produce very bad knife marks in the final thin sections. If the conditions described here are followed, selecting counterpieces that are <0.1 mm in thickness and using a slow break, then the striations usually are absent from the knife edge. The use of these methods and criteria will ensure a much larger, defect-free knife edge for thin sectioning. Under these optimum conditions, the large Wallner line (stress line) normally observed with fast breaks will mostly disappear resulting in a constant actual knife angle across the width of the glass knife edge.

12. An additional process that can be performed after all of the knives have been made is the coating of the knife edges with a discontinuous film of tungsten metal in a vacuum evaporator *(6)*. This process has been used by Griffiths et al. *(4)* and they have described marked improvement of the knife performance particularly with multiple uses of the same knife *(1)*. When possible, we use this light coating of tungsten to improve the number of times a knife may be cleaned and reused. There is a note of caution when using this tungsten coating procedure; if the tungsten becomes a continuous film, i.e., electrically conductive, then the sections will stick irreversibly to a continuous film of tungsten. It is therefore very important to use a discontinuous film of tungsten produced as described in the reference.

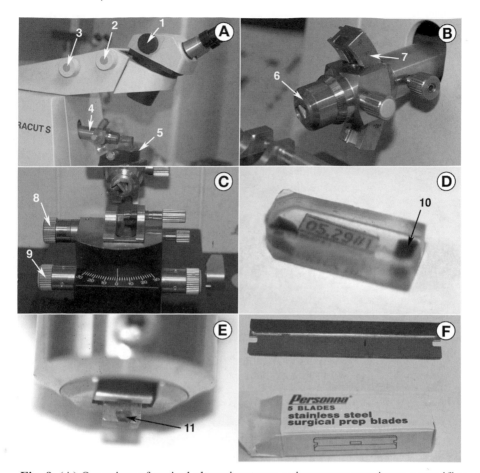

Fig. 9. (A) Overview of typical ultramicrotome, microscope eyepieces, magnification (1), focus (2), and positioning knobs (3), relationship between the microscope, specimen (4) and knife holders (5). **(B)** Specimen arc holder allowing the specimen holder (6) to rotate about the specimen arm axis and tilt about the knife edge axis (7). **(C)** Knife stage showing clearance angle adjustment (8) and ± 30° knife rotation (9). **(D)** Spurr's epoxy block with specimen (10) embedded in each end. **(E)** Specimen block (11) clamped in ultramicrotome specimen holder. **(F)** Surgical prep blades preferred for block trimming operations.

3.2. Ultramicrotomy

3.2.1. Ultramicrotomy Conditions

The ultramicrotome is an instrument designed to use a very high-quality knife (glass or diamond) to produce reproducible, thin sections (40- to 150-nm thickness) of plastic-embedded specimens. The instrument is very stable to provide the reproducible production of these sections. **Fig. 9A–C** show an overview of the major controls of a Leica Ultracut S ultramicrotome. Details of operation of a typical ultramicrotome are described in **Note 2**.

The following factors on environmental control are critical for successful, routine ultramicrotomy:

Stable room temperature: in the past, some designs used the controlled heating of the specimen arm in the ultramicrotome to provide the specimen advance during sectioning. The latest commercial instruments are mechanical advance ultramicrotomes and varying environmental temperature changes can cause problems with reproducible sectioning in these instruments.

Air currents: air currents can make ultramicrotomy extremely difficult. These sources can include a nearby open window, an overhead air conditioning duct blowing onto the work area or against a wall behind the ultramicrotome, or the operator exhaling onto the knife/specimen region.

External vibrations: vibrations can come from a variety of sources, including the operator touching the ultramicrotome during operation, setting the ultramicrotome on a table or bench that is attached directly to the building walls, or using a weakly constructed standalone table. In general, an active air table or balance table will provide the best results under all conditions. The proximity of building elevators, heavy foot traffic in a hallway, the opening and closing of doors, building air handling equipment or pumps, or even nearby bus and rail lines have been noted to seriously disrupt sectioning when the ultramicrotome is not on a stable or anti-vibration table.

The rest of this section will describe the manual trimming of a Spurr's plastic-embedded specimen, the facing of the specimen block face, the cutting of thick sections (800 nm) using a glass knife with a water-filled boat, the pickup of the thick sections and transfer to a microscope slide for light microscopy, the approach to the block face with a diamond knife, the filling of the boat of the diamond knife with water, the cutting of ultra-thin sections (40–150 nm), and the pickup of the thin sections using a copper EM grid. With practice, all of these steps can be completed within 15 min. A skilled microtomist can trim the specimen block and cut thick sections and thin sections at the rate of four to six blocks per hour. The beginning microtomist can expect to take a few weeks practice to achieve this level of skill.

3.2.2. Trimming Specimen Block

The shapes and sizes of plastic-embedding molds are many, and they can be purchased to suit almost any type of specimen. The specimen block shown in **Fig. 9D** is one in which pieces of specimen are embedded in each end of the block. One end is chosen and mounted in a specimen holder as shown in **Fig. 9E**.

Our laboratory prefers to use the more substantial and easy to hold surgical prep blades as pictured in **Fig. 9F**. Using the prep blade, a line is scribed in the

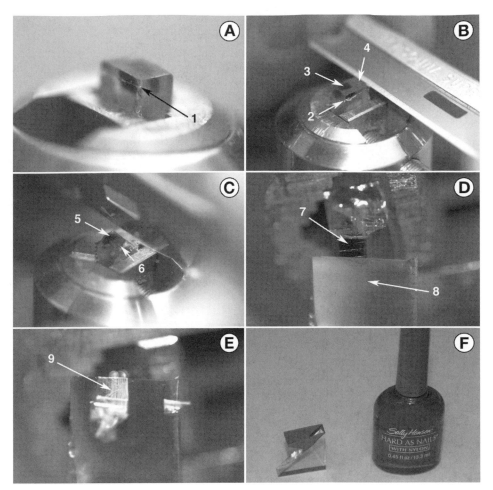

Fig. 10. (A) Scribe mark (1) on the near side of block indicating the depth to trim in order to reach the level of the specimen. **(B)** Prep blade shown in position for trimming away excess plastic (2 and 4) so the specimen (3) reaches the block face. **(C)** Trimming away the excess plastic (6) on one of the four sides of the specimen (5). **(D)** Block face (7) containing specimen ready for rough facing of the block face using a dry, boat-less glass knife (8). **(E)** Sections (9) on the dry glass knife after rough trimming of the block face. The rough trimming is continued into the block until the entire block face is "polished or shiny" in appearance. **(F)** Fingernail polish is used to seal the truf style boat onto a good glass knife.

side of the block indicating the amount of plastic that has to be removed from the block face to expose the specimen (**Fig. 10A**, number 1). **Figure 10B** illustrates the prep blade being used to shave into the block face until the specimen is exposed completely in the block face.

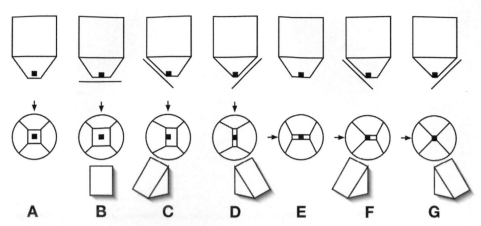

A   B   C   D   E   F   G

Fig. 11. Steps for using the ultramicrotome to trim and face the specimen block. **(A)** Specimen block mounted in the ultramicrotome; **(B)** dry glass knife is used to cut into the block until the specimen appears in the sections; **(C)** the knife is rotated left 30° to 45° and the block is trimmed until the left edge of the specimen appears in the sections; **(D)** the knife is rotated right of center 30° to 45° and the block trimmed until the right edge of the specimen appears in the sections; **(E)**. The specimen block is rotated 90° counter-clockwise; **(F)** the knife is rotated left 30° to 45° and the block is trimmed until the left edge of the specimen appears in the sections; **(G)** the knife is rotated right of center 30° to 45° and the block trimmed until the right edge of the specimen appears in the sections. There are variations on this that are equally correct. An alternate method is to rotate the specimen block for the angled cuts with the knife. The block also may be trimmed into a trapezoidal shape and sectioned with the largest base of the trapezoid positioned to be the edge of the block to make the first contact with the knife edge.

Figure 10C illustrates the cutting away of the plastic on one of the four sides of the specimen, removing all plastic that does not contain the specimen. With practice, a microtomist can adequately prepare the four sides of the block using a prep blade and not have to spend the extra time trimming with the ultra-microtome or other mechanical trimming device. There are cases in which it is desirable to trim certain specimens using the ultramicrotome itself. These steps are depicted as shown in **Fig. 11**. There are many variations on this, but a typical series of steps are illustrated which produce a perfect square or rectangle depending on the original shape of the specimen block.

3.3.3. Facing Specimen Block With Dry Glass Knife

Figure 10D illustrates the trimmed specimen (7) aligned with a glass knife for rough facing of the block face. The block is sectioned until sections containing the entire block face are achieved (*see* **Fig. 10E**, number 9). These sections are usually cut using a 800-nm thickness setting on the ultramicrotome and a sectioning speed of 1 mm/s.

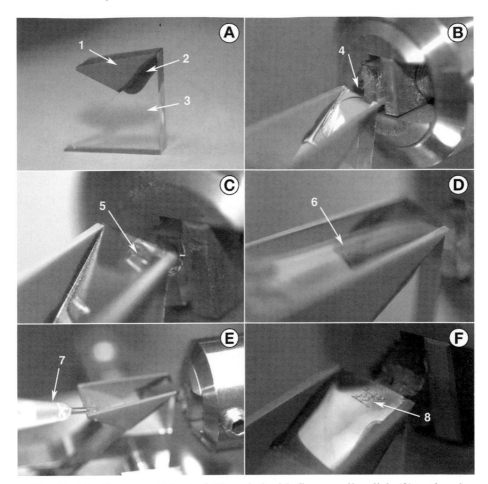

Fig. 12. (A) Close-up of the truf (1) sealed with finger nail polish (2) to the glass knife (3), forming a boat that will be filled with water. **(B)** Glass knife with dry boat shown approaching the polished block face (4) of the specimen. **(C)** Several thick (800 nm) sections (5) are taken with the knife dry. These sections are removed with a blast of canned air. **(D)** Glass knife boat shown filled with water, the curved meniscus (6) of water overfilling the boat is shown reflecting the ultramicrotome light. **(E)** The bulb micro-pipette (7) is shown being used to reduce the level of the water so the surface of water is level with the glass knife edge. **(F)** Thick (8) sections (800 nm) shown floating adjacent to the knife edge that are ready to be picked up. Note the wrinkled appearance is caused by the compression of the sections in the direction of cutting.

3.3.4. Cutting Thick Sections

Thick sections 800 nm in thickness are cut using a wet glass knife. **Figure 10F** and **Fig. 12A** illustrates the truf (1) and nail polish (2) used to attach the boat to a good glass knife *(3)*.

The glass knife with boat attached is mounted in the knife holder of the ultramicrotome and the block face (*see* **Fig. 12B**, number 4) is aligned with the knife edge following the guidelines detailed in **Figs. 14** and **15**. Approach is made, and a few 800-nm thick sections are cut with the knife prior to filling it with water. This step is a final polishing step and sections are shown in **Fig. 11C**, number 5. The dry sections are removed with a short blast of compressed or canned air.

The boat is filled with water as detailed in **Fig. 12D**. (6) Points to the meniscus of the water above the level of the sides of the boat. In **Fig. 12E**, a micropipet (number 7) is used to remove the excess water down to the level of the knife edge leaving a slightly concave meniscus surface. The microtome is used to cut several thick sections, which can be seen floating on the water surface in **Fig. 12F**, number 8. These thick sections (800-nm thick) are wrinkled and show no interference colors in the sections. The lack of color is the result of their thickness, i.e., only thin sections show colors in the sections. The surface tension of the water is not sufficiently strong to flatten these thick sections on the water.

The thick sections are transferred from the boated glass knife using a micro pipette (*see* **Fig. 13A**, number 1). The 1.5-mm diameter tip is inserted under the floating thick sections (*see* **Fig. 13B**, number 2) and transferred to a labeled glass slide which has a drop of water on its surface (*see* **Fig. 13C**, number 3). The glass slide has been marked on the underside with a black permanent marker so the sections can be found after the water has been dried on a 70°C hot plate. After the sections have been dried on the hot plate, they can be stained using a mixture of 1% toluidine blue in 1% sodium borate. Toluidine blue stain solution is applied to the dried section(s) using a filtered syringe such that the sections are surrounded with a small amount of the stain. The slide is placed on the same hotplate that is used for the drying of the sections until the edges of the sections begin to turn green. At that point the excess stain is washed away using distilled water and then the sections are allowed to air dry. This procedure usually adds sufficient contrast to see the morphology of the specimen using the light microscope.

3.3.5. Cutting Thin Sections

Thin sectioning (40- to 150-nm thickness) can be performed using a water-filled diamond knife or one of the balanced-break glass knives with attached truf boat as described previously. Of course a glass knife that would be used for ultrathin sectioning would never be used for cutting the thick sections before thin sectioning. If the glass knife used for ultrathin sectioning is cared for the same as a diamond knife, then it can be reused many times, as described previously. In the following description a balanced-break glass knife can be used as

Fig. 13. (A) Close-up view of the plastic micropipet tip. The tip is about 1.5 mm in diameter (1) and makes an ideal pickup tool for transferring the thick sections to the glass slide. **(B)** Micropipet being used to pickup the floating thick sections (2). **(C)** Micropipet transferring thick sections (3) to a drop of water on a glass slide for light microscopy. **(D)** Diamond knife (4) with integral boat for water. **(E)** Diamond knife mounted in the knife holder of the ultramicrotome, the clearance angle adjustment (5) is shown again to remind people to change this angle depending on the recommended setting of the knife manufacturer. **(F)** Diamond knife edge (7) is aligned with the block face (6). The details of this alignment procedure are shown in **Fig. 14.**

an alternative if one does not have a diamond knife available. A very important point to remember for successful thin sectioning is that the smaller the trimmed block face, the easier it will be to produce thin sections. This is especially impor-

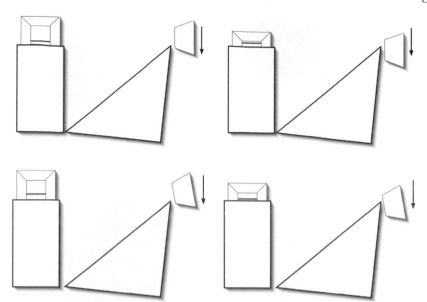

Fig. 14. Block tilt may make it necessary to align the block face with the vertical plane of the knife edge. The top row shows the knife edge reflection in the block face. As the specimen is moved down, the relative space between the knife edge and its reflection are parallel and the same vertical distance whether the reflection of the knife edge is viewed using the bottom of the block (**left side**) or the top of the block (**right side**). The top row indicates proper alignment of the block face with the knife edge The bottom row depicts the situation that occurs when the top of the block is tilted closer to the knife edge than the bottom of the block. When the reflection of the knife edge in the block face is observed, the distance between the knife edge and the reflected edge is large and gets smaller as the block face is moved down. To correct this, the block must be rotated clockwise in order to bring the block face parallel to the vertical plane of the knife edge. If the opposite is seen, that is the knife edge to reflected edge distance is smaller at the bottom of the block and larger at the top, then the block must be rotated anti-clockwise to bring the face of the block parallel to the vertical plane of the knife edge.

tant for the beginner in ultramicrotomy. A good rule is when you think that you have the block face as small as possible, then use one third of that size.

A suitable diamond knife with integral boat as shown in **Fig. 13D**, number 4, is mounted in the knife holder of an ultramicrotome as shown in **Fig. 13E**. The knife clearance angle is adjusted (*see* **Fig. 13E**, number 5) to the clearance angle recommended by the knife manufacturer. When using the excellent glass knives produced as previously described a clearance angle of 6° is recommended. **Fig. 13F**, illustrates the approach of the specimen (6) to the diamond knife (7). The same approach and alignment steps are followed as illustrated in **Figs. 14**

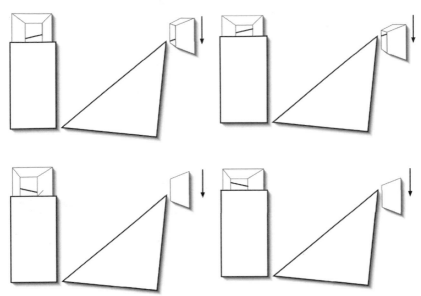

Fig. 15. Shown are the alignment of the knife edge and block face in the horizontal plane. The top row illustrates what the reflection of the knife edge in the block face looks like in the situation that the left edge of the block is closer to the knife than the right edge. To correct this situation, the knife is rotated to the right until the reflected edge in the block face is parallel to the knife edge. The bottom row illustrates the opposite condition, i.e., the right edge of the block is closer to the knife than the left edge. This condition is corrected by rotating the knife to the left until the reflected edge in the block face is parallel to the knife edge. It is not uncommon to have to adjust the knife as shown in this figure and in **Fig. 14** before sectioning can be performed.

and **15**. After the approach is complete a few ultrathin dry sections are cut from the block. The sections are then removed with a short blast of compressed or canned air.

Figure 16A illustrates filling the diamond knife with water until a marked meniscus is observed as shown in **Fig. 15B**, number 1). The excess water is removed using a micropipet, as shown in **Fig. 16C**, with the meniscus adjusted so the surface is concave. The ultramicrotome is set to automatic mode after adjusting the cutting window (detailed in **Note 2**). Cutting speeds of 0.7 to 1.0 mm/s usually are appropriate for the cutting of thin sections. **Fig. 16D–F** illustrate sections that have chatter (2), which was introduced by touching the ultramicrotome while a section was being cut and shows as light and dark bands in the floating section. Two free floating sections (3) are seen in **Fig. 16E** and a very nice ribbon of three sections (4) is shown in **Fig. 16F**. These sections all exhibit a light gold interference color indicative of sections approx 100 nm in

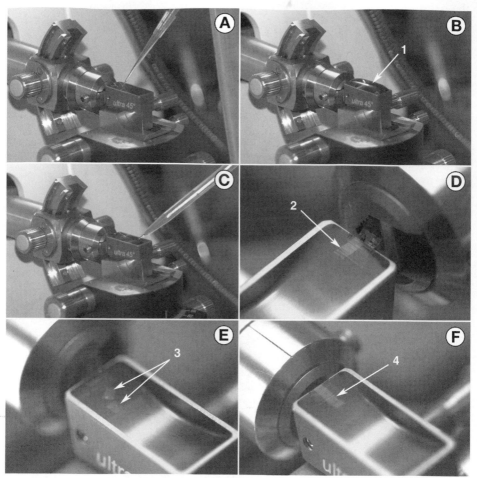

Fig. 16. Illustration of the procedure of filling the diamond knife with water. (**A**) micropipet used to overfill the diamond knife boat. (**B**) meniscus (1) of the water over-filling the boat. (**C**) micropipet is used to reduce the level of water so the water is horizontal next to the knife edge. (**D**) two thin sections in a ribbon, chatter is evident in the section labeled (2). (**E**) two sections floating free (3) on the surface of the water in the boat. (**F**) a ribbon of three sections (4) that have the same color and are chatter free.

thickness. At least they are light gold in the original color photos! It is recommended that the beginning microtomist obtain one of the color thickness cards which give the color vs thickness of sections floating on a water surface. These are available for free from the diamond knife manufacturers and distributors.

A common problem that is occasionally encountered is the water wetting the surface of the specimen block. When this occurs it also transfers water to the back surface of the knife. Sectioning will not be possible when this occurs and the specimen block and knife must be dried off before sectioning can be resumed. Sometimes the water can be removed by gentle blotting with tissue paper, however be very careful if this is tried with a good knife, diamond, or glass, because it is possible to damage the knife edge in this process. It is safer to remove the knife, dry it with canned or compressed air and repeat the filling and approach steps from the beginning.

The use of Spurr's embedding medium has been described in this chapter. The fundamental principles apply to all embedding mediums; however, the sectioning characteristics relating to block size and thickness of sections that can be cut will depend on the formulation of epoxy plastic that is used. The other factor is how hydrophilic or hydrophobic the various formulations available will exhibit. Some epoxies will be more difficult than others with regards the propensity for the block to attract water to its surface during the sectioning process. The use of antistatic guns or other devices may be helpful in these cases. The full description of this is beyond the scope of this chapter.

If the sections have compression that is not removed by the surface tension of the water, the sections can be stretched by exposing them to chloroform vapor. **Figure 17A** illustrates the use of a scintillation vial containing chloroform to treat the sections as they are floating on the water filled boat. The sections can be directly observed in the ultramicrotome microscope during this procedure and seen to expand and change from a gold interference color to a silver color which indicates a change in thickness from approx 100 nm to 70 nm. The cap is briefly removed allowing the vapors of chloroform to settle over the surface of the water. Because chloroform is heavier than air, it is easy to use without exposure of the microtomist to the vapor. It is very quick (less than 5 s) and results in good spreading of the sections.

Just before use, several EM grids are immersed in the dilute nitric acid solution described in the **Section 2** for 2 to 3 min. This brief acid etch of a copper grid ensures that the grid surface is clean prior to picking up sections. A clean EM grid as shown in **Fig. 17B** is picked up using a pair of self closing cross-action EM forceps. The grid is then transferred to just under the surface of the water-filled diamond knife as shown in **Fig. 17C**, number 1). A Dalmatian dog hair glued to a diagonally cut plastic soda straw. **Figure 17D** shows two different types of plastic soda straws. The dog hair is held in place using fingernail polish.

As illustrated in **Fig. 17E**, the dog hair (2) is used to gently push a floating section over the EM grid, which is held just below the surface of the water. The EM grid is gently elevated through the surface of the water, thus capturing the section on the dull side of the EM grid. The grid with sections up is then transferred to a clean filter paper where the section dries down and attaches to the EM grid.

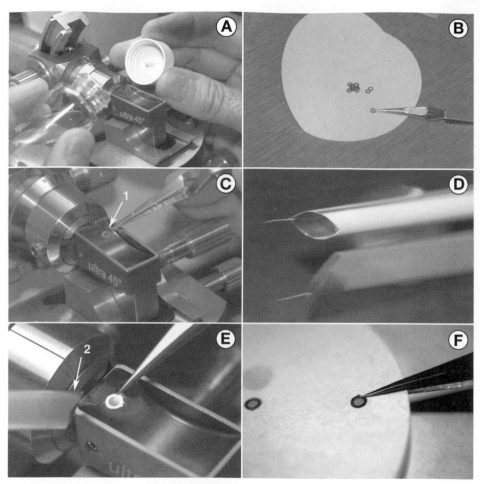

Fig. 17. (A) illustrates using chloroform vapor to cause a spreading of the thin sections as they float on the surface of the water, which relieves compression in the sections. A small heat wand can also be used for spreading the sections. In **(B)**, a grid is shown being picked up using the cross-action EM forceps. In **(C)**, the grid (1) is submersed below the water surface in the boat. **(D)** Our laboratory uses Dalmatian dog hairs on soda straws as a very durable microprobe for handling sections. **(E)** shows the dog hair (2) being used to gently guide a floating section (corners of the section are just visible beyond the edge of the grid) onto the submersed EM grid, once the section is in place the grid is gently lifted vertically from the water. In **(F)**, the wet grid is transferred to dry filter paper to remove the water from the underside of the grid.

The EM grid with sections can then be stored in a grid box for later staining with various chemical (*see* Chapter 6 in this volume) and immunological stains and probes (*see* Chapter 13 in this volume). The thin section for EM has very low contrast without these heavy metal stains and/or immunological stains using gold-labeled antibodies. Many examples of these special stains are documented in the scientific literature, and it is best to do a little research in the library or talk to other experts about results concerning a particular tissue.

It is critical to ensure that the diamond or glass knife is cleaned immediately after use and that the knife edge, glass or diamond, is kept free of sections that could dry down on the knife edge and make subsequent sectioning difficult to impossible. The diamond or glass knives are rinsed using a stream of distilled water and then briefly dried using several short blasts of compressed air. If more extensive cleaning becomes necessary please follow the diamond knife manufacturer's recommended cleaning procedure. In the case of the glass knives with attached trufs, the same procedure(s) that are recommended for diamond knives can be followed.

The steps that have been discussed apply to all types of thin and thick sectioning using an ultramicrotome. In the absence of a diamond knife an excellent glass knife with a truf attached can be substituted in the previous steps describing thin sectioning. It is hoped that these fundamental steps have been presented in enough detail to enable the beginner microtomist to get a good start into the exciting world of preparing specimens for the light and electron microscope. No discussion of ultramicrotomy would really be complete without some comments regarding common sectioning artifacts that are seen. **Note 3** provides a brief overview of the major sectioning artifacts that will be observed in the binoculars as well as some tips on things that can go wrong and result in poor sectioning.

4. Notes

1. Modifications and description of the old LKB glass knife maker, which is no longer in production, are still readily available to most EM laboratories worldwide. This instrument is described in **Fig. 18**. The items that are removed from this knife maker to permit the use of the balance break method are a metal shelf that extended across the knife maker at position (8), which was used to support one end of a glass strip using the incorrect method of making glass knives. A protractor plate was removed at position (10). The protractor was used to make glass knives with angles less than 45°. This protractor also prohibited getting a good balance on the strips of glass so it was removed. The last items removed from this knife maker were two spring loaded measuring pins (16), which applied a great deal of upward force when trying to balance the glass on the balancing pins (not visible). In this example the pins were completely removed, alternatively they can be taped down flush with the top of the knife maker so that they stay out of the way.

   The one critical alignment, which must be checked is the placement of the score directly over the center of the balance pins as shown in **Fig. 19**. This figure describes

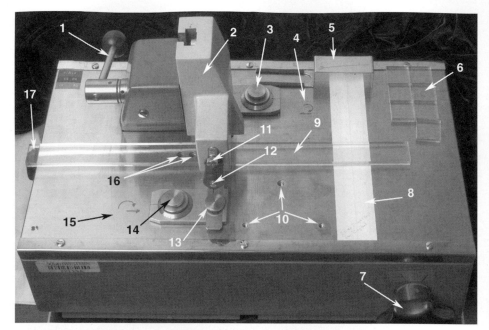

Fig. 18. LKB model 7800 Glass Knife Maker description of parts and modifications. The major parts are (1) locking lever shown in the raised position, it is used to raise and lower the anvil (2) which has the breaking pins that clamp the top of the glass. Upper fork cam (3) is used to position the upper fork left and right for alignment of the score with the upper apex of the glass squares. A pictogram is drawn at (4) to help remember that turning the cam clockwise moved the upper apex of the glass square to the left. The pickup fork (5) is used to easily move the broken knives out from under the breaking anvil. The glass squares (6) that are ready to be made into knives are shown stored on the right. The breaking force is applied with the knob (7), which causes the breaking force to increase as it is turned in a clockwise direction. The original support shelf was removed from area (8) to allow the glass strips to be balanced without obstruction. A strip of glass (9) is shown balanced on the balancing pins, which are below the glass and not seen in the picture. The area (10) is where a protractor that was originally on the knife maker has been completely removed. The score cam (11) and score knob (12) are attached to the anvil. The centering of the glass over the balancing pins is accomplished using the locking screw (13). The lower fork cam (14) for centering the lower glass apex under the scoring wheel is shown and a pictogram (15) is drawn to remind the operator that turning this knob clockwise causes the fork to move the lower glass square apex to the right, which helps when the lower apex has to be moved closer to the score to get the required break near the lower apex. (16) Points to the location of two measuring pins that were removed because they interfere with the balancing of the glass strips. The control (17) that releases or applies the spring tension which holds the upper fork against the glass strip or holds the squares during breaking. This control is released to insert or remove the glass strips and squares.

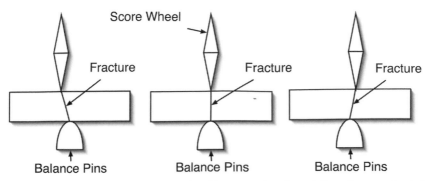

Fig. 19. It is critical that the score wheel is properly aligned directly over the balance pins. The reproducible production of excellent glass knives requires that all of the fracture surfaces be perpendicular and very flat. This illustrates the adjustment of the score wheel to achieve the required alignment. The left diagram shows the score wheel to the left of center. It is simple to test this alignment by standing a strip of glass vertically on the fresh fracture. The glass strip will lean to the right if the score wheel is to the left of center. The right hand diagram shows the score wheel too far to the right. If a freshly broken strip of glass is stood on end with this condition, the strip will lean to the left. The score wheel alignment is adjusted until a freshly broken strip stands perpendicular to the support surface. There is an adjustment to achieve this alignment on the back of the score device consisting of a spring held with two machine screws. When the screws are released a half turn, the score wheel can be positioned until it is aligned as shown in the center of the drawing. It is a trial and error adjustment but once achieved it does not change with use. It is usually only disturbed when the score wheel is replaced.

how to judge whether the score wheel is adjusted properly. If the score does not produce a perpendicular fracture, then the alignment of the score wheel left or right must be performed before reproducible excellent knives can be produced using this instrument.

2. A description follows of the basic components and operation of a typical, modern ultramicrotome used to produce thick sections for light microscopy and thin sections for electron microscopy. A general understanding of the construction of the ultramicrotome will help with understanding the basic principles of operation of the instrument.

A schematic diagram of a typical modern ultramicrotome is shown in **Fig. 20** and describes the basic components that make up the system. The knife stage is used to hold the glass or diamond knife and can be rotated about the vertical axis and translated in the x and y horizontal plane. This translation is controlled using mechanical controls or stepper motors to achieve very fine control of the movements of the stage. The micro advance of the specimen block toward the knife edge is controlled by the use of a lever arm style of advance where a large motion

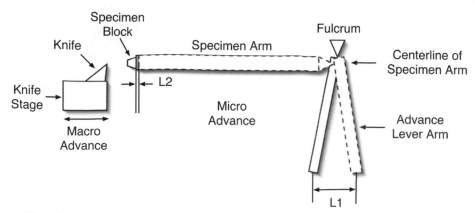

Fig. 20. This diagram illustrates the basic operating principle of a typical ultramicrotome. The Macro advance is achieved using a mechanical x-y knife stage. The movement along the y axis is depicted in the diagram shown above. The microfeed is achieved by using a lever arm arrangement with about a 100:1 ratio (the ratio will vary depending on the manufacturer). If we assume this 100:1 ratio then to achieve a specimen advance of 100 nm at position L2, a 10.0-μm advance would occur at position L1. This advance at L1 is typically achieved by using a very fine pitch threaded screw driven by a stepping motor to control the advance.

at position L1 is converted to a very small incremental advance shown at L2. In this example a 100:1 ratio is described, but it will vary depending on the manufacturer's requirements of the ultramicrotome.

The cutting cycle is divided into the cutting stroke shown on the left in **Fig. 21** and the retraction cycle shown on the right. This type of cycle provides precise control of the cutting speed when an ultra thin section is cut within the cutting window while the retraction part of the cycle is at a higher speed. This improves the ratio of sections cut per unit of time. Thus, the microtomist spends less time waiting on the return of the block to the top of the cutting window. All ultramicrotomes that are manufactured today give very good reproducibility of the sectioning process. These instruments are utilizing more computer control with each evolutionary model that is produced.

3. Things to check if difficulties are identified with the sectioning process include the following: (1) chatter is fairly common and can be caused by a number of factors, including external vibrations (solution: isolate the microtome from the source of vibration); a loose specimen (solution: tighten the holding screws); a loose knife (solution: tighten the knife in the holder); dull knives (solution: change to a sharp knife); incorrect clearance angle for the hardness of the specimen block being cut (solution: adjust the clearance angle); and a speed that is too fast (solution: slow the cutting speed; the harder specimens require slower sectioning speeds). (2) compression of the section in the direction of cutting: for a soft block,

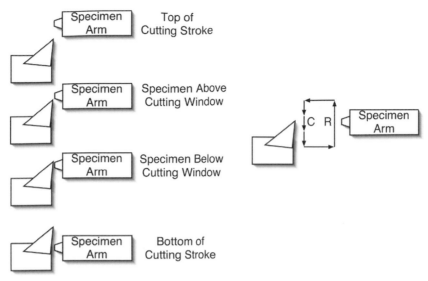

Fig. 21. Illustrated is the cutting (C) and retraction (R) cycle of a typical ultramicrotome. Details of the cutting stroke are shown in the left side of the figure. When the specimen reaches the top of the cutting stroke the micro advance has occurred depending on the thickness setting of the ultramicrotome (i.e., 90 nm). The specimen continues down to a region above the cutting window, where the motor driving the cutting stroke slows down to the desired cutting speed (i.e., 1.0 mm/s). This slow speed continues until the specimen block has traveled below the cutting window (below the knife edge). At this point the motor speeds up to the return speed (10-100 mm/sec), reaches the bottom of the cutting stroke and begins a fairly large (tens of micrometers) retraction. The specimen block then travels on the back of the retraction cycle during which a micro advance takes place. At the top of the retraction cycle, the retraction is removed and a thin section cutting stroke is repeated. The cutting window is always adjusted just before sectioning and is dependent on the height of the specimen block. The cutting speed, thickness advance and, in some cases, the retraction speed also are under control of the operator.

return to oven for additional curing time so the block becomes harder; for an actual knife angle too large, choose a better knife with a smaller angle; for a dull knife, choose a better knife; and for a cutting speed that is too fast, try a slower cutting speed. (3) striations or scratches in the sections: for edge imperfections, choose a better knife; for dirt and debris in the block, check the processing for contamination; and for a dull knife, choose a better knife. These are the common causes that usually are identified as causing the most problems for thin sectioning. You will note that a lot of the problems are traceable to the quality of the knife that is being used for sectioning.

Acknowledgments

Thanks to Dennis Bellotto for his helpful reading of the text and the many years that he has done super work for me in my laboratory. I also deeply appreciate the contact with the wonderful instructors and students that I have had the pleasure of learning from at the EMBO and FEBS courses over the last few years.

References

1. Griffiths, G. (1993) *Fine Structure Immunocytochemistry* Springer, Berlin.
2. Yamamoto, T. (1963) Method of staining with toluidine blue for epoxy resin-embedded tissues for light microscopy. *Kaibogaku Zasshi* **38,** 124–128.
3. Tokuyasu, K. and Okamura, S. (1959). A new method for making glass knives for thin sectioning. *J. Biophys. Biochem. Cytol.* **6,** 305–308.
4. Griffiths, G., Simons, K., Warren, G., and Tokuyasu, K. (1983) Immunoelectron microscopy using thin, frozen sections: application to studies of the intracellular transport of Semliki Forest virus spike glycoproteins. *Methods Enzymol.* **96,** 466–485.
5. Webster, P. (1993) A simple modification to the LKB 7800 series Knifemaker and a balanced-break method to prepare glass knives for cryosectioning. *J. Microsc. (Oxford)* **169,** 85–88.
6. Roberts, I. M. (1975) Tungsten coating-a method of improving glass microtome knives for cutting ultrathin sections. *J. Microsc. (Oxford)* **103,** 113–119.

6

Poststaining Grids for Transmission Electron Microscopy

Conventional and Alternative Protocols

E. Ann Ellis

Summary

Poststaining ultrathin sections on grids is an essential process in preparing specimens for examination using transmission electron microscopy. This process commonly consists of staining in aqueous solutions of uranyl acetate followed by lead citrate and can result in good consistent quality staining if certain precautions are taken. This chapter covers conventional and alternative methods for poststaining ultrathin sections, including microwave-assisted procedures and corrective measures when problems are encountered.

Key Words: Poststaining; uranyl acetate; lead citrate; microwave staining; precipitates.

1. Introduction

Poststaining is necessary to improve differential contrast on ultrathin sections for examination using transmission electron microscopy (TEM). A number of factors contribute to a consistent staining quality of grids. Some of the most important factors are cleanliness of the working environment, the tools used for handling grids during the process of poststaining, and the fluids that come in contact with the sections. Two other important considerations are the time interval between picking sections up on grids and staining, which is especially important when dealing with Spurr's low-viscosity embedding medium *(1)*, and whether or not unstained sections have been exposed to the electron beam. Better contrast is achieved by poststaining grids immediately after the sections are picked up and without exposure to the electron beam. If it is necessary to check the quality of sections before poststaining, only one grid from a batch should be examined by TEM.

From: *Methods in Molecular Biology, vol. 369*
Electron Microscopy: Methods and Protocols, Second Edition
Edited by: J. Kuo © Humana Press Inc., Totowa, NJ

Although there are numerous methods in the literature and various commercial devices for poststaining grids, the methodology presented here is based on keeping the procedures simple and reproducible and using common laboratory materials. The most commonly used methodology consists of poststaining grids with aqueous uranyl acetate followed by Reynolds' lead citrate *(2)*. Uranyl and lead ions are general or nonspecific stains; however, it is known that certain chemical groups react with these ions. Uranyl acetate reacts strongly with phosphate and amino groups such that nucleic acids and phospholipids in membranes stain. Lead binds to negatively charged components such as hydroxyl groups and areas that react with osmium tetroxide, such as membranes. Phosphate groups may be involved in lead staining because the use of phosphate buffers often enhances overall staining intensity.

Many problems associated with low contrast and stain precipitation are related to the age of the stains. Old stains loose reactivity, and uranyl acetate deteriorates when exposed to light. Prolonged staining times (longer than 10–15 min for uranyl acetate and longer than 3–5 min for lead citrate) can result in stain precipitates because of the drying at the surface of the drops of stain. Staining at increased temperatures also contributes to precipitate formation as the result of increased evaporation of the stain solvent. These precipitates are not easily removed by washing with deionized water. A number of salvage methods have been developed for removing stain precipitates and restaining the grids; however, it is better to understand the staining parameters and prevent the formation of these precipitates. Thicker sections (gold interference colors) stain faster and more intensely than thinner sections (gray-to-silver interference colors) because there are more sites for the stains to react in a thicker section. In addition, contrast can be improved in lightly contrasted sections by using a smaller objective aperture and/or working at a lower accelerating voltage. The trade-off in working at a lower accelerating voltage is decrease in resolution; however, decreased resolution will probably not be an issue when doing lower magnification, survey work.

There are several alternative staining procedures that are useful with difficult specimens. Marinozzi rings *(3)* originally were proposed for staining sections that required oxidation before specialized stains such as silver methenamine or phosphotungstic acid-chromic acid; however, these rings are useful for staining large sections and/or preventing buildup of dried stain near grid bars. Use of the microwave for staining with aqueous uranyl acetate results in improved contrast and shorter staining times.

Contamination of sections during staining can and does occur when precautions in the avoidance of collection of "dirt" stain precipitates are not followed. Salvage techniques have been developed over the years and offer some relief for these problems when it is not practical to start over and do the procedures correctly the first time.

2. Materials

2.1. General

1. Freshly prepared and filtered solution of aqueous uranyl acetate.
2. 5-mL Plastic syringe with 0.22-µm filter, (one syringe with filter for uranyl acetate and another syringe with filter with deionized water).
3. ParafilmM®™ (Parafilm) or porcelain spot plates.
4. Electron microscopy grade forceps.
5. Glass Petri dishes.
6. Four 100-mL glass beakers.
7. Freshly boiled deionized or distilled water.
8. Filter paper points.

2.2. Preparation of Staining Solutions

2.2.1. Aqueous Uranyl Acetate

1. Prepare 2% (w/v) solution of uranyl acetate [$UO_2(CH_3COO)_2 \cdot 2H_2O$] in 5 mL of freshly boiled warm deionized or distilled water. Use a magnetic stirrer and stir until there are no uranyl acetate crystals visible in the bottom of the beaker. It is a good idea to cover the top of the beaker with Parafilm (*see* **Note 1**). The pH of a freshly prepared 2% aqueous solution of uranyl acetate is approx 4.
2. Filter the solution through Whatman #42 or #50 filter paper or equivalent.
3. Store the filtered solution in a dark colored container or wrap the container with aluminum foil to exclude light since uranyl acetate is light sensitive.
4. Filter an appropriate volume of stain through a 0.22-µm micro filter when ready to stain grids.

2.2.2. Methanolic Uranyl Acetate
(Acetone Free Methanol, Ethanol Can Be Substituted for Methanol)

1. Prepare 5% (w/v) solution of uranyl acetate in 50% (v/v) methanol. Cover the beaker with Parafilm and stir until there are no uranyl acetate crystals visible in the bottom of the beaker.
2. Filter and store the solution as described above for aqueous uranyl acetate.

2.2.3. Lead Citrate

1. In a 50-mL volumetric flask with glass stopper, place the following components: 1.33 g of lead nitrate [$Pb(NO_3)_2$], 1.76 g of sodium citrate [$Na_3(C_6H_5O_7) \cdot 2H_2O$], 30 mL of freshly boiled and cooled deionized or distilled water.
2. Stopper the volumetric flask and shake vigorously for 1 min; continue intermittent shaking for 30 min. The solution should be milky.

3. Add 8 mL of 1 N NaOH solution, carbonate free (solutions purchased from a reliable vendor work better than freshly prepared NaOH solutions made in the laboratory; *see* **Note 2**). The solution will become clear.

4. Dilute the solution to 50 mL with freshly boiled deionized or distilled water. The solution of lead citrate is ready for use and should be stored in the stoppered bottle or any closed container such that CO_2 can be excluded. The pH of the solution should be approx 12.0.

2.3. Marinozzi Rings

1. Pieces of unexposed TEM film which have been cleared in photographic fixer and then washed and dried as one does for TEM negatives.
2. Standard-size hole punch.
3. Smaller-size hole punch or 3-mm grid punch.
4. Ethanol.
5. Alcohol lamp.

3. Methods

3.1. Standard Staining Procedure

1. Place several drops of uranyl acetate on a clean piece of Parafilm. (Clean porcelain spot plates also can be used).
2. Fill several (3–4) small (100 mL) beakers with warm, freshly boiled deionized water.
3. Clean a pair of forceps with acetone or alcohol followed by rinsing with deionized water.
4. Pick up the grid with clean forceps and float the grid, section side down, on a drop of uranyl acetate stain. Leave the grids on the stain for 5 to 10 min. The staining time must be determined empirically for each specimen and thinner sections (silver to gray interference colors) require longer staining time than thicker sections (pale gold interference colors).
5. While the grids are staining in uranyl acetate, prepare a CO_2 free chamber (*see* **Note 3**) for staining with lead citrate. In a glass Petri dish, place a piece of Parafilm or dental wax. Place pellets of NaOH or KOH around the edge of Parafilm to absorb CO_2. Place the top on the Petri dish. Use a long Pasteur pipet and remove a small volume of lead citrate solution from the volumetric flask by going below the surface of the lead citrate solution (*see* **Note 3**). Immediately, place several drops of lead citrate on the Parafilm in the CO_2 free chamber by opening the top of the Petri dish only enough to insert the Pasteur pipet. Discard the first two to three drops of lead citrate on the NaOH pellets, then place several drops (one drop for each grid) of lead citrate on the Parafilm. Close the dish immediately.
6. After 5 to 10 min in uranyl acetate, remove the grid from the uranyl acetate, dip the grid in the first beaker of warm, freshly boiled deionized water and swirl the grid around. Repeat the process in the three other beakers of warm, freshly boiled deionized water. Keep the grid wet.

7. Place the grid on the drop of lead citrate; be careful in opening the Petri dish top only enough to insert the grid and float it section side down on the stain.
8. Fill several small (100 mL) beakers with warm, freshly boiled deionized water.
9. After 2 to 3 min, open the Petri dish only enough to remove the grid and swirl the grid in the beakers of water as done in **step 6**.
10. Dry the grid using filter paper points (cut triangles of filter paper) by placing the paper point between the points of the forceps (*see* **Note 4**).

3.2. Alternative Staining Methods

If staining times beyond 5 to 10 min do not give adequate contrast, then alternative staining techniques may give better results than prolonged staining times. In addition, prolonged staining in lead citrate may result in regressive staining due to the high pH. Several methods have been reported in the literature and each has its merits.

3.2.1. Alcoholic Uranyl Acetate

Uranyl acetate has limited solubility in water, is more soluble in alcohols than in water, and increases in solubility with increased temperature *(4)*. In addition, alcohol is absorbed by epoxy resins and improves penetration of uranyl acetate into sections. When using methanol as the solvent for uranyl acetate, acetone-free methanol should be used. Alcohol based stains are not recommended for staining epoxy resins based on Luft's Epon formulation *(5)* but can be used safely with Spurr's low-viscosity resin and Mollenhauer's Epon-Araldite *(6)* formulations. Nadic methyl anhydride in Luft's formulation is probably the source of the problem.

3.2.2. Marinozzi Rings (3)

This method is very useful when handling serial sections or sections from large block faces for which conventional methods are more likely to result in folds in the sections and or stain precipitate near grid bars.

1. Unexposed TEM film that has been cleared in photographic fixer and then washed and dried as one does for TEM negatives.
2. Punch discs from pieces of cleared TEM film using the larger hole punch and then use the small hole punch or grid punch to punch a hole in the center of the disc.
3. Wash the discs with holes in the center in ethanol and then rapidly flame each one in the flame of the alcohol lamp. This makes the surface hydrophobic so that the disc (Marinozzi ring) will float on the surface of water or staining solutions.
4. Raise the water level in the knife boat slightly and then arrange sections as you would arrange sections to pick up on grids. Hold the Marinozzi ring with forceps and go down on the sections as you would in picking sections up from above. The sections will now be in a drop of water in the Marinozzi ring and can be transferred

from one fluid to another in depressions of spot plates or small plastic weighing dishes.

5. Sections in the Marinozzi rings may be stained and washed at elevated or at room temperature as long as precautions are taken to minimize evaporation of the staining or washing fluids. The usual precautions should be taken in preparing and filtering all fluids, which come in contact with the sections.

6. Pick up sections on the type of grid that is appropriate for the specimen and study. This method is very useful when picking up sections on single slot grids in order to maximize the area viewed and find specific sites in the specimen.

3.3. Double Staining With Lead Citrate (Triple Staining)

Contrast in specimens with inherently low contrast can be improved by staining first with lead citrate for 1 min followed by uranyl acetate for 5 to 10 min and then a final 1-min stain with lead citrate. This method was first proposed by Daddow *(7)* and has been endorsed by Bell for skin sections *(8)* and Maunsbach and Afzelius *(9)* for general specimens with low contrast.

3.4. Microwave-Assisted Staining

Chien et al. *(10)* have used microwave-assisted staining to not only speed up the staining process but also to improve the final contrast of the section. Staining in the microwave must be performed in a moist chamber and the times and powers (wattage) can be determined empirically. The section shown in **Fig. 1** was poststained for 5 s in a moist chamber in 2% aqueous uranyl acetate at room temperature at a power setting of 250 watts in a PELCO® Coldspot™ Laboratory microwave, washed four times in freshly boiled water, and then stained in lead citrate for 1 min on the bench. Shorter staining times (10 s or less) at low wattages give better results. Freshly prepared stains must be used; use of old uranyl acetate and lead stains will result in formation of extensive precipitates (**Fig. 2**).

3.5. Contamination and Stain Precipitates

Sometimes there is serious contamination of grids caused by "dirt" on the section and/or stain precipitates that render the grid useless. In the case of "dirt" on the sections, which renders micrographs unacceptable for publication, it is necessary to cut new sections and take precautions to prevent the collection of dirt on the sections as outlined by Mollenhauer *(11,12)*. "Dirt" adheres well only to a dry section and "dirt" on most sections comes from the surfaces of staining and wash fluids *(11)*. Keep grids wet throughout the staining process; never blot the grids. Dry the grids after the staining is finished by touching filter paper points to the edge of the grid on the forceps side of the grid. Touching filter paper to the grid opposite the side of forceps results in pulling any form of "dirt" onto

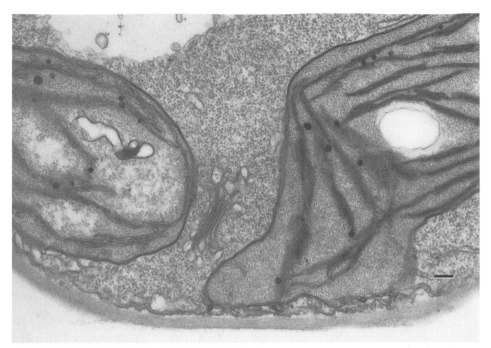

Fig. 1. Section through a cotyledon of *Arabidopsis* sp. stained for 5 s in 2% aqueous uranyl acetate in the microwave at 250 watts followed by 1 min in lead citrate on the bench. The grid was examined and photographed at 100 kV accelerating voltage. Bar = 200 nm.

the grid. Capillary fluid often collects between the tips of the forceps as grids are being held; remove this fluid by placing a filter paper point between the tips of the forceps before placing the grid on a piece of filter paper or in a grid box.

Another source of section contamination is bacteria, which can grow on 0.22-μm filters or in wash bottles used with deionized water. This problem often happens with water used in the boat of the knife as well as with water used for the final wash after staining. Using freshly boiled water and replacing the filters on a monthly basis can help to eliminate this potential problem.

Uranyl acetate precipitate often is recognized as needle-like objects on the section and can result from not filtering the stain properly immediately before use or from evaporation during prolonged staining. Many problems with poor staining are associated with the use of old solutions of uranyl acetate. Freshly prepared uranyl acetate stains more intensely and can eliminate prolonged staining times, a factor contributing to uranyl acetate precipitate. Alcoholic uranyl acetate stains are used to improve the contrast of some difficult specimens. Care

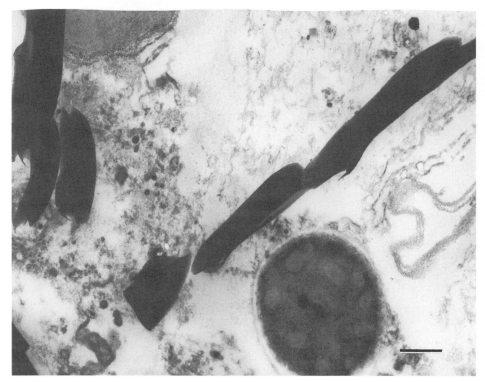

Fig. 2. Section from a yeast mitochondrial preparation stained in the conventional manner using old stains. Note the large and small electron dense precipitates. Bar = 500 nm.

should be taken to wash with the same concentration of alcohol that the stain is dissolved in and going through a decreasing series of alcohol concentrations down to water helps to prevent rapid drying and possible precipitate formation.

A common, unrecognized cause of lead precipitates and poor staining intensity with lead citrate is incorrect pH. A pH lower than 12 causes precipitation of lead carbonate which appears as electron dense particles of various sizes and shapes. As the pH approaches 14, staining is severely reduced to the point that there is loss of contrast (2). The use of carbonate free, commercially prepared 1 N NaOH solutions is a simple, practical way to achieve the correct pH when preparing lead citrate solutions. Another possible source of problems in preparing lead citrate is the quality of the deionized or distilled water. Proper maintenance of the deionized or distilled water source is essential.

In the event that there is severe stain precipitation that cannot be worked around, there are methods which can be used to salvage grids with important ultrastructural information. Avery and Ellis *(13)* proposed the use of oxalic acid, a chelator, dissolved in methanol for removal of uranyl acetate precipitate in *en bloc* stained sections and on sections stained with hot methanolic uranyl acetate. Oxalic acid removes only uranyl acetate precipitate. Several methods *(14–16)* have been developed to destain sections with both uranyl acetate and lead precipitates. Treatment of sections with freshly prepared 2% (w/v) aqueous uranyl acetate for 6 to 8 min or 10% (v/v) acetic acid for 1 min followed by several rinses in distilled water can remove both uranyl acetate and lead precipitates. The dilute acetic acid treatment can be used not only for precipitate removal, but also for removing the original stain from sections. These salvage techniques are not always successful with sections that have been exposed extensively to the electron beam. It is better to remove grids from the beam and destain, wash and restain as quickly as possible.

Sometimes a fine, electron dense, "peppery" precipitate appears in sections after post staining; specimens which exhibit this fine osmium related precipitate should be recut and picked up on nickel grids or Marinozzi rings. These sections can then be oxidized with 1% to 2% (w/v) freshly prepared periodic acid for 5 to 10 min at room temperature, washed with deionized water and stained in the conventional manner with uranyl acetate and lead citrate *(17,18)*. Dilute hydrogen peroxide (1–3% v/v) may be used as an oxidizing agent but is much harsher on the sections; periodic acid is the preferred oxidizer.

4. Notes

1. Prepare small volumes (5 mL) of stain on a weekly (more often if needed) basis for more consistent results. This also eliminates problems of disposing of large volumes of old uranyl acetate stain. Check with your local environmental health and safety office for appropriate local regulations for disposal of uranyl acetate and other chemical wastes generated in the laboratory.
2. The correct pH for Reynolds's lead citrate is 12. The cause of incorrect pH in preparing lead citrate is the use of NaOH pellets to prepare a 1 N NaOH solution instead of using a commercially produced, carbonate free 1 N NaOH solution. It is not possible to weigh NaOH pellets accurately due to their size and hygroscopic nature.
3. Lead citrate reacts with CO_2 in the atmosphere to form lead carbonate precipitate, which appears as electron dense deposits on the grid. Every effort should be made to reduce the formation of lead carbonate by avoiding CO_2 contamination. Spreading fresh pellets of NaOH around the Parafilm in the Petri dish helps to absorb CO_2. Because NaOH is very hygroscopic, it is better to buy small bottles (100 g/bottle) of NaOH to insure that the reagent is dry.
4. Grids should be kept wet throughout the staining process and only allowed to dry after all staining is complete.

References

1. Spurr, A. R. (1969) A low-viscosity epoxy resin embedding medium for electron microscopy. *J. Ultrastruct. Res.* **26,** 31–43.
2. Reynolds, E. S. (1963) The use of lead citrate at high pH as an electron-opaque stain in electron microscopy. *J. Cell Biol.* **17,** 208–212.
3. Marinozzi, V. (1961) Silver impregnation of ultrathin sections for electron microscopy. *J. Biophys. Biochem. Cytol.* **9,** 121–133.
4. Hayat, M. A. (1975) *Positive Staining for Electron Microscopy,* Van Nostrand Reinhold Co., New York.
5. Luft, J. H. (1961) Improvements in epoxy resin embedding methods. *J. Biophys. Biochem. Cytol.* **9,** 409–414.
6. Mollenhauer, H. H. (1964) Plastic embedding mixtures for use in electron microscopy. *Stain Technol.* **39,** 111–114.
7. Daddow, L. Y. M. (1983) A double lead stain method for enhancing contrast of ultrathin sections in electron microscopy: a modified multiple staining technique. *J. Microsc. (Oxford)* **129,** 147–153.
8. Bell, M. (1988) Artifacts in staining procedures, in *Artifacts in Biological Electron Microscopy* (Crang, R. F. E. and Klomparens, K. L. eds.), Plenum Press, New York, pp. 81–106.
9. Maunsbach, A. B. and Afzelius, B. A. (1999) *Biomedical Electron Microscopy: Illustrated Methods and Interpretations,* Academic Press, New York.
10. Chien, K., Van de Velde, R. L., Heusser, R. C., Shiroishi, H., and Cohen, A. H. (1994) A rapid phosphotungstic acid staining method on ultra-thin sections. *Proc. Ann. MSA Meeting* **52,** 318–319.
11. Mollenhauer, H. H. (1974) Poststaining sections for electron microscopy. *Stain Technol.* **49,** 305–308.
12. Mollenhauer, H. H. (1975) Poststaining sections for electron microscopy: an alternate procedure. *Stain Technol.* **50,** 292.
13. Avery, S. W. and Ellis, E. A. (1978) Methods for removing uranyl acetate precipitate from ultrathin sections. *Stain Technol.* **53,** 137–140.
14. Kuo, J. (1980) A simple method for removing stain precipitates from biological sections for transmission electron microscopy. *J. Microsc. (Oxford)* **120,** 221–224.
15. Kuo, J. and Husca, G. L. (1980) Removing stain precipitates from biological ultrathin sections. *Micron* **11,** 501–502.
16. Kuo, J., Husca, G. L., and Lucas, L. N. D. (1981) Forming and removing stain precipitates on ultrathin sections. *Stain Technol.* **56,** 199–204.
17. Mollenhauer, H. H. and Morré, D. J. (1978) Contamination on thin sections, cause and elimination. *Proc. Ninth Internat. Congress on Electron Microscopy,* vol. II, 78–79. Toronto.
18. Ellis, E. A. and Anthony, D. W. (1979) A method for removing precipitate from ultrathin sections resulting from glutaraldehyde-osmium tetroxide fixation. *Stain Technol.* **54,** 282–285.

7

Negative Staining of Thinly Spread Biological Samples

J. Robin Harris

Summary

Negative staining is widely applicable to isolated viruses, protein molecules, macromolecular assemblies and fibrils, subcellular membrane fractions, liposomes and artificial membranes, synthetic DNA arrays, and also to polymer solutions. In this chapter, techniques are provided for the preparation of the necessary support films (continuous carbon and holey/perforated carbon). The range of suitable negative stains is presented, with some emphasis on the benefit of using ammonium molybdate and of negative stain-trehalose combinations. Protocols are provided for the single-droplet negative staining technique (on continuous and holey carbon support films), the negative staining-carbon film technique, for randomly dispersed fragile molecules, 2D crystallization of proteins, and for cleavage of cells and organelles. The newly developed cryonegative staining procedure also is included. Immunonegative staining and negative staining of affinity labeled complexes (e.g., biotin-streptavidin) are discussed in some detail. The formation of immune complexes in solution for droplet negative staining is presented, as is the use of carbon-plastic support films as an adsorption surface on which to perform immunolabeling or affinity experiments, before negative staining. Dynamic biological systems can be investigated by negative staining, where the time period is in excess of a few minutes, but there are possibilities to greatly reduce the time by rapid stabilization of molecular systems with uranyl acetate or tannic acid.

Key Words: Negative stain; carbon support film; holey carbon support film; cryonegative staining; immunolabeling; affinity-labeling; the single droplet technique; negative staining-carbon film (NS-CF) technique; dynamic negative staining, uranyl acetate; ammonium molybdate; trehalose, polyethylene glycol (PEG).

1. Introduction

The negative staining of virus particles for the study of transmission electron microscopy (TEM) was introduced in the late 1950s, following the establishment of a standardized procedure by Brenner and Horne in 1959 *(1)*. In a matter

From: *Methods in Molecular Biology, vol. 369*
Electron Microscopy: Methods and Protocols, Second Edition
Edited by: J. Kuo © Humana Press Inc., Totowa, NJ

of time, this staining technique was applied to other thinly spread biological particulates, usually when a purified or semipurified aqueous suspension of freely suspended (i.e., nonaggregated) material was available. In addition to viruses, these methods have been applied to purified enzymes and other soluble protein molecules and components of molecular mass in the range of approx 200 kDa up to several MDa (such as the annelid hemoglobins, molluscan hemocyanins, and ribosomes), isolated cellular organelles, membrane fractions, bacterial cell walls and membranes, filamentous protein structures of many types, also liposomal and reconstituted membrane systems, and even synthetic synthetic deoxyribonucleic acid (DNA) arrays and polymers (as aqueous or organic solvent solutions).

The physical principle behind negative staining may at first glance appear to be rather simple but, on closer inspection, it is found to be somewhat more complex. Essentially, a water-soluble heavy metal-containing negative staining salt is used to surround and permeate within any aqueous compartment of a biological particle. After air-drying, a thin amorphous film or vitreous glass of stain (i.e., noncrystalline) supports and embeds the biological material and at the same time generates differential electron scattering between the relatively electron-transparent biological material and the electron-opaque negative stain. In this way, electron images are generated, which represent an approximation to the molecular envelope or solvent excluded surface of the particle. Simple air-drying undoubtedly leaves a good quantity of water bound to the biological material and within the surrounding negative stain, but this will be rapidly removed *in vacuo* within the electron microscope, unless the specimen is cooled in liquid nitrogen before cryotransfer of the specimen. Some of the subtleties and hazards of this oversimplified description will be expanded upon herein and recently have been examined in some detail elsewhere *(2)*.

After many years of relative stability and lack of progress, negative staining is currently undergoing significant technical development and, hopefully, improvement, to overcome some of the inherent undesirable aspects, such as excessive particle flattening and drying artifacts *(3)*, which should in turn lead to a better understanding of the hazards that can sometimes be generated after particle-stain interaction and drying. In addition, the combination of the established techniques of cryoelectron microscopy for the study of unstained biological materials with those of negative staining has opened up new and exiting possibilites through the cryonegative staining technique *(4)*.

It is the aim of this chapter to present some of the well-established and newer procedures *(5,6)* for air-dry negative staining using samples supported on continuous thin carbon support films and spread unsupported across small holes in holey carbon films *(7)*, with indication of the varying possibilities, applications, and technical limitations. Specimens prepared in this manner can be studied at room temperature in any conventional TEM, with or without low-electron dose study. Best-quality, high- resolution data will, however, be obtained only after specimen cooling (e.g., to −175°C) with image recording from mechanically and electronically stable specimens (i.e., with no specimen movement/drift), with minimal image astigmatism and under low electron dose conditions.

As with many electron microscopical preparative procedures, there can be a number of alternative ways to achieve the same end result from negative staining. For instance, sample and stain can be applied to a thin support film by a fine spray (nebulizer) individually or mixed, sample and stain can be applied direct to a support film from a pipet tip, or sample and stain can be transfered from small droplets on a clean Parafilm, Polythene, or Teflon surface. All these approaches and others can work, but it is the opinion of the author that the last, the single-droplet technique, is the simplest and most reliable. Thus, this procedure will be presented in full, followed by protocols for the negative staining-carbon film procedure, which is useful for freely spread single molecules, two-dimensional (2D) protein crystal formation, and for cleavage of cells and organelles. The newly developed, now standardized, technique of negative staining across the holes of holey carbon support films and of cryonegative staining (also on holey carbon or Quantifoil support films) also will be included. The ability to perform immuno- and affinity-labeling experiments in combination with negative staining will be expanded upon, and the use of negative staining for the study of dynamic biological systems will be mentioned.

A number of alternative and more specialized negative staining techniques have some value for the investigation of specific biological or molecular systems (for full coverage, *see* Harris *[2]*). Because a range of different negative stain salts are available, each with slightly different chemical properties and varying interaction with biological material, individuals often tend to favor the routine use of only one or two of these, to the exclusion of others that may be perfectly suitable and usable. Thus, although several useful negative stains are mentioned in **Table 1**, emphasis will be placed upon the use of ammonium molybdate, the stain that the author currently finds to be most reliable for studies with biological and artificial membranes, protein molecules, and virus particles. Nonetheless, the use of uranyl acetate and Na- or K-phosphotungstate as high-contrast negative stains for the preliminary TEM assessments remain useful. The reader should, however, bear in mind that slightly different opinions may well be expressed elsewhere. Also, it must be mentioned that even with ammonium molybdate, care should be taken to assess the possibility of deleterious effects resulting from sample-stain interaction *(2,3)*. Once appreciated and understood, such interactions can be of biochemical value in some instances.

Table 1
Negative Stain Solutions

Commonly used negative stain solutions

These stains are generally prepared as 1% or 2% w/v aqueous solutions,[a,b] but for
negative staining across holes and for cryo-negative staining a higher concentration
is usually required (*see* protocols).
Uranyl acetate
Uranyl formate
Sodium/potassium phosphotungstate
Sodium/potassium silicotungstate
Ammonium molybdate
Methylamine tungstate
Methylamine vanadate (Nanovan)

Negative stain-carbohydrate combinations

All of the above negative stains can be prepared as 2% to 6% w/v aqueous solutions
containing 0.1% to 1% w/v carbohydrate (e.g., glucose or trehalose).[c]

Negative stain-PEG combinations

The inclusion of 0.1 to 0.5% w/v polyethylene glycol (PEG) M_r 1000 in 2% w/v
ammonium molybdate creates a solution that potentiates 2D virus and protein crystal
formation.[d]

[a]A low concentration (e.g., 0.1 mM to 1.0 mM) of the neutral surfactant n-octyl-β-D-gluco-
pyranoside (OG) can be added to any of the above negative stain solutions to improve the spreading properties and assist permeation within biological structures.

[b]The pH of negative stain solutions can usually be adjusted over a wide range; this does not
apply to the uranyl negative stains, which readily precipitate if the pH is significantly increased
above pH 5.0. By complexing uranyl acetate with oxalic acid, an ammonium hydroxide neutral-
izable soluble anionic uranyl-oxalate stain can be created, but this possesses an undesirable granularity after drying and increased sensitivity to electron beam damage.

[c]Glucose and in particular trehalose can provide vitreous protectection to the biological material
during air-drying. The presence of trehalose also creates a slightly thicker supportive layer around
the sample, thereby reducing flattening. Electron beam instability of these carbohydrates neces-
sitates minimal routine or low electron dose irradiation conditions, assisted by specimen cooling
where possible. The inclusion of 1% trehalose reduces the net electron density of the negative stain
solution; this is why a higher stain concentration (e.g., 5% w/v) is used.

[d]When mixed with a purified viral or protein solution and spread as a thin layer on mica or
across the holes of a holey carbon support film (also with trehalose present), this AM-PEG
solution can induce 2D crystal formation (*see* text). Variation of the concentration of the PEG
and the pH of the solution is always required, to obtain the optimal conditions for 2D crystal
formation (*2,4*).

2. Materials

2.1. Equipment

1. The principal large item of equipment that is needed to prepare negatively stained EM specimens is a vacuum coating apparatus (e.g., the Edwards model Auto 306, the Bal-Tec model BAE 080 T, the Emitech model K405X, or Agar TEM Turbo Carbon Coater), together with facility to perform glow-discharge treatment. The latter may be an attachment within the vacuum coating apparatus or a separate item of equipment. Carbon coating, to produce thin continuous carbon, carbon-plastic, or perforated (holey) carbon support films (*see* **Subheading 2.2.** and also Harris *[2]*), can be performed using carbon rods, carbon fiber, or a carbon electron beam source, as described by the equipment manufacturer concerned. A carbon thickness monitor can be useful, but is not essential. Brief (30 to 60 s) glow discharge treatment of support films is particularly useful to combat the inherent hydrophobicity of the carbon surface, which interferes with the spreading and attachment of biological materials and the smooth spreading of a thin film of aqueous negative stain over and around the biological particles (i.e., satisfactory embedment in a high contrast medium). The use of prolonged ultraviolet irradiation has also been shown to render the surface of carbon support films hydrophilic.
2. For cryonegative staining, a plunge-freezing apparatus and associated consumables (e.g., liquid nitrogen, ethane gas) will be required. Some time will need to be devoted to master the use of this equipment and the cryotransfer of frozen specimens to a TEM *(2)*.
3. Numerous smaller items of equipment are needed (*see* **Fig. 1**), such as Parafilm, fine curved forceps (with rubber or plastic sliding closing ring, or use reverse-action forceps), fine straight forceps, a range of fixed volume automatic pipets (e.g., 5 µL, 10 µL, 20 µL; and variable volume pipettes, up to 1000 µL), plastic tips, scissors, metal needle/finely pointed probe, mica strips, filter paper wedges (e.g., cut from Whatman No. 1), Petri dishes with filter paper insert, 300 or 400 mesh electron microscope (EM) specimen grids (usually copper, but nickel or gold for immunonegative staining), grid storage boxes, a microcentrifuge, tubes, and tube racks. Last, and most important, tissue paper or other absorbent types of tissues need to be available to regularly wipe the tips of the forceps and needles, immediately after use, to avoid sample cross-contamination.

2.2. Support Films

Although support films can be readily purchased as consumables from the various EM supplies companies, it is still usual for individuals to prepare their own. Some time needs to be devoted to the perfection of these ancillary techniques, to make available a ready supply of the necessary continuous or holey carbon support films for conventional negative staining and for the preparation of thin frozen-hydrated/vitrified cryo-negatively stained specimens.

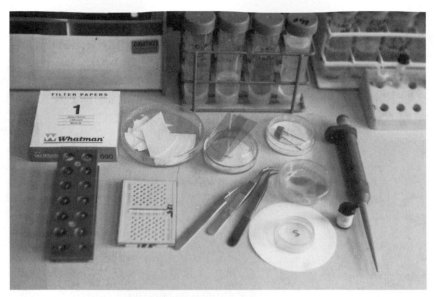

Fig. 1. An example of some of the small equipment and consumables needed for the routine production of negatively stained specimens.

Thin carbon support films can routinely be prepared by *in vacuo* carbon deposition onto the clean surface of freshly cleaved mica, with subsequent floatation of the carbon layer onto a distilled water surface followed by lowering on to a batch of EM grids (300 or 400 mesh) positioned beneath, i.e., in a Buchner funnel or glass trough with controlled outflow of the water *(2)*. The thickness of the carbon can be assessed by a crystal thickness monitor during continuous carbon evaporation, but this assessment is not esential. With a little experience, repeated short periods of evaporation from pointed carbon rods readily enable the desired thickness (e.g., approx 10 nm) to be achieved, based upon the faint gray color of a piece of white paper placed alongside the mica.

Carbon-plastic (e.g., colloidon, formvar, butvar, and pioloform) support films can be produced by first making a thin plastic film on the surface of a clean glass microscope slide, from a chloroform solution (0.1 to 0.5% w/v). This plastic film is then released from the glass slide after scoring of the edges with a metal blade, with floatation onto a distilled water surface. An array of EM grids can then be positioned individually on the floating plastic sheet, or the plastic sheet can be lowered on to an array of pre-positioned EM grids at the bottom of the funnel or trough, when the water level is reduced. After drying, the batch of grids can be carbon-coated *in vacuo*.

The presence of a thin plastic layer provides an increase in strength to the support, which is desirable for the extended sequence of steps required for immunolabeling (*see* **Subheading 3.4.**), but the extra support thickness does inevitably slightly reduce image detail and the maximal level of resolution. However, if desired, for both continuous carbon-plastic and holey/perforated carbon-plastic films, the plastic can be dissolved by washing grids singly in an appropriate solvent such as chloroform or amyl acetate, before use.

Most would agree that the production of holey/perforated carbon-plastic or carbon support films (also termed micro-grids) still remains something of a difficult art. Some therefore resort to the purchase of holey carbon films, or Quantifoil carbon support films with regularly sized and spaced holes, produced by semiconductor lithographic techniques. The simplest procedure is to initially perforate a drying film of plastic on the surface of a precooled (4°C) clean glass microscope slide by heavily breathing on to the surface. The small water droplets in the breath perforate the plastic film during evaporation of the chloroform. Alternatively, a more reproducible way of producing perforated plastic films is to use a glycerol-water (0.5% v/v each) emlusion in chloroform containing 0.1% or 0.2% w/v formvar. With vigorous shaking, an emulsion of small droplets is readily produced. On dipping a clean glass microscope slide in and out of the emulsion, allowing it to drain-off and dry, the thin plastic film so produced will be found to contain an array of small holes of varying size. A phase contrast light microscope can be used to quality control the production of holes in the drying plastic film. After wiping the glass edges and the edge of the film on the surface of the slide (i.e., approx 2 mm) with a paper tissue, the perforated plastic film can then released from the slide by floating the film onto a water surface and lowered on to a batch of EM grids previously placed beneath, as described previously. After drying, the grids should be carbon-coated, with *in vacuo* deposition of an additional layer of gold or gold/pladium, if desired. The inclusion of the metal layer enables the quality of holey carbon grids to be more easily assessed by bright field or phase contrast light microscopy, but its prime advantage is for rapid TEM focussing at higher magnfications and the fact that the presence of a metal layer improves the spreading properties of the surface, so no glow discharge treatment is necessary before use. Attempts have been made by some researchers to standardized this procedure *(7)*, but most do tend to include their own individual variations *(2)*.

2.3. Reagents and Solutions

1. Buffer: 20 mM Tris-HCl at pH 7.0 or 8.0 is most useful for diluting biological sample solutions. 5 mM Tris-HCl can be used as an on-grid washing solution.
2. Glutaraldehyde: make a 1% working solution in distilled water from the 25% w/v stock solution. Add an aliquot directly to biological sample.
3. Trehalose: make a 10% w/v stock solution in distilled water. Add an aliquot to negative stain solution, to give 1.0% or 0.1%, as desired.

4. Polyethylene glycol (PEG), M_r 1,000: make a 10% w/v stock solution. Adjust to desired pH with a 0.01 N NaOH. Add an aliquot to negative stain and protein solution, as required.
5. Uranyl acetate/formate: use as 1% or 2% w/v aqueous solution, without pH adjustment.
6. Ammonium molybdate Na/K-phosphotungstate and -silicotungstate: use as 1% or 2% (w/v) aqueous solution, or as 5 % (w/v) solution when 1% trehalose present. Adjust pH as desired with 1.0 N Na/KOH, usually to 7.0.
7. Methylamine tungstate: use as 1% or 2% w/v solution, with pH adjustment as desired.
8. Sodium/methylamine vanadate: use as a 2% w/v solution. Useful for low contrast negative staining.
9. Preimmune rabbit serum: dilute to 0.1% v/v with phosphate-buffered saline (PBS).
10. Bovine serum albumin (BSA): make up 0.1% w/v solution in PBS.
11. Rabbit anti-mouse IgG/Fab': dilute with PBS, as required.
12. Protein A: 5/10-nm gold conjugate: dilute with PBS.
13. Streptavidin: 5/10-nm gold conjugate: dilute with PBS. Possible stability problem.
14. Ammonium acetate: use as 0.155 M solution.
15. Glycerol: add as a protectant to protein/viral solutions at 50% v/v. Add to 0.155 M ammonium acetate solution at 30% v/v.
16. Alcian blue: use as 0.01% w/v aqueous solution.

2.3.1. Some Further Comments on the Use of Reagents and Solutions

Although prefixation/chemical crosslinking generally is not required for the negative staining of biological particulates, if it emerges that a sample is exceptionally unstable in the available negative stains, previous fixation with a low concentration of buffered glutaraldehyde (e.g., 0.05% or 0.1% v/v) may be included. This fixation can be performed in solution or by direct on-grid *droplet* treatment of material already adsorbed to a carbon support film. It must, however, be borne in mind that the chemical attachment of glutaraldehyde to the available basic amino acid side groups, producing intraprotein crosslinkage and stablization, may at the same time produce structural alterations at the higher levels of resolution. Also, in solution, care must be taken to avoid protein-protein crosslinkage/aggregation induced by glutaraldehyde.

The more commonly used negative staining salts are listed in **Table 1** (for a more detailed listing, including some of the less commonly used negative stains, *see* Harris *[2]*).These negative stains are generally used as 2% w/v aqueous solutions, but there is always the possiblility of increasing the concentration to provide greater electron density for small proteins or reducing the concentration for excessively thick biological samples that retain a greater volume of surrounding fluid. If the stain concentration is too low, on air-drying the thin layer of fluid that surounds a biological particle may not leave a sufficiently thick layer of amorphous salt to completely embed and support the particle, thereby resulting in partial-depth staining and undesirable sample flattening.

The adjustment of negative stain pH to a value close to that of the sample buffer is standard. This is not usually possible for the uranyl negative stains, which precipitate at pH values greater than approx pH 5.5. It should also be borne in mind that the presence of even traces of phosphate buffer are incompatible with the use of uranyl negative stains.

The addition of a protective carbohydrate such as trehalose (e.g., 0.1% to 1.0% w/v) to the negative stain solution has the advantage of creating a thicker supportive layer of dried stain, while at the same time helping to actually preserve the biogical sample during air-drying. Trehalose has uniquely beneficial properties in this respect because of the retention of sample structure within a vitreous carbohydrate-negative stain layer. Trehalose is thought to replace or protect protein-bound water and has been shown to have widespread value as a protectant in bioscience. The inclusion of a greater concentration of negative stain (e.g., 5 or 6% w/v) is then usually required to create an optimal electron density *(5,6)*. Specimen-bound water, with or without the presence of a carbohydrate, will be greatly reduced once a specimen is inserted into the high vacuum of the TEM unless direct cooling in liquid nitrogen is performed first and followed by cryotransfer to the electron microscope (with the specimen then maintained and studied at low temperature). As specimen lability within the electron beam initially is related to the presence of vitreous water, the early and continued success of conventional negative staining would appear to be the result of the rapid *in vacuo* removal of almost all loosely bound water from the biological material and amorphous stain, prior to electron irradiation. With the current widespread availability of TEM low-dose systems, image recording from frozen-hydrated negatively stained specimens (produced following air-drying or by rapid plunge freezing), can be successfuly pursued, thereby creating the possibility of improved image resolution because of sample hydration maintained at low temperature. Although negative staining and cryoelectron microscopy do now appear to have some significant overlap *(2,7,8)*, it is likely that the two separate technical approaches will often be maintained for the forseeable future. Indeed, for high-resolution low-temperature negative stain studies, the phrase high-contrast embedding media has been introduced to avoid the negative connotations and undesirable limitations of conventional room temperature air-dry negative staining on continuous support films *(8)*, that can at least in part be overcome when holey carbon support films are used (*see* **Subheading 3.1.2.**).

2.4. Sample Material

For conventional on-grid negative staining, it is desirable to have purified sample material in the form of a free suspension (i.e., without any large aggregates) in water or an aqueous buffer solution, at a concentration of ca 0.1 to 1.0 mg protein/mL. For negative staining on holey carbon support films *(7)* and for the negative staining-carbon film technique (when used to produce 2D crystals *[2,9]*) the optimal concentration will need to be in the order of 0.5 to 2.0 mg/mL protein. For lipid suspensions, lipoproteins, nucleic acids, nucleoproteins, and viral particles, these concentration figures provide only a general guide; the main aim in all cases is to avoid overloading the specimen with sample material since particle superimposition will always obliterate structural detail. The presence of a high concentration of sucrose, urea, or other solute in the sample suspension will introduce some problem for negative staining, so these reagents must be removed. These reagents can be removed by previous dialysis or gel filtration using a dilute buffer solution or by carbon adsorption and on-grid washing with distilled water or a dilute buffer solution, immediately before negative staining.

3. Methods

Protocols will be presented below for negative staining using the single-droplet procedure, applied to samples adsorbed to a continuous carbon film or spread across the small holes of a holey/perforated carbon film *(7)*. It is the considered opinion of the author, that the use of holey carbon films with ammonium molybdate-trehalose as the negative stain, although technically slightly more difficult, is preferrable for detailed negative stain studies. The negative staining-carbon film (NS-CF) technique, for 2D crystallization or proteins and viruses *(2,9)*, for randomly dispersed molecules, and for cleavage of cells and organelles *(10)* will be given. In addition, application of the droplet negative staining procedure for the production of antibody- or affinity-labeled specimens also will be given *(11,12)*, together with comments on the possibilities for dynamic negative staining experiments (where the time-period is generally in the range of a few minutes to hours, rather than milliseconds or seconds). Finally, a procedure for the preparation of cryonegatively stained specimens with be presented *(4)*.

Despite the precedural listings given herein, the user should, having gained some limited experience, be prepared to freely introduce small technical variations to suit any local requirements determined by the biological sample, the equipment available and the aims of any individual study. Thus, an overall scientific/technical awareness that improvements at the grid level can continually be sought and included is highly desirable *(2,3)*.

3.1. The Single-Droplet Negative Staining Technique

This general procedure is applicable to samples adsorbed to continuous carbon films and freely spread across the holes of holey carbon films. With slight modification, it is applicable to immuno-negative staining and dynamic negative staining experiments (*see below*, **Subheadings 3.3.** and **3.4.**).

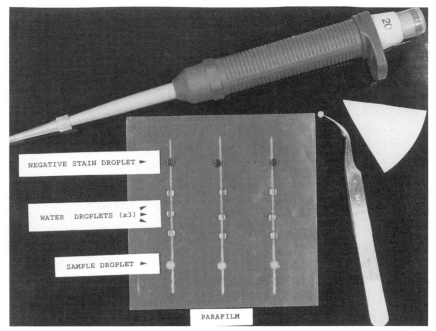

NEGATIVE STAIN DROPLET ▶

WATER DROPLETS (x3) ◀

SAMPLE DROPLET ▶

PARAFILM

Fig. 2. A representative example of the layout of material for the single-droplet negative staining procedure. The number of water-wash droplets is variable, depending upon the buffer and salt concentration in the sample.

3.1.1. Conventional Negative Staining on Continuous Carbon Films

1. Cut off a piece of Parafilm from a roll (length depending upon the number of samples and grids to be prepared), so that individual samples can be spaced by approx. 1.5 cm, as shown in **Fig. 2**. (In general, it is best not to attempt to prepare more than six to eight specimen grids as a single batch, unless this is absolutely necessary).

2. Place the Parafilm, paraffin-wax-down, on to a clean bench surface and before removing the paper overlay produce a number of parallel lines by scoring with a blunt object across the paper. Then remove the paper overlay, leaving the Parafilm loosly attached to the bench surface.

3. Place 20-µL droplets of sample suspension, distilled water (or dilute e.g., 5 mM buffer solution) and negative stain solution on to the Parafilm, as in **Fig. 2**. The number of water droplets can vary, depending upon the concentration of reagent to be removed from the sample, with the proviso that each successive wash may introduce additional breakage of the fragile carbon support film.

4. Take in a pair of curved forceps an individual specimen grid coated with a thin continuous carbon support film. (Brief glow discharge treatment should be applied in advance to increase hydrophilicity of the carbon surface and thereby improve the sample and stain spreading). Touch the carbon surface to the sample droplet. After a period of a time, ranging from 5 to 60 s (*see* **Note 1**), remove almost all the fluid by touching the edge of the grid carefully to a filter paper wedge.

5. Before the sample has time to dry, wash the adsorbed sample with one or more droplets of distilled water, and each time carefully drain away the excess water from the grid, as in **step 4**.

6. Touch the grid surface to the droplet of negative stain and likewise remove excess fluid. Then allow the thin film of sample and negative stain to air-dry, before positioning the grid in a suitable container (Petri dish or commercially available grid storage box; *see* **Note 2**).

7. After drying at room temperature, grids are immediately ready for TEM study, under either conventional electron dose conditions at ambient temperature or under low electron dose conditions at either ambient temperature or after specimen cooling in the TEM (e.g., to approx −180°C; *see* **Note 3**).

3.1.2. Negative Staining on Holey Carbon Support Films

This variant of the droplet negative staining technique is particularly useful for detailed studies on samples that have already been assessed by conventional negative staining on continuous carbon support films and justify more detailed investigation under conditions where carbon adsorption and specimen flattening are avoided. The protocol given in **Subheading 3.1.1.** should be followed, with incorporation of the following variations, as described by Harris and Scheffler *(7)*:

1. For holey carbon films, glow discharge usually is not necessary when a thin microcrystalline layer of gold or pladium metal has also been deposited.

2. It is necessary to use one of the negative stain-carbohydrate combinations (e.g., 5% w/v ammonium molybdate or sodium phosphotungstate + 0.1% w/v trehalose; *see* **Note 4**) because an unsupported air-dried film of sample and stain alone breaks very readily. This situation is considerably improved by the presence of carbohydrate, but concentrations in excess of 1.0% w/v trehalose lead to rapid damage in the electron beam unless low electron dose conditions are employed.

3. For specimens prepared across holey carbon films, sample concentrations of 1.0 to 2.0 mg/mL protein generally are desirable, as a greater quantity of free material is lost from the holes during the washing and staining steps. Biological material, however, often is retained at the fluid/air interface at the side of the grid opposite to where the filter paper is touched as the result of surface tension forces.

4. The sample washing, to remove interfering buffer and saline, can be done either with 20-µL droplets of 0.1% w/v trehalose or the 5% w/v negative stain + 0.1% w/v trehalose solution.

5. Because there is a tendency for the film of stain across the holes to be too thick, at the final stage of the procedure the filter paper should be held in contact with the edge of the grid for approx. 10 to 20 s, to remove as much fluid as possible.

6. Inclusion of PEG M_r 1000 (e.g., 0.1 mg/mL to 0.5 mg/mL) in the specimen solution, the trehalose washing solution and/or in the negative stain-trehalose solution can promote 2D crystal formation of isometric viruses and macromolecules across

the holes (*see* **Note 5**). In this case, longer times of incubation before the intial removal of excess sample and stain with the filter paper are desirable, to allow sufficient time for 2D crystal formation at the air-fluid interface, before drying.

7. Dry the grid at room temperature and study in a TEM within approx 1 d because recrystallization of the vitreous ammonium molybdate-trehalose film may occur.

3.2. NS-CF Technique

This procedure intitially was developed for the production of 2D arrays/crystals of viral particles *(9)*. The technique also has been adapted for the study of randomly dispersed fragile macromolecules and directly cleaved cells and organelles *(10)*, with the samples dried *in vacuo* on mica from a protective glycerol-volatile buffer solution, before carbon coating and negative staining.

3.2.1. Two-Dimensional Crystallization of Protein Molecules and Virus Particles

1. Prepare small pieces of mica, approx 1.5×0.5 cm, with one end pointed (*see* procedural outline in **Fig. 3**). Cleave the mica with a needle point to expose two untouched perfectly clean inner surfaces.
2. On a small piece of Parafilm, mix a 10-µL volume of sample (e.g., purified virus or protein solution, approx 1.0–2.0 mg/mL) with an equal volume of 2% ammonium molybdate (AM) containing 0.1% or 0.2% w/v PEG M_r 1,000. The pH of the AM solution can vary between pH 5.5 to pH 9.0. (Some AM precipitation may be encountered during storage at pH 6.5 and lower, over a period of months.)
3. Apply 10-µL quantities of the sample-AM-PEG to the clean surface of two pieces of mica (held by fine forceps) and spread the fluid evenly with the edge of a plastic pipet tip. Hold each piece of mica vertically for 2 s to allow the fluid to drain towards one end, remove most of the pooled fluid, and then hold horizontally, creating an even very thin film of fluid. Allow the fluid to slowly dry at room temperature within a covered Petri dish (*see* **Note 6**). A clearly visible zone of progressive drying towards a final deeper pool can usually be defined. 2D crystal formation will occur at this stage of the procedure, in all probability at the fluid-air interface, because adsorption of biological material to the untreated mica surface does not occur.
4. Coat the layer of dried biological sample on the mica surface *in vacuo* with a thin film of carbon (5 to 10 nm).
5. Float off the carbon film + adsorbed biological material (randomly dispersed and as 2D arrays or crystals) onto the surface of a negative stain solutution (e.g., 2% w/v uranyl acetate or ammonium molybdate) in a small Petri dish (*see* **Note 6**). Recover pieces of the floating film directly onto uncoated 400 mesh EM grids from beneath, with careful wiping on a filter paper to remove excess stain and any carbon that folds around the edge of the grid. Often, the freshly deposited carbon film does tend to repel the aqueous negative stain, leading to understaining rather than

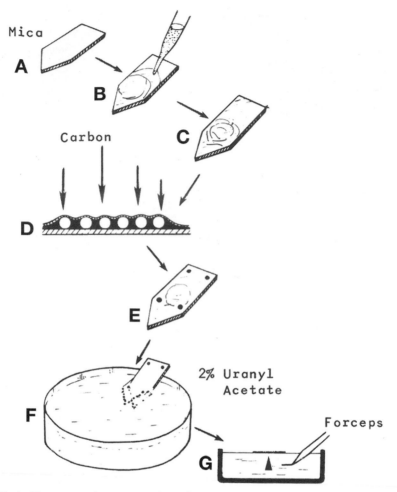

Fig. 3. A diagrammatic presentation of the succesive stages of the negative staining-carbon film (NS-CF) procedure (*9*).

overstaining; attempt to avoid this if encountered, by a shorter contact time with the filter paper. Allow the grid to air dry before positioning on a filter paper in a Petri dish or placing into a grid storage box.
6. Study stained grids in a TEM.

3.2.2. Negative Staining of Glycerol-Containing Solutions of Fragile Proteins

1. For this procedure (*10*), prepare small pieces of freshly cleaved mica, as in **Subheading 3.2.1.**
   Pipet approx 10-µL volumes of glycerol-containing solutions of protein or virus (0.1 to 0.2 mg/mL) onto the mica pieces. Remove excess fluid by touching to a filter paper (as in **Fig. 3**).

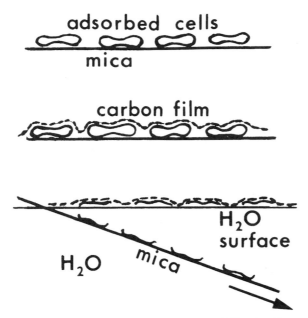

Fig. 4. A diagrammatic representation of the immobilized cell wet-cleavage by the negative staining-carbon film procedure *(10)*. The example shown uses erythrocytes, but cultured cells and isolated subcellular organelles also can be used.

2. Dry the glycerol-protein solution on the mica surface *in vacuo* at 10^{-5} Torr for 1 h, or longer (if a buffer is present, it should be volatile, i.e., ammonium acetate)
3. Coat the mica surface with a thin layer of carbon (10 nm) *in vacuo*.
4. Float off the carbon film + adsorbed protein onto either a negative stain solution or distilled water. Recover small pieces of the floating film with 400 mesh EM grids, from beneath the fluid surface. Remove excess stain solution by touching to the edge of a piece of filter paper. (If floated onto distilled water, finally add negative stain from a 20-μL droplet of stain on Parafilm).
5. Invert the EM grid (i.e., stain and protein then uppermost) and position horizontally on a filter paper to dry at room temperature.
6. Study stained grids in TEM.

3.2.3. Immobilized Cell and Organelle Cleavage by the NS-CF Procedure

This procedure *(10)* can be used for the study of wet-cleaved red blood cells, white blood cells, and cultured cells in suspension. It also has possibilities for the study of isolated cellular organelles, such a mitochondria, nuclear envelope, membrane vesicles, liposomes, and reconstituted vesicular membranes (*see* outline in **Fig. 4**), and can be combined with immuno- or affinity-labeling.

1. Prepare small pieces of freshly cleaved mica, as in **Subheading 3.2.1.**

2. To promote cell attachment, impart a positive charge on the mica surface by immersion in 0.01% w/v aqueous Alcian blue solution for 30 s, followed by thorough washing in distilled water and air drying.
3. Pipet 10-μL volumes of cell suspension onto the mica surface and spread evenly over the whole area. Place the mica horizontally for a few minutes, for the cells to become attached.
4. Washed away excess unbound cells by vertical immersion in three changes of PBS. Attached cells can be monitored by light microscopy at this stage. Brief cellular fixation with 0.1% v/v glutaraldehyde can be applied at this stage, if desired, but this may interfere with the cleavage **step 8** of this **Subheading**.
5. Remove the PBS by flooding the mica surface three times with 30% v/v glycerol-0.155 *M* ammonium acetate.
6. Wipe the under surface of the mica with tissue and remove the glycerol-ammonium acetate on the upper surface by drying *in vacuo* at 10^{-5} Torr for 1 h or longer.
7. Coat the dried cells with a thin layer (10 nm) of carbon *in vacuo*.
8. Float off the carbon film onto the surface of distilled water in a small Petri dish (*see* **Fig. 4**). At this stage, the cells are physically split, by a wet-cleavage process, leaving half the cell attached to the mica, the other half being removed with the carbon film. Note that over-stabilization of the cells with too much glutaraldehyde will prevent this cleavage.
9. Pick up small pieces of the floating carbon film on 400 mesh EM grids, from beneath the fluid surface, and immediately touch the membrane-containing surface of the carbon onto a 20-μL droplet of negative stain solution (e.g., 2% w/v uranyl acetate or ammonium molybdate) on a Parafilm surface.
10. Remove excess negative stain solution by touching the edge of the grid to a filter paper wedge, invert the grid and allow to dry at room temperature.
11. Study stained grids in TEM.

3.3. Cryonegative Staining

This procedure is a variant of the standard plunge freezing/vitrification procedure for the production of unstained cryospecimens (*see* Chapters 17 and 20). In the present instance, the technique is designed to produce frozen-hydrated specimens where a thin aqueous film of *negative stain* + the biological sample is spread across the small holes of a holey carbon support and is rapidly vitrified, as presented by Adrian et al. (*see* **Note 7** *[4]*). The equipment and materials required are essentially the same as for cryoelectron microscopy *(2)*. An additional procedural stage is included, within which the biological sample is mixed and incubated with a high concentration (16% w/v) ammonium molybdate solution prior to blotting and rapid freezing. The possibility also exists to create monolayer arrays and 2D crystals of isometric viral particles and macromolecules immediately before freezing by including PEG in the sample and negative stain solution (*see* **Note 8**). A diagrammatic description of the overall cryoneg-

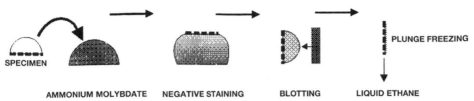

SPECIMEN   AMMONIUM MOLYBDATE   NEGATIVE STAINING   BLOTTING   LIQUID ETHANE   PLUNGE FREEZING

Fig. 5. An outline of the cryonegative staining procedure *(4)*. Immediately after the blotting stage, a short period of time (a few seconds) will enable some evaporation to occur with concentration of the sample and stain, prior to plunge freezing/vitrification. This procedure can also incorporate polyethylene glycol within the sample prior to freezing.

ative staining procedure is given in **Fig. 5**, taken from Adrian et al. *([4]; see* also Harris *[2]* for further details).

1. Apply 5 µL of sample solution to a holey carbon support film held by a pair of straight fine forceps. This is followed by short period of time, ranging from 10 to 60 s, during which some concentration of sample at the fluid/air interface will occur.
2. Invert the holey carbon grid + sample solution onto a 100-µL droplet of 16% w/v ammonium molybdate solution and allow it to float for a period of time, usually ranging between 10 s to 3 min. The forceps + grid and sample mixed with ammonium molybdate are then positioned in the clamp of the plunge freezing apparatus.
3. Carefully blot the grid face-on with a filter paper, which is held in direct contact with the grid for approx 1 to 2 s and gently lifted off. After a further 1 to 2 s, during which time some evaporation of water will occur to produce an optimal thickness of the aqueous film spanning the holes (achieved by personal experience only), the release catch of the *guillotine* is instantly activated (by a hand or foot mechanism) and the grid rapidly plunged into liquid nitrogen-cooled liquid ethane. A thin vitreous layer is produced, containing negatively stained sample material, across the small holes and surface of the holey carbon support film.
4. The forceps are then released from the holding mechanism and very quickly lifted from the liquid ethane into the liquid nitrogen, with the removal of as much adhering liquid ethane as possible (any remaining will solidify as a contaminant on the grid and can obliterate the specimen and/or interfere with the grid-holding screw or clip of the cryoholder).
5. At this stage, one or more grids can be stored in the small circular plastic transfer holder. Such holders, with a screw-on cap, can be stored in ice-free liquid nitrogen, using a modification of the system routinely used for storing aliquots of viable cultured cells. Repeat specimens can be prepared, with the precaution that the tips of the forceps holding a new holey carbon grid must not contain any moisture, because it will prevent the easy removal of the grid after the rapid freezing sequence (light coating of the forceps tips with Teflon will help, if a problem is encountered).

6. The TEM cryotransfer holder should meanwhile have been precooled, either when within the electron microscope or when inserted into the cryo-workstation. The anticontaminator system of the electron microscope must also be precooled. The small holder for the cryospecimens is then rapidly moved from the polystyrene box of liquid nitrogen to the liquid nitrogen within the cryostation. From there, a grid is positioned in the space at the end of the cryoholder, the holding-clip or -screw inserted, the sliding protective shield moved over the specimen and the specimen holder rapidly transferred to the electron microscope. The exact sequence of actions relating to the individual electron microscope air-lock evacuation, insertion and rotation of the cryoholder need to be performed as descibed by the manufacturer.

7. After insertion into the transmisison electron microscope, the small Dewar flask of the cryoholder should be rapidly filled or topped-up as necessary and time then allowed for the specimen holder temperature to stabilize (usually −170°C to −185°C) and the electron microscope to achieve its high vacuum.

8. After removal of the sliding shield of the specimen holder, low electron dose TEM study and image recording can commence.

3.4. Negative Staining
of Immunolabeled and Affinity Labeled Samples

The combination of immunological labeling of protein molecules, viruses, intact cytoskeletons, isolated cellular membrane fractions, fibrillar, and cyto-skeletal proteins in combination with negative staining offers considerable pos-sibilities for antigen/epitope localization. For plant viruses, this approach was elegantly demostrated by Roberts as early as 1986 *(11)*, yet the the the use of nega-tive staining in combination with immunolabeling of proteins and isolated cel-lular components and animal viruses has been relatively small, when compared to postembedding immunolabeling of cells and tissues (*see* Chapters 12 and 13). All the negative staining protocols presented herein can readily be adapted to study antibody-labeled samples, with the exception of the 2D crystallization procedure (*see* **Subheading 3.2.1.**).

Two main approaches can be followed. The first approach requires previous preparation of the biological material in combination with a defined epitope-specific polyclonal antibody, monoclonal antibody (IgG or Fab' fragment) in solution, using antibody alone or antibody conjugated to colloidal gold or a smaller gold cluster probe. IgG molecules will crosslink soluble protein mol-ecules, often creating specific linkage patterns that can assist the definition of epitope location. The use of Fab' fragments does in theory present the possibil-ity for higher resolution definition of epitope localization, but TEM study is then inherently more difficult and does require some confidence that all avail-able epitopes have a bound Fab'. With a satisfactory biological preparation, held in suspension as small immune complexes/immunoprecipitates (IgG) or soluble complexes (Fab'), the standard single-droplet negative staining procedure can be followed (using gold or nickel grids), as described in **Subheading 3.1.** (*see* also Harris *[12]*). An equivalent approach using the biotin-streptavidin system also can be used when a component within the biological sample is biotinylated. Both these forms of in-solution labeling can also be used to prepare samples for cryo-negative staining.

The second approach uses the fact that particulate biological material can be adsorbed to a robust carbon-coated plastic support film to create an immobilized thinly spread layer, which can then be allowed to interact successively with blocking solution, primary monoclonal antibody solution unlabelled, or colloidal gold-conjugated (usually with 5-nm or 10-nm colloidal gold), wash solutions, colloidal gold-conjugated secondary polyclonal antibody or protein A-gold wash solutions, distilled water, and finally a negative stain solution (sequential labeling using a secondary visualization marker is usually best). This does, in general, follow the pattern of on-grid postembedding immunolabeling of thin sectioned material, with the caution that somewhat greater handling precautions need to be adopted, particularly if the support film is carbon alone; carbon-plastic support films have superior handling properties. If dissociation is encountered during the prolonged incubation and washing sequence, then brief fixation of the sample with 0.05 or 0.1% v/v glutaraldehyde can be useful.

The location of primary or secondary antibody-gold conjugates, rather than unlabeled primary antibodies alone, has been found to be most successfully pursued by this approach. Thus, with an unconjugated primary monoclonal antibody the reaction can be performed in combination with a conjugated secondary antibodies and/or the streptavidin/biotin reaction (with biotinylated primary or secondary antibody, followed by streptavidin-conjugated gold probe).

Site-specific labeling of functional groups using derivatized gold cluster probes (Nanogold and Undecagold) also presents increasing possibilities for the future. In this instance, because of the small size of the gold probe, the use of a low-contrast negative stain, such as 2% w/v sodium vanadate or 3% ammonium molybdate containing 1% trehalose, is desirable. In general, with colloidal gold labeling, negative staining can be performed with any appropriate negative stain. As with immunolabeling of thin sectioned material, gold or nickel grids must be used and all appropriate controls included.

An experimental lay-out of sequential 20-μL droplets on Parafilm can be adopted, for example: sample solution, blocking solution, primary monoclonal antibody (can be biotinylated), washing solution, secondary gold-conjugated polyclonal antibody (streptavidin-gold, if the first step uses a biotinylated antibody), washing solution or distilled water and negative stain, essentially similar to that shown in **Fig. 2**. For prolonged incubation times at room temperature, individual blocking or antibody droplets, or the whole Parafilm sheet, can be covered with a Petri dish or other container, containing a piece of water-moistened tissue paper to minimise evaporation.

A standard protocol is given below (*after* Petry and Harris *[13]*), which can readily be modified to suit the requirements of almost any experimental conditions:

1. Take a glow discharge-treated carbon-formvar/butvar film on a nickel or gold 400 mesh EM grid and touch to the surface to a 10-μL droplet of sample solution (usually glutaraldehyde-fixed).
2. After a short time, remove excess solution by carefully touching to the grid edge to a filter paper. Select the sample adsorption time depending upon the concentration of material in solution. Avoid overloading the carbon surface with biological material.
3. Block nonspecific reactions by floating the grid for 10 min on a 20-μL droplet of 1.0% preimmune rabbit serum in PBS to remove free antibody. If Protein A-gold is to be used, avoid serum; use instead 0.1% w/v BSA in PBS. Remove excess fluid with a filter paper wedge.
4. Float the grid on a 20-μL droplet of hybridoma culture supernatant or purified primary IgG antibody for 2 h (*see* **Note 9**), at room temperature in a moist environment.
5. Wash the grid surface with five sucessive 20-μL droplets of 10% preimmune rabbit serum in PBS (or 0.1% w/v BSA in PBS; *see* **Note 10**).
6. Incubate the sample for 2 h by floating the grid on a 20-μL droplet of secondary antibody (e.g., rabbit anti-mouse IgG- or Protein A 5-nm gold conjugate), at room temperature.
7. Wash the grid surface with three 20-μL droplets of PBS and three droplets of deionized water.
8. Negatively stain the grid surface by touching to a 20-μL droplet of 1% or 2% w/v stain solution. Ammonium molybdate or sodium phosphotungstate (pH 7.0) will generally be found to be superior to uranyl acetate (pH 4.5).
9. Air-dry the grid at room temperature and study in a TEM.

This procedure can be combined with cellular permeablization procedures or surfactant extraction to reveal cellular cytoskeletons (*see* **Note 11** and also Hyatt *[14]*).

3.5. Dynamic Negative Staining

The peroid of time required to produce conventionally negatively stained specimens, on continuous or holey support films, requires approx 30 s. The drying time for the thin aqueous film of stain, at room temperature, is 2 or 3 min. With the cryo-negative stain procedure, the situation is similar during the initial stages, but the plunge freezing is extremely rapid (in ms).This provides the possibility that negative staining can be usd to monitor progressive or dynamic biological events, as long as they occur at time perods in excess of approx 2 min. However, Zhao and Craig *(15)* have performed time-resolved studies using an accurate rapid grid application system, whereby previous stabilization of myosin filaments for only 10 ms with uranyl acetate or tannic acid influenced the effect of a variety of interactive solutions. This system could probably be modified to trap rapid structural changes in other biological systems directly "on the grid," by applying the uranyl acetate rapidly after a very short time after a biological agent of functional interest. As some viruses react rapidly to a low pH and shed surface proteins, pH treatment of carbon-adsorbed viruses could provide a means to assess dynamic structural changes.

The PEG interaction time and drying time during the negative staining–carbon film procedure and during the holey carbon negative staining procedure greatly influences the nucleation of 2D crystals and molecular association to produce supramomolecular assemblies. Although such times are likely to minutes rather than seconds, the procedure can provide useful molecular information (*see* examples below).

3.6. Selected Examples of TEM Data
Produced Using the Negative Staining Protocols

Negative staining using the droplet procedure on continuous carbon support films is widely applicable, indeed to almost all isolated biochemical and biological samples, and indeed solutions of synthetic polymers. Numerous examples of data could be given; for further possibilities the reader is directed to the extensive scientific literature and the author's book *(2)*. The *Escherichia coli* chaperone GroEL complexed with the lower mass GroES to form the symmetrical/ellipsoidal complex is shown in **Fig. 6**. This specimen is negatively stained with 2% uranyl acetate on a continuous carbon support. Some chains of GroES-linked GroEL molecules are present, together with free GroES molecules. When trehalose is included in the negative stain solution, a thicker layer of dried stain and carbohydrate supports the biological material. This layer reduces flattening and enables the molecules to be more freely supported in varying angular orientations. This is shown in **Fig. 7** for the elongated multidecamers and didecamers of keyhole limpet hemocyanin type 2, negatively stained with a mixture of 5% ammonium molybdate and 1% trehalose (pH 7.0). There is some evidence, by comparison with uranyl acetate, that ammonium molybdate might release carbon-adsorbed molecules, thereby increasing their freedom immediately before stain dries. When the single-droplet negative staining procedure is used with holey carbon support films, greater care has to be taken at the final stage to remove enough stain solution in order to produce a thin dried film. When this is achieved the holes of the specimen grid will be evenly spread with the biological sample embedded in the thin layer of dried stain + trehalose. **Figure 8** shows an example of the decameric hemocyanin from *Octopus dolfleini*, spread across a holey carbon support film in the presence of 5% ammonium molybdate and

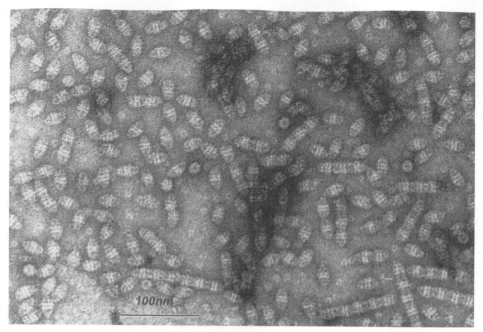

Fig. 6. The symmetrical complex of GroEL and GroES, negatively stained with 2%
uranyl acetate after adsorption to a continuous carbon support film. Note the presence
of some longer GroEL/GroES chains and free GroES molecules.

0.1% trehalose (pH 7.0). Inclusion of polyethylene glycol often promotes 2D
crystal formation across the holes, achieved by the author for many different
molecules and viruses. For tomato bushy stunt virus, the formation of 2D arrays
and crystals is shown in **Fig. 9**, negatively stained with 5% ammonium molybdate containing 1% trehalose and 0.1% PEG M_r 1000 *(7)*.

The negative staining-carbon film procedure is particularly successful when
used for the production of 2D protein and virus crystals. The example given in
Fig. 10 is of the 20S proteasome from *Thermoplasma acidophilum*, in this case
negatively stained finally with 2% ammonium molybdate rather than the more
usually used uranyl acetate. Note the large 2D crystal with a few proteasome
molecules missing from the lattice. For the *E. coli* chaperone GroEL, when 1 m*M*
Mg-ATP is present the molecule adopts a unique 2D hexagonal lattice with a
large central hexagonal space, that in the example shown (*see* **Fig. 11**) is filled
with negative stain (uranyl acetate). The inset in **Fig. 11**, shows the 2D image
reconstruction from this 2D crystal, revealing the manner in which the individual GroEL molecules are positioned within the hexagonal lattice.

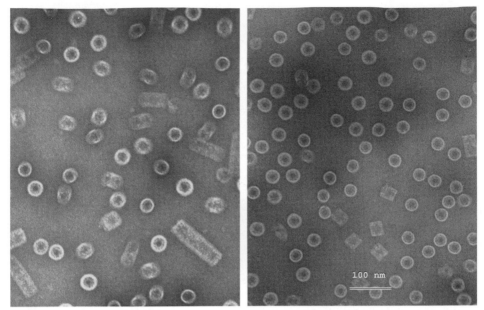

Fig. 7. Keyhole limpet hemocyanin (elongated KLH2 multidecamers and didecamers) negatively stained with 5% ammonium molybdate containing 1% trehalose on continuous carbon support films. This stain combination creates a relatively thick embedding layer that can support the molecules in varying orientations, as shown by the presence of the indistinct images of tilted molecules (left-hand panel). Bars,100 nm.

For fragile molecules that require the presence of a high concentration of glycerol for stability or are dramatically aggregated by uranyl acetate during the conventional droplet negative staining, the negative staining-carbon film procedure offers an alternative possibility. The enzyme tripeptidyl peptidase-II is in this category. In solution, tripeptidyl peptidase-II forms oligomers that generate elongated arc and double-bow complexes. The oligomeric intermediates that lead to the formation of these complexes are shown in **Fig. 12**.

The negative staining-carbon film procedure also can be used to produce cellular cleavage. In **Fig. 13**, an example is shown of a human erythrocyte wet-cleaved to reveal a single membrane layer, with the internal/cytoplasmic surface negatively stained with 2% uranyl acetate.

Cryonegative staining is likely to become an increasingly used technique for intermediate resolution molecular and virus studies (i.e., slightly above 1 nm), as it provides the benefits of freeze-preservation combined with the benefits of increased image contrast imparted by the negative stain. **Figure 14** shows an electron micrograph of cryo-negatively stained keyhole limpet hemocyanin

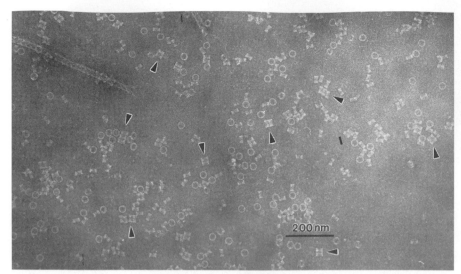

Fig. 8. Octopus hemocyanin negatively stained with 5% ammonium molybdate + 1% trehalose, spread unsupported across a hole in a holey carbon support film. Note the presence of the single ring-like end-on hemocyanin decamers and small stacks of side-on decamers (arrowheads).

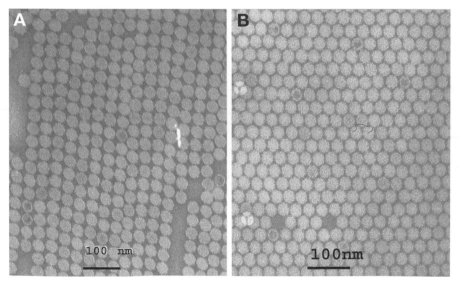

Fig. 9. Two examples of tomato bushy stunt virus negatively stained with 5% ammonium molybdate containing 1% trehalose and 0.1% polyethylene glycol (M_r 1000) across holes in a holey carbon support film. The presence of the polyethylene glycol has induced the formation of a disorganized 2D array (**A**) and a 2D crystal with a hexagonal lattice (**B**). Modified from Harris and Scheffler (7).

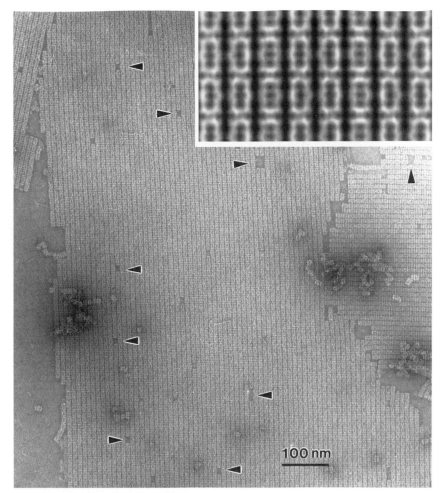

Fig. 10. A 2D crystal of the 20S proteasome from *Thermoplasma acidophilum* produced by the negative staining-carbon film procedure, with 2% ammonium molybdate (pH 7) as the final negative stain. Image processing suggests that the unit cell may contain six molecules because of the specific rotational orientation of the heptameric proteasome within the crystal (modifed from Harris *[2]*). Note the absence of individual and small groups of molecules from the 2D crystal lattice (arrowheads).

type 1 (KLH1) reconstituted from the subunit, with many didcecamers present at different angular orientations. Also present are single KLH1 decamers and tubular polymers. The somewhat-superior molecular detail here should be compared with that shown in **Figs. 7** and **8**.

Negative staining of immunologically and affinity-labeled biological samples is likely to be of increasing significance in the years ahead, as more epitope-

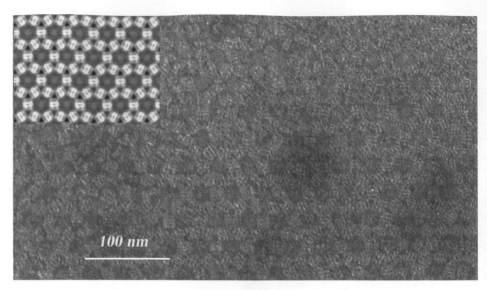

100 nm

Fig. 11. A complex honeycomb 2D hexagonal lattice the *E. coli* protein GroEL in the presence of 1 m*M* Mg-ATP, prepared by the negative staining-carbon film procedure. The inset shows a reconstruction from the 2D crystal, which reveals the complex linkage of the individual cylindrical GroEL molecules forming the hexagonal lattice. This hexagonal 2D lattice was never encountered in the absence of ATP, which is in accord with evidence from others showing that a significant shape change is induced when ATP binds to GroEL.

specific and residue-specific labels become available. Two examples of "in solution" immunonegative staining will be given. An immune complex of KLH2 with a monoclonal IgG directed against an epitope on the end of the molecule is shown in **Fig. 15**, negatively stained on a continuous carbon support film by 5% ammonium molybdate containing 1% trehalose. The molecules are linked to form chains, with one or more IgG molecules forming the bridges between the ends of the molluscan hemocyanin didecamer. When a KLH2 decamer ends a chain, further extension is not possible as the required epitope is not located at this position. If a biotinylated component is present within a biological molecule, subsequent labeling with streptavidin-gold is possible. **Figure 16** shows an example of cholesterol microcrystals with a biotinylated mutant streptolysin-O molecule bound onto their surface. In solution, streptavidin conjugated 5-nm gold particles bind to the biotin, and negative staining with 2% ammonium molybdate shows convincing labeling at the edges and surface of the cholesterol microcrystals. A similar approach can be performed at a higher-resolution level if a His-tagged component is present by using a nickel-complexed gold cluster. When immunolabeling is performed on biological particles already adsorbed onto a

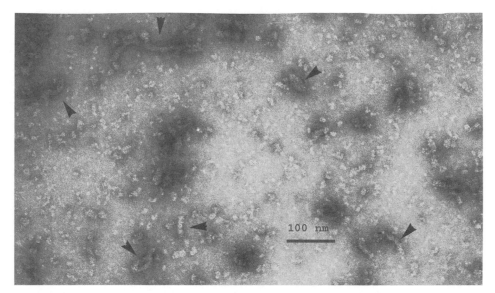

Fig. 12. Human erythrocyte tripeptidyl peptidase-II negatively stained with 2% uranyl acetate, by the negative staining-carbon film procedure *(10)*. Conventional droplet negative staining of this enzyme complex with uranyl acetate on a carbon support film induced unacceptable aggregation. This specimen was produced directly from a protective 30% glycerol solution, by *in vacuo* drying followed by carbon coating and floating off onto negative stain. The high molecular mass tripeptidyl peptidase complex is revealed as elongated arc-like oligomers of the protein (arrowheads), alongside smaller oligomers.

carbon-plastic support film, considerable possibilities exist for antigen localization. Micronemes isolated from the parasite *Cryptosporidium parvum* are very fragile, and even after stablization with 0.05% v/v glutaraldehyde some show damage. In **Fig. 17**, micronemes are shown, following on-grid labeling with a mouse monoclonal antibody and a secondary IgG complexed to 5 nm gold, negatively stained with 2% ammonium molybdate. This IgG is directed to a protein epitope that is present within the micronemes. Thus, only the content of damaged micronemes is labelled. The smooth-surfaced intact stain-excluding micronemes show no labeling.

Negative staining of dynamic biological systems has considerable potential, as long as the time period under consideration is in excess of approx 2 min. The formation of the dodecahedral macromolecular complex containing 12 toroidal 218-kDa human erythrocyte peroxiredoxin-2 decamers has been attempted by the author using negative staining on holey carbon support films in the presence of PEG. If the enzyme sample is incubated with PEG for a few minutes and then EM specimens produced with PEG present at every washing and negative staining

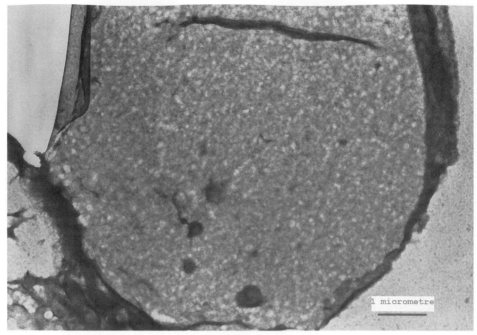

Fig. 13. A single layer of human erythrocyte membrane produced by cell cleavage during the negative staining-carbon film procedure *(10)*. The membrane has been negatively stained with uranyl acetate, revealing the uneven spectrin network on the cytoplasmic surface.

stage, dodecahedral peroxiredoxin-2 complexes can be shown by TEM to have formed (*see* **Fig. 18**). The complete standardization of this sterically interesting macromolecular assembly, which is induced instead of 2D crystal formation in this instance, is currently under investigation.

The interaction time of antigen-antibody and biotin-streptavidin can often be assessed using negative staining. For the biotin-streptavidin reaction, this has been found to be rapid, by showing the formation of streptavidin-labeled tubules of synthetic DNA formed from oligonucleotides, one of which was biotinylated *(16)*, after an incubation period of only 30 min (*see* **Fig. 19**). This specimen was spread across a holey carbon support film and was negatively stained with 5% w/v ammonium molybdate containing 0.1% w/v trehalose. Similarly, the interaction time of the *Vibrio cholerae* cytolysin with cholesterol microcrystals occurs at the crystal bilayer edges within a few minutes, but extends onto the planar surface over a period of hours *(17)*. The images of this material, shown in **Fig. 20**, were negatively stained with 2% ammonium molybdate on a carbon support film.

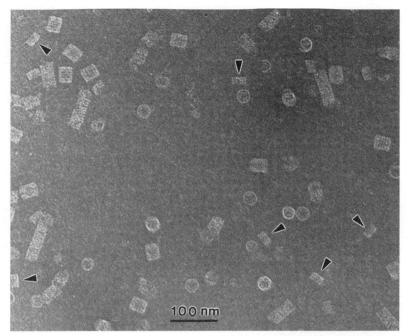

Fig. 14. Reassociated keyhole limpet hemocyanin type 1 revealed by cryo-negative staining with 15% w/v ammonium molybdate. Note the presence of decamers (arrowheads), didecamers and longer tubular structures (courtesy of Marc Adrian).

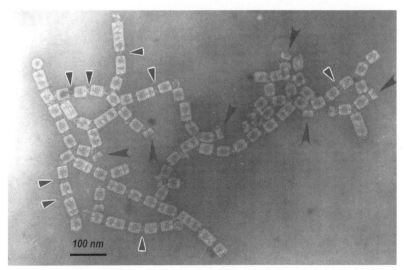

Fig. 15. An immune complex containing IgG-linked keyhole limpet hemocyanin type 2 molecules. Note the single decamers (large arrowheads) and multiple IgG molecules linking the hemocyanin molecules (small arrowheads). Negatively stained on a continuous carbon film with 5% ammonium molybdate containing 1% trehalose.

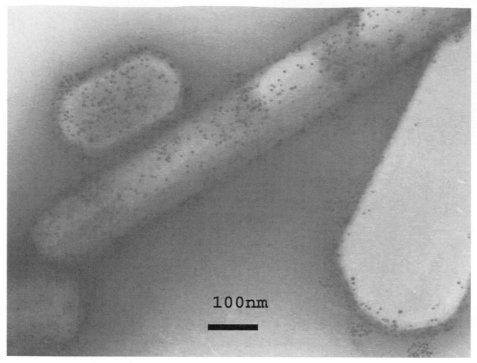

Fig. 16. Cholesterol microcrystals with bound biotinylated streptolysin-O (SLO) mutant (N402C), which does not form the characteristic SLO pores but retains an affinity for cholesterol. The cholesterol-bound biotinylated streptolysin-O was labelled with streptavidin-conjugated 5-nm colloidal gold particles. The specimen was negatively stained on a continuous carbon support film with 5% ammonium molybdate containing 1% trehalose.

A useful survey of the application of negative staining and image classification is given in the recent report by Ohi et al. *(18)*, which complements the examples given above.

4. Notes

1. Specimen grids with a low concentration of biological material require a longer carbon-adsorption time and those with high concentration of material require shorter time.
2. Negatively stained specimens can usually be stored for many weeks. However, for specimens containing a mixture of trehalose and ammonium molybdate negative stain, after a period of a few days, the initially amorphous/vitreous glass from the air-dried mixture of stain and carbohydrate (which still contains a consider-

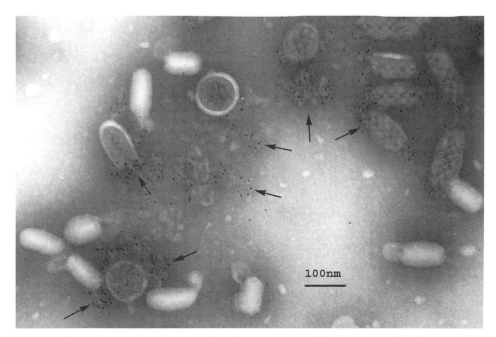

Fig. 17. Immunonegative staining of micronemes isolated from *Cryptosporidium parvum* sporozoites after adsorption onto a continuous carbon-formavar support film, labeled with the the anti-microneme monoclonal antibody ABD and detected with anti-mouse IgG conjugated to 5-nm gold particles. Labeling only occurs on the damaged micronemes (arrows), where the content of the microneme is accessible or has escaped onto the background. Intact micronemes (smooth surfaced elongated electron transparent particles) are unlabelled. The specimen was negatively stained with 2% ammonium molybdate. Modified from Petry and Harris *(13)*.

able quantity of bound water) will exhibit signs of undergoing recrystallization of the negative stain. Such specimen grids should therefore be studied in TEM as soon as possible after preparation (i.e., within approx 2 days) or stored with desiccation.

3. If air-dried negatively stained specimens are subjected to rapid freezing and cryo-transfer, the remaining specimen-bound water will be retained within the thin film of sample + stain. They must then maintained at low temperature (–180°C) and require low electron dose study. Specimens can also be cooled following room temperature insertion into the TEM; this will produce maximal dehydration of the specimen before cooling to a low temperature. In general, for all specimen conditions, a low-dose study is desirable to obtain the best quality data, particularly for those specimens containing carbohydrate and is essential for those containing vitreous water.

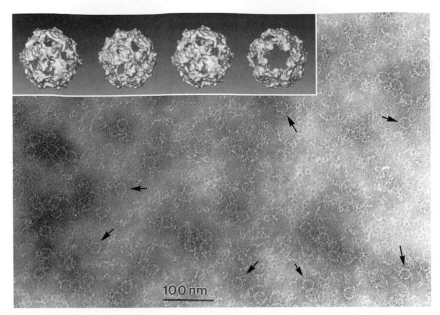

Fig. 18. The higher-order dodecahedral complex of human erythrocyte peroxired-oxin-2, formed during negative staining with 5% ammonium molybdate, 0.1% trehalose (pH 6.5) across a holey carbon support film, in the presence of 0.2% PEG M_r 1000. Note the presence of many single M_r 220,000 kDa ring-like Prx-2 decameric molecules scattered on the background, as well as small chains and clusters of molecules. Depending upon the orientation of the dodecahedra, 2-, 3-, and 5-fold symmetry features may be revealed. Some distortion of the dodecahedra often occurs, together with aggregation. The inset shows a 3D low-resolution (approx 3 nm) of the Prx-2 dodeca-hedron (diameter approx 20 nm), in varing orientations. Previously unpublished data of the author and Ulrich Meissner.

4. Although the use of ammonium molybdate is described, the possibility also exists of using an alternative negative stain solutions such as sodium phosphotungstate, sodium silicotungstate, or methylamine tungstate in combinaton with trehalose for negative staining across holes. Uranyl acetate solution is not considered to be suitable in this instance because of instability in the electron beam and excessive image granularity.

5. The PEG concentration and molecular weight of the PEG can be varied within a reasonable range, for instance, the concentration can range from 0.05% to 0.5% w/v and the molecular weight of the PEG from M_r 1,000 to M_r 10,000. If protein or viral aggregation is encountered, they usually will indicate that the PEG concentration is excessive or that the pH is too low. At higher PEG concentrations, individual microcrystals of PEG may be produced during air-drying.

6. One possible disadvantage of the NS-CF procedure is the fact that the biological sample is air-dried twice and subjected to direct *in vacuo* carbon-coating, which

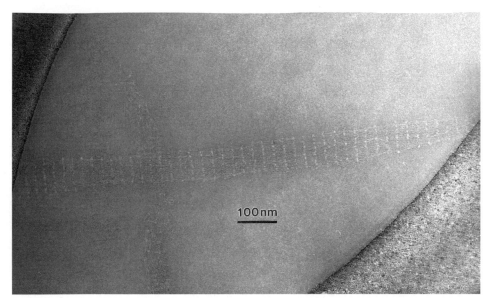

Fig. 19. A synthetic helical DNA tubule containing biotinylated oligonucleotide with labeling of the the tubules in-solution by streptavidin. The streptavidin molecules are bound periodically along the length of the tubule (*see* Mitchell et al. *[16]*). The specimen was prepared on holey carbon support film and negatively stained with 5% ammonium molybdate, 0.1% trehalose (pH 7.0).

might then restrict the penetration of negative stain at the zones of carbon-protein attachment. In general, this does not appear to be a problem, but some tendency for the final negative stain film to be rather too thin may be encountered because of the hydrophobicity of carbon. The drying time on the mica during 2D crystal formation can be varied by placing in a humid environment or a refrigerator. At the final stage, any of the conventional negative stain solutions, with or without the carbohydrate trehalose can be used. If a very thin carbon film is deposited *in vacuo* to obtain best-possible resolution, this fragile carbon layer can be supported on a holey carbon film. It is also possible to float the carbon film + attached sample on to distilled water, with recovery and removal of excess water, followed by rapid plunge-freezing, as for the preparation of unstained or negatively stained frozen-hydrated specimens.

7. The 16% w/v ammonium molybdate solution used for cryonegative staining may induce biochemical instability with some samples *(4)*. pH variation should then be explored, as can previous fixation with a low concentration of glutaraldehyde (e.g., 0.05 to 0.1% v/v). The higher-than-usual concentration of ammonium molybdate (e.g., 16% w/v here, compared with 2% to 5% w/v for air dry negative staining) is necessary because a very thin aquous film of sample + stain is produced by direct blotting across the holes of the grid. Note the similarity of this cryo-procedure to

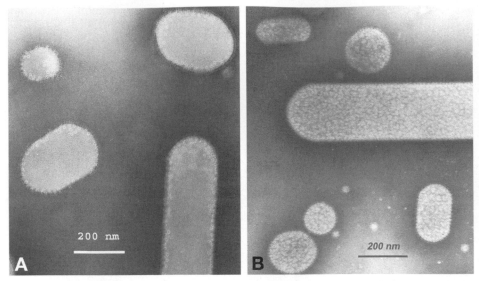

Fig. 20. Cholesterol microcrystals after interaction with *Vibrio cholerae* cytolysin (VCC) for **(A)** 15 min and for **(B)** 1 h at room temperature. The specimens were prepared on continuous carbon support films and negatively stained with 2% ammonium molybdate (pH 7.0; *see* Harris et al. *[17]*). Note that in (A) the VCC oligomer has bound only to the exposed edge of the stacked cholesterol bilayers, whereas in (B) the planar surface is also coated with VCC oligomers.

    air-drying of negatively stained specimens across holes (**Subheading 2.1.**, and also cf Adrian et al. *[4]* and Harris and Scheffler *[7]*).

8. For 2D crystallization of icosahedral and filamentous viral particles, macromolecules and enzyme complexes during the cryonegative staining procedure, the optimal concentration of the start material, the pH of the ammonium molybdate solution, the concentration of PEG and the time period of incubation of the sample with PEG and ammonium molybdate all need to be determined by the experimenter for each sample under investigation.

9. The necessary dilution of antibody and gold probe solutions and adequacy of blocking can only be determined by the individual experimenter, with the comment that the higher the antibody dilution and more efficient the blocking, the greater the labeling specificity is likely to be and the lower background non-specific labeling. Generally, prefixation of the biological sample should not be necessary for immunonegative staining. Thus, loss of sample antigenicity should not apply within this system. However, a brief on-grid fixation with a low concentration of glutaraldehyde (e.g., 0.05% to 0.1% v/v) can be included, if sample instability is encountered during the blocking, labeling, washing and negative staining sequence, or there is prior knowledge regarding sample instability in-solution. Appropriate negative controls should always be included in all immunological studies.

10. With Protein A-gold, it is be recommended that blocking with rabbit preimmune serum should not be used, instead use 0.1% w/v bovine serum albumin or casein in PBS.
11. A more detailed presentation of immunogold labeling techniques in relation to negative staining can be found in Hyatt *(14)*, and useful information on the use of gold cluster probes can be obtained in the promotional literature from Nanoprobes Inc.

References

1. Brenner, S. and Horne, R. W. (1959) A negative staining method for high resolution electron microscopy of viruses. *Biochim. Biophys. Acta* **34,** 60–71.
2. Harris, J. R. (1997) *Negative Staining and Cryoelectron Microscopy: the Thin Film Techniques.* RMS Microscopy Handbook Number 35, BIOS Scientific Publishers Ltd., Oxford.
3. Harris, J. R. and Horne, R. W. (1994) Negative staining: a brief assessment of current technical benefits, limitations and future possibilities. *Micron* **26,** 5–13.
4. Adrian, M., Dubochet, J., Fuller, S. D., and Harris, J. R. (1998) Cryo-negative staining. *Micron* **29,** 145–160.
5. Harris, J. R., Gebauer, W., and Markl, J. (1995) Keyhole limpet hemocyanin: negative staining in the presence of trehalose. *Micron* **26,** 25–33.
6. Harris, J. R., Gerber, M., Gebauer, W., Wernicke, W., and Markl, J. (1996) Negative stains containing trehalose: application to tubular and filamentous structures. *J. Microsc. Soc. Am.* **2,** 43–52.
7. Harris, J. R. and Scheffler, D. (2002) Routine preparation of air-dried negatively stained and unstained specimens on holey carbon support films: a review of applications. *Micron* **33,** 461–480.
8. Orlova, E. V., Dube, P., Harris, J. R., et al. (1997) Structure of keyhole limpet hemocyanin type 1 (KLH1) at 15 Å resolution by electron cryomicroscopy and angular reconstitution. *J. Mol. Biol.* **272,** 417–437.
9. Horne, R. W. and Parquali-Ronchetti, I. (1974) A negative staining-carbon film technique for studying viruses in the electron microscope. I. Preparation procedures for examining isocahedral and filamentous viruses. *J. Ultrastruct. Res.* **47,** 361–383.
10. Harris, J. R. (1991) Negative staining-carbon film technique: New cellular and molecular applications. *J. Electron Microsc. Techn.* **18,** 269–276.
11. Roberts, I. M. (1986) Immunoelectron micrisocopy of extracts of virus-infected plants, in *Electron Microscopy of Proteins*, vol. 5, *Viral Structure* (Harris, J. R. and Horne, R. W., eds.), Academic Press, London, pp. 293–357.
12. Harris, J. R. (1996) Immunonegative staining: epitope localization on macromolecules. *Methods* **10,** 234–246.
13. Petry, F. and Harris, J. R. (1999) Structure, fractionation and biochemical analysis of *Cryptosporidium parvum* sporozoites. *Int. J. Parasitol.* **29,** 1249–1260.
14. Hyatt, A. D. (1991) Immunogold labelling techniques, in *Electron Microscopy in Biology: A Practical Approach* (Harris, J. R., ed.), IRL Press, Oxford, pp. 59–81.

15. Zhao, F.-Q. and Craig, R. (2003) Capturing time-resolved changes in molecular structure by negative staining. *J. Struct. Biol.* **141,** 43–52.
16. Mitchell, J. C., Harris, J.R., Malo, J., Bath, J., and Turberfield, A. J. (2004) Self-assembly of chiral DNA nanotubes. *J. Am. Chem. Soc.* **126,** 16342–16343.
17. Harris, J. R., Bhakdi, S., Meissner, U., et al. (2002) Interaction of the *Vibrio cholerae* cytolysin (VCC) with cholesterol, some cholesterol esters, and cholesterol derivatives: a TEM study. *J. Struct. Biol.* **139,** 122–135.
18. Ohi, M., Li, Y., and Walz, T. (2004) Negative staining and image classification—Powerful tools in modern electron microscopy. *Biol. Proced. Online* **6,** 23–34.

8

Recent Advances in High-Pressure Freezing

Equipment- and Specimen-Loading Methods

Kent L. McDonald, Mary Morphew, Paul Verkade, and Thomas Müller-Reichert

Summary

This chapter is an update of material first published by McDonald in the first volume of this book. Here, we discuss the improvements in the technology and the methodology of high-pressure freezing (HPF) since that article was published. First, we cover the latest innovation in HPF, the Leica EM PACT2. This machine differs significantly from the BAL-TEC HPM 010 high-pressure freezer, which was the main subject of the former chapter. The EM PACT2 is a smaller, portable machine and has an optional attachment, the Rapid Transfer System (RTS). This RTS permits easy and reproducible loading of the sample and allows one to do correlative light and electron microscopy with high time resolution. We also place more emphasis in this article on the details of specimen loading for HPF, which is considered the most critical phase of the whole process. Detailed procedures are described for how to high-pressure freeze cells in suspension, cells attached to substrates, tissue samples, or whole organisms smaller than 300 μm, and tissues or organisms greater than 300 μm in size. We finish the article with a brief discussion of freeze substitution and recommend some sample protocols for this procedure.

Key Words: High-pressure freezing; high-pressure freezer; freeze substitution; cryofixation; EM specimen preparation; cryoEM; electron tomography; vitreous cryosections; correlative light; electron microscopy.

1. Introduction

"It is not the pressure-freezing itself, but the preceding preparation, which is the limiting factor also for this type of cryofixation..." Hans Moor *(1)*

Since the last review of high-pressure freezing (HPF *[2]*), much has happened in the field. There are new HPF machines (Leica EM PACT2, this chapter and *[3]*),

From: *Methods in Molecular Biology, vol. 369*
Electron Microscopy: Methods and Protocols, Second Edition
Edited by: J. Kuo © Humana Press Inc., Totowa, NJ

new techniques and holders for loading samples, and new applications of the method. There also have been repeated confirmations of the observations of Hans Moor, the developer of the original HPF machine *(4)*, that the geometry of the specimen holders, the types of material surrounding the specimen in the holder, and the other aspects of specimen loading are crucial to success in cryofixation by HPF *(1)*. Accordingly, in this chapter we focus on the process of specimen loading more than the HPF instrument itself. Because it is also possible to undo good freezing by poor handling after the samples are frozen, we will say a few words about freeze substitution, perhaps the most common way to process rapidly frozen samples.

Different types of samples often require different specimen carriers and/or loading strategies. Therefore, we will organize part of the text according to the various specimen types, starting with single cells and proceeding to more complex tissues and even whole organisms. Within each category we will discuss the appropriate type of carriers, external cryoprotectants, and handling tips that we believe will result in optimal freezing. Because newcomers to the field often want to know the difference between the types of freezing machines, we will offer a brief explanation of these differences. Our overall goal is to provide the reader with enough information that they can use HPF equipment and share our excitement about this technique, which we believe to be the state of the art for preserving cellular ultrastructure in many samples, especially those too large to be frozen by other methods. We also will provide ample references for further reading and some illustrations of the newest instruments and accessories. For those who have not read previous material on HPF, we will briefly reiterate here what high pressure freezing is and why it is better than conventional methods of specimen preparation for electron microscopy, in addition to listing some of the current applications of the technique.

HPF freezes specimens up to several hundred microns in thickness under pressure of more than 2000 Bar. This pressure retards the nucleation and growth of ice crystals and effectively lowers the freezing point of water by approx 20°C *(1)*. The consequence of these physical effects is to allow freezing to greater depths than is possible by plunge freezing or impact against a cooled metal block. In the latter methods, an optimistic average depth of good freezing is on the order of 10 μm. With any method, the actual depth of good freezing will vary with the chemistry of the particular cell type. In general, cells with less free water will freeze better than those that have more aqueous cytoplasm. Likewise, some cells also have natural cryoprotectants, such as high sugar and/or protein contents. Once frozen, cells can be treated in many different ways, such as freeze-fracturing, vitreous cryosectioning, freeze drying, or freeze substitution. After freeze drying or freeze substitution, they can be processed for scanning electron microscopy or embedded at low or room temperatures for subsequent sectioning.

By conventional fixation, we mean immersion of cells or tissues into fixatives such as glutaraldehyde made up in buffer solutions, rinsing in buffer, then postfixation with osmium tetroxide and dehydration at room temperature with organic solvents such as ethanol or acetone. Comparison studies of tissues processed this way and by fast freezing–freeze substitution show that conventional methods result in more obvious distortions of cell structure *(5–10)*. Although these distortions usually are referred to as fixation artifacts, we know from other studies *(11)* that the driving force for these distortions is actually the dehydration steps. During freeze substitution at temperatures down to −90°C, there is insufficient thermal motion to create distortions on this relatively large scale.

For studies that do not require high resolution, it is not essential to use low-temperature methods. For example, one may only want to know whether there are certain organelles in particular regions of cells and tissues and light microscopy may be inconclusive on the subject. Conventional methods will answer this question, although the ultrastructure will be less accurate than if cryomethods were used. However, there are times when optimal preservation of ultrastructure is necessary. Vitreous cryosectioning or morphometric analysis and three-dimensional modeling by tomography or serial sections are good examples. EM immunolabeling is another. Loss of proteins, lipids, or other molecules by the extraction that is typical of room temperature dehydration processing *(12,13)* is unacceptable for accurate immunolabeling. The structural biology community is becoming increasingly interested in HPF for cell biological studies because they recognize it as an essential specimen preparation method for the applications just mentioned, especially vitreous cryosectioning and tomography.

Space does not permit a detailed description of the specimen loading for every type of cell and tissue that can be frozen using high pressure. However, a search of the HPF literature may turn up an appropriate reference. To facilitate this search, we have provided a literature survey that can be accessed at http://mpi-cbg.de/HPF or at http://biology.berkeley.edu. For the purposes of this chapter, we will discuss the following categories for HPF: cells in suspension; cultured cells grown on substrates; whole organisms that are small enough to be loaded into HPF specimen carriers; and tissue samples taken by biopsy needle, dissection, or other methods. For each of these sections, we will provide a list of materials needed, followed by a detailed procedure for HPF for one to two different sample types. Next, however, we will give a brief general introduction to HPF machines, specimen carriers, and fillers used as external cryoprotectants during HPF.

2. Materials

2.1. HPF Machines

In the previous version of this article *(2)*, we focused exclusively on the BAL-TEC HPM 010 high-pressure freezer because it was the most generally used machine at the time. However, now there is a new high-pressure freezer, the Leica EM PACT *(3)*, that is being used by an increasing number of laboratories. Because the BAL-TEC HPM 010 is unchanged since the last writing, we will mostly comment on the EM PACT in this chapter. In fact, the version of the EM PACT released several years ago is now being replaced by the EM PACT2, so we will explain how this machine works and how it differs from the BAL-TEC machine as well as the original EM PACT. There is another commercially available machine (the HPF Compact 01) from Engineering Office M. Wohlwend (Sennwald, Switzerland) similar to the Leica EM HPM *(14)*, but it is not widely used at this time and we have no direct experience with it so we will not discuss it in this chapter (*see* **Note 1**).

2.1.1. The EM PACT2

The EM PACT2 is a fundamentally different design from the BAL-TEC HPM 010. In the BAL-TEC machine, the pressure and cooling systems are "in-line," whereas in the EM PACT they are separate. Detailed explanations of the differences can be found in the articles of Moor *(1)* and Studer et al. *(3)*. The practical consequence of this difference is that the EM PACT is considerably smaller, uses a fraction of the liquid nitrogen, is quieter, and perhaps most importantly, it is portable, that is, it can easily be moved to different sites for freezing rather than having to bring the samples to the HPF. In many cases, this is not a problem; however, there are times when it is an advantage to bring the HPF machine to the material, or to other equipment that cannot be moved readily, such as sophisticated light microscopes. For example, the EM PACT2 has a Rapid Transfer System (RTS, in **Fig. 1**) developed by one of us (P.V.) that permits a transfer from a light microscope to the freezer within 5 s. Here, we will cover the new features of the machine itself, including RTS.

The new EM PACT2 is available in two versions; with and without the RTS. The major difference between the EM PACT1 and the standard EM PACT2 is the pressure at which the LN_2 cools the sample. Where the old machine used 8 Bar, the new one uses 10 Bar, which gives even better cooling rates. As an add-on feature, the RTS is available. Besides being constructed for rapid transfer it is also very well suited for easy and reproducible, that is, standardized, loading of samples, a key issue in good freezing! Using the EM PACT2 with RTS, we have reached much greater freezing efficiencies (>90% well-frozen cells) for almost all samples tested than with the old EM PACT1 (*see* e.g., Manninen et

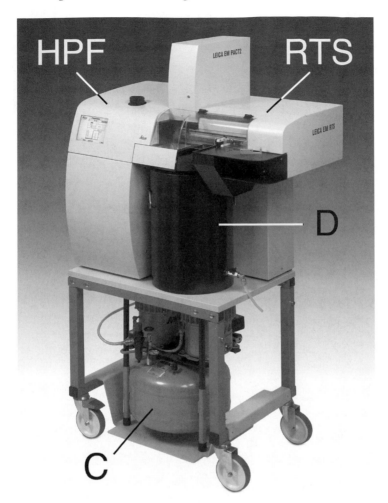

Fig. 1. The new Leica EM PACT2 high pressure freezer. The system consists of the main high pressure freezing unit itself (HPF), the optional Rapid Transfer System (RTS), the dewar (D) for holding liquid nitrogen, and the compressor (C) for generating the high pressure. Note that the machine is on a cart with wheels so that (although it has a total weight of 145 Kg) it still can be moved easily from lab to lab if necessary. Also, the machine will fit next to a light microscope when correlative light and electron microscopy experiments are performed. (Courtesy of Leica Microsystems, Inc.)

al. *[15]*). Whether this improvement in freezing efficiency is due to the result of the change in LN_2 pressure or the loading with the RTS is unknown, but it is our belief that both factors contribute. What, then, does the RTS do?

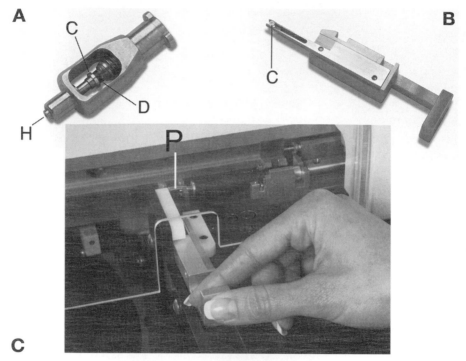

Fig. 2. Elements of the Rapid Transfer System. (Images courtesy of Leica Micro-systems, Inc.) **(A)** The so-called "pod" for holding EM PACT specimen carriers (C), which fit up against a diamond surface (D). The sample is pressurized through the high-pressure tube (H). In the actual preloading stage of use in the RTS, the specimen carrier (C) is not in the pod. **(B)** The Rapid Transfer System loading device (rapid loader) with a specimen carrier (C) inserted in the tip. This device can be fitted into a modified light microscope stage to facilitate rapid transfer from the microscope to the HPF. **(C)** Insertion of the loading device into the Rapid Transfer System. A preloaded pod (P) is in position to receive the cup and automatically tighten it against the diamond, insert it into the HPF, where it will freeze and drop into liquid nitrogen. The time from releasing the loading device from your fingers to having a specimen in liquid nitrogen is approx 2.5 s.

First, the RTS is preloaded with a freezing pod (*see* **Fig. 2A**). Next, a specimen carrier in which the sample will be placed is premounted in the so-called rapid loader (*see* **Fig. 2B**). As soon as the sample is loaded in the carrier the rapid loader is transferred to the RTS (*see* **Fig. 2C**), where it exactly fits in the pod. The rapid loader is released, and this triggers the following events: the specimen carrier is automatically put into the pod, the pod is inserted into the EM PACT, and the sample is frozen and dropped into liquid nitrogen. As described this whole procedure only takes a few seconds and gives very reproducible cryofixation.

Another feature of the EM PACT machines is that they give you pressure and cooling rate information on each freezing event. In this regard, it differs from the BAL-TEC machine where you can get the information only from a special test holder that is different from the holder used for freezing material. By having the information for the actual freezing of your material, it is possible to reject any sample where the temperature and cooling curves are not optimal. In the EM PACT machines, this information can also be stored in the machine and subsequently downloaded to a computer for permanent records of all the freezing events.

2.2. Specimen Carriers

The size, shape, and mass of the specimen carriers are critical factors in achieving good freezing. Moor (1) suggested that if the specimen carriers were not well-suited to the specimen, then the researcher should consider making custom carriers. Unfortunately, this customization is not always easy or inexpensive to do, and the number of options commercially available from vendors is limited. Yet, working within these constraints, it has been possible to achieve good freezing for a wide variety of samples. However, if your samples are not freezing well, then it is worth considering custom carriers as one option for improvement.

2.2.1. BAL-TEC and Wohlwend Machines

As illustrated in McDonald (2), the basic specimen carriers for the BAL-TEC system are simple wells 2 mm in diameter that can be arranged in different combinations to produce chambers of varying depth between 100 and 600 μm. The so-called Type A carrier has a 100-μm deep well on one side and a 200-μm deep well on the other. The Type B carrier has a 300-μm deep well on one side and is flat on the other. These are also the basic design for the Wohlwend machine. However, Wohlwend has designed some different specimen carrier holders that he also makes available for the BAL-TEC instrument (*see* **Note 2**). One of these holds low-mass copper carriers that were designed to be used with the Balzers JFD 030 jet freezing device (*see* **Fig. 3** and also Walther *[16]*). After freezing, these sample carriers can be used in the BAL-TEC freeze fracture machines, or they can also be used to prepare samples for freeze substitution. These specimen carriers are approx 100-μm thick, or approx half the thickness of the walls of the regular aluminum wells. Having thinner walls should improve the rate of heat transfer in the sample (17). Finally, BAL-TEC offers other specimen carrier geometries in their catalog, but these have rarely been used to

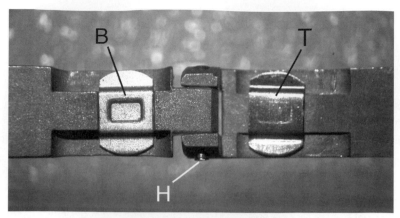

Fig. 3. A new specimen carrier holder for the BAL-TEC and Wohlwend HPF machines. A bottom piece (B) with a rectangular well is loaded into the holder, filled with material, and covered with a flat top piece (T) when the hinge (H) is swung closed.

our knowledge. For one thing, they are quite expensive compared with other available carriers, which may be a limiting factor in their use.

2.2.2. EM PACT2

This machine offers a wide variety of specimen holder geometries *(14)*. There are general cup carriers that have a well width of 1.2 or 1.5 mm and a depth of 100, 200, and 400 µm. These are generic carriers that can be used for a wide variety of materials similar to the BAL-TEC basic carriers. In a variation on this theme, the EM PACT also has carriers that do not have a hole for the pressure tube, but instead use a thin "membrane" of copper to separate the compartment that holds the material from the pressure (*see* **Fig. 4**). This membrane seems to reduce the problem that occurs with some materials, for example, a plant leaf, or cartilage tissue, in which the tissue opposite the pressure hole may be damaged by direct contact with the pressurizing fluid. Another specimen carrier geometry for the EM PACT is the tube system *(14)*, which is popular with researchers performing vitreous cryosectioning of cell suspensions *(18,19)*. Cells are pulled into the tube by capillary action, frozen, then the tube is trimmed away with a trimming tool and the contents sectioned at −160°C. It is possible to do the same thing with the cup systems of any HPF machine, but you have to trim away much more metal which will reduce the life of the trimming tool. There is a freeze fracture specimen carrier for the EM PACT that leaves a dome of material above the carrier that can be fractured in a freeze fracture machine. This machine has a carrier that accepts material from a microbiopsy punch. The carrier is unloaded from the biopsy needle in a special transfer station (*see* **Fig. 5**) that facilitates very rapid loading of the carrier before freezing, thereby

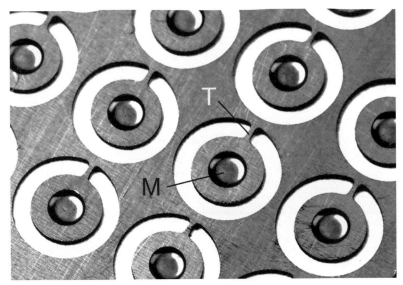

Fig. 4. A sheet of new membrane carriers for the EM PACT HPF machines. The center well is a thin membrane (M) that transmits the pressure without direct contact onto the sample. The 2.8-mm wide carriers are removed from the sheet by cutting the tab (T) with a razor blade or scalpel. (Courtesy of Leica Microsystems, Inc.)

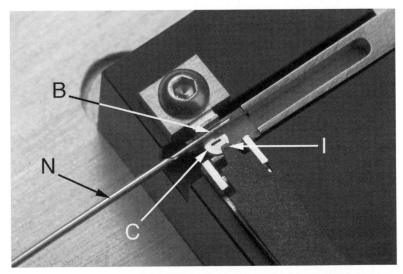

Fig. 5. A biopsy specimen carrier (C) in position to receive the biopsied material (B) at the end of the biopsy needle (N). The insertion tool (I) moves forward to the needle and presses the biopsy material into the slot of the specimen carrier. The whole operation from taking the biopsy to freezing can be done in approx 15 s or less with practice. (Courtesy of Leica Microsystems, Inc.)

Table 1
Size and Thickness Measurements of EM Grid Spacers

Avg. Thickness (µm)	Range (µm)	Size	Cat. No.[a,b]
27	27–28	"Chien" 2 mm hole	9GC20H[a]
35	33–37	0.4 × 2 mm slot	1GC42S[a]
48	47–50	1 × 2 mm slot	1GC12H[a]
62	54–65	1 × 2 mm slot Ni grid	1GN12H[a]
27	26–29	0.4 × 2 mm slot Ni grid	1GN42S[a]
100	100–102	Synaptek DOT 0.5 × 2 mm slot	4511[a]
50*	*	1 × 2 Copper-Rhodium oval slot	M2010[b]

[a]Ted Pella catalog.
[b]Electron Microscopy Sciences catalog.
*Not measured, estimated size only.
(Courtesy of Dr. Rick Fetter, Rockefeller University, unpublished)

reducing some of the damage because of more lengthy times between excision and freezing. Finally, a very specialized version of a carrier is the so-called "live cell carrier" designed for correlative light and electron microscopy experiments. It has a wide pressure hole for visualization in the light microscope and comes with a finder grid for re-localization of a particular cell.

2.2.3. "Custom" Carriers and Other Strategies

As mentioned previously, it is sometimes necessary to go beyond what is commercially available for specimen carriers. This led Craig et al. (20) to develop carriers for the BAL-TEC HPM 010 that would allow them to do freeze fracture more conveniently. Likewise, work in Martin Müller's laboratory resulted in sapphire disks as substrates for growing tissue culture cells (21). Sawaguchi et al. (22) grew or attached cells directly to the flat surface of BAL-TEC specimen carriers as a way to get good thermal contact to the cells and thereby improve the freezing quality. Richard Fetter of the Rockefeller Institute, New York (unpublished results) used EM slot grids as variable depth spacers to create shallower wells than were readily available for BAL-TEC carriers. The impetus for this approach was unpublished work by Eyal Shimoni (Weizmann Institute, Tel Aviv, Israel) that modeled rates of heat transfer as a function of well depth. Reducing the well depth had an even greater effect on heat transfer rates than reducing the thickness of the metal in the carriers (17). Using slot grids, it is possible to adjust the depth of the well to just fit the size of some specimens (see **Table 1**).

2.3. Fillers and Cryoprotectants

Although the size and geometry of the specimen holder should be optimized to provide the best freezing rate possible, the medium surrounding the sample is probably even more critical to the quality of freezing obtained. The entire volume of the inner cavity of the specimen holder must be filled. Air bubbles within the holder act as insulators and collapse during pressurization that can deform a sample milliseconds before freezing *(23)*. For most samples, the best choice of "filler" for the specimen is one which has some cryoprotective ability. Work on cryoprotectants for HPF has never been systematically studied, and much work remains to be done. Therefore, many of the suggestions that follow in the detailed methods are empirical and should not be regarded as applicable in all situations.

Cryoprotectants improve the quality of freezing directly by suppressing the formation of extracellular ice crystals and indirectly by reducing the amount of heat released by the crystallization process, thereby increasing the overall rate of cooling of the sample *(23)*. Because the premise of these procedures is to preserve samples without perturbing their natural processes, a number of cryoprotectants can be used to accomplish this goal. Depending on their interaction with cells, they are classified as either extracellular (nonpenetrating) or intracellular (penetrating).

2.3.1. Extracellular

This class of compounds includes nonpenetrating and/or hydrophobic substances with low osmotic activity. These compounds work by binding up water outside the cell, thus the name "extracellular." If your cells are surrounded by liquid water (or dilute salt buffer) during HPF, it will turn to ice before your cells are frozen and drastically reduce the heat transfer out of your sample. At liquid nitrogen temperatures, the heat transfer coefficients for sapphire are 960 $Jm^{-1}s^{-1}K^{-1}$, for copper, and aluminum they are 460 $Jm^{-1}s^{-1}K^{-1}$ and 410 $Jm^{-1}s^{-1}K^{-1}$, respectively, but for water they are negligible *(24)*. Ice is an insulator not a conductor of heat. Some examples of extracellular fillers include:

1. 10% to 20% Bovine serum albumin (BSA).
2. Yeast paste made up with 5% to 10% BSA or other solutions.
3. 15% to 25% Dextran (MW 39,000); 5-15% Ficoll (MW 70,000).
4. 1-Hexadecene.
5. Other: 0.5% to 2.0% low-melting point agarose, 15% polyvinylpyrolidone, 50% to 100% cold water fish gelatin, sucrose, and 10% to 20% fetal calf serum.

Recent results in our laboratory have shown that 20% BSA will give reproducibly good freezing results for a diverse array of organisms, including tissue

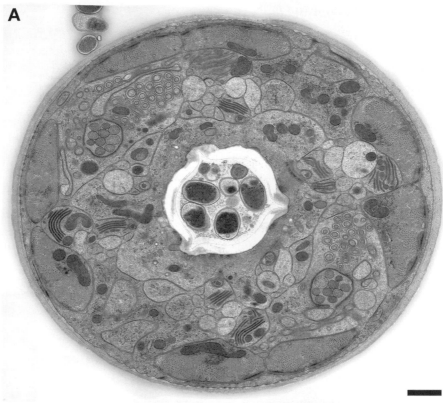

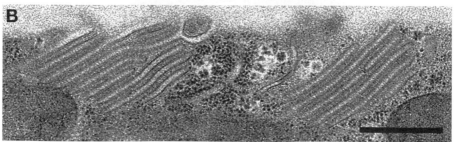

Fig. 6. Electron microscopy of whole-mount *C. elegans*. (**A**) Low-magnification cross-section view of high-pressure frozen hermaphrodite, freeze-substituted in acetone containing 1% OsO$_4$ and 0.1% uranyl acetate and embedded in Epon/Araldite. The image illustrates the structural organization of the head region. Bar = 1 μm. (**B**) Higher magnification of *C. elegans* morphology showing details of characteristic surface structures. Bar = 0.5 μm.

culture cells, cell suspensions, *Drosophila* and wasp embryos, *Caenorhabditis elegans* (*see* **Fig. 6**), isolated mouse kidney glomeruli, marine sponges (*see* **Fig. 7**), and choanoflagellates. It is our opinion that 20% BSA is a very good first choice as a cryoprotectant for HPF. Just make it up in a solution that is compatible with the cells that are being frozen. For example, with the marine sponges (*see* **Fig. 7**), we used seawater, for tissue culture cells we used growth medium, with kidney we used a Ringer's solution, and so on (*see* **Note 3**).

2.3.2. Intracellular

Cryoprotective compounds that are added exogenously and penetrate cells can be referred to as intracellular cryoprotectants. These are less desirable as fillers because they have the potential to affect the physiology of the cells by causing osmotic and other changes. However, there are situations in which the addition of these compounds seems acceptable. Sometimes cells can adapt to these changes and continue to grow normally. In other cases, the cells are surrounded by a thick diffusion barrier, and immersion in a potentially penetrating chemical seems to have no effect, probably because it does not penetrate the cells in such a short exposure. Some examples of commonly used intracellular fillers include (1) dimethyl sulfoxide; (2) 8% to 10% methanol; and (3) 10% to 20% glycerol.

2.4. General Equipment

1. High-pressure freezer (*see* **Subheading 2.1.**).
2. Specimen carriers for HPF (*see* **Subheading 2.2.**).
3. Dissecting microscope with fiber optic light source.
4. Specimen loading stations.
5. 4-L Dewar, or larger, for refilling LN_2 in HPF.
6. 1-L Widemouth Dewar for temporary storage or transfer of samples.
7. LN_2 refrigerator or storage Dewar.
8. 2-mL Cryotubes.
9. Cryotube rack with locking bottoms to match cryotubes.
10. Long tweezers (Ted Pella, Inc., cat. no. 5306) with foam or Velcro insulation.
11. Soft (number 2 or softer) pencil for writing on cryotubes.
12. Gloves for handling cold samples.
13. Liquid nitrogen source.

2.5. Cell Suspensions

2.5.1. Centrifugation

1. Centrifuge with appropriate speeds not to damage cells.
2. Centrifuge tubes, preferably plastic.
3. Dog toenail clippers.
4. Paper wicks (Pella, cat. no. 115-28).
5. 0.5–10-µL Micropipetor and tips.
6. Fine-tipped paint brush, for instance, size 0 red sable or finer.
7. Toothpicks or other small wooden sticks.

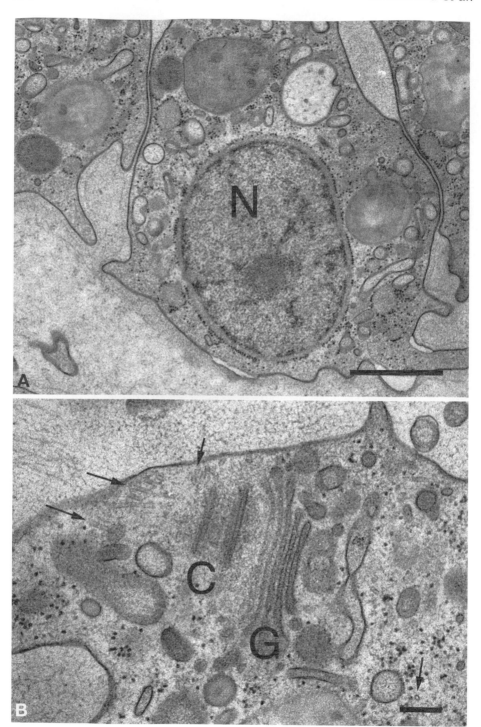

8. Cryoprotectant solution of your choice, for instance, 1-hexadecene, 20% BSA, or 10% glycerol.

2.5.2. Filtration

1. Vacuum source (pump or house line).
2. Vacuum filtration apparatus, typically a 15-mL Millipore set-up.
3. 0.4-μm-Pore-size polycarbonate filters, 25 mm in diameter.
4. Toothpicks or other implement to scrape cells from filter.
5. Syringe needle bent at 45° near the tip.

2.6. Cell Monolayers on Sapphire Disks

1. Sapphire disks (*see* **Note 4**).
2. Diamond marking pen.
3. Apparatus and materials for growing cells.
4. 20% BSA made up with growth medium or other solutions that are compatible with the physiology of the cells.
5. Fine tweezers.
6. Micropipetor and tips, or fine-tipped paint brush to apply BSA solution to cups.
7. 1-Hexadecene.
8. For the BAL-TEC HPF: (a) Chien-style slot grids (*see* **Table 1**) and (b) flat-sided (Type B) specimen planchettes.
9. For the EM PACT HPF: membrane specimen holders (*see* **Fig. 4**).

2.7. Tissues or Whole Organisms Smaller Than 300 μm

2.7.1. Whole Caenorhabditis elegans Worms

1. Healthy worms growing at the proper temperature, preferably on agar plates seeded with *Escherichia coli*. If liquid culture conditions are used, try to transfer worms to agar and remove excess fluid before freezing.
2. A worm pick and "sticky" *E. coli* for picking worms (*see* **Note 5**).
3. 20% BSA made up with M-9 (*see* **Note 6**) or PBS solution in a small tube.
4. A size 0 red sable paint brush, or micropipetor and tips able to deliver a small amount of 20% BSA solution.
5. A fine-tipped tungsten or other needle.

Fig. 7. (*Opposite page*) Cells of the marine sponge *Oscarella carmela*, high pressure frozen using 20% BSA in seawater as the filler. Courtesy of Drs. Scott Nichols and Nicole King, University of California, Berkeley). (**A**) A cell with a prominent nucleus (N) and different types of membranous organelles. Note the smooth membrane profiles and the parallel membranes where the cells make contact. Bar = 1 μm. (**B**) Higher magnification view of a cell containing a Golgi apparatus (G) in close proximity to a centriole (C), with numerous microtubules (arrows) nearby. Bar = 200 nm.

6. Specimen carriers: (a) Leica–membrane carriers; (b) BAL-TEC – Type B (flat-sided) carriers; and (c) slot grids of known thickness, appropriate to the size of the worms (*see* **Table 1**).
7. Paper points (Pella, cat. no. 115-18), or filter paper wedges for wicking fluids.
8. 1-hexadecene-saturated filter paper (5.5 cm) in a small Petri dish.

2.7.2. Correlative LM-EM on Single Caenorhabditis elegans Embryos

1. Stereomicroscope with light source.
2. Fluorescence light microscope.
3. Dialysis tubing with an inner diameter of 200 μm (*see* **Note 7** and Hohenberg et al. *[25]*).
4. Micropipetor (0.5- to 10-μL size) and gel loader tips.
5. Nail polish.
6. *C. elegans* strain expressing either GFP::histone, GFP::γ-tubulin, or both *(26)*.
7. Worm buffer: M9 containing 20% BSA (w/v; Sigma).
8. Tools, for instance, two syringe needles, for cutting open worms to release embryos.
9. Small (2.5–6 cm) plastic Petri dishes.
10. Microscope slides.
11. Fine forceps.
12. Freezing planchettes for HPF. Membrane carriers for the EM PACT machines, and Type A, 200-μm deep wells for the BAL-TEC machines.
13. Crimping tool for sealing dialysis tubing (*see* **Note 8**).

2.8. Tissues or Whole Organisms Larger Than 300 μm

2.8.1. Soft Animal Tissues

1. Appropriate equipment and permits for killing and dissecting live animals.
2. Microbiopsy gun (This is available from Leica, though it can be used for either HPF machine).
3. Microbiopsy transfer station. This station is designed to position the biopsied material directly over the hole of the biopsy specimen carrier (*see* **Fig. 5**) and to quickly press it into the hole. The BAL-TEC system has no such transfer device, although one could perhaps make something similar so that the material rests just over the specimen carrier after the biopsy is taken.
4. Biopsy specimen carriers (EM PACT).
5. Type A (200-μm deep side) or Type B (300-μm deep side) specimen carriers (BAL-TEC).
6. Filler material, either 20% BSA or 1-hexadecene.

2.8.2. Plant Tissues

1. Leaf or root tissue.
2. Razor Blades or 2-mm disposable biopsy punch (Stiefel Laboratories, sold by Technotrade International, cat. no. LH01847KN).

3. Humid chamber.
4. Filler solution: 8% to 10% methanol, 20% BSA, 1-hexadecene, 0.13 *M* sucrose, or 0.1 *M* mannitol.
5. Specimen carriers of the appropriate depth.
6. Vacuum chamber if freezing leaf tissue.

3. Methods

The following are examples of successful methods for given samples. These procedures may be combined or altered as fits the need of the cells and the system. For the following, a dissecting microscope setup in close proximity to the freezer is invaluable for sample loading. Such diversity exists in the types of cells and tissues that can be frozen that it is impossible to list specific procedures for each. For purposes of this chapter, we will consider cell suspensions, cells growing on substrates, whole cells or organisms smaller than 300 μm in the thinnest dimension, and whole cells or tissues larger than 300 μm. Be aware that, for your own material, you may have to modify these basic procedures. When doing so, keep in mind the general guidelines that inform all loading strategies.

3.1. General Loading Strategies

1. Always start with the smallest volume of cells you can. Use grid spacers in the BAL-TEC system (The EM PACT doesn't allow this strategy) if possible. If you find you get good freezing but want more cells try the next larger volume. Repeat until you find the most suitable size that still gives reliable freezing.
2. Concentrate the cells as much as you can without harming them by physical distortion. If using solid tissues, try to fill the cavity with as much as you can and reduce the amount of filler solution to a minimum.
3. In the BAL-TEC system, coat the top planchette with l-hexadecene whenever possible. This will make it much easier to separate the two planchettes under liquid nitrogen (LN₂). In the Leica or BAL-TEC system, you also can coat the well of the cup specimen holders, or the sides of the biopsy slots with hexadecene if you want the tissues to release from the cups before embedding; however, it is sometimes difficult to get cells loaded into hexadecene-coated wells.
4. Work quickly to prevent drying out of tissues. Work in a moist chamber if necessary.

3.2. Suspensions of Cells

This category includes a vast array of cell types, including bacteria, algae, protozoa, and cultured cells from all classifications. Regardless of the source, the cells should be growing in optimal physiological conditions just before freezing. For many cell types this means log phase growth, not stationary phase. Particular requirements of temperature, aeration, agitation, and so forth must be met for each cell type. It is not recommended that cells be harvested, concentrated, and stored on ice before freezing. There are five common ways to handle cells in suspension: by centrifugation, filtration, adherence to a substrate, sedimentation, and by growing them in small dialysis tubes.

3.2.1. Centrifugation

1. Grow or harvest cells under optimal conditions.
2. Spin down cells at appropriate speeds.
3. Remove excess solution above pellet with fine-tipped plastic transfer pipet.
4. If you think the cells will not be harmed, resuspend the pellet in an equal volume of cryoprotectant solution (except 1-hexadecene, which is immiscible with aqueous solutions). Spin again and remove excess solution as detailed previously.
5. If using Eppendorf or 15-mL conical plastic tubes, cut off the tip with the dog toe-nail clippers. This modification will make it easier to access the pellet.
6. Transfer material in pellet with micropipettor, toothpick, or paint brush to the HPF planchette.
7. If the cells are not concentrated enough (they should be as closely packed as possible without distortion), they can be allowed to settle and the excess fluid wicked off with paper points. The best way to do the wicking is to use a paper point on either side of the planchette. Usually the fluid will come off, and the cells will remain in the well. With only one paper point or wick, the cells have a tendency to climb onto the wick.
8. If using a BAL-TEC machine, coat the top planchette with 1-hexadecene and cover bottom planchette.
9. Freeze.
10. Transfer frozen cells to cryotubes for storage, or better yet, to cryotubes with fixative for freeze substitution. Work in a small Styrafoam box with a cryotube rack firmly wedged into it and LN_2 in the bottom. Cryotubes (with or without fixative) can be opened with one hand because the rack will hold it firm. Put the opened vial under LN_2 and transfer the specimen carrier with frozen cells into it, making sure to use pre-cooled forceps and not let the cells warm up in any way. Screw down the cap finger tight. If the vial is for storage only, make sure that there are holes in the lid to let trapped LN_2 escape. If the cells are in vials with fixative, place in precooled freeze substitution device, or store in LN_2 until ready to freeze substitute (*see* **Note 9**).

3.2.2. Filtration, e.g., of Yeast Cells

1. Set up filtration apparatus (*27*).
2. Pour 5 to 15 mL of cells into filtration column.
3. Apply suction to pull down cells onto filter, using care not to let them get too dry.
4. Remove column from apparatus, and scrape cells from filter with toothpick tool.
5. Fill well of specimen carrier with cells (*see* **Note 10**).
6. Freeze and process as in **Subheading 3.2.1., step 10**.

3.2.3. Other Methods for Cell Suspensions

Space does not permit additional detailed descriptions of some other methods, but several of these are available in published works. The following are a few examples: (1) sedimentation of plant protoplasts *(28)*; (2) adherence to a substrate with poly-L-lysine *(22)*; (3) growth in capillary tubes *(25)*; and (4) plant cells in culture *(29)*.

3.3. Monolayers

3.3.1. Tissue Culture Cells on Sapphire Disks

1. Use a diamond marking pen to scribe an asymmetric symbol on top of sapphire disk, so you can tell which side the cells are on. Alternatively, you can carbon coat the disk and scratch the symbol in the carbon *(21,30)*.
2. Sterilize sapphire disks by alcohol or other method and inoculate cells for growth.
3. Grow cells to desired confluency.
4. For BAL-TEC HPF, (a) place one flat-sided planchette (flat side up) in tip of specimen holder; (b) place Chien grid on top of planchette; (c) remove the cells from the culture dish with fine forceps, noting which side the cells are on; (d) dip the disk/cells briefly in BSA solution; (e) insert disk, cell side down onto the Chien grid spacer; (f) cover the disk with the flat side of another flat-sided planchette that had been previously coated with l-hexadecene. This will facilitate separation of the planchettes from the disk under liquid nitrogen; and (g) freeze.
5. For EM PACT2 HPF: (a) preload a membrane specimen carrier into the specimen loader; (b) fill the membrane carrier with 20% BSA made up with culture medium; (c) place the sapphire disk with cells into the cup, cells facing up; (d) load the cup into the freezing pod; and (e) freeze.
6. Separate cells from substrates in polymerized resin as described in one of the next two sections.

3.3.1.1. BAL-TEC HPM 010

During resin infiltration, the disk can be separated from the specimen carrier. This usually happens by itself as a result of agitation in resin mixtures, but if that doesn't happen, the disk can be removed from the carrier manually. Using the asymmetric numeral on the disk as a guide, embed it in a flat-bottomed embedding capsule *(22)* such that the cells are facing into the body of the capsule. After polymerization, trim away the resin around the disk so there is a clean border between sapphire and resin, hold in LN_2 vapors for a minute, then pull off the disk with forceps or pliers. It is important to note that the carrier will more easily detach from the resin block if the resin is a hard formulation, e.g., Epoxy 812 (*see* **Note 11**).

3.3.1.2. EM PACT2

When cells are grown on sapphire and frozen in a carrier, the cells on the sapphire disk cannot be removed during the freeze substitution. When doing so,

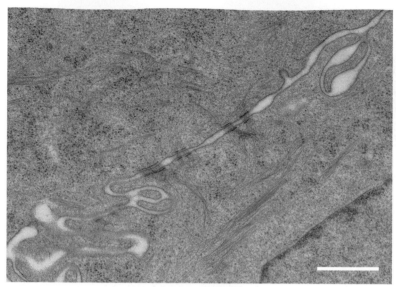

Fig. 8. Electron micrograph of part of a MDCK cell, grown on a sapphire disk. The sample was frozen with the EM PACT2 + RTS. Junctional complexes with actin cables connecting the cells are nicely visualized. After HPF the sample was freeze substituted then embedded in Epon. Bar = 200 nm.

the orientation of the cells on the sapphire (on which side are the cells) would be lost. Therefore, the carrier and sapphire disc are freeze substituted and embedded in resin as a whole. Use a flat embedding mold and put the carrier straight up in the mold or use a flat-bottomed capsule *(22)*. Make sure the cell side is facing into the body of the resin block. After polymerization, the resin around the carrier is thoroughly trimmed away with a scalpel. The resin block is dipped into liquid nitrogen and the carrier and disk either jumps off or can be removed right after taking the block out of the LN$_2$ with a pair of pliers or forceps. The cells remain in the resin and are right at the surface, which is optically smooth so you can look at the cells in a light microscope and pick out the one that was observed prior to freezing. **Figure 8** shows an example of semipolarized MDCK cells grown on a sapphire disk, frozen according to the protocol described for the EM PACT2.

3.3.2. Other Methods

Other methods of freezing adherent cultured cells include growing them on membrane filters *(15,31)* or, if one is using a BAL-TEC machine, growing them directly attached to the flat side of a type-B specimen carrier *(22)*. If cells are grown in plastic culture dishes and you want to scrape them off the substrate and freeze them, see Schlegel et al. *(32)*.

3.4. Tissues or Whole Organisms Smaller Than 300 µm

The types of material that are routinely frozen in this category include embryos of various organisms, whole *C. elegans* worms *(33)*, isolated pieces of tissue, or plant materials. Most topics on the processing of these samples were covered in the previous version of this chapter *(2)*, and therefore we will not cover them again here, except to say that 20% BSA may be an alternative to 1-hexadecene as a choice for filler/cryoprotectant. However, because there is increasing interest in freezing whole *C. elegans* worms, we will include here detailed instructions for freezing them. Although this topic was covered in a previous article *(33)*, the article did not include methods for the EM PACT2 HPF or the use of BSA as a filler. Furthermore, the new method given here for freezing worms in the BAL-TEC gives a much higher yield of well-frozen worms. We also describe how the EM PACT2 can be used to do correlative light and electron microscopy on single *C. elegans* embryos.

3.4.1. EM PACT

1. Preload a membrane specimen carrier (*see* **Fig. 4**) into the specimen carrier holder.
2. Add a small amount of 20% BSA solution to the well of the carrier.
3. Pick enough worms to fill the carrier.
4. Load holder into the EM PACT and freeze.

3.4.2 BAL-TEC

1. Preload a Type B specimen carrier flat side up into the tip of the specimen carrier holder.
2. Wet the bottom side of a slot grid with 1-hexadecene and put on top of the flat specimen carrier (*see* **Fig. 9A**).
3. With the paint brush or micropipetor add a small volume (0.3–0.5 µL) of 20% BSA solution to the slot of the slot grid.
4. Pick about 20 worms with the pick and transfer them to the BSA in the slot (*see* **Fig. 9B**). Sometimes it helps to use the needle to push the worms off the pick into the liquid.
5. Remove excess BSA with paper points or filter paper until the level is even with the top of the slot.
6. Wet the flat side of another Type B specimen carrier with 1-hexadecene on the filter paper and blot off excess. Place on top of the slot grid, flat side down.
7. Close the specimen carrier holder, insert into HPF machine and freeze.
8. Split the two specimen carriers under liquid nitrogen before proceeding to freeze substitution. The 1-hexadecene on the top carrier should help it to release easily. The worms and slot grid will usually stick to the bottom carrier.

3.5. Correlative LM-EM on Single Caenorhabditis elegans Embryos

We briefly describe here the use of capillary tubing to perform correlative light and electron microscopy of early *C. elegans* embryos in mitosis (*see* **Fig. 10**).

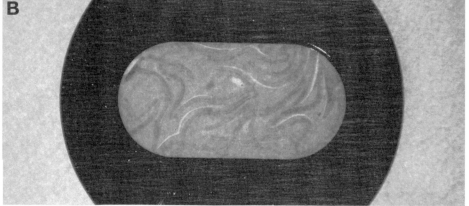

Fig. 9. (A) A 3.05-mm slot grid (**S**) (EMS cat no. M2010-CR) in the tip of a BAL-TEC HPM 010 specimen carrier holder, sitting on top of the flat side of a Type B specimen carrier. The slot will be filled with *C. elegans* worms and another flat sided specimen carrier put on top to create a shallow well for more optimal freezing. **(B)** The same slot grid as seen in **Fig. 7A**, now filled with *C. elegans* worms in a paste of *E. coli*. This sample has been through HPF, freeze substitution, and infiltration with resin and the worms are still in the slot. Preparing worms this way also helps to keep them flat in one plane, which is an advantage for some kinds of sectioning analysis.

Previously, capillary tubing was described for freezing of either cell suspensions *(25)* or *C. elegans* hermaphrodites *(33)*. Here, tubing is used to stage isolated mitotic embryos by light microscopy before HPF and freeze-substitution. This correlative approach takes advantage of excellent ultrastructural preservation of previously staged embryos and can be combined with three-dimensional studies using electron tomography *(26)*. This correlative approach should be applicable to all HPF machines (*see* **Note 12**).

1. Prepare "loading device" for collecting isolated early embryos into capillary tubes. Mount a piece of dialysis tubing about 2 cm long into gel loader tips using nail polish for sealing, let dry, and mount the tip on a pipetman.
2. Cut worms open in small Petri dishes using M9 buffer containing 20% BSA and select an early *C. elegans* embryos under a dissecting scope. Suck the selected early embryo into the capillary tubing.
3. Submerge tubing into BSA-containing M9 buffer in a small plastic Petri dish. Using the crimping tool, cut the tubing away from the pipet tip, then cut the region of the tubing containing the early embryo into a length that will fit into the specimen carrier. The ends should be crimped so that the early embryo will not leak out (*see* **Note 13**).
4. Transfer specimen carrier from the Petri dish to droplet of BSA-containing buffer onto a glass slide and observe early developmental events using either DIC or fluorescence microscopy. In the latter case, a line expressing like GFP::histone;GFP::γ-tubulin can be used for live cell imaging *(34)*.
5. At an appropriate developmental stage, use forceps to transfer the embryo-containing tube from the microscope slide to a pre-filled specimen carrier (*see* **Note 14**), freeze, and store in LN_2. Perform freeze-substitution and thin-layer embedding of the specimen as described *(33)*.

3.6. Tissues or Whole Organisms Larger Than 300 μm

The challenge in freezing tissues is to minimize the physical and physiological damage when removing a piece for loading and freezing. Some tissues are so soft and/or delicate (e.g., brain and other nerve tissues) that this is nearly impossible to achieve and other strategies must be employed. In these cases, we recommend fixing the material with 2% glutaraldehyde before freezing. For these soft animal tissues, perfusion fixation is strongly recommended rather than dissection and immersion fixation. The biopsy punch system developed by Hohenberg et al. *(35)* is another option for freezing fresh tissue and it will work for either the BAL-TEC or EM PACT systems. However, the EM PACT system has a very convenient transfer station (*see* **Fig. 5**) that facilitates rapid transfer of the excised tissue into the specimen carrier and, if the EM PACT2 is used, very rapid transfer into the freezer. Such speed is essential for most soft animal tissues. Fortunately, not all tissues are this sensitive and there

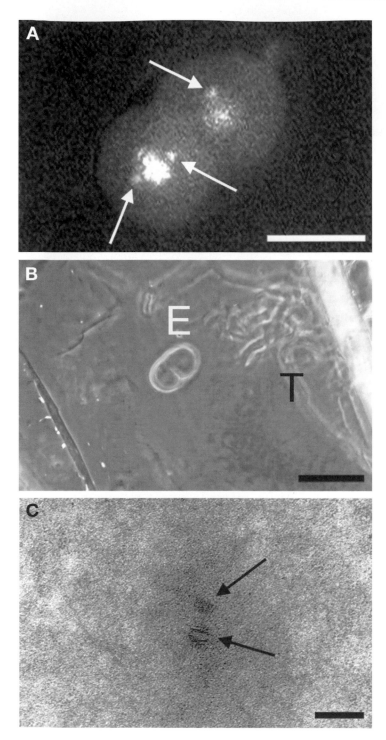

are some that are sufficiently homogeneous and/or hard (e.g., cartilage and some plant tissues) that removing a small piece without damage does not seem to be a problem. Nonetheless, speed of transfer and freezing is still important to minimize physiological damage from dissection.

3.6.1. Soft Animal Tissues Sampled With a Biopsy Needle

1. Ensure that all materials are ready prior to dissection and biopsy.
2. Inject biopsy needle into tissue.
3. Transfer to the Microbiopsy Transfer Station (EM PACT), or manually position over the specimen carrier filled with cryoprotectant/filler.
4. Transfer material to specimen carrier.
5. Freeze.

3.6.2. Plant Tissues

1. In a humid chamber, cut small pieces of root/leaf tissue with a razor blade or biopsy punch. If leaves have air in them, transfer to filler material and place in a vacuum until the sample sinks.
2. Fill well of specimen carrier with filler solution. Use the shallowest carrier possible. Leaf tissue can withstand some compression in the carrier.
3. Freeze.

3.7. Procedures After HPF

Freezing is only the first step in a series of operations between living cells and viewing them in an electron microscope. Great care must be taken with frozen cells to ensure that they do not warm up before all the water is removed, usually by freeze substitution. Transfer them under liquid nitrogen from the HPF machine into the vials with frozen freeze substitution fixative. Always use precooled forceps when handling the specimen carriers and do not lift them out of the nitrogen at any time. Once the material is in the frozen freeze substitution fixative, transfer that to a precooled freeze substitution device set at −90°C. There are some who believe that it is better to transfer to a freeze substitution device set to −155°C then warm it up slowly to −90°C *(36)*; however, we have not found this to be necessary. If in doubt, try the experiment yourself.

Fig. 10. *(Opposite page)* Correlative microscopy of an isolated *C. elegans* embryo. **(A)** Fluorescence microscope image (using a GFP::histone;GFP::-tubulin line, *see* Oegema et al. *[34]*) of a two-celled embryo collected in a cellulose capillary tube. The spindle poles are marked by arrows. Bar = 25 µm. **(B)** Low-magnification light microscope image of a resin-embedded two-celled *C. elegans* embryo (E) in the cellulose capillary tube (T) after high-pressure freezing and freeze-substitution. Bar = 50 µm. **(C)** Low-magnification electron microscope image of one of the centrosomes of the same mitotic cell shown in (B). The centrioles (arrows), and the PCM are clearly visible. Bar = 500 nm.

The details of how to do freeze substitution vary widely in the voluminous literature on the subject. We recommend a standardized approach to freeze substitution, using the same fixative combinations for all material, and then varying them as needed after initial observations. For observations of morphology only, we use 1% osmium tetroxide plus 0.1% uranyl acetate in acetone as the substitution fluid. For immunocytochemistry we use 0.2% glutaraldehyde plus 0.1% uranyl acetate in acetone. The recipes for making these fixative combinations are given in detail in (2). Although osmium in acetone alone is a fixative often used for freeze substitution, we strongly urge users to add the uranyl acetate. It provides improved membrane contrast (37) and seems to protect some samples from the destructive actions of osmium. We also recommend that acetone be used as the organic solvent and not methanol, even though some tissues look okay after methanol-based freeze substitution. For other tissues, methanol has been shown to be too aggressive during freeze substitution and extraction of cytoplasm and membranes can result (7). The acetone should be free of water, but we do not recommend using molecular sieves as an agent to dry the acetone. There are compounds in some molecular sieves that seem to dissolve in the acetone and react with the osmium (5). We buy EM-grade acetone from EM supplies vendors in 100-mL bottles and use new bottles for making up fixatives and rinsing during resin infiltration. One sign that there may be problems with the freeze substitution fixative is if your samples turn black. In our hands, using the fixatives as prepared in McDonald (2), samples are light-to-medium brown in color after freeze substitution.

The timing of the freeze substitution procedure also varies considerably and may depend to some extent on the size or nature of the tissues being processed. There are a few systematic studies of the freeze substitution process, and these are reviewed in the excellent article by Steinbrecht and Müller (5). However, the results of these studies have not led to a widely accepted single-freeze substitution protocol. The following method has worked well in our laboratories, and it is an intermediate between relatively rapid procedures (e.g., Hess [38]) and those taking many days (39,40). Freeze substitution starts at −90°C for 8 h, warms up to −60°C at 5°C/h, remains at −60°C for 8 to 12 h, then is warmed to 0°C at a rate of 5°C/h. Cells are held at 0°C until time for further processing, when they are warmed to room temperature, rinsed in pure acetone. and processed for embedding. If your cells are too low in contrast when you look at them in the microscope, hold them at −20°C for 12 h during the warm up instead of −60°C.

3.8. Summary and Prospectus

HPF is one of the most significant advances in specimen preparation techniques for biological electron microscopy since the introduction of glutaraldehyde for primary fixation *(41)*. It extends the well-known advantages of cryo-fixation to cells and tissues up to several hundred micrometers in size instead of the several micrometers of good freezing possible with most other cryofix-ation techniques. However, to perform HPF effectively, it is essential to learn how to load the sample properly into the machine. Specimen loading, includ-ing the choice of specimen carrier and material to pack around the cells, is the key to success with this method.

Electron tomography and immunoelectron microscopy (including correla-tive light-electron microscopy) are described as being at the forefront of modern electron microscopy techniques *(42)*. Likewise, vitreous cryosections, and elec-tron tomography of vitreous cryosections *(43)*, also are applications that promise a look at biological ultrastructure in a "close-to-native" state. It is significant that nearly all recent publications in these fields use high pressure freezing as the method of fixation. We look forward to the day when all EM studies of whole cells and tissues use some sort of cryofixation to preserve cellular fine structure. For cells larger than approx 10 µm, this is best accomplished by HPF.

4. Notes

1. Information about the Wohlwend HPF machine HPF Compact 01 can be found at www.technotradeinc.com. Technotrade, Inc. is the U.S. distributor for the Wohl-wend instrument and accessories. Wohlwend is based in Switzerland and can be reached by writing or faxing requests to: Engineering Office M. Wohlwend GmbH, CH-9466, Sennwald, Switzerland. He does not maintain a website or accept tele-phone inquiries.
2. Engineering Office M. Wohlwend GmbH is a good source of specimen carriers or specimen carrier holder tips for the BAL-TEC machine. They are of excellent qual-ity and considerably cheaper than if you order them from BAL-TEC. The best way to order is by Telefax (081/7572243). Write or fax a request for the price list.
3. We typically only use a small amount of BSA solution; therefore, we weigh out 1 g and add solution to make up 5 mL of final volume. To speed up the dissolution of the BSA, you can warm it up (a 37°C oven works well) for a few minutes.
4. Sapphire disks for the Leica EM PACT2 can be ordered from Leica Microsystems, Inc. For the BAL-TEC machine, disks can be ordered either from BAL-TEC, Tech-notrade International, Inc., or Engineering Office Wohlwend.
5. If you are not familiar with picking worms, have someone from a *C. elegans* labor-atory show you how, or have them pick the worms during the freezing run. Using a flattened piece of platinum wire with some *E. coli* on the tip, it is possible to "pick" 20 or more worms from a plate in a minute or so. But this is only possible if the *E. coli* paste is not too liquid. Practice will tell you the right consistency.
6. M-9 buffer has the following composition: 22 mM potassium phosphate monobasic (KH_2PO_4), 19 mM NH_4Cl; 48 mM sodium phosphate dibasic (Na_2HPO_4), 9 mM NaCl.
7. Dialysis tubing in the U.S. is available from Spectrum, 23022 La Cadena Dr., Suite 100, Laguna Hills, CA 92653, USA. Ask for Spectra/Por Hollow Fiber Bundles,

200 µm inner diameter, cat. #132 290. In the United States and Europe, dialysis tubing can also be ordered from Leica Microsystems, Inc.

8. The crimping tool is key to success with dialysis tubing. You can make one from a number 11 scalpel tip by cutting or filing off the final millimeter or so of the blade, then honing it with a whetstone to the shape of a chisel.

9. We do not recommend using cryotubes that seal with o-rings because they are likely to trap LN_2. We use Nalgene cryovials (cat. no. 5000-0020) and although they are not recommended for use with LN_2, we have used them for more than 15 yr without problems.

10. If you overfill the cup, you can adjust the level to be even with the rim by scraping away excess yeast with the syringe needle tool. Place the needle across the middle of the cup and scrape toward the outside. Repeat for the other side. Freeze immediately or the cells will dry out.

11. Our routine Epoxy 812 formulation is: 23.5 g of Epon 812 substitute, 12.5 g of DDSA, 14 g of NMA, and 0.37 mL of DMP-30.

12. Although applicable for all three HPF machines, it should be noted that stage-defined freezing is best when using the EM PACT2 equipped with the rapid loading device (RTS). The time window between light microscopic observation and freezing can be considerably reduced.

13. To crimp and cut the tubing, place it (with contents inside) in a small plastic Petri dish with enough BSA solution in the bottom to just cover the tubing. Crimp one end by pressing down the tubing with the flat side of the chisel end, then cutting the tubing with the sharp part of the chisel. Repeat with the other end of the tubing, making sure to cut a length that will fit into the specimen carrier cup.

14. The carrier for the BAL-TEC machine should be a 200-µm deep well, the carrier for the EM PACT should be a membrane carrier. In both cases, prefill the well with 20% BSA made up with worm buffer M-9.

Acknowledgments

The authors wish to thank all of our collaborators over the years who have provided material for HPF and have been patient with us as we experimented with their samples. We also wish to thank Leica Microsystems, Vienna, for providing access to the EM PACT2 machine and images for this publication prior to the official release date of the instrument. Work by M.M. was supported in part by grant RR00592 from the National Center for Research Resources of the National Institutes of Health to J.R. McIntosh. T. M.-R and P.V. wish to thank Jana Mäntler and Susanne Kretschmar of the MPI Dresden EM Facility for help with electron microscopy.

References

1. Moor, H. (1987) Theory and practice of high-pressure freezing, in *Cryotechniques in Biological Electron Microscopy* (Steinbrecht, R. A. and Zierold, K., eds.), Springer-Verlag, Berlin, pp. 175–191.

2. McDonald, K. L. (1999) High pressure freezing for preservation of high resolution fine structure and antigenicity for immunolabeling, in *Electron Microscopy Methods and Protocols*, 1st ed (Hajibagheri, M. A. N., ed.), Humana Press, Totowa, NJ, pp. 77–97.

3. Studer, D., Graber, W., Al-Amoudi, A., and Eggli, P. (2001) A new approach for cryofixation by high-pressure freezing. *J. Microsc. (Oxford)* **203**, 285–294.

4. Moor, H. and Riehle, U. (1968) Snap-freezing under high pressure: a new fixation technique for freeze-etching. *Proc. 4th Eur. Elect. Microsc.* **2**, 33–34.

5. Steinbrecht, R. A. and Müller, M. (1987) Freeze substitution and freeze drying, in *CryoTechniques in Biological Electron Microscopy* (Steinbrecht, R. A. and Zierold, K., eds.), Springer-Verlag, Berlin, pp. 149–172.

6. Kiss, J. Z. and Staehelin, L. A. (1995) High pressure freezing, in *Techniques in Modern Biomedical Microscopy; Rapid freezing, freeze fracture, and deep etching* (Severs, N. J. and Shotton, D. M., eds.), Wiley-Liss, New York, pp. 89–104.

7. McDonald, K. (1994) Electron microscopy and EM immunocytochemistry, in *Drosophila melanogaster: Practical Uses in Cell and Molecular Biology* (Goldstein, L. S. B. and Fyrberg, E., eds.), *Methods in Cell Biology* 44, Academic Press, San Diego, pp. 411–444.

8. McDonald, K. L., Sharp, D. J., and Rickoll, W. (2000) Preparing thin sections of *Drosophila* for examination in the transmission electron microscope, in *Drosophila: A Laboratory Manual,* 2nd ed. (Sullivan, W., Ashburner, M., and Hawley, S., eds.), CSHL Press, Cold Spring Harbor, NY, pp. 245–271.

9. Kaneko, Y. and Walther, P. (1995) Comparison of ultrastructure of germinating pea leaves prepared by high-pressure freezing-freeze substitution and conventional chemical fixation. *J. Electron Microsc.* **44**, 104–109.

10. Murk, J. L., Posthuma, G., Koster, A. J., et al. (2003) Influence of aldehyde fixation on the morphology of endosomes and lysosomes: quantitative analysis and electron tomography. *J. Microsc. (Oxford)* **212**, 81–90.

11. Small, J. V. (1981) Organization of actin in the leading edge of cultured cells: influence of osmium tetroxide and dehydration on the ultrastructure of actin meshworks. *J. Cell Biol.* **91**, 695–205.

12. Plattner, H. and Bachmann, L. (1982) Cryofixation: A tool in biological ultrastructural research. *Int. Rev. Cytol.* **79**, 237–304.

13. Gilkey, J. C. and Staehelin, L. A. (1986) Advances in ultrarapid freezing for the preservation of cellular ultrastructure. *J. Electron Microsc. Tech.* **3**, 177–210.

14. Studer, D., Michel, M., Wohlwend, M., Hunziker, E. B., and Buschmann, M. D. (1995) Vitrification of articular cartilage by high-pressure freezing. *J. Microsc. (Oxford)* **179**, 321–332.

15. Manninen, A., Verkade, P., LeLay, S., et al. (2005) Caveolin-1 is not essential for biosynthetic apical membrane transport. *Molec. Cell. Biol.* **25**, 10087–10096.

16. Walther, P. (2003) Recent progress in freeze-fracturing of high-pressure frozen samples. *J. Microsc. (Oxford)* **212**, 34–43.

17. Shimoni, E. and Müller, M. (1998) On optimizing high-pressure freezing: from heat transfer theory to a new microbiopsy device. *J. Microsc. (Oxford)* **192**, 236–247.

18. Al-Amoudi, A., Chang, J. J., Leforestier, A., et al. (2004) Cryoelectron micros-copy of vitreous sections. *EMBO J.* **23,** 3583–3588.
19. Zhang, P., Bos, E., Heymann, J., Gnaegi, H., Kessel, M., Peters, P. J., and Subramaniam, S. (2004) Direct visualization of receptor arrays in frozen-hydrated sections and plunge-frozen specimens of *E. coli* engineered to overproduce the chemotaxis receptor Tsr. *J. Microsc. (Oxford)* **216,** 76–83.
20. Craig, S., Gilkey, F. C., and Stähelin, L. A. (1987) Improved specimen support cups and auxiliary devices for the Balzers high pressure freezing apparatus. *J. Microsc. (Oxford)* **148,** 103–106.
21. Hess, M. W., Müller, M., Debbage, P. L., Vetterlein, M., and Pavelka, M. (2000) Cryopreparation provides new insight into the effects of Brefeldin A on the struc-ture of the HepG2 golgi apparatus. *J. Struct. Biol.* **130,** 63–72.
22. Sawaguchi, A., Yao, X., Forte, J. G., and McDonald, K. L. (2003) Direct attach-ment of cell suspensions to high-pressure freezing specimen planchettes. *J. Microsc. (Oxford)* **212,** 13–20.
23. Dahl, R. and Staehelin, L. A. (1989) High-pressure freezing for the preserva-tion of biological structure: theory and practice. *J. Electron Microsc. Tech.* **13,** 165–174.
24. Robards, A. W. and Sleytr, U. B. (1985) Low temperature methods in biological electron microscopy, in *Practical Methods in Electron Microscopy,* vol. 10 (Glauert, A. M., ed.), Elsevier, Amsterdam.
25. Hohenberg, H., Mannweiler, K., and Müller, M. (1994) High-pressure freezing of cell suspensions in cellulose capillary tubes. *J. Microsc. (Oxford)* **175,** 34–43.
26. O'Toole, E. T., McDonald, K. L., Mäntler, J., et al. (2003) Morphologically dis-tinct microtubule ends in the mitotic centrosome of *Caenorhabditis elegans. J. Cell Biol.* **163,** 451–456.
27. McDonald, K. and Müller-Reichert, T. (2002) Cryomethods for thin section elec-tron microscopy, in *Guide to Yeast Genetics and Molecular and Cell Biology, Parts B and C* (Guthrie, G. and Fink, G. R., eds.), *Meth. Enzymol.* 351, Academic Press, San Diego, pp. 96–123.
28. Lonsdale, J. E., McDonald, K. L., and Jones, R. L. (1999) High pressure freezing and freeze substitution reveal new aspects of fine structure and maintain protein antigenicity in barley aleurone cells. *Plant J.* **17,** 221–229.
29. Samuels, A. L., Giddings, T. H. Jr., and Staehelin, L. A. (1995) Cytokinesis in tobacco BY-2 and root tip cells: a new model of cell plate formation in higher plants. *J. Cell Biol.* **130,** 1345–1357.
30. Reipert, S., Fischer, I., and Wiche, G. (2004) High-pressure freezing of epithelial cells on sapphire coverslips. *J. Microsc. (Oxford)* **213,** 81–85.
31. Morphew, M. and McIntosh, J. R. (2003) The use of filter membranes for high-pressure freezing of cell monolayers. *J. Microsc. (Oxford)* **212,** 21–25.
32. Schlegel, A., Giddings, T. H., Ladinsky, M. S., and Kirkegaard, K. (1996) Cellular origin and ultrastructure of membranes induced during poliovirus infection. *J. Virol.* **70,** 6576–6588.

33. Müller-Reichert, T., O'Toole, E. T., Hohenberg, H., and McDonald, K. L. (2003) Cryoimmobiliztion and three-dimensional visualization of *C. elegans* ultrastructure. *J. Microsc. (Oxford)* **212,** 71–80.

34. Oegema, K., Desai, A., Rybina, S., Kirkham, M., and Hyman, A. A. (2001) Functional analysis of kinetochore assembly in *Caenorhabditis elegans. J. Cell Biol.* **153,** 1209–1226.

35. Hohenberg, H., Tobler, M., and Müller, M. (1996) High pressure freezing of tissue obtained by fine needle biopsy. *J. Microsc. (Oxford)* **183,** 133–139.

36. Monaghan, P., Perusinghe, N., and Müller, M. (1998) High-pressure freezing for immunocytochemistry. *J. Microsc. (Oxford)* **192,** 248–258.

37. Giddings, T. H. (2003) Freeze-substitution protocols for improved visualization of membranes in high-pressure frozen samples. *J. Microsc. (Oxford)* **212,** 53–61.

38. Hess, M. W. (2003) On plants and other pets: practical aspects of freeze substitution and resin embedding. *J. Microsc. (Oxford)* **212,** 44–52.

39. Rensing, K. H. and Samuels, A. L. (2004) Cellular changes associated with rest and quiescence in winter-dormant vascular cambium of *Pinus contorta. Trees* **18,** 373–380.

40. Segui-Simmaro, J. M., Austin II, J. R., White, E. A., and Staehelin, L. A. (2004) Electron tomographic analysis of somatic cell plate formation in meristematic cells of *Arabidopsis* preserved by high-pressure freezing. *Plant Cell* **16,** 836–856.

41. Sabatini, D. D., Bensch, K., and Barnett, R. J. (1963) Cytochemistry and electron microscopy. The preservation of cellular structures and enzymatic activity by aldehyde fixation. *J. Cell Biol.* **17,** 19–58.

42. Koster, A. J. and Klumperman, J. (2003) Electron microscopy in cell biology: integrating structure and function. *Nat. Rev. Mol. Cell Biol.* **Suppl,** SS6—SS10.

43. McIntosh, J. R., Nicastro, D., and Mastronarde, D. (2005) New views of cells in 3-D: an introduction to electron tomography. *Trends Cell Biol.* **15,** 43–51.

9

Cryoultramicrotomy

Cryoelectron Microscopy of Vitreous Sections

Dimitri Vanhecke, Luca Studer, and Daniel Studer

Summary

Cryoultramicrotomy allows the sectioning of vitrified biological samples. These biological samples are preserved at the atomic level and represent the real structure at the moment of freezing. Cryoultramicrotomy produces ultra-thin cryosections that are investigated in a cryoelectron microscope. The necessity of working during the whole preparation at temperatures less than −140°C results in some difficulties, including the cryosection transfer from the knife-edge to the electron microscropy grid; the grid handling in the cryochamber and the grid transfer into the cryoholder of the electron microscope. Furthermore, ice crystal contamination (from air humidity) can obscure the structures of interest in the sections. It is mainly know-how and experience that will prevent the contamination of ice crystals and the recrystallization of the sections during the manipulations. Here, we describe the tips, tricks, the tools, and methods that help to overcome these burdens and pave the path for successful cryoultramicrotomy.

Key Words: *Saccharomyces cerevisiae*; bovine articular cartilage; frozen hydrated sectioning; vitrification; cryotransmission electron microscopy.

1. Introduction

The future of ultrastructural investigations of biological specimens in electron microscopy (EM) is to localize proteins and protein complexes in such way they fit atomic resolution maps generated by X-ray or nuclear magnetic resonance analysis *(1,2)*. To obtain a bulk sample with the potential to be preserved at an atomic level, we have to investigate cryosections of vitreous biological samples at a temperature less than −140°C. This method is known as CEMOVIS (i.e., Cryo Electron Microscopy of Vitrified Sections, introduced by Al-Amoudit et al *[1]*). The difference between the cryosectioning technique

From: *Methods in Molecular Biology, vol. 369*
Electron Microscopy: Methods and Protocols, Second Edition
Edited by: J. Kuo © Humana Press Inc., Totowa, NJ

and other specimen-preparation methods in EM is the water status in the biological sample, which remains unchanged, reflecting the living structure of the cell at the time of freezing. On the downside, high-pressure freezing, cryosectioning, and the beam-sample interaction in the microscope can introduce artefacts. High-pressure freezing and beam-sample interaction artefacts are both inherent to the method. However, there is a chance to circumvent cryosectioning artefacts when the sectioning process is further improved *(3)*. Remember that future developments always begin with achievements in the past: Fernandez-Moran initiated cryosectioning more than 50 yr ago *(4)*, and the first acceptable cryosections of vitreous samples were been shown 20 yr ago by Dubochet and colleagues *(5)*. Whole-cell cryoelectron tomography recently was performed successfully *(6)*.

In this chapter, we describe how high-pressure frozen biological samples are transferred to the cryoultramicrotome and how cryosections are brought on an EM grid and mounted on a cryoholder for cryomicroscopy. The sections are investigated at −170°C. All aforementioned manipulations have to be conducted at less than −140°C, which is the recrystallization temperature of pure water.

2. Materials

2.1. Equipment

1. Cryoultramicrotome (Leica UCT plus Leica EMFCS; Leica-Microsystems, Vienna, Austria).
2. Ionizer (Diatome Antistatic Line II; Diatome Ltd, Nidau, Switzerland).
3. Knives: Trimming diamond knife and cryodiamond knife 35° (both knives: Diatome Ltd, Nidau, Switzerland).
4. Cryotransfer system Gatan Model 626.DH (Gatan Inc., Pleasanton, CA).
5. Electron microscope Zeiss EM912 Omega with cryoequipment (Zeiss, Oberkochen, Germany).

2.2. Tools

1. Small containers (*see* **Fig. 1A**).
2. Self-made platelet sample clamp (*see* **Fig. 2A**).
3. Forceps.
4. Self-made eyelashes glued to wooden sticks (*see* **Fig. 1B,C**, and also **Note 1**).
5. Copper EM grids: mesh size 600 (hexagonal meshes) up to 1000 (square meshes).
6. Grid box.
7. Section pressing tool (*see* **Fig. 1D,E**).
8. Partially self-made grid table for pressing the sections.
9. Cryostorage unit (−196°C) for the long-term storage of vitrified samples.
   All other tools (not listed here) are part of the equipment listed above (*see* **Subheading 2.1.**).

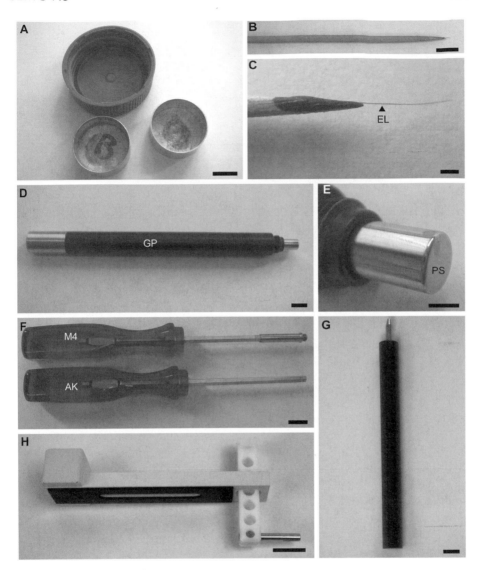

Fig. 1. Tools necessary for manipulating samples, sections, and grids. (**A**) Small cups for sample transfer (small ones: aluminium, large one: plastic). Bar = 1 cm. (**B**) and (**C**) An eyelash (EL) mounted with nail polish on a wooden stick for section transfer from the knife-edge to the grid. Bar = 1 cm in (B), 1 mm in (C). (**D,E**) The grid pressing tool (GP) with its polished surface (PS) for pressing the sections to the grid. Bar = 1 cm in (D), 5 mm in (E). (**F**) Two types of screwdrivers: the Allen Key (AK) for tightening screws and the M4 thread (M4) for pick up and transport of tools in the cryo-chamber. Bar = 1 cm. (**G**) A sample fastener used for closing of the sample clamp from outside the cryo- chamber. Bar = 1 cm. (**H**) The positioning tool, for the positioning of small objects (e.g., ion gun and grid holder) behind the knives. Bar = 1 cm.

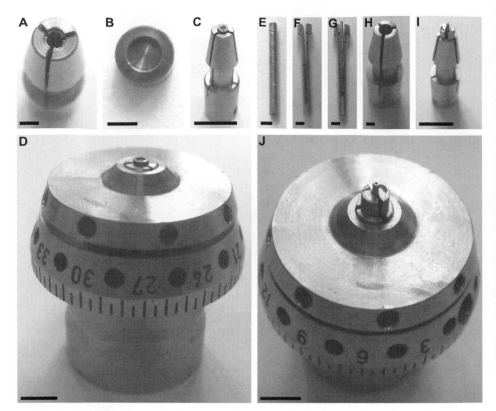

Fig. 2. Sample clamps for the sample holders. **(A–D)** show the tools for the copper platelets, whereas in **(E–J)**, the necessary tools for the copper tubes are depicted. **(A)** The platelet clamp. A three-toothed self-made clamp with a mouth diameter of 2.2 to 1.8 mm can successfully clasp a copper platelet. **(B)** An empty copper platelet in top view. **(C)** The copper platelet positioned in the jaws of the sample clamp. **(D)** The situation as encountered during sectioning. The clamp is embedded in the tightening system, which adjusts the mouth diameter. **(E)** The copper tube sample holder. **(F)** An empty tube clamp. **(G)** The two-toothed clamp can contain almost the entire copper tube. **(H)** The sample clamp for fixation of the commercially available tube clamp (note: different from the one described in **A**). **(I)** The tube sample clamp has a mouth diameter of 2 mm, which can hold the filled tube clamp. **(J)** The sample clamp and tube clamp in the tightening system. Bars: A, F, G, H = 2 mm; B, E = 1 mm; C, D, I, J = 1 cm.

2.3 Samples

We use samples, high-pressure frozen in a Leica EMPACT (Leica-Micro-systems, Vienna, Austria). In the case of unicellular organisms, such as yeast cells (*Saccharomyces cerevisiae*), the vitrification (*see* **Note 2**) of the sample takes place in a copper tube (*see* **Fig. 2E**; outer diameter 650 µm, inner diameter 300 µm, 5-mm long). In the case of tissues, such as cartilage, a round copper platelet (*see* **Fig. 2B**; diameter 2 mm, with a central inner cavity of 1.2 mm diameter and a depth of 200 μm) is used. As a consequence, copper always surrounds the biological sample during high pressure freezing, and it has to be trimmed away to get cryosections. For further information on the technique of high-pressure freezing and/or the equipment; please consult *(7)* and also the Chapter 8 by McDonald et al. in this volume.

3. Methods

3.1. Cryosectioning

3.1.1. Preparation of the Cryochamber

1. If the samples are enclosed in the copper tubes, install the original sample clamp (*see* **Fig. 2H–J**) and tube clamp (*see* **Fig. 2F,G**) for copper tubes (*see* **Fig. 2E**). If you are planning to section tissue samples in platelets (*see* **Fig. 2B**), install the platelet sample clamp (*see* **Fig. 2A–D**). The most convenient is to install these tools when the chamber is still at room temperature.
2. Fill the 25-L reservoir with liquid nitrogen (LN$_2$). Install and connect the hose and cool down the cryochamber to −160°C (*see* **Fig. 3A–D**).
3. Transfer the frozen sample from its storage place in LN$_2$ at −196°C to the cryochamber (*see* **Note 3**) with the help of aluminium caps or plastic lids obtained from commercially available mineral water bottles (*see* **Fig. 1A**).
4. Introduce the sample in the sample clamp (CL; *see* **Fig. 4**; and also **Note 4**).

The temperature in the cooled down cryochamber has to be lower than −140°C to prevent recrystallization of the sample. Samples in a cooled-down cryochamber do not have to be submersed in LN$_2$ because the gaseous phase of LN$_2$ in the cryochamber is sufficiently cold.

- For samples in the copper tubes: use a pair of LN$_2$ precooled forceps to pick up the tube (*see* **Fig. 2E**) and insert it in the precooled tube clamp (*see* **Fig. 4B**). The bulk of the tube must be inserted in the clamp, and only a small part should obtrude to guarantee a stable fixation (vibration free) of the copper tube.
- For the platelets, a similar procedure must be conducted. However, place the precooled tightening system in upright position on the precooled knife stage (*see* **Fig. 4A**) and fix the platelets in the sample clamp as shown in (*see* **Fig. 2D**). Reinstall the tightening system to the sectioning arm.

5. Fix the sample in the sample clamp using the tightening system (TS, in **Fig. 4A,B**) and the sample fastener (**Fig. 1G**). Turn clockwise to tighten, turn anticlockwise to loosen.
6. Place the knives in the knife holder as shown in **Fig. 5** (*see* **Note 5**). This can be performed at room temperature. Introduce the knife holder with the knives in the precooled cryochamber using the screwdriver equipped with a M4 thread (*see* **Fig. 1F**). Allow a cool-down of 2 min on the preparation bench (PB) of the cryochamber (*see* **Fig. 4C** and inset).

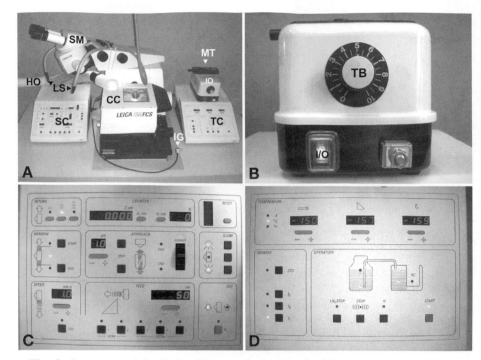

Fig. 3. Overview of the Leica Ultracut S, equipped with the Leica EM FCS cryo-chamber. (**A**) The cryogen is introduced via a hose (HO) running from a 25-L dewar (not shown) to the cryo-chamber (CC). A stereoscopic microscope (SM) and extra light sources (LS) are mounted on a turnable arm above the cryochamber. A strong electric field is generated at the tip of the ion gun (IG), ionizing air molecules. The ionizer can be tested with the multitester (MT). The two controllers besides the cutting equipment manage the sectioning process. The controller on the left covers the parameters for sectioning (SC), whereas the right one regulates the temperature (TC). (**B**) Frontal view on the ionizer. The green on/off tumble button (I/O) is the main switch. The strength of the electric field for ionization can be adjusted by the turning button (TB). (**C**) The sectioning controller with settings for section thickness, window, speed, illumination and more. Please find more detailed information in the Leica Ultracut S manual. (**D**) The temperature controller. The temperature of three separate locations is measured: the sample, the knife and the gaseous phase in the chamber. The "operation" field includes the level of LN_2 in the dewar (four LED indication), cryo-chamber tank (two-LED interface), and the control for activating the dewar LN_2 pump and heating.

7. Once cooled down, tighten the knife holder in front of the sectioning arm using an Allen key (*see* **Fig. 1F**).
8. Connect the ion gun (*see* **Fig. 3A,B**) to the positioning tool (*see* **Fig. 1H**) and aim it to the edge of the trimming knife (*see* **Fig. 4D;** and also **Note 6**).
9. Load a grid onto the grid holder (*see* **Fig. 6**).

3.1.2. Loading a Grid Onto the Grid Holder (to Be Carried Out at Room Temperature)

1. At room temperature, fix the grid holder (GH) using the Allen key into the grid loader (GL, in **Fig. 6A,B**).
2. Place a grid (GR) on the surface of the grid loader (*see* **Fig. 6B**).
3. Open the grid clamps of the grid holder (GC, in **Fig. 6C,D**) using an Allen key.
4. Move the edge of the grid (GR) between the two jaws of the grid clamp (GC) using an eyelash (*see* **Fig. 1B,C**) or a pair of forceps.
5. Close the grid holder (GH) using an Allen key. The clamps close and the grid is fixed in the grid holder.
6. Release the grid holder (*see* **Fig. 6F**).

3.1.3. Trimming

The sample holders are made from copper. In the case of unicellular organisms, such as yeast, the sample holder is a copper tube. In the case of tissues, such as cartilage, the sample holder is a copper platelet. In any case, copper surrounds the sample. Copper is soft enough to be sectioned with a trimming diamond knife without causing damage to the knife.

Details concerning the relative position of the knife to the sample are as follows:

1. Tilt angle of the specimen holder: the cryochamber is not equipped for tilted sectioning.
2. Clearance angle: the clearance angle can be altered on the side of the knife holder using an Allen key. The angle range is confined between 3° and 9° (*see* **Fig. 5**).
3. Sectioning angle: Can be adjusted using the handle at the front bottom of the cryochamber (on the outside of the cryochamber).

During trimming, the ionizer is helpful for blowing away section debris. Place the gun close to the knife-edge, with the ionizing force at full power.

In both sample systems, copper surrounding the vitreous sample must be removed by trimming. By doing so, a rectangular sample surface is shaped. In the case of the copper tubes (inner diameter 300 µm), a copper-free surface of 200 µm × 200 µm can be produced. In the case of the platelets (sample diameter 1 mm) a copper-free surface of about 850 µm × 850 µm can be achieved, but better sectioning conditions will be obtained if the sample is reduced to dimensions comparable to the copper tubes (200 µm × 200 µm).

For convenient trimming, select a section thickness of 500 nm. Making thinner sections will be more time demanding without bringing extra benefits, whereas thicker sections induce higher forces on the knife and trimming can damage the knife. High speed and automatic sectioning are the proper settings for convenient trimming.

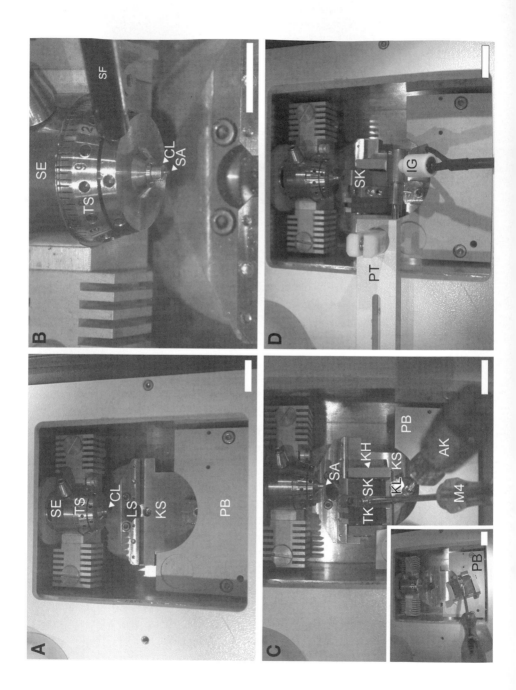

182

Trimming is explained for the copper tube sample holder. In the case of platelets, a similar procedure can be used, but more sample and copper must be trimmed away.

The aim is to produce a surface of approximately 200 μm × 200 μm on a 50-μm high mesa.

1. Trim the entire sample surface (copper tube up to 200-μm deep; platelets up to 30-μm deep). An entirely dark and shiny sample surface is an indication for its vitreous state (proving vitrification, however, is only possible by electron diffraction of the cryosections in the electron microscope). White samples illustrate a crystalline state of water.
2. Position the knife-edge parallel to the pretrimmed sample surface, then move the knife to the right until the left outer edge of the knife is about 70% (if 100% is the sample radius) away from the sample centre. In this setting, the entire copper wall on the right side (of the tube) can be trimmed away. This is the easiest way to produce a mesa with minimal loss of biological sample (in the case of the copper tube).
3. After trimming 50 μm, stop and retract the knife (>50 μm). Move the knife to the other (left) side of the sample (right knife-edge is at 70% to the left of the sample centre) and repeat. Do not change any rotational settings, only the horizontal movement of the knife (the knob on the right-hand bottom on the outside of the chamber). In this way, two perfect parallel sides, each 50 μm high, can be trimmed.
4. Rotate the sample 90° and repeat the process described previously. In doing so, one will obtain a 50-μm high mesa with a rectangular surface of 200 μm × 200 μm, devoid of the copper encasement.

Humidity of the ambient atmosphere will crystallize as soon as it enters the cryochamber and small ice crystals will form a dust-like layer on the interior

Fig. 4. *(Opposite page)* Preparation of the cryochamber before trimming and sectioning. **(A)** The sectioning arm (SE) ends in a tightening system (TS) with a radial closing mechanism for the clamp (CL). A light source (LS) throws light from below (switched off). The knife stage (KS) is shown here without knives. The preparation bench (PB) is located at the proximal end of the chamber and is elevated. **(B)** Enlargement of the sample tightening system. The sample previously high pressure frozen in a small copper tube (SA) is clasped in the clamp (CL), which was introduced in the tightening system of the sectioning arm (SE). By moving the outer disc clockwise, by means of the sample fastener (SF), the sample is tightened. Anti-clockwise rotation will open the tightening system. **(C)** Mounting of the knife set: the trimming knife (TK) on the left and the sectioning knife (SK) on the right in the knife holder (KH). Before mounting the warm knives on the knife stage (KS), cooling down for at least 2 min, away from the sample on the preparation bench (PB) is essential. The knife holder is tightened to the knife stage by closing the knife lock (KL) using the Allen key (AK). **(D)** Mounting the ion gun. Assure the gun is not switched on during manipulation. Use the positioning tool (PT) described in **Fig. 1H** to position the cooled-down gun close to the knives in the cryo-chamber. Point the gun towards the knife-edge. Bars = 2 cm.

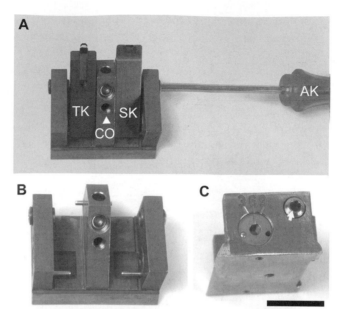

Fig. 5. The knife holder with knives. **(A)** The knife holder with a trimming knife (TK) and a sectioning knife (SK). The knives are fixed in their positions by closing the screw (arrow) on the side (see also **Fig. 5C**) with the Allen key (AK). A connection (CO) for a M4 thread equipped screwdriver allows manipulation of the knife holder in the cryo-chamber. **(B)** The knife holder without the knives. **(C)** The side of the knife holder, showing the lock for the knife (top right) and the clearance angle modulator (left). Both are Allen key operated. The numbers above the modulator represent the clearance angle in degrees. Bar = 5 mm.

of the cryochamber. The electron beam cannot penetrate ice crystals. Therefore, ice crystals on the surface of the sections are worsening their quality because the crystals obscure ultrastructural details of the specimen (*see* **Note 7**). Introduction of water molecules, for instance, through fingers touching, breathing, in the cryochamber must be kept to a minimum. Use forceps or with M4 thread equipped screwdrivers to transfer samples and tools in and out the cryochamber. Close the cryochamber with the cover whenever possible. Unfortunately, the formation of ice crystals is not entirely preventable.

3.1.4. Sectioning

The knob on the right-hand bottom on the outside of the chamber controls the movement of the knife stage along the horizontal axis. Use this knob to position the cryoknife in line with the trimmed sample. If necessary, slightly unscrew the knife holder and use the Allen key with the M4 thread to adjust the cryoknife.

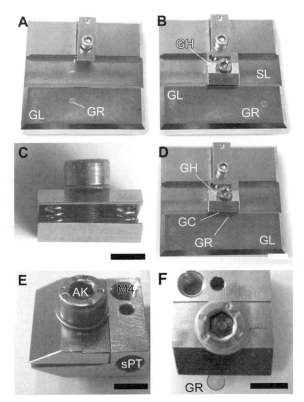

Fig. 6. Loading a grid. The loading of the grid is done at room temperature. (**A**) Starting position with the grid loader (GL) and an empty grid (GR). (**B**) The grid holder (GH) can be placed in the grid loader (GL) and moved back and forth in the slot (SL). An Allen screw in the middle of the slot fixes the grid holder. (**C**) The grid holder seen from the front, showing the grid clamp. (**D**) The situation in the grid loader (GL) before the pick up of the grid (GR) in the grid holder (GH). The grid is positioned at the edge, between the grid clamps (GC). (**E**) As viewed from the side, three connectors can be seen on the grid holder: an Allen screw (AS) on top for opening and closing the grid clamp, a M4 thread (M4), tangential from the Allen screw, for transfer and manipulation in the cryo-chamber and a sideway entry for the positioning tool (sPT), allowing positioning of the grid holder behind the knives. (**F**) The grid holder holding a grid (GR). Bars = 5 mm.

The magnitude of ionization on the knife-edge is an important aspect of cryo-sectioning. It depends both on the strength of the ionizer (changeable on the source) and on the distance of the ion gun to the knife-edge. Too little ionization at the knife-edge will result in heavily compressed sections, whereas too much ionization will blow away the sections from the knife-edge. Therefore, the optimal settings have to be found empirically according to the microenvironmental climate (important factor: relative air humidity). High amounts of ionisation on the knife-edge are advantageous. In our experience, stronger ionisation allows higher sectioning speeds, which are beneficial in the reduction of crevasse formation (*see* **Note 8**; Al-Amoudi et al. *[8]*).

1. If necessary, use an eyelash (*see* **Fig. 1B,C**) to remove ice and section debris from the knife-edge, in addition, use the ionizer to blow away the debris.
2. Position the sample a few micrometers away from the knife-edge. The sample should be parallel and equal-distant to the knife-edge. To do so, use the light source from below to create a reflection of the knife-edge on the sample. The reflection should be the same over the entire block surface (akin to plastic sectioning alignment). A clearance angle of 6° of the knife provides the best results for sectioning.
3. Manually move the knife toward the sample until the light from below almost disappears between the knife-edge and the sample.
4. The ultramicrotome is set to automatic mode and cutting section thickness to 50 nm. Use maximum cutting speed until the sectioning starts. Optimal cutting speeds can vary enormously due to the sample and the atmospheric conditions around the microtome. They have to be found by trial and error. However, a cutting speed of approx 1 mm/s or lower is fine for most samples. Adjust the ion gun parameters until smooth (transparent) sections are obtained.
5. The aim is to make a ribbon (RI) with five or six sections, as shown in **Fig. 7D**: head-tail (with the top edge of the last section still attached to the knife-edge; KE in **Fig. 7**). Start the first ribbon on the left hand side of the knife.

Fig. 7. (*Opposite page*) The handlings of the transfer of sections to the grid. The transfer of sections is the most challenging procedure of all cryosectioning-related manipulations. (**A**) Switch off the ionizer and remove the ion gun (IG) from the positioning tool (PT) outside the cryo-chamber. Let the positioning tool warm up and dry. Mount the grid holder and cool it above the preparation bench (PB). The grid table (GT) is mounted on the preparation bench (PB) and grid presser (GP) is introduced in the cryo-chamber to adjust its temperature to −160°C. If the grid presser is impeding the manipulations during section transfer, it can be placed in the chamber after the transfer. (**B**) Zoom in on the grid holder (GH) loaded with one grid (GR). The frontal edge of the precooled grid is positioned as close as possible towards the knife edge of the sectioning knife (SK), thereby making the transfer distance of the ribbons as short as possible. Ensure not to touch the knife-edge with the grid. (**C**) Situation in the cryochamber ready for the transfer of sections to the grid. (**D**) Close up of the knife-edge (KE). Note the mesa form of the sample (SA) after trimming. The ribbons (RI) are attached by their last section at the knife-edge. Inset: grid (GR) with 3 ribbons (RI outlined by the fine rectangular bars). The vitreous samples reflect almost no light (hence the black colour of the sample). Bar = 250 μm. Ribbons (RI) of 50 nm thin sections appear highly transparent at the knife-edge (the ribbon right from the indication RI; white ribbons show heavily compressed sections).

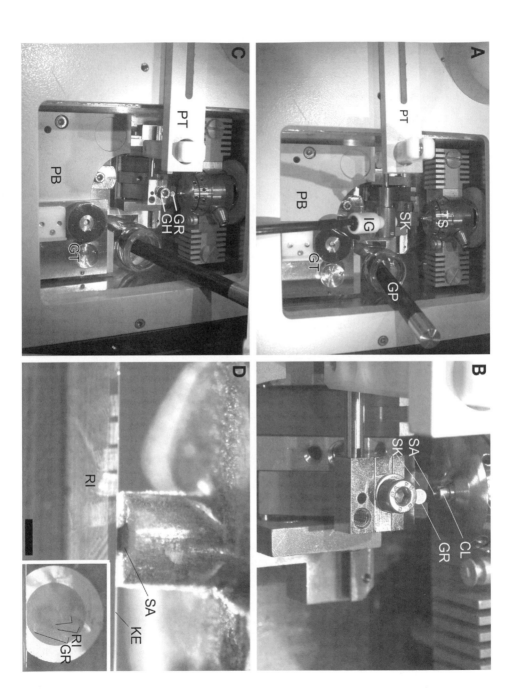

187

6. When such ribbon is obtained, stop the sectioning and retract the knife stage. Move it to the right of the ribbon already produced. Repeat **steps 2–5** to produce a second ribbon.

7. Proceed until there is no more space for the ribbons on the knife-edge (*see* **Fig. 7D**). This is depending of the width of the block face and the knife width. As a thumb rule, six to eight ribbons will fit easily on a grid.

3.1.5. Pick-up of the Sections and Transfer to the Grid

This step might turn out to be the most challenging step in the entire process of CEMOVIS. Training, patience, and a steady hand are helpful.

1. Switch off the ionizer and remove the ion gun from the positioning tool (*see* **Fig. 7A**). Attach the grid holder (*see* **Fig 7B**).

2. Cool down the grid holder by placing it on the preparation bench (see PB in **Fig. 7**) of the cryochamber. Do not bring the grid holder in the vicinity of the knife edge before it is cooled down to less than −140°C. Allow at least 2 min for the cool down.

3. Arrange the grid horizontally in the same focal plane as the knife edge, and as close as possible to the knife-edge (*see* **Fig. 7B,C**).

4. Use two eyelashes to transfer the ribbons to the grid (*see* **Fig. 7D, inset**). The easiest way is to disconnect the ribbon top edge from the knife-edge with one eyelash while using the other to keep the ribbon stretched. Then position both eyelashes under the ribbon and make the transfer. It is possible that the ribbons will fold, curl up or fly away (*see* **Note 9**).

3.1.6. Section Pressing (see **Fig. 8**)

This step assures that the sections do not get lost during the subsequent procedures. The sections must stay at temperatures less than −140°C to avoid recrystallization, hence the procedure must be conducted in the cryochamber.

1. Place a small jar filled with LN_2 in the cryochamber. Cool down the polished surface of the grid presser (*see* **Fig. 1D,E**) to LN_2 temperature (*see* **Fig. 7C**).

2. Arrange the grid holder on the self-made grid holder frame (HF, in **Fig. 8A**). The grid, loaded with sections, is orientated toward the round polished surface (*see* **Fig. 8B**). To avoid ice crystal adsorption on the polished surface, cover it during cooling with a coverslip, and remove the cover moments before the transfer of the grid takes place.

Fig. 8. (*Opposite page*) Final steps in the cryochamber which are carried out on the grid table: the unloading of the grid from the grid holder, the grid pressing and the temporary storage in a grid box. (**A**) Situation of the grid table before the grid transfer. The polished surface is sheltered by a coverslip which is removed just before the grid is transferred. The grid holder frame (HF) receives the grid holder with the grid. Via the polished

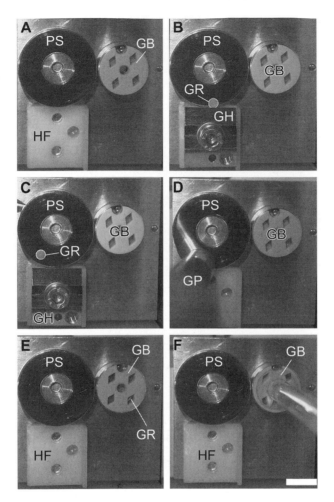

Fig. 8. *(Continued)* surface, the grid will end up vertically in the grid box. **(B)** When the grid holder (GH) is placed in the grid holder frame, the grid (GR) automatically hangs a few micrometers above the rotating polished surface (PS, with an axis in the middle). By loosening the Allen screw of the grid holder, the grid will be released and fall on the polished surface just below. **(C)** After the release of the grid, the grid holder can be removed safely with the positioning tool. **(D)** With the grid presser (GP, *see* **Fig. 1D–E**), the sections are pressed onto the grid. The polished surface below (PS) prevents damage of the sections during pressing. Keep in mind that the grid presser should only make movement in the Z direction, and no slanted, tilted or skewed movements to avoid distortion of the sections. Remove the grid presser in a quick and fluent movement. **(E)** Using precooled tweezers; the grid (GR) can be transferred from the polished surface (PS) into the grid box (GB). Once the sections are pressed onto the grid, there is only a small risk that the sections will fall from the grid. **(F)** The grid box (GB) can be closed with a handle and transferred into a vessel filled with LN_2. Bar = 1 cm.

3. Detach the grid from the grid holder by loosening the screw of the grid clamp with the Allen key. The grid lands on the polished surface below (*see* **Fig. 8C**).
4. Use an eyelash (*see* **Fig. 1B,C**) to position the grid away from the edges of the round polished surface.
5. Use the grid presser (*see* **Fig. 1D,E**) to apply a vertical force on the sections and the grid (*see* **Fig. 8D**). Ensure the sample press is precooled and not dripping wet with LN_2 (which may lead to sections washed away from the grid). Only apply pressure in the Z-axis and not in the X-Y axes.
6. Transfer the grid to one of the empty pockets of the grid box (*see* **Fig. 8E**).
7. Close the grid box with the precooled handle (*see* **Fig. 8F**).
8. Keep the grids at temperatures below −140°C all times, that is, in the cryochamber, in a grid box submerged in LN_2 or in a cooled-down cryotransmission electron microscope.

3.2. Transfer to the Microscope

3.2.1. Preparation of the Cryotransfer Workstation

1. Cool down the cryoequipment of the electron microscope and the cryoholder (inserted into the electron microscope) with LN_2 (fill the dewars completely until the LN_2 stops boiling). When the cryoholder reaches a temperature of −165°C or lower, one can start the transfer of the grid.
2. Close the side entry port of the cryotransfer workstation with the plug.
3. Fill the insulated vessel (IV, in **Fig. 9A**) of the Gatan cryotransfer workstation with LN_2.

3.2.2. Preparation of the Gatan Cryoholder

1. Place the shutter of the cryoholder in the "closed" position with the shutter control knob (SC, in **Fig. 9A, 9C**).
2. Remove the Gatan cryoholder from the microscope: (a) pull away until halfway out and (b) turn 90° anticlockwise and catch the excess LN_2 in a recipient (<1 L) in a fast-but-gentle movement.
3. Remove the plug from the side entry port (SeP, in **Fig. 9A**) of the cryotransfer workstation. LN_2 starts to gush out.
4. Draw out the cryoholder completely from the microscope and immediately insert the cryo holder in the workstation (as shown in **Fig. 9A**).
5. Refill the vessel with LN_2.

Fig. 9. (*Opposite page*) Loading of a grid into the Gatan cryo-holder. (**A**) Small dewar (DE) assures the low temperature of the cryospecimen holder (SH), whereas a shutter mechanism (SC = shutter control) can cover the grid and protect it against ice crystal contamination. The transfer occurs in the cryo-transfer workstation (WS). The specimen holder enters via the side entry port (SeP) into the LN_2 filled insulated vessel (IV). Minimize the transfer time from the TEM to the cryo transfer workstation and vice versa! (**B**) Tip of the specimen holder with the shutter (SR) in opened position: grids are fixed by a clip ring (CR). (**C**) Tip of the specimen holder with the shutter (SR)

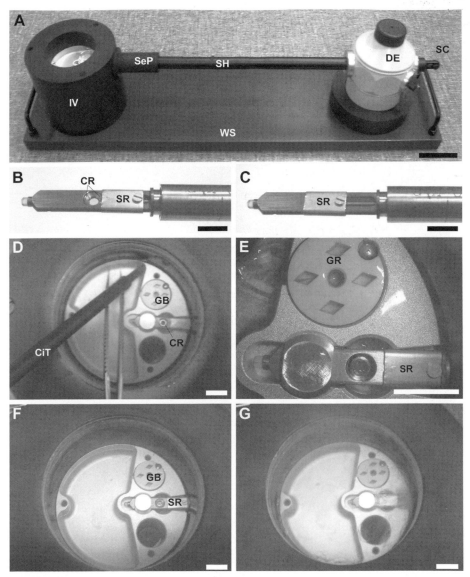

Fig. 9. *(Continued)* in the closed position: the position of the shutter can be changed using the shutter control (SC; **Fig. 9A**). **(D)** Top view in the insulated vessel with all tools set: clip ring insertion tool (CiT), which allows locking of the clip ring (CR), tweezers for the grid transfer and the grid box (GB) with the grid. The picture may appear blurred as the result of the LN_2 in the vessel. **(E)** Magnification of the tip of the specimen holder in the insulated vessel: the shutter is in opened position and the clip ring is removed. **(F)** The insulated vessel in top view after the transfer of the grid: the shutter is still opened. **(G)** The specimen holder, with grid, ready to transferred to the TEM: the shutter is closed. Bars = 1 cm.

3.2.3. Grid Transfer

1. Precool tweezers and the clip ring insertion tool key (*see* CiT in **Fig. 9D**).
2. Place the grid box (GB, in **Fig. 9**) with the grids in the isolated vessel (*see* **Fig. 9E**): (a) open the grid box and (b) open the shutter (SR, in **Fig. 9E**) with the shutter control knob (SC, in **Fig. 9A**).
3. Remove excess LN_2 as much as possible, but ensure the grid stays just above the LN_2 level when placed in the holder (doing so guarantees low temperature but avoids floating away of the grid in the LN_2).
4. Open the clip ring (CR, in **Fig. 9B, 9D**) with the clip ring insertion tool (CiT, in **Fig. 9D**).
5. Transfer the grid (*see* **Fig. 9E**).
6. Close the clip ring (*see* **Fig. 9F**). Keep the tool submerged in LN_2 during grid transfer. Refill the LN_2 to submerge the grid.
7. Close the shutter (*see* **Fig. 9G**).

3.2.4. Gatan Transfer to the Microscope

1. Remove the cryo holder from the workstation (WS, in **Fig. 9A**).
2. Swiftly place the cryo holder into the microscope until halfway inserted (the dewar will again release its LN_2 content).
3. Wait until the electron microscope airlock is evacuated.
4. Insert the cryoholder into the microscope completely.
5. Refill the dewar of the cryoholder; check the cold finger of the electron microscope.
6. Open the shutter (SC, in **Fig. 9A**).

3.3. Microscopy

Vitreous sections of yeast in overview are depicted in **Fig. 10**. Before inserting the cryoholder, adjust the electron microscope settings. Beam damage of cryosections has to be avoided. Therefore, make sure that you can take micrographs in the minimal dose mode. In this mode you focus close to, but not in the region of interest to minimize beam damage. As you will experience the sample shows only good contrast after a certain electron dose (some electrons/nm^2); however, too high electrons dose leads to bubbling (*see* **Fig. 11B**). **Figure 11** depicts cryosections of a yeast cell. Remind that the contrast of such pictures is not rendered by heavy metal staining but by natural mass distributions. After successful cryosectioning and electron microphotography, high resolution micrographs can be taken as shown in **Fig. 12 inset** and Al-Amoudi et al. *(1)*.

Figure 12 shows vitreous cartilage high pressure frozen by the platelet system. A chondrocyte is partially shown (*see* **Fig. 12**). The fine-structured cytoplasm reflects real-life mass distribution (including water), not heavy metals binding sites. Collagen fibres are abundant in the extracellular matrix (*see* **Fig. 12 inset**). The striation of the fibres (arrow) shows the 64 nm repetitive pattern of collagen.

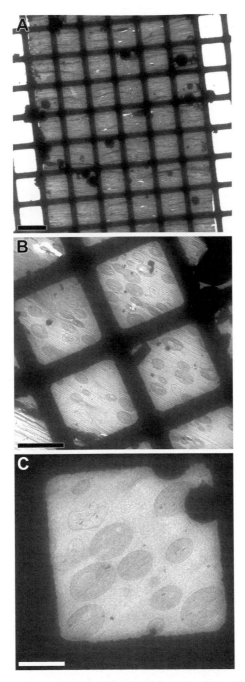

Fig. 10. Overview of vitreous sections in the cryo transmission electron microscope. The images show progressively higher magnification of the same region. Vitreous yeast cells are shown. Bars A = 25 µm, B = 10 µm, C = 5 µm.

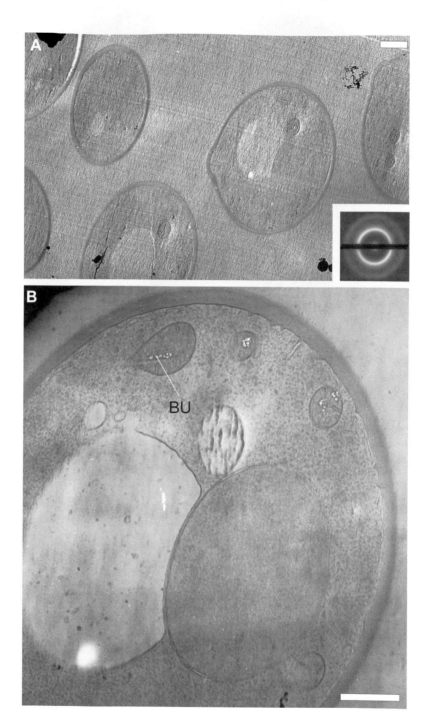

4. Notes

1. Eyelashes are so fine that they are cooled down almost instantly in the cold cryo-chamber. Just hold them for a few seconds in the cryochamber to adjust their temperature.

2. Ice crystals are formed when biological samples are immobilized by cooling. However, when cooling is fast, approx 200-μm thick samples may vitrify under high-pressure (2000 Bar) conditions. In vitreous samples, water does not form ice crystals and the water molecules cease their movements, resulting in the preservation of the ultrastructure in the amorphous specimen. Vitrification in transmission EM is only shown by electron diffraction. Diffraction patterns showing only blurred rings prove that the status of water is vitreous. In contrast, crystalline ice diffraction patterns show sharp spots (hexagonal ice) or sharp rings (cubic ice).

3. Forceps always should be precooled when handling cryofixed samples. When frozen samples are handled with warm (room temperature) tweezers, recrystallization of ice in the sample may occur, resulting in severe deterioration of the sample. Place the tips of the forceps in liquid nitrogen until LN_2 boiling stops.

4. Cryofixed samples must be stored and maintained at temperatures less than $-140°C$ at all times. The transport of samples and grids loaded with cryosections to and from the cryochamber is conducted out in LN_2-filled vessels, for instance, plastic caps. Ensure that the necessary supplies are ready before starting cryosectioning: (a) Styrofoam box filled with LN_2 (circa 0.5 L with the samples submerged), (b) transport vessels, and (c) forceps. Plastic lids obtained from commercially available mineral water bottles or 50-mL centrifuge tubes come in handy for the transfer of the samples because they cool down fast, sink in LN_2, don't break at low temperatures, are easy to handle with forceps, and can hold an optimal volume of LN_2 (enough to make the transfer, but not too much to flood the cryochamber).

5. With the exception of the positioning tool, all tools are fixed with an Allen key (Hex key) 3-mm socket. Transfer from and to the cryochamber of tools (such as the knife holder) is performed with a M4 threaded screwdriver (a 3-mm Allen key equipped with a M4 thread). Only 3-mm Allen and M4 keys are used.

6. Be aware to switch on the ionizer only when the gun is mounted and in position. Switch it off before any manipulation with the ion gun. Aiming an operational ion gun on metal surfaces can cause heavy static electricity streaks. Highly charged

---

Fig. 11. (*Opposite page*) Frozen hydrated yeast cells. This sample was prepared using the copper tube system. (**A**) An overview with a few cells is shown. (**B**) Magnification of one of the cells. The bubbles (BU) in the mitochondria illustrate beam damage. It is obvious that not all structures are at the same amount sensitive to beam damage; beside the mitochondria, all structures are well preserved. That there are almost no sectioning artifacts visible is attributable to the fact that the applied electron dose was quite large. The beam "irons" the sections. Note that membranes and cell organelles are well preserved. The diffraction pattern with its blurred rings (insert in **A**) shows that the section derived from a vitreous sample. Bars: A = 1 μm; B = 500 nm.

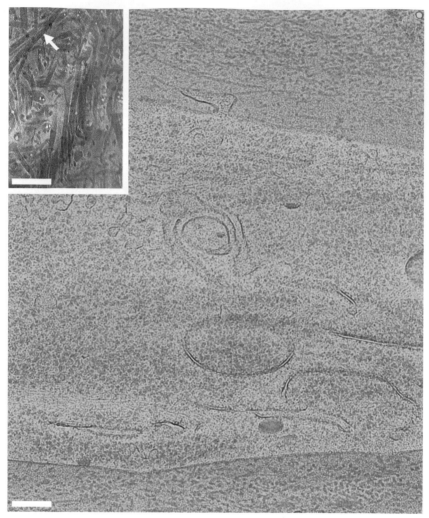

Fig. 12. Vitreous cartilage prepared and sectioned in the platelet system. The cytoplasm is fine-structured and shows several membrane bound structures dispersed over the cell. The extracellular matrix is abundant in collagen proteins. In the inset the approx 64-nm striation of collagen (arrow) can be seen. Bars = 1 μm.

sample surfaces (tribo-electric effects) occur in insulated samples and are generated by the high cutting forces of ultrathin sectioning. Ion flow discharges the sample, ensuing optimal section gliding on the knife edge. The optimal conditions (distance from knife and strength of ionisation) have to be determined empirically (depending on the microclimate in the room).

7. When copper debris remain on the sample, one can use a paper napkin soaked with LN_2, which is brought in a vessel to the preparation bench (PB in **Fig. 4A**) and drop

LN$_2$ over the sample; the copper debris float away. Do not use pipets to blow away debris because the used air will be warm (risk of recrystallisation of the sample) and plenty of moisture (condensation of ice crystals everywhere) will be released.

8. In frozen hydrated sections, cutting artifacts are apparent in three forms. There are knife marks, running parallel but nonrepetitive with the cutting direction. There is chatter, running perpendicular to the cutting direction. Chatter is repetitive and is visible as darker and lighter waves in the section. Chatter is formed as the result of changing adsorption forces of the section to the knife. Crevasses are small cracks, more or less perpendicular to the cutting direction. Details are described in Al-Amoudi et al. *(8)*.

9. In our experience, cryosectioning of vitreous samples seems to be subject to environmental parameters, such as the weather. On warm humid days, it has been nearly impossible to transfer the sections from the knife-edge to the grid without a folding effect of the ribbons. The sections curled up and became useless. We see the humidity as the main player in this process. The only practical remedy is to install a cabinet with humidity control.

References

1. Al-Amoudi, A., Chang, J. J., Leforestier, A., et al. (2004) Cryo-electron microscopy of vitreous sections. *EMBO J.* **23,** 3583–3588.
2. Lucic, V., Förster F., and Baumeister, W. (2005) Structural studies by electron tomography: from cells to molecules. *Ann. Rev. Biochem.* **74,** 833–865.
3. Al-Amoudi, A., Dubochet, J., Gnaegi, H., and Lüthi, W. (2003) An oscillating cryo-knife reduces cutting-induced deformation of vitreous ultra thin sections. *J. Microsc. (Oxford)* **212,** 26–33.
4. Fernandez-Moran, H. (1960) Low temperature preparation techniques for electron microscopy of biological specimens based on rapid freezing with liquid helium II. *Ann. NY Acad. Sci.* **85,** 689–713.
5. Dubochet, J., McDowall, A. W., Menge, B., Schmid, E. N., and Lickfeld, K. G. (1983) Electron microscopy of frozen-hydrated bacteria. *J. Bacteriol.* **155,** 381–390.
6. Medalia, O., Weber, I., Frangakis, A. S., Nicastro, D., Gerisch, G., and Baumeister, W. (2002) Macromolecular architecture in eukaryotic cells visualized by cryoelectron tomography. *Science* **298,** 1209–1213.
7. Studer, D., Graber, W., Al-Amoudi, A., and Eggli, P. (2001) A new approach for cryofixation by high-pressure freezing. *J. Microsc. (Oxford)* **203,** 285–294.
8. Al-Amoudi, A., Studer, D., and Dubochet, J. (2005) Cutting artefacts and cutting process in vitreous sections for cryo-electron microscopy. *J. Struct. Biol.* **150,** 109–121.

10

Cell-Free Extract Systems and the Cytoskeleton

Preparation of Biochemical Experiments
for Transmission Electron Microscopy

Margaret Coughlin, William M. Brieher, and Ryoma Ohi

Summary

Cell-free systems can be used to reconstitute complex actin or microtubule-based phenomena. For example, a number of extracts support actin-dependent propulsion of *Listeria monocytogenes*, whereas *Xenopus laevis* extracts support formation of a microtubule-based meiotic spindle. Working *in vitro* opens these complex processes to biochemical dissection. Here, we describe methods to view these in vitro preparations by thin-section electron microscopy.

Key Words: *Listeria monocytogenes*; *Xenopus laevis*; cell-free extract; actin; microtubule.

1. Introduction

The structural organization of the cytoskeleton is crucial to its function. Understanding the morphogenesis of the actin and microtubule cytoskeletons is difficult, however, because they are biochemically and spatially complex. Studying the two systems in vitro allows easier access to their underlying biochemistry. Here, we describe methods to view the ultrastructure of the reaction products by electron microscopy (EM). In this way, cytoskeletal organization can be related to specific protein factors.

We use *Listeria monocytogenes* as a model system for understanding the dynamic organization of actin and *Xenopus laevis* egg extracts to understand the behavior of microtubules in the mitotic spindle. Despite both being well-developed in vitro systems, their organizations are not yet fully understood. We have developed EM techniques to gain a better understanding of fundamental

From: *Methods in Molecular Biology, vol. 369*
Electron Microscopy: Methods and Protocols, Second Edition
Edited by: J. Kuo © Humana Press Inc., Totowa, NJ

mechanisms underlying cytoskeletal morphogenesis. Moreover, the overall methods described in this chapter can be extended to other in vitro microtubule and actin systems.

2. Materials

2.1. General Equipment

1. Parafilm.
2. Pipettemen (P1000, P200, P20, P2) and appropriate tips.
3. Aclar film for coverslips and mounting (*see* **Note 1**).
4. Spearhead Power Punch and Die Set (*see* **Note 2**).
5. Rubber mallet.
6. Glow discharge apparatus.
7. High molecular weight poly-L-lysine hydrobromide (Sigma, cat. no. P-1524).
8. 50-mL tri-pour polypropylene beakers.
9. Tongue depressor.
10. Vacuum desiccators with desiccant.
11. Polystyrene Petri dishes.
12. Styrofoam shipping containers.
13. Digital low-temperature thermometer.
14. Fluorescence microscope.
15. Dissecting microscope.
16. Polypropylene transfer pipettes.
17. Forceps.

2.2. Equipment Specific for Listeria Experiments

1. Perfusion chambers (*see* **Fig. 1**). These are constructed by adhering two 25-mm × 2-mm strips of Scotch 3M double-stick tape to a 25-mm × 75-mm glass microscope slide parallel and distant to one another by 10 mm at the inside edges. Place symmetrically on the strips a 12.7-mm round Aclar coverslip. Tap lightly with forceps handle to secure. The perfusion chamber will hold a volume of 8 to 10 µL.
2. Whatman 3 MM filter paper cut into 62.5-mm × 12.5-mm slivers.
3. Razor blades.
4. Humidified Petri dish with support for glass slide.
5. Aluminum inserts from heating block (used to conduct cold in ethanol bath).

2.3. Equipment Specific for Xenopus Experiments

1. Wide-bore (0.75–1 mm) P200 tips prepared by cutting them at the ends.
2. 250-mL polypropylene beakers.
3. Vacuum aspirator.
4. Ceramic coverslip holder with handle (*see* **Note 3**).
5. 150-mm Petri dish humidification chamber.
6. 15 mL of modified Corex centrifuge tubes with removable chocks (*see* **Fig. 5** and also Evans et al. *[1]*). These tubes have a small layer of silicone at the bottom,

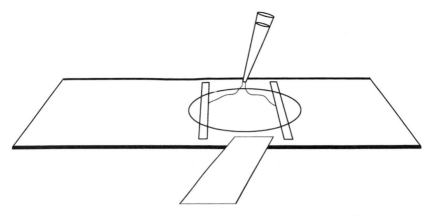

Fig. 1. Perfusion chamber for *Listeria* actin comet tail reactions. The chamber consists of a standard, glass microscope slide and an Aclar coverslip separated by two strips of double-stick tape. Fluid is introduced into one end of the chamber and wicked out from the other with filter paper.

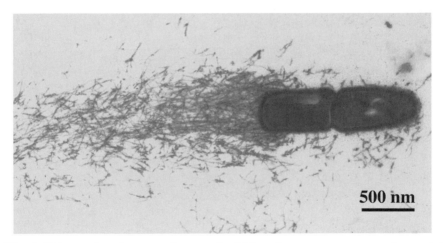

Fig. 2. Thin section electron micrograph of *Listeria monocytogenes* and associated actin comet tail assembled in bovine brain extract.

followed on top by a fixed Plexiglass layer, which enables the placement of a removal chock at the bottom of the tube that can be taken out by a large, unfolded paperclip. We usually mount a coverslip on top of the chock, onto which structures contained within extract can be centrifuged.

2.4. Buffers

1. Assay buffer (for actin experiments): 20 mM PIPES pH 6.8, 100 mM KCl, 2 mM MgCl$_2$, 2 mM adenosine triphosphate, 2 mM b-mercaptoethanol.
2. 1 M PIPES, pH adjusted to 6.8 using solid potassium hydroxide.

3. 5XBRB80: 400 mM K-PIPES, pH 6.8, 5 mM MgCl$_2$, 5 mM K-ethylenebis(oxy-ethylenenitrilo)tetraacetic acid (K-EGTA) pH 7.7.
4. Sperm dilution buffer: 10 mM K-N-hydroxyethylpiperazine-N'-2-ethanesulfonate (HEPES), pH 7.7, 100 mM KCl, 1 mM MgCl$_2$, 150 mM sucrose, 10 µg/mL cytochalasin D.
5. "Frog fix": 60% glycerol, 1X MMR, 1 µg/mL Hoechst 33342, 10% formaldehyde.
6. Hooking solution: 500 mM K-PIPES, pH 6.8, 1.4 mM MgCl$_2$, 1 mM K-EGTA, pH 7.
7. 1 mM GTP: 2.5% dimethyl sulfoxide, 1.5 mg/mL tubulin.
8. Cacodylate buffer: make 0.2 M solutions of sodium cacodylate trihydrate (Electron Microscopy Sciences, cat. no. 12300) and 0.2 N HCl. Mix the two stocks to obtain the appropriate pH, and then dilute with water to 50 mM.

2.5. Proteins and Protein Conjugates

1. G-actin, fluorescently labeled, and biotinylated G-actin are prepared by established techniques *(2)* or are available commercially (Cytoskeleton Inc., Boulder, CO).
2. 5 nm of gold-conjugated streptavidin (EY Laboratories, San Mateo, CA).

2.6. Miscellaneous Reagents

1. Uranyl acetate.
2. Lead citrate (Ted Pella, cat. no. 512-26-5).
3. Ethanol.
4. Propylene oxide (Electron Microscopy Sciences, cat. no. 20412).
5. 1 M MgCl$_2$.
6. 500 mM K-EGTA, pH 7.7.
7. Electron microscopy grade 25% glutaraldehyde (Ted Pella, cat. no. 18426).
8. Free base lysine powder (Sigma, cat. no. L 5501).
9. Paraformaldehyde powder (Electron Microscopy Sciences, cat. no. 19200).
10. 4% aqueous osmium tetroxide (Ted Pella, cat. no. 18465).
11. Potassium ferricyanide (Sigma, cat. no. P8232).

2.7. Fixatives

1. Actin hard fix: 1:1 mixture of 100 mM lysine and 6% glutaraldehyde in 0.05 M cacodylate buffer, pH 7.0 made immediately before use.
2. Actin soft fix: 1% glutaraldehyde in assay buffer.
3. Microtubule "hard fix": a 1:1 mixture of 100 mM lysine and 6% glutaraldehyde in BRB80 buffer for a final concentration of 50 mM lysine and 3% glutaraldehyde.
4. Reduced osmium: 1% osmium tetroxide with 0.8% potassium ferricyanide in 0.05 M cacodylate buffer, pH 7.0.

2.8. Embedding Resins

Epon araldite, a formula that lends itself to serial sectioning. For 30 mL: 8.4 mL Epon 812 (Polysciences, cat. no. 08791), 5.4 mL araldite (Polysciences, cat. no. 00552), 16.2 mL DDSA (Polysciences, cat. no. 00563). Mix these three components with a tongue depressor. Add 1.2 mL BDMA (Polysciences, cat. no. 00141). Evacuate bubbles in vacuum desiccator.

3. Methods

3.1. Preparation of Listeria monocytogenes Tissue Extracts, and Proteins

Cultivation, chemical inactivation, and storage of *L. monocytogenes* have been described previously *(4)*. A number of cell and tissue extracts support actin comet tail assembly off *L. monocytogenes* including outdated platelets *(4)*, *Xenopus* eggs *(5)*, bovine calf brain *(6)*, and Hela cells *(7)*. Comet tail assembly can also be performed with a mixture of pure proteins *(8)*. Comet tail assembly can frequently be seen by phase contrast with a 40X objective but is most easily followed by fluorescence microscopy using fluorescently labeled actin.

3.1.1. Listeria *Actin Comet Tail Assembly*

1. Rapidly thaw an aliquot of tissue or cell extract.
2. Thaw an aliquot of chemically inactivated *L. monocytogenes*. Dilute the stock 1:100 to 1:1000 in assay buffer and store on ice.
3. Fill the perfusion chamber with the diluted suspension of *L. monocytogenes* in assay buffer. Place the perfusion chamber upside down in a humidified chamber and incubate for 10 min at room temperature. Inverting the chamber facilitates the binding of *Listeria* to the coverslip over the glass side.
4. Prepare 25 µL of the actin comet tail assembly mixture for each perfusion chamber. This mixture contains any of the extracts described previously diluted into assay buffer and supplemented with 4 µ*M* G actin and 1 µ*M* tetramethylrhodamine-labeled G actin. The exact dilution of extract required for comet tail formation depends on the tissue source and method of preparation and, therefore, needs to be determined empirically using light microscopy. For a detailed example of the biochemistry for assembling actin comet tails in perfusion chambers, *see* Brieher et al. *(2)*.
5. Flip the perfusion chamber right side up. Perfuse in all 25 µL of the extract + actin solution while wicking from the other end of the chamber (*see* **Note 1**). Incubate the reaction at room temperature. Check the extent of the reaction periodically by fluorescence microscopy. In our experience, comet tails assembled in brain extract typically form and extend to 10 µm in 7 to 10 min.

Once the reaction has progressed to the desired extent, proceed to lysine glutaraldehyde fixation (*see* **Subheading 3.1.3.**). **Fig. 2** shows a representative thin section electron micrograph of an actin comet tail prepared by this method.

3.1.2. Localizing Sites of Actin Polymerization With Biotinylated Actin and Streptavidin Gold

Actin (and actin binding proteins) can be labeled with biotin, introduced into the chamber, and detected with streptavidin-coated gold particles. This tech-

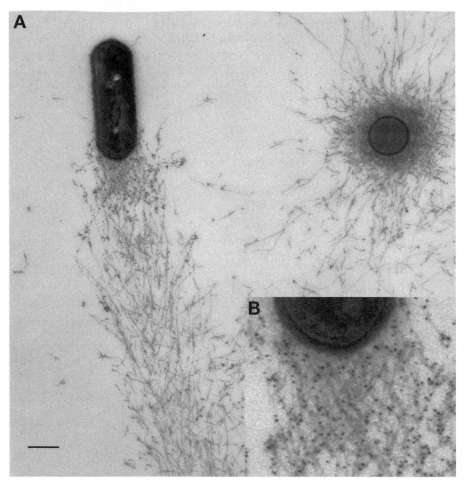

Fig. 3. Streptavidin gold labeling of biotinylated actin incorporation during a 10-s period in bovine brain extract. Comet tails initially were assembled in brain extract. The solution was then replaced with brain extract containing biotinylated actin and fixed 10 s later. Biotinylated actin was detected with 5 nm of streptavidin gold. Bars, **A** = 500 nm and **B** = 167 nm.

nique allows a snapshot of the incorporation of components into the comet tail. The experiment is best performed with two people to minimize time between addition of the biotinylated protein and fixation. **Figure 3** shows a representative example of a pulse of biotinylated actin labeled with streptavidin gold.

1. Prepare a 25-μL solution of the comet tail assembly mixture with the extract of choice. Add biotinylated actin to 1 μ*M* final concentration. Incubate on ice for 30 min.
2. Assemble actin comet tails in perfusion chambers as described above using a mixture lacking biotinlylated actin. Follow the extent of the reaction by light microscopy.

3. While the first comet tail reaction is proceeding, prepare the soft fix and hold at room temperature.
4. Once the reaction has progressed to the desired extent, replace the contents of the chamber with 25 µL of the comet tail assembly mix containing biotinylated actin.
5. After 10 s, wash the chamber once with 10 µL of assay buffer.
6. Fix immediately by flushing the chamber twice with 15 µL of soft fix. Incubate in the humidified chamber for 15 min at room temperature.
7. Wash three times with assay buffer.
8. Flush the chamber twice with a 1:20 dilution of 5 nm of streptavidin gold in assay buffer. Incubate 60 min at room temperature in a humidified chamber.
9. Wash the chamber three times with 15 µL of assay buffer.
10. Proceed to lysine-glutaraldehyde fixation.

3.1.3. Fixing and Embedding Actin Samples

1. Fixation solutions are 100 mM lysine and 6% glutaraldehyde made separately in 0.05 M cacodylate buffer pH 7.0, then mixed in equal parts immediately before use *(9)*. Fix for 6 min, then exchange with 3% glutaraldehyde alone in buffer for another 12 min at room temperature.
2. Rinse twice with cacodylate buffer.
3. Postfix with 1% osmium tetroxide with 0.8% potassium ferricyanide in cacodylate buffer *(10)* for 15 min on ice in the hood. Important: It is important to work with osmium in a well-ventilated hood. The fumes are toxic. Dispose of appropriately as hazardous waste.
4. Rinse twice with cacodylate buffer, then twice with distilled water.
5. Stain in a solution of 1% aqueous uranyl acetate (filtered before use) overnight or for a minimum of 2 h at 4°C in the dark. Use additional wet Kimwipes in humidification chamber to make sure sample does not dry out overnight.
6. Rinse three times with distilled water.
7. Dehydration by progressive lowering of temperature method *(11)*. To take the samples through a graded ethanol series while lowering the temperature has required that we devise a setup that cools the alcohols and provides a similarly cool stage on which to place the glass slide perfusion chamber. We modified the humidification chamber by wetting the filter paper with 95% EtOH instead of water and eliminating the plastic support strips for the slide. This will then conduct the appropriate degree of cold without freezing the slide to the paper.For the first two steps, we set the humidification chamber directly on ice and cool bottles of 35% EtOH and 50% EtOH in it as well. Fluid transfers are performed as described previously, introducing cold ethanol on one side with a Pipetman, while simultaneously wicking fluid from the opposite side of the chamber with a strip of Whatman filter paper. Allow 5 min for each step. For the subsequent steps, super-cooled ethanol baths are constructed out of two Styrofoam shipping containers into which 95% EtOH is poured to a level of about one inch. Aluminum blocks from a bench-top heating block are inverted into the ethanol, flat side on top. An electronic thermometer probe is placed in the ethanol bath. Dry ice is dropped into it until the temperature

reaches −20°C in one box and −40°C in the second. Cool bottles of 50% EtOH and 80% EtOH in the first bath and 95% EtOH and 100% EtOH in the second. The humidification chamber with microscope slide perfusion chamber is moved from one super-cooled aluminum block to the next for the appropriate temperature for each ethanol. Exchange the 100% EtOH twice, then move the humidification chamber to room temperature in a laboratory fume hood. Avoid fluid evaporation and sample drying.

8. Prepare two small cups, one with fresh 100% EtOH and one with 100% propylene oxide in the hood.
9. Remove the perfusion chamber from the humidification chamber and use a no. 3 Dumont forceps to dismantle it by gently pulling the Aclar coverslip from the tape strips.

3.1.4. Infiltrating

These steps are the same for coverslips from spin down and perfusion chamber experiments.

1. Line two 60-mm polystyrene Petri dishes with Parafilm for infiltration steps.
2. Prepare 7.5 ml of 2:1 propylene oxide:epon araldite, and 7.5 ml 1:2 propylene oxide: epon araldite, cap and invert to mix thoroughly.
3. Decant the 2:1 into one of the Parafilm lined Petri dishes.
4. Dip the coverslips into the cup of fresh 100% EtOH, then into the 100% propylene oxide, and lastly into the dish of 2:1 propylene oxide:epon araldite.
5. Infiltrate 45 min, then pipet off and replace with the 1:2 solution and infiltrate 1 h. Dispose of waste plastic in appropriately labeled hazardous waste container.
6. Drain coverslips on folded paper towel, then place in second Parafilm lined Petri dish and apply 100% epon araldite. Infiltrate at least 1 h, longer if possible. Drain coverslips on a folded paper towel.

3.1.5. Mounting Assembly

The mounting method here has been devised to result in the embedded sample as a small plastic disk of uniform thickness that can be taped to a microscope slide facilitating the identification of areas of the sample or structures of a specific phenotype for sectioning using a standard inverted cell culture microscope. A diagram of the mounting assembly is given in **Fig. 4**. Written instructions continue here:

1. Use the Spearhead Power Punch and Die Set to cut Aclar "washers." Punch a circle with the die one size larger than the coverslip, and then cut the middle out with one size smaller.
2. Cut two Aclar squares (one slightly smaller than the other) large enough to accommodate the number of coverslips abutted next to each other.
3. Clean washers and squares with acetone.
4. Dip the washers in epon araldite and position on the larger of the Aclar squares, placed in the lid of a 100-mm Petri dish.

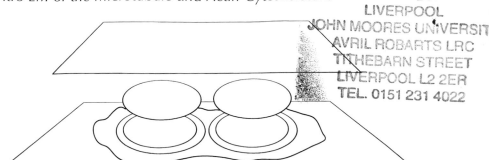

Fig. 4. Aclar sandwich coverslip mounting assembly. The sandwich consists of Aclar washers resting on a square of Aclar in a puddle of epon araldite. Aclar coverslips containing the sample are placed on top of the washers. The sandwich is sealed with a second square of Aclar.

5. With a transfer pipet, apply enough epon araldite inside and around the washers to make a thin, continuous pool. Remove bubbles with a narrow point transfer pipet.
6. With forceps, grasp coverslips with sample side face down and lower over washer.
7. Pipet small drops of epon araldite on the backs of each coverslip and lower second Aclar square over the assembly to complete "sandwich".
8. Place the bottom of the Petri dish on top and weight for uniform sample thickness.
9. Polymerize at 65°C for 48 h.

3.1.6. Sample Selection and Remounting for Sectioning

1. Peel top square of Aclar away from "sandwich."
2. Score around periphery of washers with razor blade and remove plastic surrounding samples.
3. Score around coverslip and remove plastic layer over washers.
4. Peel coverslip off of sample.
5. Use forceps to lift washer with sample disk inside and tape to microscope slide. Avoid fingerprints.
6. View on an inverted cell culture microscope to identify structures of interest and mark with the needle of a tuberculin syringe.
7. Use fresh razor blade to excise individual structures.
8. Remount on blank Beem capsule stub with two-part epoxy glue and harden.
9. Trim block face and section.
10. Stain sections with 1% uranyl acetate in 50% methanol solution, then 0.4% lead citrate.

3.2. Preparation of Xenopus *Egg Extracts, and Proteins*

3.2.1. Spindle Assembly in Xenopus *Egg Extracts*

Cytoplasm from unfertilized *Xenopus* eggs and demembranated *Xenopus* sperm nuclei are prepared as described (*12*), and used to assemble bipolar spindles.

1. To 20 µL of extract, add sperm nuclei to a final concentration of approx 400 µL and calcium chloride in sperm dilution buffer to a final concentration of 0.4 mM. Multiple reactions can be generated by adding sperm nuclei and calcium to a larger volume of extract, but small (<35-µL) aliquots should always be made after the addition of calcium because spindle assembly occurs more robustly in small volumes. For visualization of microtubules, approx 50 µg/mL fluorophore-labeled tubulin (3) can be added at any step throughout the spindle assembly time course. Incubate the extract at 20°C for 80 min to allow chromosome replication to occur.

2. Add 0.75 volumes (e.g., 15 µL) of untreated, cytostatic factor-arrested extract to each tube containing extract and replicated DNA. Intact cytostatic factor in native extract will reimpose a metaphase arrest on the cycled reaction, and spindle assembly is frequently complete 60 min after the addition of cytostatic factor extract addition. We monitor the efficiency of spindle assembly using conventional wide-field microscopy after mixing 1 µL of extract with 4 µL of fixative directly on a approx 1-mm thick, 25- × 75-mm glass slide.

3.2.2. Spindle Microtubule Polarity Determination Using Hook Decoration

Incubation of microtubules with pure tubulin in a buffer that promotes polymerization results in the addition of tubulin protofilament "hooks" along the walls of pre-existing microtubules (13). The handedness of the hooks, observed in serial cross section electron micrographs, indicates the orientation of the microtubule, and follows the "right-hand rule"; curve the fingers of your hand in the direction of the hooks, and your thumb will point towards the minus end. Hook decoration has primarily been used for observing microtubule polarity in intact cell model systems (14,15), although it has been applied to determine microtubule polarity distributions in DNA bead spindles assembled in *Xenopus* extracts (16). What follows is an adapted method to hook microtubules in sperm chromatin-derived bipolar spindles assembled in *Xenopus* extracts. **Figure 6** shows a representative cross section through a spindle that has been treated to generate hook decorated microtubules, as well as a cross section through a control untreated spindle.

1. Aclar coverslips for samples are prepared as follows: Using a tuberculin syringe with needle, scratch numbers on 12.7-mm Aclar coverslips to distinguish individual samples. If care is taken to keep the surface with the scratched number on it upwards, it will eventually be transferred to the plastic in which the sample is embedded.

2. Aclar coverslips are highly hydrophobic. Just before use, glow discharge for 10 to 12 min and coat with 1 mg/mL poly-L-lysine. This step has made a significant difference in the number of spindles retained on the coverslip through the many processing steps. Rinse with distilled water and place on the removable chocks from the spin down tubes and gently lower to the Plexiglass support surface at the bottom

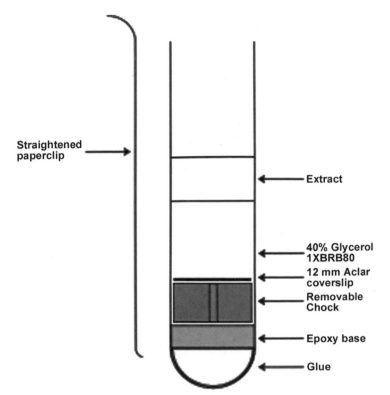

Straightened paperclip

Extract

40% Glycerol 1XBRB80

12 mm Aclar coverslip

Removable Chock

Epoxy base

Glue

Fig. 5. Modified Corex tubes used to sediment spindles assembled in *Xenopus* egg extracts. The tubes are made by solidifying glue and epoxy at the bottom of 15-mL Corex tubes. A Plexiglass chock with a sidelong groove can be mounted on top of the epoxy base, and removed using a straightened, medium-sized paperclip. Aclar coverslips are seated on top of the chock, which together are placed at the bottom of the Corex tube. A glycerol cushion is overlaid onto the chock and coverslip, and fixed extract is pipetted over the glycerol cushion.

    of the tube. While pinning the coverslip down with an unfolded paperclip, pipet a cushion of 30% glycerol in BRB80 into the tube.

3. Using a cut-off pipet tip (bore size of approx 1 mm), transfer 20 µL of extract containing spindles to 200 µL of freshly made hooking solution in a 5-mL Falcon 2063 tube. Incubate at room temperature for 30 to 60 s (*see* **Note 4**).

4. Add 800 µL of spindle dilution buffer and mix by inversion two to three times.

5. Fixation solutions are 100 mM lysine and 6% glutaraldehyde made separately in 0.05M BRB80, mixed in equal parts immediately before use. One milliter of this is added to the extract in dilution buffer for a final concentration of 25 mM lysine and 1.5% glutaraldehyde. The tube is capped and gently inverted three times to mix. Fix for 10 min at room temperature.

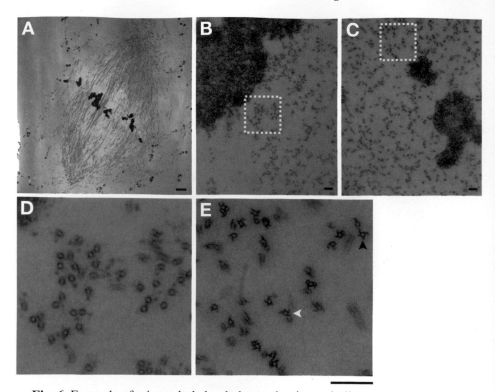

Fig. 6. Example of microtubule hook decoration in a spindle assembled in *Xenopus* egg extract. (**A**) Low-magnification, thin-section view of a bipolar spindle assembled *in vitro*. Scale bar = 2 μm. (**B**) and (**C**) show 85-nm thin cross sections of a region near the spindle equator without (**B**) and after (**C**) hook decoration. (**D**) and (**E**) 5-fold magnification of areas contained within the dashed boxes in (**B**) and (**C**), respectively. Note the tubulin extensions present on the microtubules in (**E**), which display both left and right handedness. Clockwise (right-handed; black arrow) tubulin hooks indicate the presence of a microtubules oriented with their minus ends distal to the cross section, whereas counterclockwise (white arrow) tubulin hooks are indicative of microtubules growing toward to the observed section. Bars in **B–E** = 200 nm.

6. Overlay the fixed spindles onto a 4 mL of 40% glycerol/1XBRB80 cushion in a 15-mL Corex tube modified to accept a 12-mm Plexiglass chock onto which a coverslip can be placed (*see* **Fig. 5** and also Evans et al. *[1]*).
7. Pellet spindles onto the coverslip by centrifugation in an HB6 or HS4 swing-out rotor, at 5500 rpm, for 20 min, at 18°C.
8. Wash the sample/cushion interface 2-3X with 1XBRB80 and then remove most of the cushion by aspiration. Keep coverslip wet.
9. Using the unfolded paperclip, withdraw chock and transfer coverslip to humidification chamber and apply buffer.

10. Fix again in 1.5% glutaraldehyde in BRB80 for 10 min.
11. Rinse in BRB80 two times and 0.05 M cacodylate buffer pH 7.0 two times.
12. Postfix in 1% aqueous osmium tetroxide with 0.8% $K_3Fe(CN)_6$ in cacodylate buffer for 15 min on ice in hood.
13. Pipet off osmium and dispose of as hazardous waste.
14. Rinse twice in cacodylate buffer and twice in distilled water.
15. Stain in 1% aqueous uranyl acetate (filtered before use) overnight at 4°C.
16. Dehydrate by progressive lowering of temperature method.

Place coverslips in coverslip rack with handle and move through beakers containing alcohols cooled to the appropriate temperatures:

35% EtOH,	50% EtOH	4°C	5 min each
50% EtOH,	80% EtOH	−20°C	5 min each
95% EtOH,	100% EtOH	−40°C	5 min each

Bring to room temperature in 100% EtOH.

Dip each coverslip in fresh 100% EtOH and fresh 100% propylene oxide. Proceed as in **Subheading 3.1.4.** for Infiltrating, **3.1.5.** for Embedding and **3.1.6.** for Sample Selection and Remounting.

4. Notes

1. Aclar is a polyester plastic that has been used as a substrate for cultured cells to be processed for electron microscopy first referenced in 1968 (Masurovsky and Bunge). Sheets 0.19 mm thick are currently available to microscopists through Ted Pella, Inc., cat. #10501-25.
2. The Spearhead Power Punch and Die Set is used to punch coverslips and the "washers'" referred to in the Mounting Assembly section. It was purchased from Zimmerman Packing and Mfg., Inc. 2768 Highland Ave., Cincinnati, OH 45212.
3. The ceramic coverslip staining rack can accommodate round and square coverslips of any size. It was purchased from Thomas Scientific, cat. #8542E30.
4. In contrast to hook decoration in cells, we have observed that hooking times of >1 min leads to the assembly of tubulin protofilament rosettes on microtubule lattices instead of hooks in *Xenopus* extract spindles. This leads to ambiguity in determining hook directionality and should be avoided.

References

1. Evans, L., Mitchison, T., and Kirschner, M. (1985) Influence of the centrosome on the structure of nucleated microtubules. *J. Cell Biol.* **100,** 1185–1191.
2. Brieher, W. M., Coughlin, M., and Mitchison, T. J. (2004) Fascin-mediated propulsion of *Listeria monocytogenes* independent of frequent nucleation by the Arp2/3 complex. *J. Cell Biol.* **165,** 233–242.
3. Hyman, A., Drechsel, D., Kellogg, D., et al. (1991) Preparation of modified tubulins. *Methods Enzymol.* **196,** 478–485.

4. Welch, M. D. and Mitchison, T. J. (1998) Purification and assay of the platelet Arp2/3 complex. *Methods Enzymol.* **298,** 52–61.
5. Theriot, J. A., Rosenblatt, J., Portnoy, D. A., Goldschmidt-Clermont, P. J., and Mitchison, T. J. (1994) Involvement of profilin in the actin-based motility of *L. monocytogenes* in cells and in cell-free extracts. *Cell* **76,** 505–517.
6. David, V., Gouin, E., Troys, M. V., et al. (1998) Identification of cofilin, coronin, Rac and capZ in actin tails using a *Listeria* affinity approach. *J. Cell Sci.* **111,** 2877–2884.
7. Noireaux, V., Golsteyn, R. M., Friederich, E., et al. (2000) Growing an actin gel on spherical surfaces. *Biophys. J.* **78,** 1643–1654.
8. Loisel, T. P., Boujemaa, R., Pantaloni, D., and Carlier, M. F. (1999) Reconstitution of actin-based motility of *Listeria* and *Shigella* using pure proteins. *Nature (London)* **401,** 613–616.
9. Boyles, J., Anderson, L., and Hutcherson, P. (1985) A new fixative for the preservation of actin filaments: fixation of pure actin filament pellets. *J. Histochem. Cytochem.* **33,** 1116–1128.
10. McDonald, K. (1984) Osmium ferricyanide fixation improves microfilament preservation and membrane visualization in a variety of animal cell types. *J. Ultrastruct. Res.* **86,** 107–118.
11. Carlemalm, E., Villiger, W., Actaruab, J.-D., and Kellenberger, E. (1986) Low temperature embedding, in *Science of Biological Specimen Preparation* (Mueller, M., Becker, R. P., Boyde, A., and Wolsewick, J. J., eds.), SEM Inc., AMF O'Hare, Chicago, pp. 24–59.
12. Murray, A. W. (1991) Cell cycle extracts, *Methods Cell Biol.* **36,** 581–605.
13. Heidemann, S. R. and McIntosh, J. R. (1980) Visualization of the structural polarity of microtubules. *Nature (London)* **286,** 517–519.
14. Euteneuer, U., Jackson, W. T., and McIntosh, J. R. (1982) Polarity of spindle microtubules in *Haemanthus* endosperm. *J. Cell Biol.* **94,** 644–653.
15. Euteneuer, U. and McIntosh, J. R. (1981) Structural polarity of kinetochore microtubules in PtK1 cells. *J. Cell Biol.* **89,** 338–345.
16. Heald, R., Tournebize, R., Habermann, A., Karsenti, E., and Hyman, A. (1997) Spindle assembly in *Xenopus* egg extracts: respective roles of centrosomes and microtubule self-organization. *J. Cell Biol.* **138,** 615–628.

11

Electron Microscopy *In Situ* Hybridization

Tracking of DNA and RNA Sequences at High Resolution

Dušan Cmarko and Karel Koberna

Summary

Electron microscopy *in situ* hybridization (EM-ISH) represents a powerful method that enables the localization of specific sequences of nucleic acids at high resolution. We provide here an overview of three different nonisotopic EM-ISH approaches that allow the visualization of nucleic acid sequences in cells. A comparison of various methods with respect to their sensitivity and the structural preservation of the sample is presented, with the aim of helping the reader to choose a convenient hybridization procedure. The post-embedding EM-ISH protocol that currently represents the most widely used technique is described in detail, with a special emphasis on the organization of the cell nucleus.

Key Words: Electron microscopy; double- and single-stranded DNA and RNA; *in situ* hybridization; post-embedding; preembedding; cryosections.

1. Introduction

In situ hybridization (ISH) techniques allow for specific nucleic acid sequences to be demonstrated in the cellular environment. ISH takes advantage of the specific pairing of complementary nucleic acid molecules through hydrogen bonds formed between bases attached to the sugar-phosphate backbone. The first light microscopy detection of nucleic acid sequences by complementary radioactive probes was performed by Pardue and Gall *(1)* and John et al. *(2)*. Later on, radioactive ISH was adopted for electron microscopy *(3,4)*. However, a relatively low spatial resolution could be achieved and a long exposure time of autoradiograms was necessary. Moreover, the use of radioactive probes required the implication of safety hazard regulations. All these obstacles were overcome by the introduction of nonisotopic labels and their application to electron microscopy-ISH (EM-ISH). The use of nonisotopic probes also enables simultaneous detections of

From: *Methods in Molecular Biology, vol. 369*
Electron Microscopy: Methods and Protocols, Second Edition
Edited by: J. Kuo © Humana Press Inc., Totowa, NJ

Post-embedding method

→

Fixation – Embedding – Sectioning – Hybridization – Labeling – Staining – Observation

Pre-embedding method

→

Fixation – Permeabilization – Hybridization – Labeling – Embedding – Sectioning – Staining – Observation

Frozen sections

→

Fixation – Cryoprotection – Freezing – Sectioning – Thawing – Hybridization – Labeling – "Embedding" – Observation

Fig. 1. Schematic view of three different electron microscopy–*in situ* hybridization methods.

more than one nucleic acid sequence or the simultaneous detection of nucleic acid and protein targets. As a consequence, ISH, including its electron microscopy variant, has become an essential method in a number of fields of biomedical research. Three basic EM-ISH approaches normally are used: post-embedding, pre-embedding, and cryo-EM-ISH (*see* **Fig. 1**).

During the postembedding approach, the tissue or cells are fixed, embedded into resin, sectioned, and detection is performed on surface of prepared ultrathin sections *(5–27)*. The postembedding hybridization on ultrathin resin sections dominates the current high-resolution EM-ISH methods. Especially in the case of locally abundant nucleic acids, when good ultrastructural preservation is required, postembedding on ultrathin acrylic sections represents the common choice (*see* **Note 1**).

During the pre-embedding ISH, cultured cells or vibratome sections are fixed, permeabilized, hybridized, post-fixed, embedded in resin and thin-sectioned *(28–35)*. A major drawback of pre-embedding techniques represents the need for permeabilization treatment before hybridization. When weak permeabilization results in low penetration of probes and antibodies into cells, especially into vibratome sections, stronger permeabilization results in a more extensive extraction of cellular components and in the loss of ultrastructural details. In contrast to lower preservation of cell ultrastructure, the pre-embedding approach could profit from the fact that no embedding material is used before ISH so that the probe can penetrate into the cell structures. Consequently, the labeling is, in principle, found in the entire volume of the section and, therefore, more intense signal is usually observed. In this respect, the pre-embedding may eventually allow for the detection of unique genes.

The cryo approach is based on the freezing of a fixed and cryoprotected sample that is afterwards sectioned and ISH then performed on thawed sections *(10, 36–41)*. Sections are then usually "embedded" in a thin layer of special water-soluble material (e.g., methylcellulose) that preserves the section structure during drying and observation. The accessibility of cell structures is usually better than in the case of sections embedded in hydrophilic resins. Some results even support the possibility that ISH and immuno-cytochemical probes penetrate into a volume of cryo-sections *(42)*. However, the conditions used during ISH could result in much higher damage of ultrastructure than in resin sections. This is also a reason why ultrathin cryo-sections are not convenient for the detection of double-stranded deoxyribonucleic acid (DNA) sequences that requires a harsh denaturation step.

Each of the EM-ISH techniques has its advantages and disadvantages. Generally, the development of ISH protocol represents the search for a compromise between cell utrastructural preservation and detection sensitivity. In this respect, the final choice of methods depends on the specific aim of the study to be performed (*see* **Note 2**).

Here, we will focus on the postembedding EM-ISH in acrylic sections. Postembedding hybridization has been successfully used to localize ribosomal ribonucleic acid (RNA) and DNA, viral RNA and DNA, as well as high copy cellular messenger (m)RNAs. This most widely used technique provides, in our opinion, a very good compromise between resolution, ultrastructure preservation, and intensity of the signal.

2. Materials

2.1. Fixatives

1. 4% Formaldehyde and 0.25% glutaraldehyde in 100 mM piperazine-N,N-bis(2-ethanesulfonic acid (PIPES) buffer, pH 7.3 (working solution): Mix the paraformaldehyde with distilled water (a third of final fixative volume) in an Erlenmeyer flask. Heat to 60°C on a stir plate. When moisture forms on the sides of the flask, add sodium hydroxide (drops of 1 N solution), and stir until the solution clears. Cool the solution under the faucet. Filter, and then add glutaraldehyde (research-grade) and PIPES buffer (200 mM, pH 7.3). Adjust pH if necessary and fill with distilled water to final volume.
2. 4% Formaldehyde and 0.25% glutaraldehyde in 100 mM PIPES buffer with heptane, pH 7.3 (working solution): Combine the formaldehyde–glutaraldehyde fixative (as previously) with an equal volume of n-heptane, shake for 1 min, and then leave to sand until the phases are separated. The heptane (upper) phase is carefully decanted and immediately used for fixation.

Formaldehyde and glutaraldehyde are toxic. Use gloves, and prepare and use fixatives under fume hood.

2.2. Postfixation Processing of Cell Samples, Embedding, and Ultramicrotomy

1. 10X Phosphate buffered saline (PBS; stock): prepare 1370 mM NaCl, 30 mM KCl, 160 mM Na$_2$HPO$_4$, 20 mM KH$_2$PO$_4$. Adjust pH to 7.4 with sodium hydroxide (10 N).

Sterilize in an autoclave. Store at room temperature (RT), after opening store the bottle in a refrigerator.

2. 2% Agarose: dissolve low melting agarose in PBS at 37°C. Alternatively 10% gelatin could be used: dissolve gelatin in PBS at 40°C. Store at −20°C.
3. Dehydrating solutions. Ethanol or methanol is usually used for dehydration. For detailed information see protocol of manufacturer of embedding medium. Methanol is highly toxic. Use gloves. If methanol is used, perform dehydration step under fume hood.
4. Embedding medium (e.g., LR White or Lowicryl). Embedding media are toxic. Use gloves. Embed under fume hood.
5. Gelatin capsules.
6. Gold or nickel EM grids covered with plastic foil. Any standard procedure for their preparation can be used. If nickel grids are used, antimagnetic tweezers are recommended.
7. Ultramicrotome.
8. Glass or diamond knives.

2.3. Probes

Four types of probes labeled with biotin or digoxigenin are used.

1. Double-stranded DNA probes.
2. Single-stranded DNA probes.
3. Oligonucleotide probes.
4. Single-stranded RNA probes.

2.4. Hybridization

1. 20X Saline–sodium citrate (SSC; stock) buffer: prepare 3 M sodium chloride plus 0.3 M sodium citrate. Adjust pH to 7 with sodium hydroxide (10 N). Sterilize in an autoclave. Store at RT; store the bottle in a refrigerator after opening.
2. 10 mg/mL RNA stock solution: dissolve yeast tRNA in diethylpyrocarbonate treated water. Store at −20°C.
3. 10 mg/mL DNA stock solution: dissolve herring sperm, salmon sperm, or *Escherichia coli* DNA in sterile water. Store at −20°C.
4. 10 mg/mL RNase A stock solution: dissolve in sterile water. Store at −20°C. Dilute in 10 mM Tris-HCl, pH 7.4 before use.
5. 10 mg/mL Proteinase K, stock solution: dissolve in sterile water. Store at −20°C. Dilute in 20 mM Tris-HCl, 2 mM CaCl$_2$ buffer, pH 7.6 before use.
6. 1 mg/mL DNase I stock solution: dissolve in sterile RNase free water. Store at −20°C. RNase free DNase should be used. Dilute in DNAse I buffer (*see* **step 9**).
7. 100–400 U/μL S1 nuclease stock solution: dilute in S1 nuclease buffer (*see* **step 10**).
8. Deionized formamide: add 5 g of ion exchange resin AG501-X8 (Bio-Rad) to 100 mL of formamide. Shake for 30 min. Filter it. Store at −20°C in dark. **Caution:** Formamide is very toxic. Use gloves during manipulation.
9. 10X DNase I buffer stock solution: 100 mM Tris-HCl, 25 mM MgCl$_2$, 5 mM CaCl$_2$, pH 7.6. Store at −20°C.

10. 10X S1 nuclease buffer stock solution: 330 m*M* sodium citrate, 500 m*M* sodium chloride, and 300 mM zinc sulfate, pH 7.5.
11. 50% Dextran sulfate stock solution: dissolve dextran sulfate in distilled water. Store at −20°C.

2.5. Labeling

1. 1X PBS:BSA (1%, working solution): dissolve BSA in PBS. Sterilize by filtering using a 0.22-μm filter. Store at −20°C.
2. Antibiotin and antidigoxigenin antibodies.
3. Colloidal gold antibody complexes. The secondary antibodies are raised against antibiotin or antidigoxigenin antibodies and coupled to colloidal gold particles.

3. Methods

3.1. From Fixation to Ultrathin Sections

Primary fixation represents the first and very important step of any EM-ISH procedure (*see* **Notes 3** and **4**). Freshly prepared fixation solution should be used.

1. Prepare fixation solution containing 4% paraformaldehyde and 0.25% glutaraldehyde.
2. Fix small pieces of tissue or cultured cells at 4°C.
   Tissue: Sink pieces of tissue to the bottom of the fixation solution. If this method does not work, use vacuum remove air bubbles. They can prevent the penetration of fixative into the tissue. Fix for 4 h.
   Cells in suspension: Centrifuge at 150*g* on an Eppendorf centrifuge for 5 min and remove medium. Overlay with a fixative and resuspend the cells. Two hours are sufficient for the fixation of cells usually.
   Cells cultured in monolayer: remove medium before fixation. Fix immediately in a fixative. Two hours usually are sufficient for fixation of cells.
   If fixative penetration is a major problem, like in plant tissues owing to the presence of cell walls and hydrophobic waxy cuticles, you can use phase-partition fixation method *(6,43)*. It accelerates fixative penetration. Immerse material in the heptane phase. Fix at RT for 10 mins. Afterward, remove the material from the heptane and place it in paraformaldehyde/glutaraldehyde fixative for a further 4 h at RT.
3. Wash three times in PBS.
4. Preparation of small pieces of material.
   Tissue: Dissect tissue into small cubes (less than 1 mm³).
   Cells in suspension: Heat agarose (2%) or gelatin (10%) and keep it warm (37°C). Overlay pellet of cells in Eppendorf tube with 200 μL of agarose or gelatin and mix the content of tube by gentle pipetting on and off. Cut the end of the pipet tip to reduce shearing forces before mixing. The cells should be evenly distributed throughout the agarose or gelatin. Centrifuge the cell suspension at 150*g* for several minutes. Carefully remove excess of agarose or gelatin and chill cell

pellet at 4°C. If gelatin is used, fix cell pellet with 4% paraformaldehyde and 0.25% glutaraldehyde for 20 min at 4°C. Afterward, wash cells with PBS several times and cut the tip of tube. Remove the embedded cells and cut them in small pieces (less than 1 mm³).

Cells cultured in monolayer: Release cells by cell scrapper. Then use the same proto- col as for cells in suspension.

5. Dehydrate and perform the infiltration, embedding and polymerization with em- bedding acrylic medium (*see* **Notes 5–8**) according to the manufacturer protocol.
6. Cut ultrathin sections.
7. Collect ultrathin sections of fixed material on plastic coated gold or nickel EM grids. Do not use copper grids. Copper reacts chemically with components of hybridiza- tion protocol and the hybridization solution turns blue.

3.2. Pretreatments of Material on the EM Grids

Pretreatments include proteolytic treatment, nuclease treatment, and denatu- ration steps. Perform this and the followings steps (the hybridization as well as the immunodetection) in a moist chamber with 2 to 20 μL drops of appropriate mixtures onto a sheet of Parafilm. Incubate the grids section-side down on top of the drop at RT (if not indicated otherwise) for given time.

3.2.1. Proteolytic Treatment

The aim of this step is to remove proteins, particularly those structurally associated with nucleic acids, to improve the ISH probe accessibility to tar- gets. This step is necessary if elimination of single-stranded DNA and RNA by nucleases is indicated (*see* **Subheading 3.2.2.**) because proteins could protect DNA and RNA sequences from nuclease digestion. To digest the proteins, incu- bate the sections with up to 100 μg/mL proteinase K for 5 to 30 min at 37°C. Optimal concentration and/or time of incubation depend on a given sample and must be determined empirically. Rinse on three drops of distilled water. Air dry.

3.2.2. Nuclease Treatment

Nuclease treatment is used to remove nucleic acids that could react with the ISH probe but not target sequences themselves. The RNase A digestion serves to remove single-stranded RNA sequences in the case of DNA sequences detec- tion. Enzyme concentration of 0.1 to 10 mg/mL is recommended. DNase I is used to degrade DNA sequences in the case of detection of RNA sequences. Concentration up to 1 mg/mL is recommended. To detect exclusively double- stranded DNA sequences, S1 nuclease digestion (1600 U/mL) is required to eliminate single-stranded RNA and DNA molecules. Incubate grids with sec- tions at drops of RNase, DNase or S1 nuclease usually for 60 min at 37°C. Then, wash at RT on three drops of distilled water. Air dry.

3.2.3. Denaturation

Denaturation of nucleic acids is the next important step that is required, particularly when ISH target is double-stranded DNA. In general, denaturation treatment may have a deleterious effect on morphology. Therefore, in practice, a compromise must be found between the level of hybridization signal and the level of preservation of morphology. Numerous denaturation approaches were implemented, including heating with or without presence of formamide and acid or alkali treatments. Among these treatments the incubation of sections with approx 0.5 *N* NaOH for several minutes seems to be most convenient for detection of double-stranded DNA in resin sections. To denature double-stranded DNA incubate grids with 0.5 *N* NaOH for 4 min. Rinse on three drops of distilled water and wash in jet of distilled water. Air dry.

3.3. Hybridization

The composition of the hybridization buffer must provide optimal conditions for the formation of probe-target hybrids. It usually contains formamide, dextran sulfate and blocking DNA or RNA molecules in SSC buffer. Formamide weakens hydrogen bonds between strands of nucleic acids and thus facilitates a work at a lower hybridization temperature. Dextran sulfate increases the hybridization speed. In aqueous solutions dextran sulfate is strongly hydrated. It results in an apparent increase in probe concentration and consequently in higher hybridization rate. Salmon sperm DNA or transfer (t)RNA represents common blocking nucleic acids that reduce nonspecific binding between ISH probes and thin sectioned sample. *E. coli* DNA is used instead of salmon sperm DNA if homology between target sequence and salmon sperm DNA is high. The detection of ribosomal genes represents such an example of high homology. tRNA is typically used in the case of oligonucleotide probes as it does not require probe denaturation step. SSC buffer contains monovalent ions that interact with nucleic acids and decrease the electrostatic repulsion between two strands of the nucleic acid duplex. The common hybridization solution consists of 50% formamide, 2X SSC, 10% dextran sulfate and 400 µg/mL blocking RNA or DNA. Although its actual composition can be altered to optimize hybridization for the sample under study, it is usually more convenient to change hybridization temperature and duration. The hybridization experiments require several controls (*see* **Notes 9–14**).

1. Prepare probe-containing hybridization mixture (2 to 10 µg/mL biotin or digoxigenin labeled probe in 50% formamide, 2X SSC, 10% dextran sulfate and 400 µg/mL RNA or DNA.
2. Heat the hybridization solution in order to denature probe if necessary (double stranded DNA probe or long RNA probes) at 92°C for 10 min or at 96°C for 5 min.

3. Chill on melting ice to prevent reassociation of the denatured molecules.
4. Place the grids on the droplets of cold hybridization solution.
5. Perform ISH in a wet chamber for 1 to 16 hr at 37°C to 65°C (*see* **Notes 11** and **12**).
6. Wash the hybridized sections 3 × 5 min with 2X SSC and 3 × 5 min with PBS.

3.4. Immunolabeling

Indirect methods using biotin- or digoxigenin-labeled probes detected by specific antibodies represent the most popular systems for detection of the hybridization reaction (*see* **Note 15**).

1. To block nonspecific bindings, incubate each grid with PBS containing 1% BSA (PBS-B) for 10 min.
2. Incubate with antibiotin or antidigoxigenin antibody diluted in PBS-B for 1 h (*see* **Note 16**).
3. Rinse the grids at drops of PBS-B for 3 × 5 min.
4. Apply a secondary antibody bound to gold particles in PBS-B for 1 h.
5. Wash in PBS-B for 3 × 5 min and in PBS 3 × 5 min.
6. Postfix the grids in 2% glutaraldehyde in PBS for 20 min.
7. Wash with distilled water 5 × 1 min.
8. Air dry and stain with 3% uranyl acetate in distilled water for 30 min.
9. Wash in water and dry.

4. Notes

1. The sensitivity of post-embedding EM-ISH is usually lower compared with light microscopy ISH performed on whole cells (*see* **Fig. 2**). In this respect, one should note that, even multicopy genes produce commonly scarce signal. The typical example represents detection of human ribosomal genes. They are localized in the specific chromosome regions as tandem repeats. Approximately 40 of gene copies are found in each such a region in human cells. Despite that the signal on section of the human cell is usually formed by a few gold particles. This should be kept in mind when considering the power of a postembedding EM-ISH approach. On the other hand, it represents a powerful tool for localization of multiple sequences such as viral DNA and RNA (*see* **Fig. 3**), or transcripts of highly active genes, all this in the context of highly preserved ultrastructural details.
2. When hesitating what ISH procedure (postembedding, pre-embedding or cryosections) represents the best choice for the tracking of the sequence of interest, determine what preservation of ultrastructure is needed and what kind of nucleic acid you are looking for. If fine preservation represents the critical step, try the postembedding technique first. It provides a convenient structural preservation of cellular content. This feature is still more important in the case of detecting of double stranded DNA sequences, as usually a harsh denaturation step is required. In the case of cryosections as well as pre-embedding approach, denaturation usually results more-or-less in an alteration of ultrastructural details, even in a compact

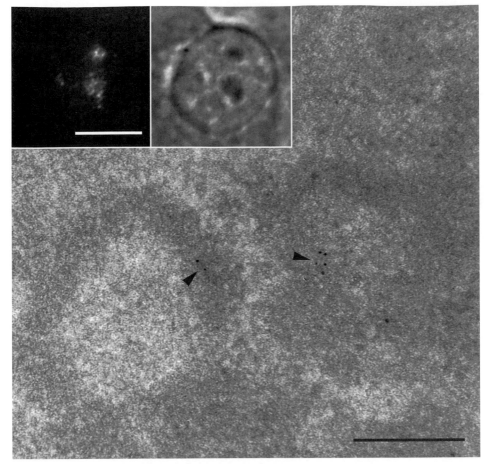

Fig. 2. *In situ* hybridization (ISH) mapping of ribosomal genes in HeLa cells using a deoxyribonucleic acid probe. Several gold particles (arrowheads) are located in specific nucleolar structures (dense fibrillar components) in the electron microscopy image. Compare with results of ISH at light microscopy level (inserts). Ribosomal genes are located in a number of fluorescent foci (left insert). The phase contrast image is shown in the right insert. Bars = (EM-ISH) 200 nm, (inserts) = 10 µm. Images are reproduced from Koberna et al. *(24)*, with the permission of the Rockefeller University Press.

compartment such as the nucleolus. However, both pre-embedding and cryosections generally can provide higher sensitivity.

3. To achieve a convenient fixation, several factors should be considered, particularly speed and temperature of fixation and pH and osmolarity of fixation solution. The speed of fixation has to be rapid, because fixation not only stabilizes the structures of cell but also stops possible autolytic processes. Therefore, the sample size should

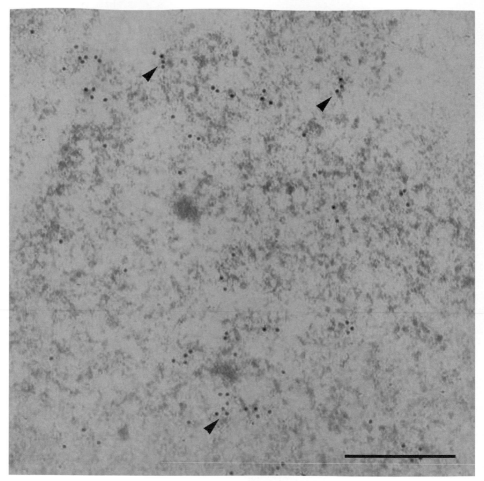

Fig. 3. Electron microscopic localization of HIV-1 RNA in a transfected COS cell. *In situ* hybridization signal for ribonucleic acid probe is revealed with colloidal-gold particles (arrowheads). The signal is confined to ribonucleoprotein containing fibrils located in nucleoplasm. Bar = 400 nm.

be as small as possible. In the case of animal models the fixation by perfusion is recommended. It represents convenient protocol with respect to high fixation speed. As this technique requires special equipment we provide here only protocol for much easier fixation by immersion. The penetration of fixatives is crucial for fixation efficiency. It generally increases with temperature elevation. On the other hand, the increased temperature can result into higher speed of autolytic enzyme activities. In this respect, fixation performed at 0°C to RT usually represents convenient compromise. pH represents another important parameter that has to be controlled.

Both acidic and basic pH could result into damage of cellular components including nucleic acids. That is why pH around neutral should be used. Osmolarity of the fixative should be equal to physiological osmotic pressure to prevent plasmolysis or cell swelling. The osmolarity is adjusted by various molecules, frequently by sodium chloride.

4. Among different crosslinking fixatives, formaldehyde and glutaraldehyde are most frequently used for EM-ISH. However, formaldehyde alone may not sufficiently preserve the ultrastructure of cells. Thus, mixture of 4% formaldehyde and up to 0.5% glutaraldehyde frequently is used. The formaldehyde has to be prepared fresh just before fixation through dissolution of powdered paraformaldehyde. Although the powdered paraformaldehyde is stable if stored at RT for a long time, the formaldehyde rapidly degrades after preparation. The EM quality glutaraldehyde is purchased as stock solution. It has to be kept at −20°C. The volume of fixative should largely exceed the volume of the specimen (at least ten times). Higher concentration of glutaraldehyde could result in the loss of sensitivity EM-ISH. In our experience it impairs preferentially detection of double stranded DNA. The detection of RNA seems to be less impaired.

5. Ethanol and methanol are commonly used for dehydration. Methanol is recommended for dehydration of samples where viral DNA is detected as ethanol dehydration could result in higher aggregation of viral DNA.

6. Several embedding acrylic resins convenient for EM-ISH are available. The most frequently used resins for *in situ* hybridization represent LR White and Lowicryl K4M. They originally were developed for protein detection but have been shown to enable the detection of nucleic acids equally well. They are polymerized by heating or ultraviolet light. The hydrophilic properties of a majority of these resins provide advantage that they may be polymerized with some contents of water. For example, as much as 5% (w/w) water can be present during polymerization in the block of material embedded in Lowicryl K4M. However, the shrinkage of sectioned material embedded in these resins is higher than in epoxy-resins. As acrylic resins exhibit low viscosity, they allow rapid tissue infiltration. The choice of resin depends rather on practice implemented in various laboratories than on well-defined rules. Generally, if the high temperature does not prevent the detection of nucleic acids, then resins polymerizing by heating (e.g., LR White) may be used, as the manipulation is easier than with resins polymerized under ultraviolet light (e.g., Lowicryl K4M).

7. All steps during Lowicryl K4M polymerization are commonly carried out at low temperatures, and polymerization is performed under ultraviolet light. Low-temperature embedding procedure usually is performed by the technique "progressive lowering of temperature." This technique is able to produce excellent morphology of nuclear fine structure, which may well approach its native state. However, Lowicryls are highly toxic, and prolonged or repeated exposure to Lowicryl K4M resin can result in strong allergic skin reaction.

8. LR White is a ready-to-use mixture of methacrylate and hardener. It must be stored at 0°C to 4°C. LR White may be polymerized by two procedures, by either a chemical

catalysts, or heat treatment. Polymerization at 4°C is carried out by the addition of the accelerator provided with the resin. Heat mediated polymerization takes place in the absence of accelerator.

9. Concerning the volume of drops of probe-containing hybridization mixture used for EM-ISH, it appears that the hybridization efficiency increases with the volume lowering of the drops. However, the others steps should be performed on the larger drops.

10. Prehybridization incubation often is necessary to lower background staining. The pre-hybridization solution contains all components of a hybridization mixture except that of the probe. It is performed at the same temperature as hybridization for 10 min.

11. A determination of temperature and time of hybridization represents the crucial step. Carry hybridization at different temperatures first. Duration should be approx 2 h. The temperature can vary substantially depending on the target, the probe, and the presence of formamide. For example, in the case of human ribosomal genes, double-stranded DNA probe and presence of 50% formamide in hybridization mixture, hybridization at 37°C for 70 min provides good result. Temperatures greater than 55°C and time approx 4 h seem to be optimal for the detection of ribosomal RNA. After optimal temperature determination, test different duration of hybridization step. Usually 1 to 6 h is sufficient time. Keep in mind that the prolongation of hybridization step could result in increasing of nonspecific signal.

12. The hybridization temperature could be estimated on the basis of melting temperature (T_m). T_m is the temperature at which 50% of double-stranded DNA is denatured. Usually, the hybridization temperature is 20 to 25°C less than T_m in the buffer without formamide. The hybridization temperature with formamide is 5°C less than T_m *(44)*. The melting temperature can be calculated for hybrids longer than 200 bp according to following equation (T_m): $T_m = 81.5 + 16.6 \log(Na+) + 0.41(\% \text{ G-C}) - 0.65 (\% \text{ F})$. Na^+ is the ionic concentration of sodium salt and F the percentage of formamide in the hybridization buffer.

13. Several control experiments must be performed to determine the hybridization specificity. As a negative control for oligonucleotide or RNA probes, samples can be hybridized with a negative probe prepared by substituting the sense probe for antisense probe. Any DNA sequence that is not present in the cell can serve as the control probe for DNA localization. Omitting of antibiotin or antidigoxigenin antibody represents common negative control of the immunodetection step. RNA (DNA) nature of ISH target should also be confirmed by RNase (DNase) digestion of sections prior the detection.

14. Perform light and EM-ISH simultaneously, at least during the initial EM-ISH assays. The failure of light microscopy ISH usually indicates that the result at EM level also will be negative, saving time.

15. The use of digoxigenin-labeled probes instead of biotin-labeled probes is recommended if the level of endogenous biotin in the sample is high. Both the liver and kidney represent typical examples of tissues containing large amounts of biotin.

16. In the case of biotin detection, avidin or streptavidin bound to gold particles can be also used.

Acknowledgments

We thank Drs. Ivan Raska, Jan Malinsky, and Helena Fidlerova for reading and helpful comments. This work was supported by grants from the Grant Agency of the Czech Republic 304/03/1121, 304/04/0692 and 304/05/0374, and from the Ministry of Education, Youth and Sports of the Czech Republic: MSM0021620806 and LC535 as well as a grant from Welcome Trust: 075834.

References

1. Pardue, M. L. and Gall, J. G. (1969) Molecular hybridization of radioactive DNA to the DNA of cytological preparation. *Proc. Natl. Acad. Sci. USA* **64,** 600–604.
2. John, H. A., Birnstiel, M. L., and Jones, K. W. (1969) RNA-DNA hybrids at the cytological level. *Nature (London)* **223,** 582–587.
3. Jacob, J., Todd, K., Birnstiel, M. L., and Bird, A. (1971) Molecular hybridization of 3H-labelled ribosomal RNA with DNA in ultrathin sections prepared for electron microscopy. *Biochim. Biophys. Acta* **228,** 761–766.
4. Croissant, O., Dauguet, C., Jeanteur, P., and Orth, G. (1972) Application of the molecular hybridization technic in situ for the demonstration with the electron microscope, of vegetative replication of viral DNA in the papillomas induced with Shope virus in cottontail rabbits. *CR. Acad. Sci. Hebd. Seances Acad. Sci. D.* **274,** 614–617.
5. Binder, M., Tourmente, S., Roth, J., Renaud, M., and Gehring, W. J. (1986) In situ hybridization at the electron microscope level: localization of transcripts on ultrathin sections of Lowicryl K4M-embedded tissue using biotinylated probes and protein A-gold complexes. *J. Cell. Biol.* **102,** 1646–1653.
6. McFadden, G. I., Bonig, I., Cornish, E. C., and Clarke, A. E. (1988) A simple fixation and embedding method for use in hybridization histochemistry on plant tissues. *Histochem. J.* **20,** 575–586.
7. Puvion-Dutilleul, F., Pichard, E., Laithier, M., and Puvion, E. (1989) Cytochemical study of the localization and organization of parental herpes simplex virus type I DNA during initial infection of the cell. *Eur. J. Cell Biol.* **50,** 187–200.
8. Le Guellec, D., Frappart, L., and Willems, R. (1990) Ultrastructural localization of fibronectin mRNA in chick embryo by in situ hybridization using 35S or biotin labeled cDNA probes. *Biol. Cell* **70,** 159–165.
9. Wachtler, F., Mosgoller, W., and Schwarzacher, H. G. (1990) Electron microscopic in situ hybridization and autoradiography: localization and transcription of rDNA in human lymphocyte nucleoli. *Exp. Cell Res.* **187,** 346–348.
10. Wenderoth, M. P. and Eisenberg, B. R. (1991) Ultrastructural distribution of myosin heavy chain mRNA in cardiac tissue: a comparison of frozen and LR White embedment. *J. Histochem. Cytochem.* **39,** 1025–1033.
11. Puvion-Dutilleul, F. and Puvion, E. (1991) Sites of transcription of adenovirus type 5 genomes in relation to early viral DNA replication in infected HeLa cells. A high resolution in situ hybridization and autoradiographical study. *Biol. Cell* **71,** 135–147.

12. Puvion-Dutilleul, F., Bachellerie, J. P., and Puvion, E. (1991) Nucleolar organization of HeLa cells as studied by in situ hybridization. *Chromosoma* **100,** 395–409.

13. Dorries, U., Bartsch, U., Nolte, C., Roth, J., and Schachner, M. (1993) Adaptation of a non-radioactive in situ hybridization method to electron microscopy: detection of tenascin mRNAs in mouse cerebellum with digoxigenin-labelled probes and gold-labelled antibodies. *Histochemistry* **99,** 251–262.

14. Olmedilla, A., Testillano, P. S., Vicente, O., Delseny, M., and Risueno, M. C. (1993) Ultrastructural rRNA localization in plant cell nucleoli. RNA/RNA in situ hybridization, autoradiography and cytochemistry. *J. Cell Sci.* **106,** 1333–1346.

15. Mosgoller, W., Schofer, C., Derenzini, M., Steiner, M., Maier, U., and Wachtler, F. (1993) Distribution of DNA in human Sertoli cell nucleoli. *J. Histochem. Cytochem.* **41,** 1487–1493.

16. Morey, A. L., Ferguson, D. J., Leslie, K. O., Taatjes, D. J., and Fleming, K. A. (1993) Intracellular localization of parvovirus B19 nucleic acid at the ultrastructural level by in situ hybridization with digoxigenin-labelled probes. *Histochem. J.* **25,** 421–429.

17. Lin, N. S., Chen, C. C., and Hsu, Y. H. (1993) Post-embedding in situ hybridization for localization of viral nucleic acid in ultra-thin sections. *J. Histochem. Cytochem.* **41,** 1513–1519.

18. Visa, N., Puvion-Dutilleul, F., Harper, F., Bachellerie, J. P., and Puvion, E. (1993) Intranuclear distribution of poly(A) RNA determined by electron microscope in situ hybridization. *Exp. Cell Res.* **208,** 19–34.

19. Visa, N., Puvion-Dutilleul, F., Bachellerie, J. P., and Puvion, E. (1993) Intranuclear distribution of U1 and U2 snRNAs visualized by high resolution in situ hybridization: revelation of a novel compartment containing U1 but not U2 snRNA in HeLa cells. *Eur. J. Cell Biol.* **60,** 308–321.

20. Raška, I., Dundr, M., Koberna, K., Melčák, I., Risueno, M. C., and Torok, I. (1995) Does the synthesis of ribosomal RNA take place within nucleolar fibrillar centers or dense fibrillar components? A critical appraisal. *J. Struct. Biol.* **114,** 1–22.

21. Morey, A. L., Ferguson, D. J., and Fleming, K. A. (1995) Combined immunocytochemistry and non-isotopic in situ hybridization for the ultrastructural investigation of human parvovirus B19 infection. *Histochem. J.* **27,** 46–53.

22. Pierron, G. and Puvion-Dutilleul, F. (1996) Localization of the newly initiated and processed ribosomal primary transcripts during the mitotic cycle in *Physarum polycephalum. Exp. Cell Res.* **229,** 407–420.

23. Melčák, I., Cermanová, S., Jirsová, K., Koberna, K., Malínský, J., and Raška, I. (2000) Nuclear pre-mRNA compartmentalization: trafficking of released transcripts to splicing factor reservoirs. *Mol. Biol. Cell.* **11,** 497–510.

24. Koberna, K., Malínský, J., Pliss, A., et al. (2002) Ribosomal genes in focus: new transcripts label the dense fibrillar components and form clusters indicative of "Christmas trees" in situ. *J. Cell Biol.* **157,** 743–748.

25. Cmarko, D., Boe, S. O., Scassellati, C., et al. (2002) Rev inhibition strongly affects intranuclear distribution of human immunodeficiency virus type 1 RNAs. *J. Virol.* **76,** 10473–10484.
26. Vazquez-Nin, G. H., Echeverria, O. M., Ortiz, R., et al. (2003) Fine structural cytochemical analysis of homologous chromosome recognition, alignment, and pairing in Guinea pig spermatogonia and spermatocytes. *Biol. Reprod.* **69,** 1362–1370.
27. Liebe, B., Alsheimer, M., Hoog, C., Benavente, R., and Scherthan, H. (2004) Telomere attachment, meiotic chromosome condensation, pairing, and bouquet stage duration are modified in spermatocytes lacking axial elements. *Mol. Biol. Cell.* **15,** 827–837.
28. Webster, H. F., Lamperth, L., Favilla, J. T., Lemke, G., Tesin, D., and Manuelidis, L. (1987) Use of a biotinylated probe and in situ hybridization for light and electron microscopic localization of Po mRNA in myelin-forming Schwann cells. *Histochemistry* **86,** 441–444.
29. Wolber, R. A., Beals, T. F., and Maassab, H. F. (1989) Ultrastructural localization of herpes simplex virus RNA by in situ hybridization. *J. Histochem. Cytochem.* **37,** 97–104.
30. Liposits, Z., Petersen, S. L., and Paull, W. K. (1991) Amplification of the in situ hybridization signal by silver postintensification: the biotin-dUTP-streptavidin-peroxidase diaminobenzidine-silver-gold detection system. *Histochemistry* **96,** 339–342.
31. Le Guellec, D., Trembleau, A., Pechoux, C., Gossard, F., and Morel, G. (1992) Ultrastructural non-radioactive in situ hybridization of GH mRNA in rat pituitary gland: pre-embedding vs ultra-thin frozen sections vs post-embedding. *J. Histochem. Cytochem.* **40,** 979–986.
32. Mitchell, V., Gambiez, A., and Beauvillain, J. C. (1993) Fine-structural localization of proenkephalin mRNAs in the hypothalamic magnocellular dorsal nucleus of the guinea pig: a comparison of radioisotopic and enzymatic in situ hybridization methods at the light- and electron-microscopic levels. *Cell Tissue Res.* **274,** 219–228.
33. Sibon, O. C., Humbel, B. M., De Graaf, A., Verkleij, A. J., and Cremers, F. F. (1994) Ultrastructural localization of epidermal growth factor (EGF)-receptor transcripts in the cell nucleus using pre-embedding in situ hybridization in combination with ultra-small gold probes and silver enhancement. *Histochemistry* **101,** 223–232.
34. Sibon, O. C., Cremers, F. F., Humbel, B. M., Boonstra, J., and Verkleij, A. J. (1995) Localization of nuclear RNA by pre- and post-embedding in situ hybridization using different gold probes. *Histochem. J.* **27,** 35–45.
35. Gonzalez-Melendi, P. and Shaw, P. (2002) 3D gold in situ labelling in the EM. *Plant J.* **29,** 237–243.
36. Tang, M., Owens, K., Pietri, R., Zhu, X. R., McVeigh, R., and Ghosh, B. K. (1989) Cloning of the crystalline cell wall protein gene of Bacillus licheniformis NM 105. *J. Bacteriol.* **171,** 6637–6648.

37. Morel, G., Chabot, J. G., Gossard, F., and Heisler, S. (1989) Is atrial natriuretic peptide synthesized and internalized by gonadotrophs? *Endocrinology* **124,** 1703–1710.
38. Dirks, R. W., Van Dorp, A. G., Van Minnen, J., Fransen, J. A., Van der Ploeg, M., and Raap, A. K. (1992) Electron microscopic detection of RNA sequences by non-radioactive in situ hybridization in the mollusk *Lymnaea stagnalis*. *J. Histochem. Cytochem.* **40,** 1647–1657.
39. Yao, C. H., Kitazawa, S., Fujimori, T., and Maeda, S. (1993) In situ hybridization at the electron microscopic level using a bromodeoxyuridine labeled DNA probe. *Biotech. Histochem.* **68,** 169–174.
40. Mandry, P., Murray, B. A., Rieke, L., Becke, H., and Hofler, H. (1993) Post-embedding ultrastructural in situ hybridization on ultrathin cryosections and LR white resin sections. *Ultrastruct. Pathol.* **17,** 185–194.
41. Li, K., Smagula, C. S., Parsons, W. J., et al. (1994) Subcellular partitioning of MRP RNA assessed by ultrastructural and biochemical analysis. *J. Cell Biol.* **124,** 871–882.
42. Takizawa, T. and Robinson, J. M. (1994) Use of 1.4-nm immunogold particles for immunocytochemistry on ultra-thin cryosections. *J. Histochem. Cytochem.* **42,** 1615–1623.
43. Zalokar, M. and Erk, I. (1977) Phase-partition fixation and staining of *Drosophila* eggs. *Stain. Technol.* **52,** 89–95.
44. Le Guellec, D. (1998) Ultrastructural in situ hybridization: a review of technical aspects. *Biol. Cell.* **90,** 297–306.

12

Correlative Light and Electron Microscopy Using Immunolabeled Resin Sections

Heinz Schwarz and Bruno M. Humbel

Summary

In correlative microscopy, light microscopy provides the overview and orientation in the complex cells and tissue, whereas electron microscopy offers the detailed localization and correlation to subcellular structures. In this chapter, we offer the detailed high-quality electron microscopical preparation methods for the optimum preservation of the cellular ultrastructure. From such preparations, serial thin sections are collected and used for comparative histochemical, immunofluorescence, and immunogold staining. In light microscopy, histological stains are used to identify the orientation of the sample, and immunofluorescence labeling is used to identify the region of interest, namely, the labeled cells expressing the macromolecule under investigation. Subsequent sections, labeled with immunogold, are analyzed by electron microscopy to identify the label within the cellular architecture at high resolution.

Key Words: Immunolabeling; immunofluorescence; immunogold; light microscopy; transmission electron microscopy; resin embedding; methacrylate; epoxy resin; correlative microscopy.

1. Introduction

Electron microscopy in combination with immunolabeling allows the localization of macromolecules at high resolution in the structural context of tissue and cells. The label can directly be bound to the antigen-containing structure. If one starts with a new antigen, this approach is definitely not the method one should use first. It can be tedious or even impossible to find the region of interest in the three-dimensional structure within the tissue by electron microscopy.

Light microscopy, on the other hand, offers the advantages of rapid screening a large area of a specimen, of determining the specificity of labeling, and of simultaneously detecting multiple antigens with ease. In bright-field or phase-

From: *Methods in Molecular Biology, vol. 369*
Electron Microscopy: Methods and Protocols, Second Edition
Edited by: J. Kuo © Humana Press Inc., Totowa, NJ

contrast light microscopy, the majority of cellular organelles cannot be directly identified because of the low-resolution power and the discernment of the color reaction products of the label. In epi-fluorescence microscopy, the signal of the label is much higher because of the dark-field imaging mode. Unfortunately, there are no reference structures visible in epi-fluorescence microscopy, so the correlation between labeled antigen and a counterstain is indispensable *(1)*, e.g., 4',6-diamidine-2-phenylindole (DAPI) to visualize the deoxyribonucleic acid (DNA), or additional labeling for assumed compartments. Still, even using confocal microscopy, colocalization cannot be unequivocally proven as a result of the poor resolution along the z-axis. Furthermore, the point of focus of the different wavelengths, for example, blue from DAPI and green from fluorescein isothiocyanate (FITC), is not at the same z-height because of the different refraction indices.

Thus, ideally one should combine the advantages of both light and electron microscopic imaging using the identical specimen. The ultimate approach would be live-cell observation and selection of the region of interest in the light microscope and subsequent high-resolution imaging of the fluorescent signal of the identical region in frozen-hydrated sections in a cryoelectron microscope *(2,3)*. Although new preparation techniques and the instrumental developments are emerging, we still prefer to exploit the advantages of already-existing advanced electron microscopy preparation techniques for correlative light and electron microscopic imaging.

An intermediate step in this direction might be the efficient conversion of a fluorescent signal in a living cell into an electron dense signal for subsequent characterization on electron microscopic level *(4,5)*. One of the drawbacks of the photoconversion method is the low contrast and the rather poor spatial resolution of deposited reaction product that may in addition obscure the structure of interest. Furthermore, although a region of interest is easy to find by light microscopy, the identical region is intricate to spot by electron microscopy. This problem is particularly troublesome if the signal is not both unequivocally identifiable and associated with a well-defined structure; consequently, one needs to collect a lot of serial sections and to screen them to find the label.

The preparation schemes of immunolabeling for light microscopy compromise the cellular ultrastructure, cause the loss of components such as lipids and nonstructural proteins, and limit the admission of the detecting antibodies through steric hindrance *(6–9)*. For pre-embedding labeling, the samples have to be opened to provide access to the intracellular antigens either mechanically by freeze-thaw methods or chemically with solvents or detergents.

In the past three decades, we have learned a lot about preservation of the cellular architecture for high-resolution electron microscopy, which raises the possibility of applying this knowledge to light microscopy. The aim of this report is to show the great potential of correlating light and electron microscopy data (**Figs. 1** and **2**), which has developed from improved specimen preparation techniques in combination with postembedding labeling for both immunofluorescence and immunogold (*see* **Note 1**).

The key observation in postembedding labeling has been the fact that antibody binding is mainly restricted to the surface of a section of properly fixed material *(10,11)*. Hence, there is no need to cut thick or semithin sections for on-section immunolabeling, because the signal would be similar to that of ultrathin sections (*see* **Note 2**). Particularly with resin-embedded material, the potential of ultrathin sectioning can be exploited to provide adjacent serial sections for light and electron microscopy *(12–14)*. The use of the identical specimen preparation via thin sections for both light and electron microscopy offers the possibility to examine the same sample at different orders of magnitude. Thus, identical sections of a specimen can be used to give an overview of labeling distribution in a total organ by fluorescence microscopy and to define precisely the fine details at the level of cellular compartments by electron microscopy. In principal, this approach is suited to Tokuyasu cryosections *(15,16)* and all methacrylate and epoxy resins, as long as antigenicity is retained.

The main advantages of Tokuyasu cryosections are twofold: the samples are not dehydrated by solvents—hence, the proteins remain in their natural aqueous environment until immunolabeling—and because of the open surface structure, more antigens are accessible than in resin sections (*see* **Note 3**). Moreover, the preparation procedure is fast and the results are obtainable within one day (*see* Chapter 13 in this volume).

Embedding in resins such as crosslinked methacrylates (Lowicryl, London Resins, Unicryl *[17–21]*) or epoxies (Epon, Araldite, Spurr's *[22–24]*) allow wet-sectioning to produce large (millimeter sized) and ultra-thin sections (≥50-nm thickness) with excellent structural preservation. Furthermore, established histological staining procedures, e.g., toluidine blue staining *(25)*, can be applied and are advantageous to differentiate between the structures within the object. Last, but not least, large collections of resin-embedded specimens can easily be stored at room temperature, requiring only a small space.

Epoxy monomers crosslink with the cellular components during polymerization; hence, they form a copolymer of resin and molecules in the biological specimen *(26)*. Thus, biological material embedded in epoxy resins is easy to section. Although this method provides excellent morphology, antigenicity often is affected. In contrast, methacrylates cures by a radical chain reaction, allowing only the monomers to polymerize *(26,27)* so that the biological components are enmeshed in a network of resin polymers. Hence, the more open structure and its higher degree of hydrophilicity than Epon provides better labeling properties.

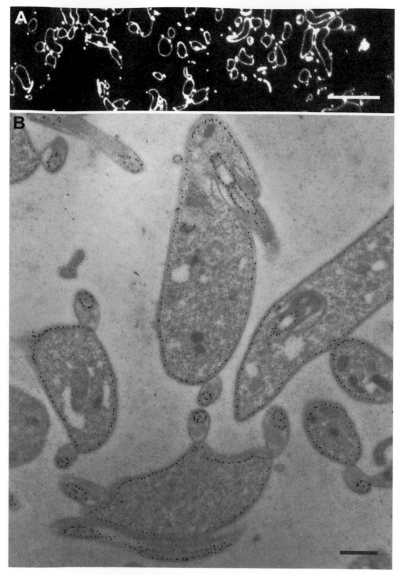

Fig. 1. Specificity and fidelity of on section immunolabeling demonstrated for a structurally visible antigen, the microtubuli cortex of *Trypanosoma brucei.* Ultrathin sections of the unicellular parasite *T. brucei,* chemically fixed, dehydrated by the progressive lowering of temperature and embedded in Lowicryl HM20. The sections are labeled for α-tubulin and detected with Cy3-tagged (**A**) or 12 nm gold particles-tagged (**B**) secondary antibodies. The fluorescence micrograph (**A**) obtained with a 100X objective (numerical aperture of 1.3) demonstrates the high resolution that can be achieved by light microscopy. Even cross-sectioned flagella are visible. The microtubuli are located directly underneath the plasma membrane and as typical 9 plus 2 structure in the flagella (**B**). Bars = 5 μm and 0.5 μm respectively.

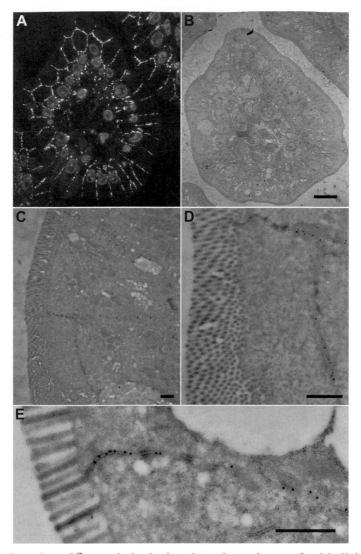

Fig. 2. Location of β-catenin in the basolateral membrane of epithelial cells in rat intestine. The intestine was immersion fixed with formaldehyde, dehydrated by progressive lowering of temperature and embedded in Lowicryl K4M. Cross section of a villus labeled for β-catenin detected with Cy3 and counterstained with DAPI (**A**). Images were taken with a 63X objective (numerical aperture of 1.4). The β-catenin is distributed in patches along the basolateral membranes of the epithelial cells. The electron micrograph (**B**), at the same magnification as the light micrograph, shows the overview structure of a cross-sectioned villus and the delineation of the individual cells. A portion of the left middle region of **B** is enlarged (**C**) and even more in (**D**). At higher magnification (**D,E**) it becomes evident that β-catenin is located mainly in the adherence junctions of the basolateral membrane but not in the desmosomes. Bars = 10 μm (A,B) and 1 μm (C–E), respectively.

Resin embedding can be combined with a large variation of fixation and dehydration procedures. Fixation can be performed chemically at ambient temperature or by cryofixation. Dehydration can be performed through a graded series of solvents at ambient temperature, by progressive lowering of temperature (PLT *[18]*) or by freeze-substitution *(28)*.

The gold standard of preparation methods is cryofixation followed by freeze-substitution and resin embedding *(28–31)*. The cryofixation method of choice is high-pressure freezing (*[32–35]*, *see* also Chapter 8 in this volume). To quote Martin Müller *(36)*, "Even the best microscopes and the most sophisticated image analysis algorithms are unable to provide high density information from samples distorted during the sampling step (i.e., specimen preparation)." Therefore, improved specimen preparation using cryotechniques in combination with postembedding labeling for both immunofluorescence and immunogold is a promising approach to bridge the gap between light and electron microscopy *(13)*.

An intrinsic property of high-resolution objectives with high numeric apertures of light microscopes is the small depth of focus (*see* **Note 4**). This results in out-of-focus blurring of the image and loss of resolution in both bright field and phase contrast illumination settings as well as in fluorescence light microscopy. This is because biological objects, even cell monolayers, are much thicker. Therefore, only thin sections not thicker than the depth of focus of objectives with a high numerical aperture (0.5–1 μm) can exploit the much higher resolution. Thus, for correlative microscopy first the region of interest is approached with the help of toluidine blue stained thin sections (**Figs. 3A** and **5A**). Second, the signal is localized on ultra-thin sections (50–100 nm) with fluorescence microscopy (**Figs. 2A,3B,3C,5B,6**) and then at high resolution with electron microscopy (**Figs. 2B–E,4**).

Fig. 3. *(Opposite page)* Distribution of prolactin in the pituitary gland of a 7-d-old zebrafish larva. The fish larvae were immersion fixed with formaldehyde, dehydrated by PLT and embedded in Lowicryl K11M. A 500-nm thick section was stained with toluidine blue (**A**), and images were taken with a 100X objective (numerical aperture of 1.3). The nuclei of the brain are clearly visible, whereas the nervous tissue is only faintly stained. The pituitary gland is directly located under the brain above esophagus. For better orientation the nuclei in an ultrathin section are counter-stained with DAPI (**B**). An ultrathin section (**C**) was labeled for prolactin and detected with Cy3. In the whole area of the tissue only a small number of cells in the anterior part of the adenohypophysis expressed prolactin *(45)*. This micrograph illustrates the difficulties to find these few cells in the electron microscope without an overview reference. Longitudinal sections, anterior to the left, dorsal to the top. Bar = 10 μm.

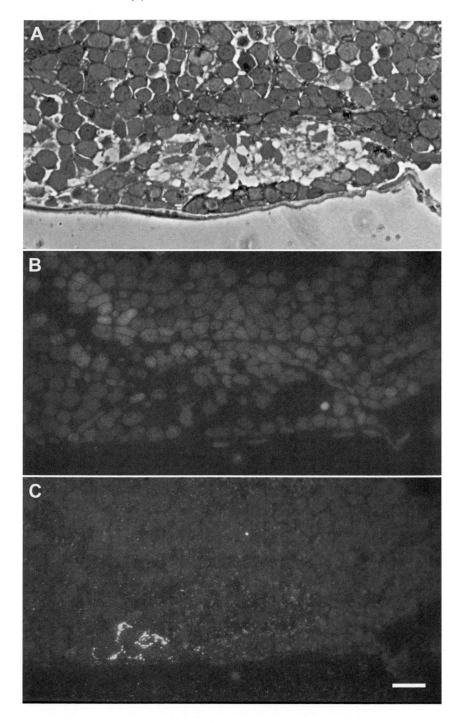

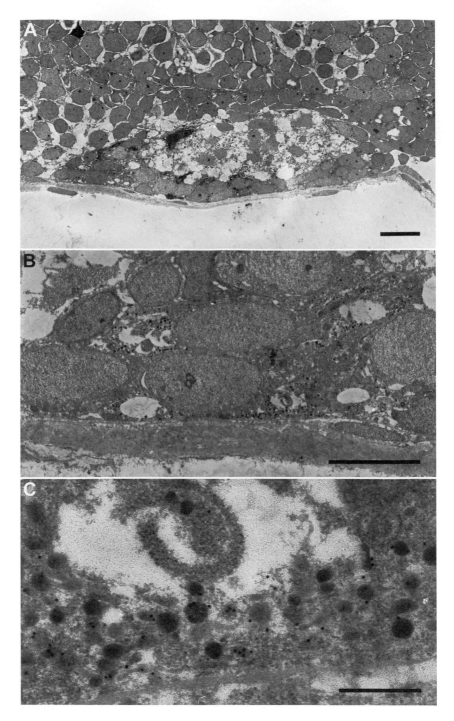

The widespread and successful application of confocal light microscopy has pushed tremendous development of a great variety of stable fluorochromes and detection systems, e.g., more precise optical filters (*see* **Note 5**). Because of the small size of the fluorochromes (a few hundred Daltons), they influence the binding properties of antibodies less than other marker systems *(8)*. Generally, if labeling with fluorescence markers fails, there will be no chance to detect the antigen with gold probes. Hermann et al. *(37)* have demonstrated the dramatic influence of even a very small gold particle (1 nm). They showed that Fab fragments directly coupled to ultra-small gold particles, detect only one copy of a homotrimer of proteins at distinct sites of the tail fibers of a T even phage, whereas, uncoupled Fab fragments detect two copies. Nevertheless, the smaller the size of the gold particle, the higher is the density of gold label *(8,38)*.

To permit optimum correlation of the results obtained by on-section labeling, the labeling conditions for immunofluorescence and immunogold have to be kept identical. The potency of this approach has been demonstrated in a series of publications *(13,39–45)*.

2. Materials

1. Poly-L-lysine stock solution. Prepare a stock solution of poly-L-lysine (Sigma, P 1274, MW 70,000–150,000 Da) at 2 mg/mL (w/v) in bidistilled water, or order a 0.1% stock solution (Sigma, P 8920). Aliquots can be stored at −20°C.
2. Methacrylates. All work has to be conducted under a fume hood. Use appropriate gloves whenever handling methacrylates because they can cause allergic reactions *(46)*. Oxygen is an inhibitor of the polymerization; therefore, avoid vigorous agitation of the methacrylates and use airtight tubes, for instance, 0.5-mL tubes (Sarstedt, cat. no. 72.699) for polymerization.
3. Lowicryl stock solutions. Weigh substances under fume hood using screw-cap glass vials (e.g., Wheaton glassware for scintillation counting). Mix until initiator has dissolved completely and store at −20°C. The mixture is stable at −20°C for several months.

Fig. 4. *(Opposite page)* Distribution of prolactin in the pituitary gland of a 7-d-old zebrafish larva visualized by electron microscopy. The fish larvae were immersion fixed with formaldehyde, dehydrated by the progressive lowering of temperature and embedded in Lowicryl K11M. Subsequent sections of the same sample as depicted in **Fig. 3** were labeled with 15 nm of colloidal gold. An overview **(A)** taken at approximately the same magnification as **Fig. 3**. Comparing the fluorescence image and this figure the labeled area can be found and slowly approached at higher magnifications **(B)**. Only at high magnification the subcellular structures, the prolactin containing vesicles the can be identified **(C *[45]*)**. Bars = 10 μm **(A)**, 5 μm **(B)** and 0.5 μm **(C)**, respectively.

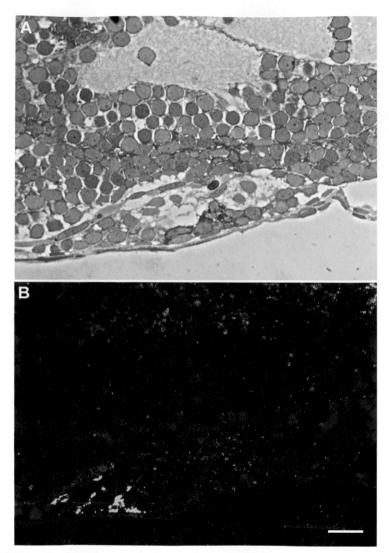

Fig. 5. Distribution of prolactin in the pituitary gland of a 7-d-old zebra fish larva. The fish larvae were immersion fixed with formaldehyde, dehydrated by the progressive lowering of temperature, and embedded in Lowicryl K11M. A 500-nm thick section was stained with toluidine blue (**A**) and images were taken with a 100X objective (numerical aperture of 1.3) in phase contrast imaging mode. The nuclei of the brain are clearly stained in blue, whereas the nervous tissue is only faintly stained. The pituitary gland is directly located under the brain above esophagus. An ultrathin section (**B**) was labeled for prolactin, detected with Cy3, and counterstained with DAPI. In the whole area of the tissue only a small number of cells in the anterior part of the adenohypophysis expressed prolactin *(45)*. Longitudinal sections, anterior to the left, dorsal to the top. Bar = 10 μm.

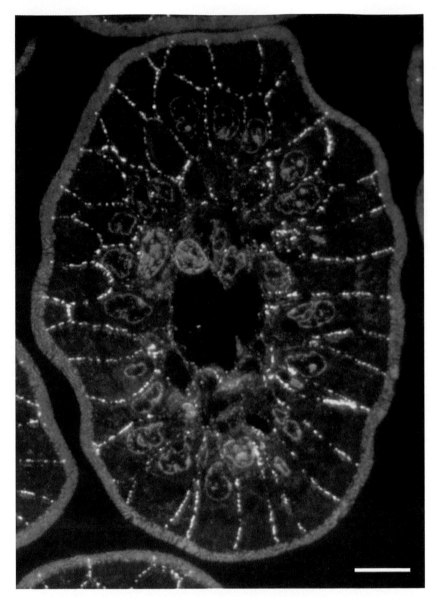

Fig. 6. Location of β-catenin and F-actin in rat intestine. The intestine was immersion fixed with formaldehyde, dehydrated by the progressive lowering of temperature, and embedded in Lowicryl K4M. Shown is a cross section of a villus labeled for β-catenin with Cy3 (yellow), F-actin with fluorescein isothiocyanate (green), and counterstained with propidium iodide (orange). Images were taken with a 63X objective (numerical aperture of 1.4). The β-catenin is distributed in patches along the basolateral membranes of the epithelial cells. The F-actin is predominantly located in the microvilli of the brush border. Bar = 10 μm.

Lowicryl K4M: 3.0 g of crosslinker A (Fluka, cat. no. 62639 or Polysciences, cat. no. 15923-1); 19.0 g of Monomer B; 110 mg of Initiator C: benzoin methyl ether (catalyst).

Lowicryl HM20: 3.3 g of crosslinker D (Fluka, cat. no. 62636 or Polysciences, cat. no. 15924-1); 18.7 g Monomer E; 110 mg Initiator C: benzoin methyl ether (catalyst).

Lowicryl K11M: 1.1 g of crosslinker H (Fluka, cat. no. 62640 or Polysciences, cat. no. 18163-1); 20.9 g of Monomer I; 110 mg of Initiator C: benzoin methyl ether (catalyst).

LR White and LR Gold are ready to use embedding media and can be purchased by any supplier for electron microscopy.

4. Perfect loop (Electron Microscopy Sciences, Fort Washington, PA, cat. no. 70944).
5. Toluidine Blue O solution. Dissolve 0.5% toluidine blue O [C.I.52040] (Serva, cat. no. 36693.02; Fluka, cat. no. 89640; Sigma, cat. no. T 0394) in an aqueous solution of 1% sodium tetraborate, pH 9.2 (Merck, cat. no. 106308; *see* **Note 6**), filter and store as aliquots, e.g., in microtubes. Centrifuge before use.
6. Coverslips, 12-mm diameter, 0.17-mm thickness (Electron Microscopy Sciences, Fort Washington, PA, cat. no. 72231).
7. PAP-Pen (Electron Microscopy Sciences, Fort Washington, PA, cat. no. 71310).
8. Counter-stains for nucleic acids: DAPI: 4',6-diamidine-2-phenylindole; for DNA staining; Hoechst 33342 and Hoechst 33258: Bisbenzimide, 2-(4-ethoxyphenyl)-5-(4-methyl-1-piperazinyl)-2,5-bi-[1]H-benzimidazole,trihydrochloride (for DNA staining); propidium iodide; for DNA and ribonucleic acid (RNA) staining.
9. Mounting media for fluorescence microscopy: Mowiol 4-88 from Hoechst (*see* Beug et al. *[47]*; Calbiochem, cat. no. 475904), or Elvanol (DuPont), or Poly-vinyl alcohol (Sigma, cat. no. P 8136, MW 30,000-70,000 Da). Commercially ready-to-use medium: Vectashield (Vector Laboratories, cat. no. H-1000).
10. Antifading agents *(48)*:1,4-Diazabicyclo[2.2.2.]octane (DABCO *[49]*); p-phenylenediamine (PPD *[50]*); n-propyl gallate (3,4,5-Trihydroxybenzoic acid n-propyl ester *[51]*).
11. Label buffer: 0.2% gelatin, 0.5% BSA, Fraction V, (PBG *[52]*); or 0.5% to 1% coldwater fish skin gelatin (Sigma, G 7765 *[53]*), 0.5% BSA; or 1% skim milk powder; or 0.1% BSAc (Aurion).
12. The proteins (*see* **Note 7**) are diluted in 1X PBS or 50 mM Tris-HCl buffer at pH 7.4.

3. Methods

3.1. Fixation

3.1.1. Chemical Fixation at Ambient Temperature

The biological specimen is fixed as close to living conditions as possible by adjusting the pH, osmolarity and temperature to those of the living conditions (*see* **Note 8**). To fix tissue of mammals, perfusion fixation is first choice. For cells or small organisms growing in medium, the addition of an equal volume of double strength fixative is recommended. For additional information, *see* Chapters 1 and 2 in this volume).

For immunolabeling, aldehydes are used as fixative. The use of osmium tetroxide is strongly discouraged (*see* **Note 9**). The typical fixation protocols applied in light microscopy (such as cold acetone, or methanol, or aldehyde in combination with permeabilization with detergents) must be avoided. They have a deleterious effect on the cellular ultrastructure, and antigens can be relocated or even extracted *(7,9)*.

Our standard fixation protocol is 2% to 4% formaldehyde or a mixture of 2% to 4% formaldehyde with 0.02% to 0.2% glutaraldehyde in HEPES or PIPES (for a comprehensive compilation *see* Griffiths *[54]*, pp 57–63) for 1 to 16 h. If necessary, aldehyde-fixed samples may be stored in 1% to 4% formaldehyde at 4°C.

3.1.2. Cryofixation

Cryofixation or cryoimmobilization is a fast physical fixation procedure that arrests cellular processes within milliseconds, thereby avoiding the typical artifacts introduced by the relatively slow diffusion-dependent chemical fixation process (i.e., minutes *[36,55–58]*). Cryofixation can be achieved in many different ways: plunging *(59,60)*, jet-freezing *(61)*, metal-mirror or slamming *(62)*, spray-freezing *(63)*, or high-pressure freezing *(32,34,64)*. In cell biology, after a long time of stagnation, high-pressure freezing has become the method of choice, because much thicker samples can be processed. For review of cryopreparation techniques, refer to (*[65–67]*, *see* also Chapter 8).

3.2. Dehydration

3.2.1. Dehydration at Ambient Temperature

Dehydration at ambient temperature is traditionally done in a graded series of increasing ethanol concentrations *(68)*. Organic solvents dissolve and extract lipids as well as denature proteins in a polarity and temperature dependent manner. Therefore, dehydration methods performed at lower temperatures, such as PLT and freeze-substitution, are preferred.

3.2.2. Ethanol Dehydration by the PLT Method and Low-Temperature Embedding

The method is as follows *(18,69)*: chemically fixed specimens are dehydrated and embedded into methacrylates according to the scheme given for Lowicryls (**Table 1**).

Either of the Lowicryl mixtures K4M, K11M, or HM20 (also the other methacrylates, LR Gold and LR White) can be used for embedding. Note that the

Table 1

Solvents/Resins	Temperature, °C	Time
30% Ethanol	0°	30 min
50% Ethanol	−20°	60 min
70% Ethanol	−20°	60 min
95% Ethanol	−35°	60 min
100% Ethanol	−35°	60 min
100% Ethanol	−35°	60 min
Ethanol/Lowicryl 1:1	−35°	60 min
Ethanol/Lowicryl 1:2	−35°	60 min
100% Lowicryl	−35°	60 min to overnight
100% Lowicryl	−35°	approx. 6 h
Embedding in Lowicryl and ultraviolet polymerization	−35°	48 h

final temperature needs to be adapted to the physical and chemical properties of the resins used (*see* **Subheading 3.3.1.1.** and **Note 10**).

3.2.3. Dehydration by Freeze-Substitution

Cryofixed tissue and cells are dehydrated by freeze-substitution for subsequent embedding into resin (*[28–31]; see* also Chapter 8 in this volume). Thus, the superior morphological preservation of cryofixation can be combined with the advantages of resin sections, for instance the ease of cutting and the possibility for on-section immunolabeling.

During the freeze-substitution process, the biological samples are dehydrated by dissolving the ice and, in addition, they can be chemically fixed. Here, the negative effects of dehydration, such as loss of lipids and proteins (*30,70–72*), are reduced greatly compared with dehydration at room temperature or by PLT.

The advantages of dehydration by freeze-substitution also may be exploited for chemically fixed specimens (*43,73,74*). After chemical fixation, the specimens are cryoprotected and frozen. This approach is particularly suited for infectious material or specimens that cannot be cryofixed by high-pressure freezing, because they are too thick or samples have to be taken in the field.

The most frequently used solvents for freeze-substitution are acetone (*28, 29*), methanol (*31,75*), or ethanol (*43,76*). They have to be liquid at −90°C (i.e., less than the recrystallization point of water in biological systems), and they have to dissolve the ice in due time, that is, within hours or a few days (*30,31*).

In principle, freeze-substitution and resin embedding can be done without additional chemical fixation (*77,78*); however, proteins can be lost during wet-sectioning (*73*). Therefore, it is suggested to add uranyl acetate and/or aldehydes to the substitution medium (*see* **Note 11**).

3.3. Resin Embedding

Some components of the resin formulations are hazardous, carcinogenic, allergenic, or toxic *(46)*. Please, follow the instructions of the manufacturer carefully. We suggest always working with appropriate gloves in a well functioning fume hood.

3.3.1. Methacrylates

Acetone is a radical scavenger and should not be used for dehydration in combination with methacrylates. Ethanol is needed as an intermediate step before resin infiltration. Polymerization of methacrylates is inhibited by oxygen; therefore, the samples must be polymerized in tightly closed tubes or in an inert gas atmosphere (*see* **Notes 10** and **12**).

3.3.1.1. LOWICRYLS

The final temperature for infiltration and polymerization is dependent on the physical and chemical properties of the Lowicryls, K4M $-35°C$, K11M $-60°C$, HM20 $-50°C$, and HM23 $-80°C$ (*[18,69]; see* manufacturer's instructions).

1. Transfer samples with a pipette into small (0.5-mL) test tubes and add fresh Lowicryl.
2. Close the tubes airtight and polymerize for 24 h at $-35°C$ using an ultraviolet (UV) lamp (366 nm).
3. The racks of tubes are placed in a stainless-steel tray filled with ethanol. This ethanol bath acts as a heat sink.
4. Cover the tubes with aluminum foil in order to obtain an indirect irradiation of the UV light to avoid irregular polymerization (*see* **Note 13**).
5. After 24 h, remove the aluminum foil and continue UV polymerization for 24 to 48 h.
6. For final polymerization, expose the tubes to sunlight (outdoors) or use a UV box under a fume hood at room temperature. Open the tubes to allow any remaining volatile Lowicryl components to escape so that the samples no longer smell like Lowicryl.

3.3.1.2. LR WHITE AND LR GOLD

Dehydrated samples can be directly infiltrated with pure resin with four to six changes during a 3-h period at room temperature. We suggest first to infiltrate in a 1:1 mixture of LR White: ethanol and then to change with pure resin several times at room temperature (*see* **Note 14**).

LR White can be polymerized either by heating between 50°C and 60°C (*see* **Note 15**) or in the cold by exposing to UV light. For UV-light polymerization (not less than $-15°C$), 0.5% (w/v) benzoin methylether (initiator C of the Lowicryls) must be added. LR Gold can be polymerized at $-25°C$ with UV light. The sectioning and labeling properties are very similar to those of the Lowicryls.

3.3.2. Epoxy Resins

Epoxy resins react covalently with the biological material antigens, so they may get masked; therefore, epoxy resins are not the best suited resins for immunolabeling. On the other hand, epoxy resins are far more stable and preserve the cellular ultrastructure better than the methacrylates. Practice shows that carbohydrates and other non-protein components are less affected by epoxy resins and can easily be detected in epoxy sections, e.g., lectin labeling (*[79, 80]; see* **Note 16**).

3.4. Coating of Microscope Slides and Coverslips

3.4.1. Gelatin-Chrome Alum Coating
of Microscope Slides for Toluidine Staining

Please also consult Romeis *(81)*.

1. Clean microscope slides and put them in a slide holder.
2. Dip the loaded slide holder in gelatin solution containing 5 to 10 g of gelatin power, 0.25 g of chromium potassium sulfate (Merck, cat. no. 101036), and 500 mL of bi-distilled water.
3. Warm the gelatin solution up to 50°C with continuous stirring for 8 to 10 min, then filtrate the solution.
4. Let the slides dry and store in a dust-free box.

3.4.2. Coating of Coverslips or Slides
to Mount Sections for Light Microscopy

The preferred coatings for immunofluorescence microscopy *(82)* are poly-lysine or alcian blue (*see* **Note 17**). We routinely use poly-L-lysine with a molecular weight between 70,000 and 150,000 Da.

1. Clean round glass coverslips or microscopy slides with either ethanol, or 1% hydro-chloric acid in 70% ethanol, or with soap.
2. Rinse with bidistilled water.
3. Dry coverslips or slides separately on a filter paper in a large dish.
4. Prepare a solution containing 0.2 mg/mL (w/v) poly-L-lysine in bidistilled water or use the commercially available 1% stock solution. Place a 30- to 50-μL drop in the center of the coverslip or cover the microscope slide with the solution.
5. Allow the coating to settle for at least 30 min in a moist chamber.
6. Rinse briefly with bidistilled water and dry face up on a filter paper for 1 h at 60°C or overnight at ambient temperature.
7. Coverslips can be coated in large batches and stored indefinitely in dust-free dishes at room temperature.

3.5. Transfer and Mounting

Sections are collected from the water trough of the knife either with a perfect loop or an uncoated electron microscopy grid and transferred onto a dry polylysine-coated coverslip or microscopy slide (*see* **Note 18**). Residual water is drained with a filter paper along the outside of the loop or grid.

3.6. Toluidine Blue O Staining

1. For toluidine blue staining *(25)*, 0.5- to 5-μm plastic sections are put on a droplet of 15% (v/v) ethanol in water on a coated slide, e.g., SuperFrost Plus, chromium gelatin, or poly-L-lysine.
2. They are dried on a hot plate (60–80°C) and immediately incubated with toluidine blue for 30 to 60 s.
3. The sections are destained using a stream of (tap) water.
4. For high-resolution light microscopy and long-term storage, the stained sections can be embedded in a thin layer of Epon under a coverslip (*see* **Notes 19** and **20**).

3.7. Immunolabeling

3.7.1. Immunofluorescence on (Ultra)thin Resin- or Cryosections

1. All incubations *(13,39)* are performed in a humid chamber on parafilm, and care must be taken to prevent the sections from drying during the whole labeling procedure.
2. Encircle sections (on coverslips) with a water-repellent silicon pen.
3. Place the coverslips on parafilm in a humid chamber with sections facing up.
4. Put a drop of approx. 25 μL of blocking buffer (*see* **Note 21**), for instance, PBG, on top of the section and incubate for 10 min.
5. Remove blocking buffer (*see* **Note 22**), add 25 μL of the primary antibody solution (*see* **Note 23**) per coverslip, and incubate for 30 to 60 min.
6. Wash five times with buffer and incubate with fluorochrome-labeled secondary antibodies, similar to the primary antibody staining conditions.
7. Wash five times with buffer and then counter-stain nuclei for 5 min either with DAPI, Hoechst, or propidium iodide (0.4–0.1 μg/mL in H_2O; *see* **Note 24**).
8. After a final wash with bidistilled water, mount coverslips on glass slides using a small drop of mounting medium for fluorescence microscopy. Slides can be stored at 4°C or −20°C. Use oil immersion objectives for imaging.

3.7.1.1. MOUNTING MEDIUM FOR FLUORESCENCE MICROSCOPY

1. Dissolve 5 g of Mowiol (Elvanol or polyvinyl alcohol *[83,84]*) in a 20-mL buffer of choice (e.g., 100 m*M* Tris pH 8.0) and stir for 16 h (*see* **Note 25**).
2. Add 10 mL of glycerol and stir again for 16 h.
3. Remove undissolved Mowiol by centrifugation. Aliquots can be stored at −20°C. To prevent fading of the fluorescence, one can add either 20 to 50 mg/mL (w/v) DABCO (that is dissolved preferentially at 60°C); or 1 mg/mL (w/v) PPD; or 10

to 20 mg/mL (w/v) n-propyl gallate to the mounting solution (*see* **Note 26**). Section can also be mounted with commercial mounting media that already contain anitfading agents, e.g., Vectashield.

3.7.2. On-Section Immunogold Labeling

This protocol can be applied to resin and cryosections. It also can be used for labeling of antigens on the surface of intact cells or labeling of small particles adsorbed onto grids. All incubations are performed in a humid chamber on parafilm, and care should be taken so that the sections do not dry during the whole labeling procedure.

1. To mask nonspecific binding sites by floating the grids, place specimens section-side down on small drops of blocking buffer (*see* **Note 21**) on parafilm for 5 min.
2. Dilute the primary antibody in the blocking buffer used in the previous step and centrifuge. The same antibody concentrations as in immunofluorescence labeling are used (*see* **Note 23**).
3. Put small drops of diluted antibody onto a clean parafilm surface (or on the drop on the coverslips for fluorescence labeling).
4. Transfer the grids and float specimens on the antibody drops for 20 to 60 min at room temperature.
5. Wash with drops of blocking buffer six times, 2 min each time.
6. Dilute the colloidal gold probe to the working concentration in blocking buffer and centrifuge to remove gold clusters.
7. Transfer the grids, section-side down, to drops of gold probe on Parafilm and incubate for 20 min.
8. Wash with three changes of blocking buffer (2 min each) and three changes of PBS (2 min each).
9. Fix the sections by floating them on 1% buffered glutaraldehyde for 5 min (optional, *see* **Note 27**).
10. Wash four times each wash for 1 min with distilled water (*see* also Chapter 13).

3.8. Contrasting Labeled Resin Sections

1. Float sections on 1–2% aqueous uranyl acetate for 2–5 min (*see* **Note 28**).
2. Rinse the sections quickly with distilled water. Check the immunolabeling in the transmission electron microscope.
3. If more contrast is required, the sections can be stained more by using lead citrate (*85*).

4. Notes

1. In pre-embedding labeling, the remaining proteins are easily accessible in three-dimensions, which often results in high label density. On the other hand, the post-embedding labeling approach (particularly of cryofixed, freeze-substituted and low-

temperature embedded samples) offers good (excellent) preservation of the cellular architecture. Consequently, only a low number of antigens are accessible at the section surface. In addition, interacting proteins remain bound to their reaction partners covering the epitope site.

2. The specific fluorescence signal is independent of the section thickness, but the autofluorescence caused by the resin molecules is. Therefore, the thinner the section, the better the signal-to-noise ratio.

3. In contrast to resin sections on Tokuyasu cryosections, the fluorescence signal is dependent on the section thickness. When only low concentrations (less than 0.2%) of glutaraldehyde were used, the autofluorescence of the cryosections is neglectable.

4. Objective lenses with a numerical aperture (NA) of 1.2 to 1.4 have a depth of focus in the range of 0.5 to 1 μm; therefore, sections of a similar thickness are recommended. Just as a reminder: a 40X-objective (NA = 0.65) has a depth of focus of 2.5 μm; a 20X-objective (NA = 0.4), of 10 μm.

5. There are numerous publications on the application of improvements of fluorochromes as well as probes for subcellular components, for instance, endoplasmic reticulum and Golgi markers. For a comprehensive commercial overview, *see The Handbook: A Guide to Fluorescent Probes and Labeling Techniques* by Richard P. Haugland, Molecular Probes, Invitrogene Detection Technologies. For filter systems, *see* the *Handbook of Optical Filters for Fluorescence Microscopy* by Jay Reichman of the Chroma Technology Corp.

6. The carbonate buffer of the original recipe *(25)* has been replaced by borate to improve pH stability.

7. In principle, any kind and concentration of proteins can be used for blocking. There effectiveness should be tested. The manufacturer advices one not to use BSAc in combination with gelatin. For notorious background labeling, increasing the salt concentration up to 0.5 *M* NaCl can be considered to counteract electrostatic interactions. Sometimes it is necessary to increase the pH of the label buffer to pH 8.0 to 8.5, for instance, when mouse IgG_1 needs to be detected with protein A gold.

8. Aldehydes do not react sufficiently a low pH *(54)*. Therefore, after an initial fixation at growing conditions (e.g., yeasts at approx. pH 4.7 *[86]*), the used fixative has to be replaced with fresh fixative with a pH greater than 7.0. The standard PBS buffer has an inadequate buffering capacity. Moreover, Tris must be avoided, because it contains primary amino groups that inactivate the aldehyde fixatives.

9. At temperatures greater than 0°C, osmium tetroxide reacts proteolytically *(87–89)*, so it should not be used in combination with immunolabeling studies.

10. The choice of methacrylates is mainly dependent on the preferences of the user or the laboratory facilities, but there are also more scientific reasons for the choice of the type of methacrylate. One major point is the sectioning properties. There is no universal resin for all types of specimens. Therefore, we suggest trying several different formulations for one distinct object and evaluating the quality of the final ultrathin section in the electron microscope. In general the biological samples

tend to shrink in hydrophobic formulations, whereas the hydrophilic sections may expand during sectioning on the water surface of the knife trough in a thickness dependent manner *(73)*. There is only a minor difference in labeling efficiency for the different methacrylates. We routinely use K11M and HM20 at –40°C, one with a hydrophilic and one with a hydrophobic surface properties, and LR White for embedding at ambient temperature with heat polymerization *(90)*.

11. Osmium tetroxide also may be used as a fixative during freeze-substitution. At low temperature, osmium tetroxide reacts differently *(91)*; therefore, the proteins are not degraded, so the deleterious effect on the antigenicity is greatly reduced. It is suggested to remove osmium tetroxide at –40°C before further processing.

12. As long as the tubes are tightly sealed, it is not essential to fill them completely with the methacrylate for a successful polymerization. The exchange of air is responsible for improper polymerization, not the small gas volume above the resin. The plasticizer of some plastic tubes also may interfere with polymerization. When this is the case, the tubes may be boiled in water and air-dried before use. Do not forget that resin monomers are very potent solvents, for example, Lowicryl HM20 dissolves polystyrene culture dishes.

13. UV polymerization is preferentially done in a freeze-substitution unit in which the polymerization temperature can be controlled precisely. Alternatively, a UV lamp is mounted into a deep freezer. For details, see the instructions of the manufacturer. The actual temperature should be monitored in a reference probe.

14. The 1:1 infiltration may be performed at –30°C after dehydrating the specimen by either PLT or freeze-substitution.

15. Sometimes the LR White resin does not contain the catalyst dibenzoyl peroxide. Therefore, it is a good idea to polymerize an aliquot of a new batch of resin prior to use.

16. The 30% Epon or Epon-Araldite:acetone mixtures may be infiltrated at –30°C after PLT or freeze-substitution dehydration (M. Müller, personal communication). The subsequent steps are conducted at room temperature. Make sure that the tubes are tightly closed during the transfer from –30°C to room temperature to avoid water condensation.

17. We do not suggest using gelatin-coated substrates or SuperFrost Plus (Menzel Gläser, Braunschweig) microscopy slides for fluorescence microscopy, because they are either too thick for high-resolution light microscopy or are autofluorescent.

18. Mounting on coverslips (12-mm diameter, 0.17-mm thickness) offers the possibility of labeling with different antibodies under different conditions without the risk of cross-contamination and is more flexible. For systematical investigations of different specimen that are labeled in a standardized manner, mounting on a microscope slide might be preferred. Mounted, air-dried resin sections can be stored for months prior to labeling.

19. The intensity of staining depends on factors such as the section thickness, the staining time, the temperature, the pH of toluidine blue solution, and the hydrophobicity of the embedding medium; for example, methacrylate sections stain much more intensely than Epon sections do. If staining is still too intense, the sections can be

destained using water or solvents, that is, ethanol. Toluidine blue will stain baso-
philic and osmiophilic structures; hence, the resulting light microscopy image will
be similar to an electron micrograph: blue stain in light microscopy corresponds
with electron density of positive stained ultrathin sections. After toluidine blue
staining, sections may be counterstained with fuchsin: Incubate the sections for 1 to
2 min on a hot plate at 60°C with a solution of 1% basic fuchsin in bidistilled water;
rinse with water and dry.

20. The stained sections are finally embedded in Epon. The use of mounting media
    containing solvents such as DPX or Permount must not be used because they destain
    the sections.

21. We recommend that all buffer and antibody solutions are centrifuged before use, for
    example, 2 min at 10,000 rpm (about 12,000*g*) in an Eppendorf centrifuge 5415.

22. Using a vacuum pump to remove the drops from the surface of the coverslips can
    speed up the whole labeling procedure for immunofluorescence. Make sure that
    the tip does not touch the sections and that fresh medium is added immediately to
    avoid drying out the sections.

23. Experience indicates using primary antibodies at a final concentration in the range
    of 1 to 5 µg/mL of specific IgG (saturating conditions). The exact concentration
    of the antibody can be determined by series of dilutions: 1/10, 1/30, 1/100, 1/300,
    etc. The criterion for the correct concentration is high labeling in absence of back-
    ground. For multiple labeling of different antigens, primary as well as secondary
    antibodies may be mixed so long as the primary antibodies were raised in differ-
    ent species. To save precious primary antibodies, the incubation buffer can be
    reduced to 10 µL (drop size). In extreme cases, 3 µL of antibody solution can be
    used. In these cases, the drop is directly placed on a piece of parafilm and the
    coverslip on top, the specimen side down. To guarantee the exact same labeling
    conditions for fluorescence and electron microscopy labeling, the ultrathin sections
    mounted on grids may be placed specimen side down on the same drop of primary
    antibodies on the coverslip. This is also an approach to save primary antibodies.

24. The DAPI is a very specific DNA intercalating stain, whereas propidium iodide
    stains both RNA and DNA. Both counterstains are very helpful for histological
    differentiation in fluorescence mode. Propidium iodide is particularly helpful in
    regions with low number of nuclei. These counter-stains easily diffuse from the
    nucleic acids with time: propidium iodide much faster than DAPI; therefore, prep-
    arations containing these stains should to be documented immediately.

25. The optimum emission signal for FITC is at pH 8.5.

26. The antifading activity of DABCO is less efficient than that of PPD, but DABCO-
    containing preparations can be stored for months in a freezer and maintain a fluo-
    rescence signal. Fresh PPD results in a brilliant signal and an absolutely black back-
    ground. It is, however, less stable than DABCO so that, within days, the embedding
    turns a brownish color, and under ultraviolet fluorescence, a red background devel-
    ops. Mounted samples can be stored frozen for months. If necessary, coverslips can
    be removed with a razor blade after the slides were submersed in buffer, for instance,
    for additional antibody labeling, for staining of precious sections with different

antibodies, for counterstaining of nucleic acids or for refreshing the embedding medium after PPD has become a brownish color.

27. For prolonged exposure to solutions with low pH (less than pH 7.0) such as aqueous uranyl acetate (pH 4.5) or some silver enhancement solutions, e.g., Danscher, pH 3.5 *(92)*, it is suggested to fix the label to the section with glutaraldehyde.

28. Staining of immunolabeled sections with alcoholic uranyl acetate is not advised, as the alcohol may precipitate proteins of the labeling solutions.

Acknowledgments

The zebrafish specimen and the anti-prolactin serum were provided by Dr. Matthias Hammerschmitt und Mrs. Gabriela Nica (MPI Immunbiology, Freiburg, Germany). We thank Mrs. Brigitte Sailer and Mrs. Ursel Müller for technical support and Mrs. Gertrud Scheer for excellent photographic work; Dr. Christopher Antos (MPI Developmental Biology, Tübingen, Germany) Dr. Garreth Griffiths (EMBL, Heidelberg, Germany) and Mrs. Elly van Donselaar (Utrecht University, Utrecht, The Netherlands) for critical reading the manuscript; and Mrs. Annegret Schwarz and Mrs. Pina Bucher for the moral and culinary support.

References

1. Griffiths, G. (2001) Bringing electron microscopy back into focus for cell biology. *Trends Cell Biol.* **11,** 153–154.

2. Brink, H. A., Barfels, M. M. G., Burgner, R. P., and Edwards, B. N. (2003) A sub-50 meV spectrometer and energy filter for use in combination with 200 kV monochromated (S)TEMs. *Ultramicroscopy* **96,** 367–384.

3. Barfels, M. M. G., Jiang, X., Heng, Y. M., Arsenault, A. L., and Ottensmeyer, F. P. (1998) Low energy loss electron microscopy of chromophores. *Micron* **29,** 97–104.

4. Grabenbauer, M., Geerts, W. J. C., Fernadez-Rodriguez, J., Hoenger, A., Koster, A. J., and Nilsson, T. (2005) Correlative microscopy and electron tomography of GFP through photooxidation. *Nat. Methods* **2,** 857–862.

5. Gaietta, G., Deerink, T. J., Adams, S. R., Bouwer, J., Tour, O., Laird, D. W., et al. (2002) Multicolor and electron microscopic imaging of connexin trafficking. *Science* **296,** 503–507.

6. Geuze, H. J. (1999) A future for electron microscopy in cell biology? *Trends Cell Biol.* **9,** 92–93.

7. Melan, M. A. and Sluder, G. (1992) Redistribution and differential extraction of soluble proteins in permeabilized cultured cells. Implications for immunofluorescence microscopy. *J. Cell Sci.* **101,** 731–743.

8. Humbel, B. M., de Jong, M. D. M., Müller, W. H., and Verkleij, A. J. (1998) Preembedding immunolabeling for electron microscopy: an evaluation of premeabilization methods and markers. *Microsc. Res. Tech.* **42,** 43–48.

9. Brink, M., Humbel, B. M., de Kloet, E. R., and van Driel, R. (1992) Evidence against the model of nuclear translocation for the glucocorticoid receptor. *Endocrinology* **130**, 3575–358.

10. Stierhof, Y.-D., Schwarz, H., and Frank, H. (1986) Transverse sectioning of plastic-embedded immunolabeled cryosections: morphology and permeability to protein A-colloidal gold complexes. *J. Ultrastruct. Molecul. Struct. Res.* **97**, 187–196.

11. Stierhof, Y.-D. and Schwarz, H. (1989) Labeling properties of sucrose-infiltrated cryosections. *Scanning Microsc.* **3**, *Suppl.* 35–46.

12. Schwarz, H. (1994) Immunolabelling of ultrathin resin sections for fluorescence and electron microscopy in *Electron Microscopy 1994, ICEM 13* (Jouffrey, B. and Coliex, C., eds.), Vol. 3, Les Editions de Physique, Les Ulis, France, pp. 255–256.

13. Schwarz, H., Hohenberg, H., and Humbel, B. M. (1993) Freeze-substitution in virus research: a preview, in *Immunoelectron Microscopy in Virus Diagnosis and Research* (Hyatt, A. D. and Eaton, B. T., eds.), CRC Press Inc., Boca Raton, FL, pp. 97–118.

14. Schwarz, H. (1998) Correlative immunolabelling of ultrathin resin sections for light and electron microscopy, in *Electron Microscopy 1998, ICEM 14* (Calderón Benavides, H. A., Yacamán, M. J., Jiménez, L. F., and Kouri, J. B., eds.), Vol. IV, Institute of Physics Publishing, Bristol, Philadelphia, pp. 865–866.

15. Tokuyasu, K. T. (1973) A technique for ultracryotomy of cell suspensions and tissues. *J. Cell Biol.* **57**, 551–565.

16. Tokuyasu, K. T. (1986) Application of cryoultramicrotomy to immunocytochemistry. *J. Microsc. (Oxford)* **143**, 139–149.

17. Acetarin, J. D., Carlemalm, E., and Villiger, W. (1986) Developments of new Lowicryl resins for embedding biological specimens at even lower temperatures. *J. Microsc. (Oxford)* **143**, 81–88.

18. Carlemalm, E., Garavito, R. M., and Villiger, W. (1982) Resin development for electron microscopy and an analysis of embedding at low temperature. *J. Microsc. (Oxford)* **126**, 123–143.

19. Newman, G. R. and Hobot, J. A. (1987) Modern acrylics for post-embedding immunostaining techniques. *J. Histochem. Cytochem.* **35**, 971–981.

20. Newman, G. R. and Hobot, J. A. (1993) *Resin Microscopy and On-Section Immunocytochemistry*, Springer-Verlag, Berlin, Heidelberg, New York.

21. Scala, C., Cenacchi, G., Ferrari, C., Pasquinelli, G., Preda, P., and Manara, G. C. (1992) A new acrylic resin formulation: a useful tool for histological, ultrastructural, and immunocytochemical investigations. *J. Histochem. Cytochem.* **40**, 1799–1804.

22. Luft, J. H. (1961) Improvements in epoxy resin embedding methods. *J. Biophys. Biochem. Cytol.* **9**, 409–414.

23. Glauert, A. M. and Glauert, R. H. (1958) Araldite as an embedding medium for electron microscopy. *J. Biophys. Biochem. Cytol.* **4**, 191–194.

24. Spurr, A. R. (1969) A low-viscosity epoxy resin embedding medium for electron microscopy. *J. Ultrastruct. Res.* **26**, 31–43.

25. Trump, B. F., Smuckler, E. A., and Benditt, E. P. (1961) A method for staining epoxy sections for light microscopy. *J. Ultrastruct. Res.* **5**, 343–348.
26. Causton, B. E. (1986) Does the embedding chemistry interact with tissue? in *The Science of Biological Specimen Preparation 1985* (Müller, M., Becker, R. P., Boyde, A., and Wolosewick, J. J., eds.), SEM Inc., AMF O'Hare (Chicago), pp. 209–214.
27. Causton, B. E. (1984) The choice of resins for electron immunocytochemistry, in *Immunolabelling for Electron Microscopy* (Polak, J. M. and Varndell, I. M., eds.), Elsevier Science Publishers, Amsterdam, pp. 29–36.
28. Van Harreveld, A., J., C. and Malhotra, S. K. (1965) A study of extracellular space in central nervous tissue by freeze-substitution. *J. Cell Biol.* **25**, 117–137.
29. Steinbrecht, R. A. and Müller, M. (1987) Freeze-substitution and freeze-drying, in *Cryotechniques in Biological Electron Microscopy* (Steinbrecht, R. A. and Zierold, K., eds.), Springer-Verlag, Berlin, Heidelberg, pp. 149–172.
30. Humbel, B. M. and Schwarz, H. (1989) Freeze-substitution for immunochemistry, in *Immuno-Gold Labeling in Cell Biology* (Verkleij, A. J. and Leunissen, J. L. M., eds.), CRC Press, Boca Raton, pp. 115–134.
31. Humbel, B., Marti, T., and Müller, M. (1983) Improved structural preservation by combining freeze substitution and low temperature embedding. *Beitr. Elektronenmikroskop. Direktabb. Oberfl.* **16**, 585–594.
32. Riehle, U. (1968) *Über die Vitrifizierung von verdünnter wässriger Lösungen.* ETH Diss Nr 4271, Federal Institute of Technology (ETH), Zürich.
33. Riehle, U. and Hoechli, M. (1973) The theory and technique of high pressure freezing, in *Freeze-Etching Techniques and Applications* (Benedetti, E. L. and Favard, P., eds.), Société Française de Microscopie Electronique, Paris, pp. 31–61.
34. Müller, M. and Moor, H. (1984) Cryofixation of thick specimens by high pressure freezing, in *Science of Biological Specimen Preparation 1983* (Revel, J. P., Barnard, T. B., and Haggis, G. H., eds.), SEM Inc., AMF O'Hare (Chicago), pp. 131–138.
35. Studer, D., Michel, M., and Müller, M. (1989) High pressure freezing comes of age. *Scanning Microsc.* **3, Suppl.** 253–268.
36. Müller, M. (1992) The integrating power of cryofixation-based electron microscopy in biology. *Acta Microsc.* **1**, 37–44.
37. Hermann, R., Schwarz, H., and Müller, M. (1991) High precision immunoscanning electron microscopy using Fab fragments coupled to ultra-small colloidal gold. *J. Struct. Biol.* **107**, 38–47.
38. Humbel, B. M. and Biegelmann, E. (1992) A preparation protocol for postembedding immunoelectron microscopy of *Dictyostelium discoideum* cells with monoclonal antibodies. *Scanning Microsc.* **6**, 817–825.
39. Albrecht, U., Seulberger, H., Schwarz, H., and Risau, W. (1990) Correlation of blood-brain barrier function and HT7 protein distribution in chick brain circumventricular organs. *Brain Res.* **535**, 49–61.
40. Bierkamp, C., Schwarz, H., Huber, O., and Kemler, R. (1999) Desmosomal localization of β-catenin in the skin of plakoglobin null-mutant mice. *Development* **126**, 371–381.

41. Fialka, I., Schwarz, H., Reichmann, E., Oft, M., Busslinger, M., and Beug, H. (1996) The estrogen-dependent c-junER protein causes a reversible loss of mammary epithelial cell polarity involving a destabilization of adherens junctions. *J. Cell Biol.* **132,** 1115–1132.

42. Hoffmann, W. and Schwarz, H. (1996) Ependymins: meningeal-derived extracellular matrix proteins at the blood-brain barrier. *Int. Rev. Cytol.* **165,** 121–158.

43. Kurth, T., Schwarz, H., Schneider, S., and Hausen, P. (1996) Fine structural immunocytochemistry of catenins in amphibian and mammalian muscle. *Cell Tissue Res.* **286,** 1–12.

44. Wilsch-Bräuninger, M., Schwarz, H., and Nüsslein-Volhard, C. (1997) A sponge-like structure involved in the association and transport of maternal products during *Drosophila* oogenesis. *J. Cell Biol.* **139,** 817–829.

45. Nica, G., Herzog, W., Sonntag, C., Nowak, M., Schwarz, H., Zapata, A. G., and Hammerschmidt, M. (2006) Eya1 is required for lineage-specific differentiation, but not for cell survival in the zebrafish adenohypophysis. *Develop. Biol.* **292,** 189–204.

46. Tobler, M. and Freiburghaus, A. U. (1990) Occupational risks of (meth)acrylate compounds in embedding media for electron microscopy. *J. Microsc. (Oxford)* **160,** 291–298.

47. Beug, H., von Kirchbach, A., Döderlein, G., Conscience, J.-F., and Graf, T. (1979) Chicken hematopoietic cells transformed by seven strains of defective avian leukemia viruses display three distinct phenotypes of differentiation. *Cell* **18,** 375–390.

48. Longin, A., Souchier, C., French, M., and Bryon, P.-A. (1993) Comparison of anti-fading agents used in fluorescence microscopy: Image analysis and laser confocal microscopy study. *J. Histochem. Cytochem.* **41,** 1833–1840.

49. Langanger, G., De Mey, J., and Adam, H. (1983) 1,4-Diazobizyklo-(2.2.2)-Oktan (DABCO) verzögert das Ausbleichen von Immunfluoreszenzpräparaten. *Mikroskopie* **40,** 237–241.

50. Johnson, G. D. and de C. Nogueira Araujo, G. M. (1981) A simple method of reducing the fading of immunofluorescence during microscopy. *J. Immunol. Meth.* **43,** 349–350.

51. Giloh, H. and Sedat, J. W. (1982) Fluorescence microscopy: reduced photobleaching of rhodamine and fluorescein protein conjugates by n-propyl gallate. *Science* **217,** 1252–1255.

52. Van Bergen en Henegouwen, P. M. P. and Leunissen, J. L. M. (1986) Controlled growth of colloidal gold particles and implications for labelling efficiency. *Histochemistry* **85,** 81–87.

53. Birrell, G. B., Hedberg, K. K., and Griffith, O. H. (1987) Pitfalls of immunogold labeling: analysis by light microscopy, transmission electron microscopy, and photoelectron microscopy. *J. Histochem. Cytochem.* **35,** 843–853.

54. Griffiths, G. (1993) *Fine Structure Immunocytochemistry,* Springer-Verlag, Berlin, Heidelberg.

55. Ebersold, H. R., Cordier, J. L., and Lüthy, P. (1981) Bacterial mesosomes: method dependent artifacts. *Arch. Microbiol.* **130,** 19–22.

56. Kaneko, Y. and Walther, P. (1995) Comparison of ultrastructure of germinating pea leaves prepared by high-pressure freezing-freeze substitution and conventional chemical fixation. *J. Electron Microsc.* **44,** 104–109.

57. Studer, D., Michel, M., Wohlwend, M., Hunziker, E. B., and Buschmann, M. D. (1995) Vitrification of articular cartilage by high-pressure freezing. *J. Microsc. (Oxford)* **179,** 321–332.

58. Studer, D., Hennecke, H., and Müller, M. (1992) High-pressure freezing of soyabean nodules leads to an improved preservation of ultrastructure. *Planta* **188,** 155–163.

59. Fernández-Morán, H. (1960) Low-temperature preparation techniques for electron microscopy of biological specimens based on rapid freezing with liquid Helium II. *Ann. N. Y. Acad. Sci.* **85,** 689–713.

60. Costello, M. J., Fetter, R., and Corless, J. M. (1983) Optimum conditions for the plunge freezing of sandwiched samples, in *Science of Biological Specimen Preparation, 1983* (Revel, J. P., Barnard, T. B., and Haggis, G. H., eds.), SEM Inc., AMF O'Hare (Chicago), pp. 105–115.

61. Müller, M., Meister, N., and Moor, H. (1980) Freezing in a propane jet and its application in freeze-fracturing. *Mikroskopie* **36,** 129–140.

62. Van Harreveld, A. and Crowell, J. (1964) Electron microscopy after rapid freezing on a metal surface and substitution fixation. *Anat. Rec.* **149,** 381–386.

63. Bachmann, L. and Schmitt, W. W. (1971) Improved cryofixation applicable to freeze etching. *Proc. Natl. Acad. Sci. USA* **68,** 2149–2152.

64. Moor, H. (1987) Theory and practice of high pressure freezing, in *Cryotechniques in Biological Electron Microscopy* (Steinbrecht, R. A. and Zierold, K., eds.), Springer-Verlag, Berlin, Heidelberg, pp. 175–191.

65. Echlin, P. (1992) *Low-Temperature Microscopy and Analysis*, Plenum Press, New York, London.

66. Zierold, K. and Steinbrecht, R. A. (eds.) (1987) *Cryotechniques in Biological Electron Microscopy*, Springer-Verlag, Berlin, Heidelberg.

67. Robards, A. W. and Sleytr, U. B. (eds.) (1985) *Low Temperature Methods in Biological Electron Microscopy* Vol. 10. Pract. Methods Electron Microsc. (Glauert, A. M., ed.), Elsevier, Amsterdam, New York, Oxford.

68. Hayat, M. A. (2000) *Principles and Techniques of Electron Microscopy. Biological Applications*, 4th ed, Cambridge University Press, Cambridge.

69. Villiger, W. (1991) Lowicryl resins, in *Colloidal Gold: Principles, Methods, and Applications*, Vol. 3 (Hayat, M. A., ed.), Academic Press, San Diego, pp. 59–71.

70. Weibull, C., Villiger, W., and Carlemalm, E. (1984) Extraction of lipids during freeze-substitution of Acholeplasma laidlawii-cells for electron microscopy. *J. Microsc. (Oxford)* **134,** 213–216.

71. Hunziker, E. B. and Herrmann, W. (1987) In situ localization of cartilage extracellular matrix components by immunoelectron microscopy after cryotechnical tissue processing. *J. Histochem. Cytochem.* **35,** 647–655.

72. Verkleij, A. J., Humbel, B., Studer, D., and Müller, M. (1985) 'Lipidic particle' systems as visualized by thin-section electron microscopy. *Biochim. Biophys. Acta* **812,** 591–495.
73. Schwarz, H. and Humbel, B. M. (1989) Influence of fixatives and embedding media on immunolabelling of freeze-substituted cells. *Scanning Microsc.* **3,** *Suppl.* 57–64.
74. Meissner, D. H. and Schwarz, H. (1990) Improved cryofixation and freeze-substitution of embryonic quail retina: A TEM study on ultrastructural preservation. *J. Electron Microsc. Tech.* **14,** 348–356.
75. Müller, M., Marti, T., and Kriz, S. (1980) Improved structural preservation by freeze substitution, in *Proc. 7th Eur. Congr. Electron Microsc.* (Brederoo, P. and de Priester, W., eds.), **2,** 720–721.
76. Grünfelder, C. G., Engstler, M., Weise, F., Schwarz, H., Stierhof, Y.-D., Boshart, M., and Overath, P. (2002) Accumulation of a GPI-anchored protein at the cell surface requires sorting at multiple intracellular levels. *Traffic* **3,** 547–559.
77. Humbel, B. and Müller, M. (1986) Freeze substitution and low temperature embedding, in *The Science of Biological Specimen Preparation 1985* (Müller, M., Becker, R. P., Boyde, A., and Wolosewick, J. J., eds.), SEM Inc., AMF O'Hare (Chicago), pp. 175–183.
78. Monaghan, P. and Robertson, D. (1990) Freeze-substitution without aldehyde or osmium fixatives: ultrastructure and implications for immunocytochemistry. *J. Microsc. (Oxford)* **158,** 355–363.
79. Tonning, A., Helms, S., Schwarz, H., Uv, A. E., and Moussian, B. (2006) Hormonal regulation of mummy is needed for apical extracellular matrix formation and epithelial morphogenesis in *Drosophila. Development* **133,** 331–341.
80. Moussian, B., Tang, E., Tonning, A., Helms, S., Schwarz, H., Nüsslein-Volhard, C., and Uv, A. E. (2006) *Drosophila* Knickkopf and Retroactive are needed for epithelial tube growth and cuticle differentiation through their specific requirement for chitin filament organization. *Development* **133,** 163–171.
81. Romeis, B. (1989) *Mikroskopische Technik*, Urban & Schwarzenberg, München, Wien, Baltimore.
82. Huang, W. M., Gibson, S. J., Facer, P., Gu, J., and Polak, J. (1983) Improved section adhesion for immunocytochemistry using high molecular weight polymers of L-lysine as a slide coating. *Histochemistry* **77,** 275–279.
83. Rodriguez, J. and Deinhardt, F. (1960) Preparation of a semipermanent mounting medium for fluorescent antibody studies. *Virology* **12,** 316–317.
84. Lennette, D. A. (1978) An improved mounting medium for immunofluorescence microscopy. *Am. J. Clin. Pathol.* **69,** 647–648.
85. Venable, J. H. and Coggeshall, R. (1965) A simplified lead citrate stain for use in electron microscopy. *J. Cell Biol.* **25,** 407–408.
86. Kärgel, E., Menzel, R., Honeck, H., Vogel, F., Böhmer, A., and Schunck, W. H. (1996) *Candida maltosa* NADPH-cytochrome P450 reductase: cloning of a full-length cDNA, heterologous expression in *Saccharomyces cerevisiae* and function

of the N-terminal region for membrane anchoring and proliferation of the endoplasmic reticulum. *Yeast* **12**, 333–348.

87. Behrman, E. J. (1983) *The chemistry of osmium tetroxide fixation* in *The Science of Biological Specimen Preparation 1983* (Revel, J. P., Barnard, T. B., and Haggis, G. H., eds.), SEM Inc., AMF O'Hare (Chicago), pp. 1–5.

88. Maupin, P. and Pollard, T. D. (1983) Improved preservation and staining of HeLa cell actin filaments, clatrin-coated membranes, and other cytoplasmic structures by tannic acid-glutaraldehyde-saponin fixation. *J. Cell Biol.* **96**, 51–62.

89. Tanaka, K. and Mitsushima, A. (1984) A preparation method for observing intracellular structures by scanning electron microscopy. *J. Microsc. (Oxford)* **133**, 213–222.

90. Humbel, B. M., Konomi, M., Takagi, T., Kamasawa, N., Ishijima, S. A., and Osumi, M. (2001) In situ localization of β-glucans in the cell wall of *Schizosaccharomyces pombe. Yeast* **18**, 433–444.

91. White, D. L., Andrews, S. B., Faller, J. W., and Barrnett, R. J. (1976) The chemical nature of osmium tetroxide fixation and staining of membranes by x-ray photoelectron spectroscopy. *Biochim. Biophys. Acta* **436**, 577–592.

92. Danscher, G. (1981) Localization of gold in biological tissue. A photochemical method for light and electron microscopy. *Histochemistry* **71**, 1–16.

13

Cryosectioning Fixed and Cryoprotected Biological Material for Immunocytochemistry

Paul Webster and Alexandre Webster

Summary

   Immunocytochemistry for transmission electron microscopy provides important information on the location and relative abundance of proteins inside cells. Gaining access to this information without extracting or disrupting the location of target proteins requires specialized preparation methods. Sectioning frozen blocks of chemically fixed and cryoprotected biological material is one method for obtaining immunocytochemical data. Once the cells or tissues are cut, the thawed cryosections can be labeled with specific antibodies and colloidal gold probes. They are then embedded in a thin film of plastic containing a contrasting agent. Subcellular morphology can be correlated with specific affinity labeling by examination in the transmission electron microscope. Modern technical advancements both in preparation protocols and equipment design make cryosectioning a routine and rapid approach for immunocytochemistry that may provide increased sensitivity for some antibodies.

   Key Words: Transmission electron microscopy; immunocytochemistry; cryosectioning; chemical fixation; antibodies; immunogold.

1. Introduction

   The application of antibodies and other affinity markers to thin sections for high-resolution immunocytochemistry is an important tool of cell biologists. Cut surfaces offer increased accessibility to antigens on sections while still preserving subcellular morphology. In the transmission electron microscope, immunolocalized molecules can be accurately correlated to the large amount of visible subcellular components in the sections. If possible, the data obtained from immunocytochemical methods should be used for a quantitative analysis of the sample (*see* **Note 1** and Chapter 15 in this volume).

From: *Methods in Molecular Biology, vol. 369*
Electron Microscopy: Methods and Protocols, Second Edition
Edited by: J. Kuo © Humana Press Inc., Totowa, NJ

Table 1
Cryosectioning and Immunolabeling Biological Material

1. Chemical crosslinking (4% formaldehyde, and 0.1% glutaraldehyde) dissolved in 100 mM sodium phosphate buffer, pH 7.2.
2. Trim tissues or scrape cells and centrifuge.
3. Prepare small specimen blocks (e.g., 3 mm^3).
4. Soak in 2.3 M sucrose (15 min to 24 h).
5. Place on specimen pins (*see* **Fig. 1**).
6. Freeze in liquid nitrogen.
7. Place in cryochamber precooled to −80°C.
8. Trim the front and sides of the block with a trim tool (*see* **Fig. 2**).
9. Cool the cryochamber to −120°C, replace trim tool with a knife.
10. Set the ultramicrotome to cut 70-nm thick sections and start the machine sectioning.
11. Retrieve sections from the dry knife surface with a wire loop filled with a 1:1 mixture of 2% methyl cellulose and 2.3 M sucrose (*see* **Fig. 4**).
12. Transfer the sections to specimen grids at room temperature (*see* **Fig. 5**).
13. Float the grids, section side down on a drop of PBS containing 0.12% glycine (5 min).
14. Transfer the grids to drops of PBS containing 10% fetal calf serum (10 min).
15. Float grids individually on 5-µL drops of diluted antibody (20 min).
16. Float grids on 500-µL drops of PBS (five drops, 2 min each).
17. Float the grids individually on 5-µL drops of diluted colloidal gold probe (20 min).
18. Wash the grids by floating on drops of PBS (five drops, 2 min each).
19. Float the grids on four drops of water, 1 min each.
20. Transfer each grid to drops of methyl cellulose containing uranyl acetate (nine parts 2% methyl cellulose plus 1 part 3% uranyl acetate), on ice for 10 min.
21. Lift off each grid in a wire loop and remove excess liquid.
22. Leave to dry, remove the grid and examine in the TEM.

Currently, two approaches to preparing thin sections for immunolocalization at the ultrastructural level are possible. However, we will only be discussing the Tokuyasu cryosectioning method in this chapter (*see* **Notes 2** and **3**). The Tokuyasu method involves infiltrating chemically fixed specimens with a sucrose solution followed by freezing, then sectioning on a dry knife while the specimens are still frozen. The sections are removed from the knife surface, thawed, and placed onto a coated specimen support grid. They are then immunolabeled at room temperature and dried in a plastic film before examination in the transmission electron microscope (*see* **Table 1**). In the other method, described in Chapter 12, resin-embedded specimens are immunolabeled.

The use of thin cryosections for immunolabeling offers two advantages over resin-embedded material: (1) cryosections can be rapidly prepared and (2) cryosections remain fully hydrated during the immunolabeling process (*see* **Notes 4** and **5**).

As with resin sections, cryosections from cryoprotected, frozen specimens can be used for light microscopy observations and electron microscopy (EM). Sequential sections can be immunolabeled and used in correlative microscopy studies (*[1]* and Chapter 12) or as a substitute for confocal microscopy (*[2,3]*; *see* also **Note 6**).

In this chapter, we describe the cryosectioning method in a way that will allow readers to produce, label, and contrast cryosections without assistance. However, the most successful way to learn how to cryosection is to either visit a laboratory where cryosectioning is routinely used or to join a hands-on cryosectioning practical course.

2. Materials

2.1. Transmission Electron Microscopy (TEM)

1. Transmission electron microscope (TEM; *see* **Note 7**).
2. Fine forceps with good-quality tips (e.g., Dumont Biologie, #5 or #7).
3. Grid boxes or grid mats for storing immunolabeled grids with dried cryosections.

2.2. EM Specimen Grids for Cryosections

1. EM specimen grids (Hexagonal 100 mesh grids).
2. Precleaned 25 × 75-mm micro slides from VWR Scientific Inc. (Media, PA, cat. no. 48312-002).
3. Thin film casting device.
4. 2% and 4% Formvar in chloroform (w/v).
5. 200 mL of glass conical flask.
6. 2% Aqueous acetic acid.
7. 100% Acetone.
8. Ross lens tissue.
9. White, 1" × 2.5" self-adhesive address labels.
10. Vacuum evaporator.

2.3. Specimen Preparation

2.3.1. Tissue Fixation

1. A solution of 4% formaldehyde, and 0.1% glutaraldehyde dissolved in 100 mM sodium phosphate buffer, pH 7.2.
2. 2.3 M Sucrose in phosphate-buffered saline (PBS).
3. Stereo microscope with overhead light source, for dissection and trimming.
4. Scalpel blades with handles.

5. Thin razor blades.
6. Metal specimen pins and forceps for holding the specimen pins.
7. Liquid nitrogen Thermos flasks with wide neck.
8. Liquid nitrogen storage tank (for specimen storage).

2.3.2. Fixing Cells Grown in Culture

In addition to the materials listed in **Subheading 2.3.1.**, the following materials are required for preparing cultured cells for cryosectioning.

1. Soft wooden stick or piece of Teflon™.
2. Kelly forceps (or artery forceps).
3. 12% Gelatin.
4. 2% Aqueous agarose.
5. Rotating mixing wheel.
6. Eppendorf benchtop centrifuge.

2.4. Cryosectioning, Section Retrieval, and Sections Storing

1. Cryoultramicrotome (an ultramicrotome equipped with cryosectioning attachment and stereo binocular eyepieces).
2. Diamond knife for cryosectioning.
3. Diamond cryotrim tool.
4. Antistatic line for ultramicrotome.
5. Liquid nitrogen.
6. 1.5-mL size Eppendorf tubes.
7. 2.3 M Sucrose in PBS.
8. Wire loops (2–3.5 mm in diameter) for section retrieval (*see* **Note 8**).
9. Eyelash probes mounted on wooden sticks.
10. Hydrophilic, Formvar/carbon-coated specimen grids (*see* **Note 9**).
11. 2% Methyl cellulose: 2.3 M sucrose mixed in a 1:1 ratio, on ice.
12. 2% Gelatin.
13. Petri dishes, 3.5 cm in diameter.

2.5. Immunolabeling

1. PBS.
2. 10% Fetal calf serum (FCS) in PBS.
3. Primary antibodies with known antigen specificity.
4. Colloidal gold probes. Protein A-gold is the recommended probe, although secondary antibodies coupled to colloidal gold also can be used.
5. Unconjugated secondary antibodies.
6. 1.5-mL size Eppendorf tubes.
7. 1% Glutaraldehyde in PBS.
8. 0.12% Glycine in PBS.
9. ParafilmM®™ (Parafilm).

2.6. Contrasting and Drying Labeled Sections

1. Large wire loops (3.5 mm in diameter) for drying the sections. These loops also can be used for transferring specimen grids from drop to drop during the washing steps of the immuno-labeling protocol.
2. Hardened filter paper (e.g., Whatman, type 5).
3. Ice in a bucket and a flat support for placing on ice (e.g., metal sheet or inverted plastic dish).
4. 3–5% Aqueous uranyl acetate.
5. 2% Aqueous methyl cellulose (viscosity 25 cps; *see* **Note 10**).
6. Parafilm.

3. Methods

3.1. Formvar Coating of EM Grids

Plastic support films over EM grids are essential for supporting the fragile thawed cryosections during subsequent handling and imaging. The ideal films should be thin, strong, and clean and should remain attached to the EM grid at all times. There are many methods described in textbooks for preparing coated grids. Here is our contribution for making coated EM grids.

1. Clean 100 mesh hexagonal EM grids (*see* **Note 11**) by placing them in a 200-mL glass flask, rinsing them with 2% aqueous acetic acid, undertaking multiple washes with distilled water, and finally with adding three changes of dry acetone. Dry the grids by placing the flask on a warm plate. Clean, dry grids will fall easily from the flask when it is inverted.
2. Briefly immerse a microscope slide in dry ethanol and dry it with Ross lens tissue.
3. Dip the dry slide into a solution of 2% Formvar in chloroform, and withdraw it carefully. Slower withdrawal will increase the thickness of the Formvar film on the glass surface. A thin film casting device will make this process reproducible.
4. Let the slide dry by propping it vertically in a clean environment.
5. Fill a deep dish with distilled water and stand it below a diffuse light source so that the light reflects off the water surface. Dirt and other particles should be easily observed if present on the water surface. Clean the water surface by passing a page of Ross lens tissue over the surface.
6. Use a razor blade or scalpel blade to score around the edge of the slide. Score around all four sides to define the limits of the film to be dislodged from the slide.
7. Hold the Formvar-coated slide at a 45° angle at the edge of the glass dish and touch the bottom of the slide to the water surface. Slowly and carefully slide the glass slide into the water while maintaining a 45° angle. The Formvar film should start to detach from the glass surface and float onto the water surface.
8. Remove the glass slide from the water when the Formvar film has completely detached from the glass surface. Examine the floating film for dirt particles and

imperfections. Interference colors, a result of the reflected light, should be present on the film. A film useful for supporting sections should have a uniform silver interference color. Yellow, purple, red, or green interference colors indicate a thicker film, whereas thinner films will have a gray color (*see* **Notes 12** and **13**).

9. Place cleaned EM grids (from **step 1**) onto the floating film using clean, sharp fine forceps. Place grids so that the shiny surface of the grid is facing up and the dull side is on the film.

10. Dry the glass slide used for casting the film and stick a 1-inch by 2.5-inch white address label on its surface.

11. Hold the glass slide vertically over one edge of the floating film and push it into the water so that the film floating on the water surface sticks to the address label on the glass slide. Forcing the film into the water will cause it to adhere to the address label, thus trapping the grids between the Formvar film and the label.

12. Once the glass slide has been totally immersed, remove it from the water making sure the grids remain attached, and leave it to dry.

13. Coat the grids with a thin layer of carbon using a vacuum evaporator. The final carbon layer should be thin enough to form a light gray coating on a sheet of white paper.

3.2. Chemical Fixation of Sample

If we are to study the location of molecules within cells, then some form of immobilization must take place before sectioning and immunolabeling occurs. Molecules that are not immobilized may be displaced from their normal location, resulting in either a false localization or a negative result if the molecules are washed away.

- The easiest immobilization method is by chemical fixation using aldehydes to cross-link cellular components. In addition to immobilizing antigens in place, fixatives also help to preserve specimen morphology (*see* **Notes 14–20**).
- For immunocytochemistry, tissues or cells can be fixed by immersion in a buffered solution of aldehyde using the same methods that are used for epoxy resin embedding. The specimen should be small (approx 1- to 3-mm cubes, if possible) for rapid penetration of aldehyde and care should be taken not to damage the sample by over-manipulation (i.e., squeezing, pulling, cutting, etc.). Large organs are best fixed by perfusing with warm fixative (*see* **Note 21**).
- The main problem when fixing tissues is the slow rate of diffusion of the chemical fixative through the specimen. When possible, it is best that perfusion methods are performed. This may vary from whole body perfusion, where the fixative is pumped through the vascular system of deeply sedated animals, to perfusion of isolated organs.
- When perfusion-fixation is not possible, then small tissue pieces should be carefully excised and immersed in fixative as soon as possible. Cutting the tissue into small pieces while in fixative will assist with penetration of fixative.

- The buffers used with fixatives will affect the morphology of biological material and although the exact role of pH in fixation is not known, it is known that a buffer must have a large buffering capacity in order to counter rapid drops in pH that occur during fixation (*see* **Note 22**).

3.2.1. Dissection and Handling of Perfusion-Fixed Tissues

1. Carefully dissect out the tissue under study once the perfusion has been completed and transfer it whole into fresh fixative (*see* **Note 23**). A stereo dissecting microscope will be of great assistance for precise dissection. Leave the tissues for 1 h in the fixative. Any pulling or squeezing at this stage may adversely affect the subcellular morphology of the tissue. Do not mince the tissue, as is sometimes recommended in other books. Damage caused by the rough cutting may result in poor morphology in the cryosections.
2. Transfer the tissue to a 100 mM phosphate buffer and slice into 1- to 2-mm thick slabs using two sharp, clean razor blades. Use one blade to immobilize the tissue by applying slight pressure and gently cut into the tissue with a second blade. To simplify the process and ensure that the cutting is not distorting the tissue, the process should be performed while watching it with the aid of a stereo microscope. It is important to avoid any squeezing or pulling because they may damage or alter subcellular morphology.
3. Use similar cutting techniques with two blades to slice the slabs into rods and then to cut the rods into small samples approx 1–3 mm^3. Dissection and handling also is discussed in Chapter 2 of this volume.

3.2.2. Fixing Cells

1. Add double strength fixative (8% formaldehyde+0.2% glutaraldehyde) to the cell culture (1:1 ratio). For best results, fixation should occur at the normal growing temperature. Cooling cells down before fixation should be avoided as this may alter morphology (e.g., depolymerize microtubules). Cell suspensions can be fixed in suspension and immediately pelleted by gentle centrifugation (we use an Eppendorf bench-top centrifuge, taking it up to maximum speed by pressing the front button and slowing down by immediately releasing the button). If possible do not centrifuge cells prior to fixation. The presence of extracellular proteins (e.g., fetal calf serum) does not normally affect intracellular labeling so washing cells before fixation is not essential.
2. For cells grown on culture dishes or in flasks, fix by the addition of double strength fixative to the culture medium. Carefully scrape the cells off the substrate using a piece of Teflon that has been beveled on the edge to form a blade and clamped in a pair of Kelly forceps. Alternatively, a soft wooden stick with a beveled edge can be used (*see* **Notes 24** and **25**).
3. Centrifuge the scraped cells into a pellet.
4. Embed the pelleted cell in a support medium (*see* **Notes 26–31**). If the pellet is loose or is falling apart, it will be necessary to embed in gelatin or agarose: (a)

remove the fixative supernatant from the cell pellet and (b) resuspend the pellet in a small volume of 12% aqueous gelatin. While the gelatin is still warm, repellet the cells. If the gelatin sets before the cells have pelleted, warm the tube until the gelatin becomes liquid again and centrifuge into a pellet. Alternatively, cells can be embedded in 2% low-melting-point agarose. Warm agarose solution is cooled to 40°C and the cells resuspended in it. The cells are centrifuged to a pellet in the warm agarose. If the agarose sets, then the agarose has to be melted again at 60°C, which may not be good for cellular morphology or antigenicity. (c) Cool the pellet in ice for 30 min or until the support medium has solidified. Cut off the tip of the tube and remove the embedded pellet. Then, slice the embedded pellet into slabs or sheets of cells and gelatin and cut away gelatin that does not contain the specimen (*see* **Note 32**). Finally, slice the slabs into rods and then into small blocks using the same method of cutting for the tissue samples. Cell fixing also is discussed in Chapters 1–4 of this volume.

3.3. Cryoprotection and Freezing

1. Transfer cubes of tissue or cell pellets to 1 mL of 2.3 *M* sucrose in PBS in an Eppendorf tube, seal the cap and leave overnight on a rotating mixing wheel at 4°C. The sucrose solution will infiltrate through the specimen (*see* **Notes 33** and **34**).
2. Transfer each infiltrated specimen block onto a clean metal specimen pin (*see* **Fig. 1**), remove excess sucrose with a damp filter paper and immerse the specimen pin and specimen in liquid nitrogen until freezing is complete (*see* **Note 35**).

3.4. Specimen Transfer to Precooled Cryochamber

1. Precool the cryochamber of the ultramicrotome to −80°C.
2. Insert the diamond cryotrim tool and allow the chamber to reach equilibration temperature. Any tools inserted into the cryochamber, and especially those that are used to manipulate the specimen, must be pre-cooled in liquid nitrogen.
3. Rapidly transfer the specimen, on the specimen pin, from the liquid nitrogen using pre-cooled forceps and quickly place it into the pre-cooled cryochamber.
4. Place the specimen pin into the specimen holder and lock it in place using minimal force.

3.5. Trimming the Specimen Block

1. Turn on the antistatic line to full power.
2. Carefully advance the front of the trim tool to the surface of the block until it is just touching the front of the block. First time users may benefit from a few lessons from more experienced ultramicrotome operators when learning this step (*see* **Note 36**).
3. Set the ultramicrotome to cut sections of 500 nm (0.5 μm) and leave the machine to section at the maximum speed setting.
4. Once a substantial portion of the block (revealing an area of between 1 and 3 mm^2) has been smoothed away, stop sectioning. This process is called "facing-off" the

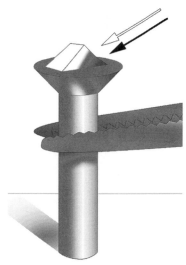

Fig. 1. After the specimen has been fixed, trimmed and infiltrated in 2.3 *M* sucrose, it is placed onto a clean specimen pin (supplied with the cryoultramicrotome). Hold the pin in a stand or with locking forceps at room temperature. Pick up the specimen from the sucrose without crushing it, remove excess sucrose and place the specimen onto the top surface of the pin. There should be a small amount of sucrose around the base of the specimen (arrow). The sucrose solution should not cover the top of the specimen. Pick up the pin with warm forceps and transfer it to liquid nitrogen. Do not pick up the pin with forceps previously cooled in liquid nitrogen. The frozen specimen can be quickly transferred to a precooled cryochamber on an ultramicrotome for trimming and sectioning.

block and should result in a shiny-smooth block face with no indentations or scratches.

5. Use the edge of the trim tool to shape the sides of the block into a "mesa" protruding approx 1 mm from the block surface, approx 1 mm wide and 2 mm long. Each side of the block is cut until a protruding mesa is formed with smooth sides (*see* **Fig. 2**). The trimming process can be performed either with a diamond cryotrim tool (shown in **Fig. 2**) or with glass knives. Because of the cost of diamond knives, we recommend that beginners practice the trimming and sectioning process with glass knives.

Semithin sections can be obtained when the cryochamber is still set at −80°C by sectioning at thicknesses of between 200 and 500 nm. The sections can be retrieved from the knife surface using a drop of 2.3 *M* sucrose in a 3.5-mm diameter wire loop (*see* **Subheading 3.8.**). Insert the loop with the drop of sucrose into the cooled cryochamber and watch it through the stereo binocular eyepieces. As it cools, a thin wisp of white "smoke" will rise from it. Use the

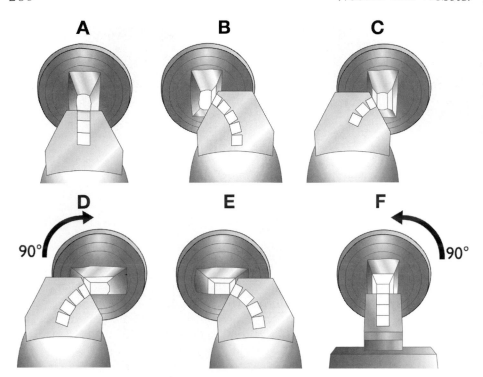

Fig. 2. Set the cryochamber at −80°C and the anti-static line set at 100% power. **(A)** Trim the front of the block until the block face is smooth and shiny. **(B)** Retract the cryotrim tool, move the trim tool to the right, advance it to the block and trim the right side of the block until the edge is smooth. **(C)** Retract the trim tool, move it to the left, and advance to trim the left side of the block. **(D)** Rotate the block 90° clockwise and trim the left side of the block. **(E)** Move the trim tool to the right and trim the right side of the block. The last two cuts will form the bottom **(D)** and top **(E)** of the specimen block. **(F)** Rotate the block 90° counterclockwise to return it to its original position, replace the trim tool with a pre-cooled knife and move the specimen to the knife for sectioning.

drop to pick up the sections when the drop stops "smoking." When mounted on glass coverslips, these sections can either be stained with any routine histology or vital stain or examined using phase-contrast or interference-contrast light microscopy and used for orientation (*see* **Fig. 3**). Alternatively, the sections can be labeled with antibodies and examined by fluorescent microscopy. Information about antibody dilutions and the relative abundance of positive labeling can be obtained this way. The high resolution of the semithin section for light microscopy makes this method of antibody screening comparable with, or better than, the results obtained using a laser scanning confocal microscope (*see* **Note 6**).

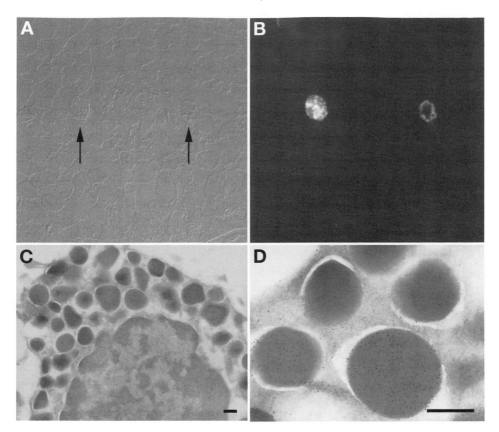

Fig. 3. Cryosections from the same specimen block can be used for light and electron microscopy. Small pieces of bovine nasal mucosa were immersed in 4% buffered formaldehyde, left overnight, cryoprotected in 2.3 M sucrose, frozen, and sectioned. The first sections removed from the specimen block were 300-nm thick and were cut at −80°C. The sections were placed on glass coverslips and immuno-labeled with primary antibody (rabbit anti-aprotinin) and then with fluorescent secondary antibody (goat anti-rabbit TRITC). The coverslips were then mounted, section-side down on a glass microscope slide. **(A)** The sections were examined by light microscopy, first using Nomarski DIC optics. **(B)** The sections were then examined under indirect epifluorescent illumination (excitation at 546 nm and emission at 590 nm wavelengths). Only two cells in the section were labeled with the antibody (arrows in A). Bar, 100 μm. The specimen block was trimmed so that it contained only the region with the aprotinin-positive cells and then thin-sectioned for electron microscopy. **(C)** From their morphology, the positive cells were identified as mast cells. **(D)** Granules in the mast cell labeled with the anti-aprotinin antibodies, not all granules labeled with the antibodies, a result that was not apparent from light microscopic observation. Bars = 0.5 μm.

3.6. Sectioning

1. Mount the diamond knife in its holder, place in the cryochamber and let it cool down and equilibrate at −120°C.
2. Carefully move the trimmed block to the knife edge and set the cutting window so that the section stroke starts just above the knife and ends just below the knife (*see* **Note 37**).
3. Set the antistatic line to 50% power or less. Variations in the antistatic line power will affect the sectioning properties of a diamond knife (*see* **Note 38**).
4. Set the ultramicrotome to cut 80-nm thick sections (*see* **Note 39**).
5. Use the rough advance to move the knife close to the block. The orientation of the knife and the block can be evaluated and, if needed, aligned to bring the knife parallel to the block.
6. Continue moving the knife to the block using the fine advance control until the two are almost touching. When advancing the knife, move the specimen up and down past the knife edge. This step will ensure that the part of the block closest to the knife will be seen. The knife can then be appropriately aligned.
7. When it seems as if the knife and the block are almost touching, continue to advance the knife to the specimen using the extra-fine advance control. This step will allow the knife to be advanced in small steps of known distance. We use 0.5-μm steps for the final adjustments.
8. When the knife and specimen block are almost touching (indicated by the presence of small amounts of debris from the specimen block on the knife after the specimen has passed over the knife edge), switch over to automatic sectioning and wait for sections appear on the knife.
9. Set the cutting speed between 0.6 mm/s and 2 mm/s. The slow cutting speed may be combined with a fast return cycle to reduce the possibility of thermal changes within the specimen.
10. If sections fail to appear, check to make sure the knife and the specimen block are aligned and almost touching. If sections appear but the sectioning is irregular (e.g., it is cutting thin and then thick sections), check the machine for a loose knife or specimen block (*see* **Note 40**).

3.7. Manipulating the Sections

1. Turn off antistatic line.
2. Pull sections from the edge of the knife with a clean eyelash probe and arrange them in groups on the knife surface (*see* **Note 41**).

3.8. Section Retrieval

1. Dip a 2-mm diameter loop in a mixture of one part 2% methyl cellulose: one part 2.3 *M* sucrose and lift a drop out on the loop *(4)*. Keep the methyl cellulose/sucrose mixture in a small tube on ice (*see* **Notes 42–44**).
2. Transfer the loop containing the sucrose/methyl cellulose drop to the cryochamber and, looking through the stereo binocular eyepieces, maneuver the loop so that it

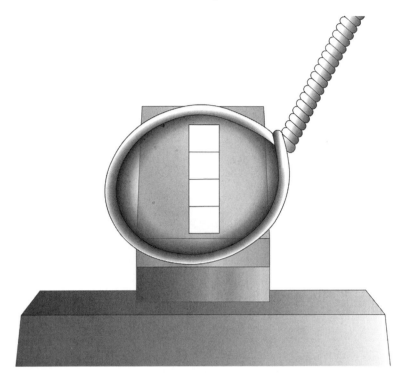

Fig. 4. The cryochamber is cooled and stabilized at −120°C, the section thickness on the ultramicrotome control is set to 70 nm, and the antistatic line is turned to 50% power. The cutting window is set and the knife is carefully advanced to the trimmed block. The ultramicrotome is left to section automatically with the cutting speed set at 0.8 mm/s. As sections start to appear on the knife surface they are pulled down using an eyelash probe. When a ribbon of sections appears on the knife surface, turn off the automatic cutting, and the anti-static line and bring a wire loop filled with one part methyl cellulose: 1 part 2.3 *M* sucrose to the sections. Pick up the sections and remove the loop from the cryochamber.

is just above a group of sections. Move the loop down until the sections touch the drop of liquid and immediately remove the loop from the cryochamber. Sections should be adhered to the bottom surface of the drop of sucrose/methyl cellulose in the loop (*see* **Fig. 4**).

3. Thaw the liquid in the loop and place it, section side down, onto a Formvar/carbon coated EM grid at room temperature. The drop can be left to dry on the grid until it is ready for labeling (*see* **Fig. 5**).

3.9. Storing Sections

1. After the sections have been placed onto the grids, they can be stored for a short time by placing them upside down onto 2% aqueous gelatin in 3.5-cm Petri dishes.

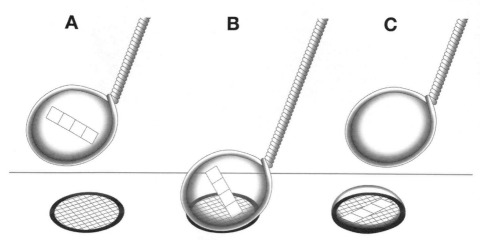

Fig. 5. Sections removed from the knife surface will thaw when brought to room temperature. (**A**) Sections on the liquid drop may be visible in a stereo dissecting microscope. (**B**) The liquid drop of methyl cellulose:sucrose with sections attached to the bottom surface is placed onto a coated specimen grid attached to a substrate. (**C**) The wire loop is carefully removed and the sections remain on the grid covered with a small drop of methyl cellulose:sucrose.

The gelatin is first solidified by cooling to 4°C and the grids are placed, section-side down onto the solid gelatin surface. To remove the grids for immunolabeling, warm the gelatin to 37°C for 30 min and then take off the grids with a 3.5-mm wire loop. The Petri dishes containing the 2% gelatin usually are stored in a moist atmosphere at 4°C to prevent them drying out.

2. Long-term storage of sections can be achieved by leaving the grids attached to the glass slide *(5)*. Sections are transferred to the grids, which are still attached to the address labels on the glass slides, with drops of 2% methyl cellulose and 2.3 *M* sucrose (mixed in a 1:1 ratio). The wire loop used for retrieving and transferring the sections to the grid is carefully removed, leaving the drop of liquid on the grid. The sections are attached to the formvar film under the liquid. When sufficient grids with sections on have been collected in this way, the whole slide is stored in a closed Petri dish at 4°C. The methyl cellulose/sucrose retrieval solution will dry over and around the sections without damaging morphology or antigenicity *(5)*. The methyl cellulose can be rehydrated and washed away by floating the grids, section-side down, on drops of cold water.

3.10. Immunolabeling of Thin Sections for TEM

Thin sections of biological material, mounted on specimen grids, can be easily labeled by floating them section-side down on small drops of polyclonal antibody. This method will work for sections of resin-embedded material as well as for thin, thawed cryosections (**Table 2**).

Table 2
Annotated Labeling Protocols Using Specific Antibodies and Protein A Gold[a]

Antibodies that bind protein A-gold (this is the same protocol described in the text):

1. 0.12% glycine in PBS	5 min
2. 10% fetal calf serum in PBS	10 min
3. Primary antibody	20 min
4. Wash with PBS	5 × 2 min
5. Protein A-gold	20 min
6. Wash with PBS	5 × 2 min
7. 1% Glutaraldehyde	5 min
8. Wash with distilled water	4 × 1 min

Antibodies that do not bind protein A-gold. A second, bridging antibody is inserted before the PAG. The bridging antibody binds to the primary antibody and will bind protein A.

1. 0.12% Glycine in PBS	5 min
2. 10% FCS in PBS	10 min
3. Primary antibody	20 min
4. Wash with PBS	5 × 2 min
5. Bridging antibody	20 min
6. Wash with PBS	5 × 2 min
7. Protein A-gold	20 min
8. Wash with PBS	5 × 2 min
9. 1% Glutaraldehyde	5 min
10. Wash with distilled water	4 × 1 min

Double labeling using two primary antibodies that bind protein A-gold *(20)*

1. 0.12% Glycine in PBS	5 min
2. 10% FCS in PBS	10 min
3. First primary antibody	20 min
4. Wash with PBS	5 × 2 min
5. Protein A-gold (1st size)	20 min
6. Wash with PBS	5 × 2 min
7. 1% Glutaraldehyde in PBS	10 min
8. 0.12% Glycine in PBS	5 min
9. 10% FCS in PBS	5 min
10. Second primary antibody	20 min
11. Wash with PBS	5 × 2 min
12. Protein A-gold (2nd size)	20 min
13. Wash with PBS	5 × 2 min
14. 1% Glutaraldehyde	5 min
15. Wash with distilled water	4 × 1 min

[a]Grids with attached sections are floated, section-side down on drops of reagent (*see* **Note 45**). When all immunolabeling steps have been performed, the grids are washed in water, and dried in the presence of a thin film of methyl cellulose containing uranyl acetate.

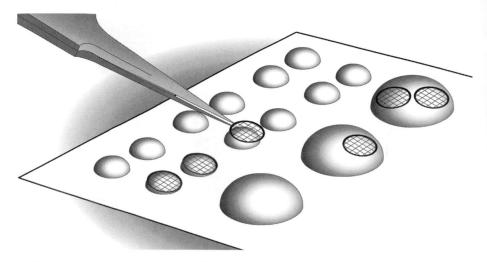

Fig. 6. Forceps are used to transfer grids to fresh drops of phosphate-buffered saline for washing after antibody incubation.

1. Attach a long piece of Parafilm to a bench surface with a drop of water (*see* **Note 45**).
2. Float specimen grids, section side down on drops (500 μL) of 0.12% glycine in PBS placed on the clean Parafilm surface. Leave for 5 min (*see* **Note 46**).
3. Transfer, and float, the grids on 10% FCS in PBS (blocking solution; *see* **Note 47**). Use the 3.5-mm diameter loop to transfer grids. Leave for 5 min.
4. Use forceps to transfer grids to 5-μL drops of rabbit antibody diluted in 10% fetal bovine serum in PBS (*see* **Note 48**). Place one grid on each drop with the section-side down. Use a wet piece of filter paper close to (but not touching) the antibody drops to create a moist atmosphere. This will prevent the antibody from evaporating away. Cover with a plastic dish and leave for 20 min.
5. Remove the grids from the antibody with forceps and float on 500-μL drops of PBS (*see* **Fig. 6**). Wash by transferring the grid to five drops of fresh PBS, leaving the grid on each drop for 3 min.
6. Use forceps to transfer the grid to a 5-μL drop of protein A-gold diluted in blocking solution (*6*).
7. Cover with a plastic dish and leave for 20 min in a moist atmosphere (*see* **step 4**).
8. Remove the grid with forceps and wash by transferring the grid to five drops of PBS leaving the grid on each drop for 4 min.
9. Float the grids to drops of 1% glutaraldehyde in PBS and leave for 5 min.
10. Wash the grids by transferring to drops of PBS, 4 × 1 min.
11. Wash with distilled water, four changes, for 1 min each. Although only a short wash, this step is important because it removes salts before incubation with uranyl acetate. Phosphates present in the sections will precipitate the uranyl salts.

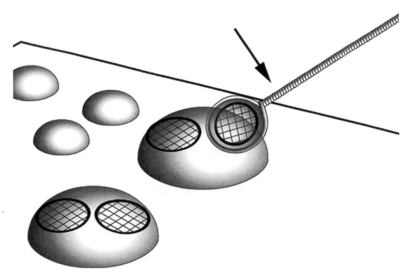

Fig. 7. Final contrasting and drying is performed after the sections have been immunolabeled with antibody and colloidal gold probes. A final wash with distilled water is followed by 10 min floating on drops of methyl cellulose:uranyl acetate on ice. The grids are first placed on small drops of the methyl cellulose/uranyl acetate mixture to remove excess water and left on a large drop for 10 min. The grids are lifted off the cold methyl cellulose/uranyl acetate mixture with a wire loop (arrow).

3.11. Contrasting and Drying Immunolabeled Cryosections

This step is essential for fine structure preservation as well as for producing enough contrast to visualize subcellular detail in the electron microscope. Many different contrasting methods exist and all can be tried using hydrated cryosections from the same specimen blocks, mounted on specimen grids. The method below was first introduced in 1984 *(7)*.

1. Mix 2% aqueous methyl cellulose with a 3% to 5% aqueous uranyl acetate solution to give a final concentration of 0.3% to 0.5% uranyl acetate (nine parts methyl cellulose to one part uranyl acetate). Mix immediately before use and keep the solution on ice at all times. More uranyl acetate in the methyl cellulose will produce more contrast in the section.
2. Put drops of the uranyl acetate/methyl cellulose solution onto a clean surface (Parafilm) on a flat surface placed on ice and float the grids, sections down, on the drops for 10 min. To wash water from the grids, transfer the grids over two small drops of the methyl cellulose/uranyl acetate mixture and onto a larger drop for final incubation (*see* **Fig. 7**). Touch the grids onto the surface of the smaller drops.
3. Loop off each grid individually from the methyl cellulose-uranyl acetate solution using a 3.5-mm diameter wire loop (*see* **Fig. 7**).

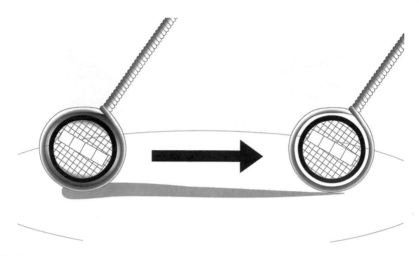

Fig. 8. Remove excess liquid from the loop by placing the loop at a 45° angle onto a clean hardened filter paper, with the section side facing down toward the filter paper. Once the methyl cellulose starts to be absorbed by the paper, gently drag the loop across the surface of the filter paper. The thickness of the dried methyl cellulose film can be varied at this stage. More methyl cellulose left on the grid will result in a thicker film. If the grid falls out of the loop it can be returned to the cold methyl cellulose/uranyl acetate mixture. The "looping" process can then be repeated.

4. Remove excess liquid from the loop by dragging the edge of the loop across a sheet of hardened filter paper (*see* **Fig. 8**).
5. Leave the grid in the loop to dry and form a thin film over the grid (*see* **Fig. 9**). Dried films of optimal thickness have gold to blue interference colors. When the film is dry, the grids can be carefully removed from the wire loop using pointed forceps to cut the film from around the periphery of the grid and lift the grid away. Store the dry grids section side up or immediately examine them in the TEM.

3.12. Examining the Immunolabeled Sections in the TEM

Production of cryosections for immunolabeling experiments has become a routine that is readily achievable by beginners (*see* **Fig. 10**). However, evaluation of sectioned specimens in the TEM is a skill that can take years to develop, and even experienced electron microscopists admit that it is a continuous learning process. The educational process can be accelerated by reading books and journals and by sharing data with experienced electron microscopists.

When the sections have been immunolabeled, the amount of information present in the section increases and adds an extra layer of complication to the evaluation process. Yet, although every antibody will have unique labeling properties, specific guidelines can be followed to aid the initial examination of sections.

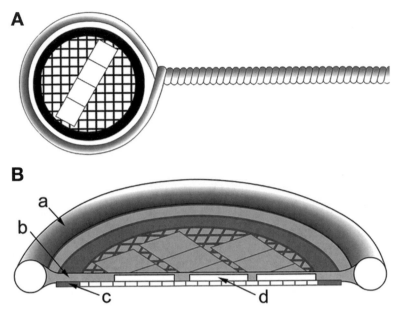

Fig. 9. (A) Each grid is dried in a loop with a thin film of methyl cellulose. Once dry, the film can be cut from the wire loop using forceps. When loose, the grid can be lifted out. **(B)** A diagrammatic cross section through the drying loop (a). The thin film of dried methyl cellulose:uranyl acetate (b) over the top surface of the grid (c), and sections embedded in the thin film (d). The thickness of the dried film will depend on the amount of methyl cellulose mixture removed from the loop. This, in turn, is influenced by the thickness of the wire used to make the loop, and the loop diameter. A thicker wire will produce a thicker film, a larger diameter loop will produce a thinner film. However, if the loop diameter is too large, the grid may fall out during drying.

1. Obtain as much information about the biological system to be examined. This may be information about the biological system, immunoblotting, light microscopy or immuno-precipitation data. Find out from others, who may have used the same specific antibodies, as much information as possible (e.g., optimal dilution, suitable blocking solutions, etc.).
2. Test the antibody by light microscopy using as many of the same reagents that will be used for EM analysis. The same specimen block can be used for EM and light microscopy (*see* **Fig. 3**). The same primary antibody colloidal gold probe that is used for EM visualization can be used for light microscopy if silver enhancement of gold is used.
3. When cryosections are first examined in the TEM the specimen morphology is strikingly different to that observed when sections of resin-embedded specimens are observed (*see* **Note 49**). The contrast of cryosections is low and often reversed with membranes appearing light (electron transparent) against a darker (electron dense) background of cytoplasm (*see* **Fig. 11**). New operators should spend time

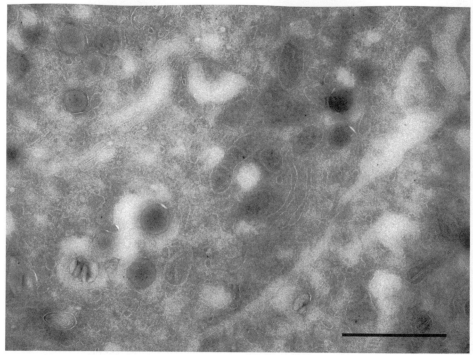

Fig. 10. Cryosectioned isolated islets of Langerhan labeled with an antibody that appears to label the mitochondria. Embedded in gelatin for cryosectioning, the section has excellent morphology and specimen contrast (courtesy of Dr. P. Nevsten). Bar = 1 μm.

     examining cryosections and perhaps comparing them with sections of resin-embedded specimens to develop an orientation map of the cryosections.

4. Once the cryosection morphology is understood, then it will be possible to evaluate immunolabeled sections. The first immunolabeling experiments with a new antibody should be with one antibody only. Understand how each antibody behaves on sections and look for specific patterns of labeling over subcellular structures. Look for a labeling pattern and develop a hypothesis from this. Quantitative methods can be used to determine the specificity of antibody labeling (*see* Chapter 15 in this volume).

5. If there appears to be no specific labeling present on the sections, it is important to determine how the gold particles are distributed. There may be too few or too many gold particles, or the colloidal gold may be present but undetectable. Detecting and differentiating specific label can be a difficult task but there are reasons why an antibody will not react on thin sections (*see* **Note 50**).

6. Understanding and documenting labeling experiments using single antibodies are an important first step. Once the patterns of single label experiments are under-

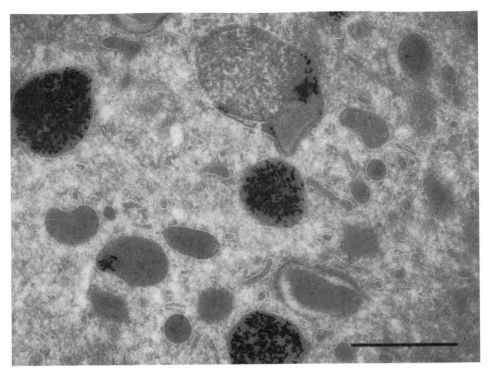

Fig. 11. A cryosectioned mouse dendritic cell containing endocytosed 5-nm BSA-gold particles. Isolated dendritic cells were incubated in culture medium containing BSA-coupled colloidal gold particles for 1 h, fixed in 4% formaldehyde and 0.01% glutaraldehyde in 100 m*M* phosphate buffer. The sections were contrasted and dried in methyl cellulose containing uranyl acetate (*see* text). Membrane-bounded structures with multiple morphologies fill with the endocytosed colloidal gold particles. Intracellular membranes have negative contrast, characteristic of being dried in methyl cellulose containing uranyl acetate. Bar = 200 μm.

stood, then it should be possible to perform multiple labeling protocols using two or more antibodies. Each antibody is applied sequentially and alternated with colloidal gold probes of different sizes (*6*).

4. Notes

1. Quantitative results and limitations of thin section immunocytochemistry: A good immunocytochemical protocol should be quantitative (*see* Chapter 15). This means that the amount of label present on the section should be directly related to the amount of antigen present in the sample. If only small amounts of antigen are present then only small amounts of signal should be present. The amount of antigen accessible to antibody on the surface of a well-fixed thin cryosection may be up to 100

less than the amount accessible in a permeabilized paraffin embedded section or a 3- to 10-μm thick section obtained using a cryostat. Evaluation of the signal obtained by light microscopy using thick sections or whole cells is recommended before progressing to a study using EM. If only small amounts of antigen are present or the antigen is located to small areas within the sample, then electron microscopic localization may not be possible without extensive sectioning.

2. Professor K. T. Tokuyasu first developed the cryosectioning method in the 1970s while working with John Singer at the University of California in San Diego *(8)*. Using a cryoultramicrotome developed by A. Kent Christensen, Tokuyasu demonstrated that chemically fixed biological material is easily sectioned at low temperature when infused with a sucrose solution *(8)*. Subsequent technical improvements of this technique and an extensive international teaching program have been the result of close collaborative efforts between Tokuyasu and his colleagues Jan W. Slot and Hans Geuze in The Netherlands, and Gareth Griffiths in Heidelberg, Germany.

3. In parallel with the technical developments introduced by Tokuyasu and colleagues, there has been an equally important series of innovations in the cryoultramicrotomes available to researchers who produce cryosections for immunocytochemistry. The newest machines on the market, when combined with the equally innovative cryodiamond knives that are currently available, make cryoultramicrotomy an easy preparation process.

4. Rapid preparation may be important in histology laboratories needing fast examination of biopsy material, and fully hydrated sections may have an increased sensitivity for antibodies that do not label resin sections. Early quantitative studies have suggested that some antibodies are able to label cryosections with a higher labeling efficiency than that observed on resin sections *(9)*, an observation that may be explained by the hydrated state of cryosections.

5. Fully hydrated sections can be contrasted, after immunolabeling, in many different ways. Sections from the same block can be treated differently to produce negative or positive contrast of varying intensity as required *(10–12)*. Labeled structures and those in close proximity to the immunolabeled structures can be imaged in different ways, allowing for increased flexibility when preparing sections for morphological assessment.

6. Cryosections can be prepared with a thickness of between 200 and 500 nm, mounted on glass coverslips and labeled with fluorescent antibodies for light microscopic observations (*see* **Fig. 3**). If sections for EM and light microscopy are prepared from the same specimen it is possible to compare labeling patterns and assess reactivity of reagents *(1)*. However, because biological tissues fixed in glutaraldehyde exhibit autofluorescence, light microscopy is best performed on specimens fixed in formaldehyde alone.

7. A basic TEM is the only requirement for imaging thawed cryosections. The TEM should be able to examine sections mounted on regular specimen grids at accelerating voltages of between 60 kV and 80 kV. No special attachments are required to visualize the dried and contrasted cryosections.

8. Although loops are commercially available for EM (e.g., Perfect loops from Electron Microscopy Sciences, PA) it is possible to easily prepare loops for use in cryosectioning using thin copper or steel wire. For reproducibility, the wire is wrapped around the stem of a drill bit of known diameter and twisted to form a loop. The twisted ends of the wire form a stem that can be inserted into a wooden stick or glued into an Eppendorf pipet tip. Loops of different diameters can be fabricated by forming them around drill bits of different diameter. For picking up sections, loops of between 2 to 3.5 mm can be used.

9. Although they are not essential, coated specimen grids can be made hydrophilic before use by placing a glow-discharge apparatus in the vacuum evaporator or in a sputter coater that has an etch function. Alternatively, a glow discharge apparatus can be constructed (*13*).

10. Methyl cellulose is a compound that is more soluble in cold water than in warm water. Use methyl cellulose with a viscosity of 25 centrepoises and add it to hot water (60°C) while mixing vigorously. When the powder has been suspended in the water, leave the solution at 4°C until it has dissolved (2–3 d). Remove dirt particles and undissolved powder by centrifuging at high speed (20,000–40,000*g*). Keep the methyl cellulose solution at 4°C at all times after it has dissolved or it will precipitate out of solution. Precipitate can be detected on sections as gray amorphous particles.

11. EM grids are made from many different materials. The most common ones are made of copper or nickel. Although we routinely use copper and nickel grids, other researchers have found disadvantages with both. If the Formvar EM grids are damaged, copper grids react with salts in the buffer to produce precipitates on the carbon film. Nickel grids are easily magnetized, making them difficult to manipulate, and they potentially can cause astigmatism when being examined in the electron microscope. Although we have not encountered these problems when using nickel grids, gold or palladium grids can be used if preferred.

12. To reduce the thickness of the Formvar film, remove the glass slide from the Formvar solution more quickly. If this does not produce the required result, the Formvar solution can be diluted with chloroform and **steps 3–9** repeated.

13. If the Formvar film is too thin, then removing the glass slide from the Formvar solution at a slower rate may help. Alternatively, the Formvar concentration in the solution can be increased by an addition of small amounts of 4% Formvar in chloroform. If necessary, excess volume can be removed.

14. The chemical processes involved when aldehydes react with biological material has been extensively reviewed in Chapter 2 of a book by Gareth Griffiths (*14*). We highly recommend this book to anyone involved with immunocytochemistry as it contains much useful information.

15. All chemical fixatives used for chemical crosslinking are hazardous materials. They will crosslink any biological material they come into contact with so they must be treated with great care. Fumes from fixatives have similar fixation properties.

16. Tissue pieces should be handled carefully throughout the whole process so that ultrastructural damage does not occur from squeezing, pulling, or shaking.

17. Preservation of morphology and retention of antigenicity are goals that are often in conflict with each other. For example, good fixation for morphology often is aimed at producing substantial extraction of cell cytoplasm to give dark (electron dense) membrane contrast on a white background. However, it is very possible that antigens of interest are part of the extracted material and important information is washed away. Conversely, too much crosslinking of cellular components may result in antigens being embedded in a dense matrix that is impenetrable to the antibody. It is for these reasons that each antibody–antigen reaction will have its own unique fixation conditions that have to be determined empirically.

18. The chemicals to be used for crosslinking and their concentration in the fixation medium can affect the antigenicity of target molecules and thus interfere with subsequent antibody binding. For example, although glutaraldehyde produces better cellular morphology, antibodies do not recognize some antigens after this fixation. Other antigens seem unaffected even by high concentrations of glutaraldehyde. So-called "insensitive" antigens can be fixed in glutaraldehyde concentrations of up to 1% for 15 to 60 min. For so-called "sensitive" antigens, formaldehyde concentrations of up to 8% can be used for exposure times of as short as a few minutes.

19. Whether using formaldehyde, glutaraldehyde, or a mixture of the two, it is important to note three things as follows: (a) Unlike glutaraldehyde, formaldehyde, when used at low concentrations (<4%) is partly reversible. It is important to avoid extensive washing after fixation with formaldehyde alone. (b) Again, unlike glutaraldehyde, the crosslinking reactions of formaldehyde occur much slower. For this reason, if preserving morphology is important, then it is best to leave the cells or tissues to be fixed for a longer time than when using glutaraldehyde. (c) Formaldehyde exists in solution as monomers and polymers and it is believed that polymers are more active at crosslinking. The polymers are present in higher numbers in more concentrated solutions of formaldehyde. The polymers also form when formaldehyde solutions are cooled to 4°C. It is possible, therefore, that short fixation times can be used with higher concentrations of formaldehyde.

20. Materials can be fixed in formaldehyde alone. We routinely use 4% formaldehyde in sodium phosphate buffer (pH 7.4), but other buffers will work. One alternative protocol is to fix in 4% formaldehyde (made from paraformaldehyde powder, w/v ratios in buffer) for 30 to 60 min and then overnight in 8% formaldehyde. Formaldehyde-fixed material should be stored and transported in formaldehyde.

21. Fixation of mammalian tissues is best performed at the temperature the cells are maintained in vivo. Once the initial fixation has occurred, the specimens can be cooled and stored at 4°C. Although physiological osmolarity is approx 360 mOsm, it does not appear that matching the osmolarity of the fixative is as important as once thought. Some of the recipes used for fixing tissues for immunocytochemistry have osmolarity values that may be more than three times higher than physiological osmolarity yet they produce useful results.

22. Factors affecting chemical crosslinking: There are reports that elevated or lowered pH of fixatives can affect morphology *(14)*. However, when fixatives are added to cell cultures growing in culture medium containing a pH indicator it is clear

that a rapid and substantial drop in pH occurs. The drop in pH has been attributed to the generation of H$^+$ ions during amine crosslinking *(14)*. The role of pH in chemical fixation has still to be fully understood.

23. If perfusion fixation of tissues is not possible, then immersion fixation is an available option. However, it is important to transfer the sample to the fixative as rapidly as possible after removal and to avoid all forms of mechanical trauma. If necessary, the tissue can be carefully sliced into small slabs as soon as it has been immersed in fixative. This will allow for rapid access of fixative to the tissue. Again care must be taken to protect fixed and unfixed tissue from any form of mechanical trauma unless this is part of the experiment.

24. An alternative way of removing cells from culture dishes is to treat them with proteinase K. Be aware, however, that this may result in the removal of surface antigens. Treat cell monolayers in ice for a few minutes with cold 20 and 50 µg/mL proteinase K in PBS. For most cell types, the lower concentration works well. Gently pipet the proteinase K solution up and down until cell free areas appear on the plastic. The cell suspension can be removed and centrifuged. Some fixative should be added to the cell suspension prior to centrifugation (to preserve the morphology) and a protease inhibitor can be added to inhibit the proteinase K activity.

25. If it is important to examine cells growing as a monolayer in their *in situ* orientation, it is possible to embed the cell monolayer in gelatin and strip the cells from the substrate while maintaining their orientation *(15)*.

26. Centrifuging small numbers of cells into a pellet is best performed in the presence of protein (BSA or FCS will work). Add a small amount (1–2%) to the suspension of cells in fixative just before centrifugation. The proteins will stop the cells from sticking to the sides of the tube during centrifugation (14,000*g* for a few seconds is sufficient to form a pellet of chemically cross-linked cells).

27. If the cell pellet can be manipulated without falling apart, then ignore this step and move onto the cryoprotection step. Gelatin is supplied by many manufacturers and comes in many grades of quality. If prepared in phosphate buffer, the salt may occasionally precipitate out of solution and the crystals cause problems during sectioning. Centrifugation of warm gelatin will remove these crystals. A good source of gelatin, which can be dissolved in phosphate buffer without forming precipitates can be obtained at low cost from food stores.

28. Gelatin embedding offers a good approach for embedding small specimens for TEM examination. There are many advantages to gelatin embedding that make it useful for even novice users. A 10% solution of gelatin will be liquid at 37°C so specimens to be embedded are never exposed to high temperatures (as may be the case with agarose embedding). If it is determined that specimens are incorrectly embedded after the gelatin has solidified, then the solution only needs to be rewarmed to 37°C for it to become liquid again. Gelatin can easily be removed from embedded specimens, or from sections, if required by washing with warm aqueous medium.

29. Unlike agarose, which encases specimens, gelatin tends to stick to biological specimens and gives extra support. This step is especially important if the specimen has to be cut into smaller pieces after being encased. Cells and tissues will fall out of

agarose when it is handled too much but gelatin remains attached to, and between, the specimens.

30. Small specimens or cell fragments can be centrifuged at 14,000g for a few seconds, in gelatin, to a known location in an Eppendorf tube. When Eppendorf tubes are used in a fixed-angle rotor, pellets always deposit at the same place—on the tube surface furthest away from the center of the rotor. If the lid hinge is placed outward, the pellet at the bottom of the tube will always be found under the hinge. Small, and even invisible pellets that have been embedded in gelatin can be easily found in this region. A few fine hairs or dust particles in the specimen make detection easier under a dissecting microscope.

31. A centrifuged pellet of gelatin-embedded specimen can be easily manipulated if it is removed from the bottom of the centrifuge tube and placed onto a glass slide wrapped in Parafilm. Warm gelatin can be poured over the pellet and a second Parafilm-coated slide pressed over the first one. Spaces placed between the two slides (folded strips of Parafilm work well for this) create a space where the specimen, embedded in a thin film of gelatin, can be found. Small fragments or tiny cell pellets can be easily seen in the gelatin film and cut out in small blocks if the gelatin is kept cool.

32. Ideally, the pieces of tissue or cell pellets should be trimmed to a pyramidal shape so that when mounted onto the specimen pin the base is wider than the top of the sample. If possible, the block face should be flat and resemble a square, rectangle or trapezoid.

33. Chemical fixation makes cell membranes permeable to the sucrose solution, and at concentrations of 1.6 *M* and greater, sucrose is frozen in a vitreous state *(7,16)*. This means that ice crystals, which can damage specimen ultrastructure, do not form even when the cryoprotected specimens are frozen slowly.

34. Although an overnight infiltration in 2.3 *M* sucrose is recommended, it is possible to leave tissue pieces and cell pellets in the sucrose for much shorter times (<30 min). However, uneven infiltration may occur in dense tissues that will make the sample more difficult to section. Some sucrose should be left around the base of the block so as to anchor the block to the specimen pin. It is also important not to allow the specimen to dry out. Not only will higher sucrose concentration produce softer blocks, but there is also a possibility of damage to subcellular morphology occurring within the sample. Sometimes during extreme drying, it is possible for small crystals of sucrose to form within the tissue. These will prevent proper sectioning and may even damage the diamond knife.

35. Start by positioning the block in front of the trim tool and cut into the block until a flat surface is produced. Then position one corner of the trim tool at one side of the block and cut at 500-nm increments into the block. Move the trim tool to the other side of the block and repeat the trimming. Turn the block one-quarter turn clockwise and repeat the side trimming process. Turn the block one-quarter counterclockwise and examine the block. Make sure the face of the block is mirror smooth and that the sides of the block are parallel to each other and perpendicular to the

knife edge. The top and bottom of the block should be parallel to the knife edge. The edges of the block should be sharp showing no signs of chipping, and clear of debris.

36. The process of sectioning cryoprotected, frozen biological material in a cryochamber is very similar to the process of sectioning plastic-embedded material at room temperature. It is worthwhile reading the chapter on sectioning resin-embedded material (*see* Chapter 5).

37. The most important factor to obtaining good cryosections is the quality of the knife. Until recently, glass knives were the preferred tools for production of cryosections. Recent improvement in diamond knife production means they can now be used routinely for the production of good cryosections. However, knife-making machines are commercially available and the theory behind glass knife making has been covered in great detail (*see* Chapter 5). A simple modification to an older knife-making machine (LKB 7800 series knife makers), commonly found in most EM laboratories, is possible *(17)*.

39. Section thickness can be varied through a wide range by manipulating the temperature of the cryochamber. At high concentrations (e.g., 2.3 *M*), sucrose improves the plasticity of frozen specimens, making them easier to section at low temperatures. For any particular sucrose concentration, cooling the block will make it harder and more brittle. Warming the block will have the opposite effect, making it softer and more plastic.

40. There are many reasons why cryosections cannot be obtained, but the most common problems are as follows: (a) loose sample: Make sure the sample is securely clamped into the specimen arm and that the sample holder is clamped to the specimen arm. (b) Loose knife: Confirm that the knife is securely fixed into its holder and that the knife holder is secure. (c) Blunt knife: It is important that the knife be sharp enough to section through the sample and produce thin sections. Good-quality diamond or glass knives are required. (d) Sucrose crystals are in the sample (this rarely occurs): Sucrose can crystallize if the sample dries before being frozen. The result is uneven sectioning qualities when thin sectioning. The unevenness is often localized to a few areas of the block. A similar effect can be observed if aldehyde, present in the specimen, cross-links gelatin used for embedding.

41. Once sections have been obtained, they will lie on the knife surface close to the edge. The sections can be carefully moved away using an eyelash mounted on a stick. It is important to separate sections that are lying on top of each other and to arrange them into small separate groups ready for picking up. When using the eyelash, it may be best to hold the sections down rather than trying to move them with the eyelash positioned under them.

42. Section retrieval using drops of methyl cellulose, first introduced by Liou et al. *(4)*, replaced a simpler retrieval method that used only a drop of 2.3 *M* sucrose. A drop of 2.3 *M* sucrose is picked up on a small wire loop and placed into the cryochamber. The drop can be used immediately to pick up the sections or left to cool down before picking up the sections. If used immediately, the sections will spread out on the drop. This step is useful in removing the compression artifact produced

during the sectioning procedure. Too much spreading, however, can damage the morphology of the lightly fixed material. The surface tension on the sucrose droplet may pull apart tissue sections or cell pellets.

43. Drops of 2% methyl cellulose alone, or 2% methyl cellulose containing 0.3% uranyl acetate can be used *(4)*. The methyl cellulose drops have a lower surface tension and will not spread out the sections when they are picked up. This results in improved morphology in the section.

44. Other pick-up solutions have been used for section retrieval with interesting results. For example, it is possible to section unfixed, uncryoprotected material and retrieve the sections from the knife using chemical fixatives *(18)*. Using this approach, the vitrified section is immediately fixed when it is thawed.

45. A layer of Parafilm on the bench will provide a clean surface on which to label and wash grids. To keep the surface clean, leave the backing paper on the Parafilm, and only expose as much of the Parafilm as needed. Do not contaminate the clean Parafilm surface by touching it or placing other contaminants on it.

46. It is important to always keep the section side of the grid wet and the back of the grid dry. Failure to keep the sections wet will result in the sample drying out. If the dry side of the grid gets wet it will not float on water droplets and will sink. Although drying the back of the grid is possible if it does get wet, salt crystals may deposit as it dries, sections may dry out or the Formvar film may break. It is best to keep the side without sections dry.

47. The antibody, diluted to a suitable concentration, is centrifuged before use. This will remove any aggregates formed during storage.

48. Blocking: Any "sticky" proteins can be used to cover nonspecific binding sites on the thawed sections. Alternative blocking agents including ovalbumin, bovine serum albumin, bovine gelatin (2%) or cold-water fish skin gelatin. Be aware that the chosen blocking proteins must not react with any of the affinity reagents to be used (e.g., rabbit serum cannot be used with protein A gold:fetal bovine serum cannot be used as a blocking agent if antibodies to serum proteins are being studied).

49. There is a misconception that cryosections produce poor morphology in biological material. If care is taken during sample preparation, sectioning and section retrieval, minimal damage will occur to the subcellular morphology. Material frozen in the presence of sucrose can be thawed out and embedded in epoxy resin, providing a simple way for evaluating morphology of frozen material. Alternatively, the cryoprotected frozen material can be embedded in resin after being dehydrated by freeze substitution using methanol as a solvent. The following are a few of the most common causes of poor morphology in cryosections: (a) Trypsinized cells: Cells growing in culture dishes are routinely passaged by treating them with a trypsin solution. When placed into fresh culture dishes, the cells will rapidly reattach to the substrate and will look normal by light microscopy within a few hours. For EM, however, the cells may require at least 2 d before returning to normal. Sometimes good morphology is obtained only after the cells have been grown for many days in the presence of excess nutrients. (b) Centrifugation: If cells are to be pelleted by centrifugation, it is best to do this after chemical fixation. If this is not

possible, then only a gentle centrifugation should be used otherwise the morphology may be adversely affected by the rough treatment given to the cells by centrifugation (c) Support film: Cryosections are usually dried after immunolabeling so that they can be examined in the electron microscope. To protect them from surface tension forces, which damage the morphology, the sections are embedded in a plastic film to protect the sections during drying. Drying artifact, caused by the plastic film over the sections being too thin, is often mistaken for poor fixation. If the morphology is better near the grid bar where the methyl cellulose film is thicker, the film may be too thin over the rest of the grid. (d) Blunt knife: Poor morphology can result from using a blunt, or suboptimal, knife for sectioning. The knife will tear the sections, pulling them apart at their weakest points. Biological materials that have only been fixed in low concentrations to form aldehyde are especially susceptible to being ripped apart in this way. In addition to producing knife marks and other obvious cutting artifacts, the damage to subcellular morphology may also be subtle and not be immediately obvious that the knife is at fault. (d) Surface tension effects: Sections of lightly fixed material also are susceptible to being ripped apart when they are first picked up from the knife surface, onto the sucrose droplet. The surface tension effects, used to spread the sections, may be too great to pull the section apart. Using a mixture of methyl cellulose and sucrose to pick up the sections, instead of sucrose alone, will avoid this artifact. (f) Mishandling trauma: Well-fixed organs and tissue from perfused animals can still show poor morphology if the fixed tissues have been dissected out carelessly. Squeezing or pulling will cause ultrastructural damage. (g) Sections too thin: Very thin sections of lightly fixed material may appear to have poor morphology. They may be the result of poor crosslinking of cellular constituents resulting in their loss from the section. However, careful examination of the sections will reveal the remaining structures to be well fixed.

50. There are many causes for poor results in immunocytochemistry, and the following is a list of common problems found when using cryosections for immunolabeling reactions, and some possible solutions. (a) No label detected. Possible causes include (1) Antigen masking or extraction: Demonstrate antibody labeling by light microscopy. When a positive result has been obtained, use the same preparative conditions for electron microscopic labeling. Some antigens are sensitive to glutaraldehyde but not to formaldehyde. Other antigens may be masked by surrounding proteins, which will result in the antigens being inaccessible to the primary antibody, protein A-gold or, in some cases, both, unless cellular material is extracted with detergents. Occasionally, antigens will not be immobilized by the chemical fixative and will either wash out of the section or be redistributed during the immunolabeling process. For such antigens, it may be best to immunolabel sections of resin-embedded material. (2) Primary antibody does not work: Causes of this problem may be the result of a wrong dilution of the antibody (perform serial dilutions to find the optimal dilution, i.e., the highest concentration that gives specific signal by low levels of non-specific binding). Make up the working dilutions fresh for each experiment and do not store diluted antibody; rather, only make up

the amounts needed for each experiment. Inadequate storage procedures producing repeated freeze-thawing (e.g., when stored in a frost-free freezer), or growth of contaminants may also result in a deactivation of the antibody. Store frozen in small aliquots or in 50% glycerol containing 0.01% sodium azide to prevent contamination. If antibodies must be stored at 4°C, then include sodium azide or chloroform to prevent microbial contamination. Prepare positive controls using other methods to demonstrate specific labeling. Change the antibody if necessary. (3) Low amounts of antigen: One of the advantages of using protein A-gold to detect antibody binding is that it can be quantitative. In some systems, the amounts of antigen present will be very low. Most often, an estimation of antigen amounts can be obtained from light microscopy and biochemical experiments. If these amounts are low, then the actual amount of labeling detected by EM will also be low. Increasing the signal will be difficult in these cases unless antigen masking is occurring. Increasing antigen accessibility by detergent treatment may help. (4) The magnification of the image is too small to detect the gold particle: Colloidal gold probes can be made in many sizes. Some multiple labeling experiments require the use of 5-nm or 4-nm gold particles. These particles are very small and require that the screen magnification in the electron microscope be approx ×18,000 to ×25,000. Gold particles less than 2 nm in particle size are extremely difficult to visualize in the transmission electron microscope without specialized detection methods (e.g., silver enhancement) (5). The image contrast is too strong: Small gold particles cannot easily be viewed on the images of resin sections that have been contrasted with large amounts of heavy metal stains. To improve the visibility of the smaller gold particles, reduce the specimen contrast as needed. (6) The protein A-gold does not bind to the primary antibody: Many antibodies, especially mouse monoclonals, do not have binding sites for protein A. If this is so, then include a bridging antibody (one that will recognize the primary antibody and bind protein A) between the primary antibody incubation and the protein A-gold step, or use secondary antibody coupled to gold particles. (7) The diluted gold preparation has been centrifuged: Unlike the primary antibodies, the gold suspensions should not be centrifuged before use. Centrifugation may remove the gold probe from the buffer, resulting in a dilution of reagent. If overcentrifuged, or if the gold particles are large, then most of the reactive probe will be removed from the buffer, which will result in lower levels of labeling than expected with the amount of label varying between experiments. In extreme cases, no labeling will be detected. (b) Background (more gold particles than expected): Background labeling is the signal produced by the primary antibody or visualization probe that is judged to be nonspecific labeling. Possible causes include the following: (1) Antibody preparation: Many background problems arise from the primary antibody. Suitable negative controls will identify this. (i) Incubate in the absence of primary antibody only and (ii) Incubate with antibodies from the same species as the primary antibody but which do not react with the cells or tissue under study. If the antibodies are producing background labeling, then it may be sufficient to treat the sections with blocking agents dissolved in PBS to prevent nonspecific binding. These include

1% gelatin (calf skin or fish skin), 2% bovine serum albumin, 5% to 10% fetal bovine serum, or 1% ovalbumin. Additionally, specimens that have been fixed with aldehydes can be treated with primary amines which will quench any free aldehyde groups with could crosslink antibodies. Useful compounds include glycine (0.12%) or ammonium chloride (50 mM) and can be dissolved in PBS. A 10 min incubation is sufficient. If background persists, then try a more dilute antibody. The best antibody dilution is that which produces the strongest specific signal possible with little or no background signal. Often, this means using the primary antibody as concentrated as possible and accepting some background labeling. Using serum instead of purified immunoglobulin fractions can also cause background labeling. The primary antibodies that usually produce the best results are affinity purified IgG fractions. If the background labeling cannot be removed, it is possible that it is a specific signal. Check the results with other labeling procedures to confirm this. Specific signal in polyclonal antibodies can often be removed using affinity adsorption techniques. (2) Background caused by blocking agents: Sometimes non-specific binding of protein A-gold can be caused by using blocking agents that bind to protein A. Rabbit or pig serum, if used as a blocking agent on the sections, will bind protein A-gold and produce a nonspecific signal. (3) Antigen migration: If the antigen has not been properly immobilized by the fixation step, it may become redistributed throughout the sample before it is washed away. Try a different fixation protocol. A good example of this phenomenon was presented by van Genderen et al. when immuno-localizing Forssman glycolipid in MDCK cells *(19)*. (4) Protein A-gold: Too high a concentration of protein A-gold can cause background labeling. Diluting the protein A-gold will remove this background. Some preparations of protein A-gold will produce background if they are not diluted in PBS containing blocking agents. The addition of a suitable blocking agent (i.e., one that does not bind protein A) will remove this background. (5) Background is specific signal: If all attempts to remove background labeling have failed, then it is possible that the signal is specific. If specific labeling is suspected, than a reexamination all the scientific data is required. (c) Contamination of the specimen: Contamination can be easily avoided by using filtered solutions and being scrupulously clean during all the handling procedures. However, even if the most careful precautions are taken, it is still possible to produce fine precipitates on sections that make evaluation of the immunolabeling impossible. Possible causes are as follows: (1) Buffers reacting with EM grids: If copper EM grids are being used, then neither the antibodies nor the gold probes can be diluted in Tris-HCl buffer. The HCl will react with the copper. If this is happening, then the antibody droplets will turn blue after the incubation. Change either the buffer in which the antibodies are stored or diluted or use nickel or gold grids on which to mount the specimens. Copper grids will react with PBS and produce small amounts of a fine precipitate over the specimen if the support film is not intact. For this, either the source of the grids can be changed or grids made from other metals substituted for the copper ones. Alternatively, grids can be covered with plastic by dipping them in Formvar solution prior to coating. (2) Uranyl acetate precipitation: Salt solutions are used for antibody dilutions and washes. If these salts are present during contrasting, then the uranyl acetate will precipitate out of solution. Wash sections with water before incubation with uranyl acetate.

Acknowledgments

The European Molecular Biology Organization (EMBO) has been generous in funding training courses that have introduced young scientists to immunocytochemical methods using cryosections. One important and unplanned result of the EMBO courses has been their effect on technique development. Bringing together a group of scientists together for 10 days each year has resulted in innovative developments and rapid spread of these innovations. This chapter was originally developed as a teaching aid for EMBO practical courses and owes much to conversations with, and encouragement from, fellow instructors. The authors have learned immunocytochemistry and cryosectioning from many people, including Professors Tokuyasu, Slot and Griffiths. This chapter also owes much to discussions with Janice Griffith (Ultrecht) and Peter Peters (Amsterdam). Participants of past EMBO courses have also helped shape the final version of this chapter. Pernilla Nevsten of nCHREM (National Center of High Resolution Electron Microscopy) Lund University, Lund, Sweden, who was a participant in a recent EMBO course kindly provided **Fig. 10**, an image from the first cryosections she produced on the course. We extend a special thanks to Ms Siva Wu and Ms Debbie Guerrero for their expert support and assistance during the preparation of this chapter. This work was supported in part by the Hope for Hearing Foundation of Los Angeles and by grants from the National Institutes of Health (5 R03 DC005335-03) and the Deafness Research Foundation.

References

1. Takizawa, T. and Robinson, J. M. (2003) Ultrathin cryosections: an important tool for immunofluorescence and correlative microscopy. *J. Histochem. Cytochem.* **51,** 707–714.
2. Griffiths, G., Parton, R. G., Lucocq, J., et al. (1993) The immunofluorescent era of membrane traffic. *Trends Cell Biol.* **3,** 214–219.
3. Takizawa, T. and Robinson, J. M. (2004) Thin is better!: ultrathin cryosection immunocytochemistry. *J. Nippon Med. Sch.* **71,** 306–307.
4. Liou, W., Geuze, H. J., and Slot, J. W. (1996) Improving structural integrity of cryosections for immunogold labeling. *Histochem. Cell Biol.* **106,** 41–58.
5. Griffith, J. M. and Posthuma, G. (2002) A reliable and convenient method to store ultrathin thawed cryosections prior to immunolabeling. *J. Histochem. Cytochem.* **50,** 57–62.
6. Slot, J. W. and Geuze, H. J. (1985) A new method of preparing gold probes for multiple-labeling cytochemistry. *Eur. J. Cell Biol.* **38,** 87–93.

7. Griffiths, G., Simons, K., Warren, G., and Tokuyasu, K. T. (1983) Immunoelectron microscopy using thin, frozen sections: applications to studies of the intracellular transport of Semliki Forest virus spike glycoproteins. *Methods Enzymol.* **96**, 466–484.

8. Tokuyasu, K. T. and Singer, S. J. (1976) Improved procedures for immunoferritin labeling of ultrathin frozen sections. *J. Cell Biol.* **71**, 894–906.

9. Griffiths, G. and Hoppeler, H. (1986) Quantitation in immunocytochemistry: correlation of immunogold labeling to absolute number of membrane antigens. *J. Histochem. Cytochem.* **34**, 1389–1398.

10. Keller, G. A., Tokuyasu, K. T., Dutton, A. H., and Singer, S. J. (1984) An improved procedure for immunoelectron microscopy: ultrathin plastic embedding of immunolabeled ultrathin frozen sections. *Proc. Natl. Acad. Sci. USA* **81**, 5744–5747.

11. Tokuyasu, K. T. (1978) A study of positive staining of ultrathin frozen sections. *J. Ultrastruct. Res.* **63**, 287–307.

12. Takizawa, T., Anderson, C. L., and Robinson, J. M. (2003) A new method to enhance contrast of ultrathin cryosections for immunoelectron microscopy. *J. Histochem. Cytochem.* **51**, 31–39.

13. Aebi, U. and Pollard, T. D. (1987) A glow discharge unit to render electron microscope grids and other surfaces hydrophilic. *J. Electron. Microsc. Tech.* **7**, 29–33.

14. Griffiths, G. (1993) *Fine Structure Immunocytochemistry*, Springer-Verlag, Heidelberg.

15. Oorschot, V., de Wit, H., Annaert, W. G., and Klumperman, J. (2002) A novel flat-embedding method to prepare ultrathin cryosections from cultured cells in their in situ orientation. *J. Histochem. Cytochem.* **50**, 1067–1080.

16. McDowall, A. W., Chang, J. J., Freeman, R., Lepault, J., Walter, C. A., and Dubochet, J. (1983) Electron microscopy of frozen hydrated sections of vitreous ice and vitrified biological samples. *J. Microsc. (Oxford)* **131**, 1–9.

17. Webster, P. (1993) A simple modification to the LKB 7800 series Knifemaker and a balanced-break method to prepare glass knives for cryosectioning. *J. Microsc. (Oxford)* **169**, 85–88.

18. Mobius, W., Ohno-Iwashita, Y., van Donselaar, E. G., et al. (2002) Immunoelectron microscopic localization of cholesterol using biotinylated and non-cytolytic perfringolysin O. *J. Histochem. Cytochem.* **50**, 43–55.

19. van Genderen, I. L., van Meer, G., Slot, J. W., Geuze, H. J., and Voorhout, W. F. (1991) Subcellular localization of Forssman glycolipid in epithelial MDCK cells by immuno-electronmicroscopy after freeze-substitution. *J. Cell Biol.* **115**, 1009–1019.

20. Slot, J. W., Geuze, H. J., Gigengack, S., Lienhard, G. E., and James, D. E. (1991) Immuno-localization of the insulin regulatable glucose transporter in brown adipose tissue of the rat. *J. Cell Biol.* **113**, 123–135.

14

Fluorescence-Integrated Transmission Electron Microscopy Images

Integrating Fluorescence Microscopy With Transmission Electron Microscopy

Paul A. Sims and Jeff D. Hardin

Summary

This chapter describes high-pressure freezing (HPF) techniques for correlative light and electron microscopy on the same sample. Laser scanning confocal microscopy (LSCM) is exploited for its ability to collect fluorescent, as well as transmitted and back scattered light (BSL) images at the same time. Fluorescent information from a whole mount (pre-embedding) or from thin sections (post-embedding) can be displayed as a color overlay on transmission electron microscopy (TEM) images. Fluorescence-integrated TEM (F-TEM) images provide a fluorescent perspective to TEM images. The pre-embedding method uses a thin two-part agarose pad to immobilize live *Caenorhabditis elegans* embryos for LSCM, HPF, and TEM. Pre-embedding F-TEM images display fluorescent information collected from a whole mount of live embryos onto all thin sections collected from that sample. In contrast, the postembedding method uses HPF and freeze substitution with 1% paraformaldehyde in 95% ethanol followed by low-temperature embedding in methacrylate resin. This procedure preserves the structure and function of green fluorescent protein (GFP) as determined by immunogold labeling of GFP, when compared with GFP expression, both demonstrated in the same thin section.

Key Words: *C. elegans*; F-TEM images; high pressure freezing; correlative microscopy; GFP; immunogold labeling; high pressure freezing; AJM-1.

1. Introduction

The pre-embedding method relies on genetic manipulation of *Caenorhabditis elegans* to create a strain with a rescuing array containing green fluorescent protein (GFP). The presence of GFP is used only to identify the genotype of embryos.

From: *Methods in Molecular Biology, vol. 369*
Electron Microscopy: Methods and Protocols, Second Edition
Edited by: J. Kuo © Humana Press Inc., Totowa, NJ

Because *C. elegans* embryos are difficult to fix with conventional chemical fixation, high-pressure freezing (HPF) is the method of choice. The agarose mount we describe for embryos is used for ultrastructural analysis, using freeze substitution with 1% osmium and 0.1% uranyl acetate.

A second postembedding method collects fluorescence images directly from thin sections. We have visualized GFP, red fluorescent protein (RFP), and fluorescent phalloidin staining in 100-nm (thin) sections after low-temperature embedding in methacrylate embedding resin *(1,2)*. This process can be accomplished using HPF or chemical fixation, but requires a cold dehydration and infiltration protocol that avoids dehydration in absolute solvent and uses a water-tolerant embedding resin. We combine immuno-transmission electron microscopy (TEM) images (anti-GFP immunogold labeling) with native AJM-1::GFP expression from the same thin section. The additional water added to the freeze substitution media to preserve GFP expression does result in some freeze damage.

1.1. Pre-Embedding Fluorescence: Laser Scanning Confocal Microscopy (LSCM), HPF, Epoxy Embedding, and TEM

Embedding embryos within a thin agarose pad allowed us to determine the genotype of embryos by LSCM and examine the same embryos by TEM *(3,4)*. We used a transgenic strain containing a lethal mutation, *ajm-1(ok160)*, rescued to viability with an extrachromosomal array containing wildtype *ajm-1* DNA fused to GFP. Animals with the array express AJM-1:: GFP and are rescued to viability. Embryos without AJM-1::GFP phenocopy *ajm-1* null mutants develop slowly and arrest as twofold embryos that are approximately twice the length of the egg.

The agarose pad is made of a thin base of high-strength agarose. The thin base layer provides the strength and toughness to keep the mount intact. An agarose pad composed only of low-melting agarose would not hold the pad together well enough to allow transfer to a HPF specimen carrier. Additionally, it is necessary for the top agarose layer to be very thin to allow imaging in a LSCM.

1.2. Postembedding Fluorescence: HPF, Acrylic Embedding, LSCM, and TEM

A complete loss of GFP fluorescence is observed in absolute ethanol *(5)*, methanol, and acetone. The addition of 1% to 5% water in the freeze substitution mixture has previously been reported to improve visualization of membranes after HPF *(6)*. Pombo et al. *(7)* viewed fluorescently labeled cryosections by LSCM and re-embedded the same sections in Epon for TEM immunogold evaluation of the same section. Luby-Phelps and co-workers *(8)* first described the detection of postembedding GFP fluorescence in one micron plastic (methacrylate) sections. We have found that GFP can be observed directly in plastic thin sections, which is more convenient than cryo-sectioning, and allows examination of the same structures by LSCM and TEM. The acquisition of a backscattered light (BSL) image, which is aligned with, and acquired at the same time as, a fluorescent image, provides surface information to align thin-section fluorescent images with TEM images from the same thin section. TEM images also can be aligned with GFP fluorescent images using patterns of anti-GFP immunogold labeling. This allows "native" GFP fluorescence to be compared with gold labeling observed by TEM. Thin sections can be viewed in a LSCM before TEM to find an area or orientation of interest or to locate a scarce antigen before TEM *(9)*. The ability to visualize antigen (GFP) in thin or thick sections is useful to confirm the presence of an antigen and to evaluate immunolabeling procedures by light microscopy.

2. Materials

2.1. General Materials Needed for Either Method

1. Glass slides.
2. Glass cover slip, 22-mm square (*see* **Note 1**).
3. High pressure freezing specimen carriers (refer to Chapter 8 in this volume).
4. 1-Hexadecene (*see* **Note 2**).
5. High-pressure freezer (we used the BAL-TEC HPM 010).
6. Petri dishes to make a humid chamber (95 and 60 mm diam. dishes).
7. 20-mL scintillation vials for resin infiltration.
8. 3-Aminopropyltriethoxy-silane, 3-APTS (Sigma, A-3648).

2.2. Pre-Embedding Method

1. Watch glass to cut open hermaphrodites.
2. Dissecting microscope.
3. Mouth pipet.
4. Microcapillary pipets (50 µL) to transfer embryos (*see* **Note 3**).
5. Scalpel to cut open gravid hermaphrodites (*see* **Note 3**).
6. Eyelash glued to a toothpick to move embryos.
7. Agarose (1% gel strength of 1000 g/cm^2 or greater for base; Invitrogen, cal. no. 15510-027; *see* **Note 4**).
8. Low melting temperature agarose (Sigma, A-9539) to immobilize embryos.
9. 2-mL Polypropylene vials with screw cap lids for freeze substitution (refer to Chapter 10 in this volume).
10. Styrofoam box, dry ice, rotary shaker in 4°C cold room (*see* **Note 5**).
11. Disposable polyethylene pipets (Fisher, cat no. 12-711-7; *see* **Note 6**).
12. Microtiter plate shaker and smaller Styrofoam box in –20°C freezer.

13. Single edged razor blade (*see* **Note 7**).
14. Sharpened tooth picks (*see* **Note 7**).
15. Rain-X (Unelko Corp., Scottsdale, AZ) to coat slides for flat embedding (*see* **Note 8**).
16. Velap (equal volumes of petroleum jelly, lanolin and paraffin).
17. Clear acetate tape (Scotch brand or similar) for use as a spacer.
18. Richardson's stain for thick sections (*see* **Note 9**).

2.3. Postembedding Method

1. Rotary mixer in −20°C freezer (*see* **Note 10**).
2. Gelatin capsules, size "00" (*see* **Note 10**).
3. Ultraviolet (UV) light (model UVL-56, long wave UV-366 nm, UVP, Inc. San Gabriel, CA) or equivalent.
4. Cardboard box to hold UV light 25 cm above gelatin capsules while in −20°C freezer (*see* **Note 11**).
5. Nickel EM finder grids, 200 mesh honeycomb (Electron Microscopy Sciences [EMS], cat. no. LH200-Ni; coat grids with formvar or other electron stable polymer; *see* **Note 12**).
6. LR-Gold embedding resin with BME (EMS kit including LR Gold and BME cat. no. 14370; *see* **Note 13**).

2.4. Buffers

Use *N*-Hydroxyethylpiperazine-N'-2-ethanesulfonate (HEPES) buffer for agarose (pre-embedding). Make agarose solutions with 0.1 *M* HEPES buffer (*see* **Note 14**).

For Immunolabeling, use the following buffers and reagents (post-embedding):

1. 0.02 *M* Phosphate buffer, pH 7.3 (*see* **Note 15**).
2. Blocking reagent, 0.5% nonfat dry milk (*[10] see* **Note 16**).
3. 20 m*M* Tris buffer (*see* **Note 17**).
4. 225 m*M* NaCl in 20 m*M* Tris buffer (*see* **Note 18**).
5. Primary antibody: Rabbit anti-GFP (Research Diagnostics Inc., cat. #RDI-GRNFP4 abr); anti-GFP antibody comparison is described in Paupard et al. *(11)*.
6. Secondary antibody: goat anti-Rabbit Lissammine Rhodamine conjugate (Jackson Immuno Research, West Grove, PA, cat. no. 111-085-003. Note: this conjuagate is no longer available).

2.5. Freeze Substitution Medium

1. The freeze substitution medium for pre-embedding method (1% osmium tetroxide with 0.1% uranyl acetate in acetone; *see* **Note 19**).
2. Freeze substitution medium for post-embedding method (1% paraformaldehyde in 95% ethanol) is prepared by adding 0.5 mL of 20% paraformaldehyde (*see* **Note 20**) to 9.5 mL 200 proof ethanol.

2.6. Preparation of 3-APTS-Coated Coverslips to Hold Half Micron Sections for LSCM

Thick sections on 12-mm round coverslips can be immunolabeling with fluorescent secondary antibodies to confirm immunoreactivity and labeling reagents and procedures prior to immunogold labeling. Thick sections are also useful to observe GFP fluorescence prior to examination of thin sections (*see* **Note 21**).

3. Methods

3.1. Pre-Embedding Method for Live Embryos

3.1.1. An Agarose Mount for Live Embryos

1. A thin (high-strength) agarose pad is formed over a standard glass microscope slide. Agarose is dissolved in 0.1 *M* HEPES (neutral pH) buffer to a final concentration of 4 to 5%. The thickness of the pad can be controlled by adding a single layer of cellulose tape over two slides on either side of the slide to be coated. Add 100 μL of melted agarose to the top of the center slide and compress the hot agarose to the thickness of a layer of Scotch tape (approximately 60 μm) with a fourth slide resting on the two adjacent tape-covered slides. Allow the agarose to solidify before sliding the top slide off. Place the slide with the agarose pad in a humid chamber. A humid chamber can be made by placing a 25-mm diameter Petri dish inside a 95-mm Petri dish and adding water to cover the bottom of the large dish.

2. *C. elegans* embryos are obtained by cutting open gravid hermaphrodites in a watch glass filled with distilled water. The desired age embryos are mouth pipetted to the agar pad and grouped together in the center of the pad using an eyelash glued to the end of a toothpick. Any visible "standing" water on the pad is pulled away from the embryos with the eyelash brush, and allowed to evaporate. Only one group of embryos, no larger than one "confocal field of view," is embedded per slide to minimize exposure to UV light. The use of a 63X objective limits the field of view to approx 10 embryos. Larger groups of embryos are impractical and difficult to navigate in the TEM.

3. Place 70 μL of 5% low-melting temperature agarose dissolved in 0.1 *M* HEPES along one edge of the agar pad. Quickly position a glass cover slip over the low melt agarose to spread it out before it solidifies. Spacers can be used on either side of the slide to obtain the correct thickness. Use the edge of a slide resting on two spacer slides on either side of the slide containing embryos to apply pressure to the cover slip and compress the agarose to a uniform thickness. Ideal mount thickness is 100 μm, which is the thickness of the smallest HPF specimen carrier configuration (*see* **Note 22**).

4. Seal the edges of the slide with hot velap to prevent dehydration; the slides are now ready for confocal microscopy.

3.1.2. Confocal Microscopy of Live Embryos

1. Acquire a focal series through the group of embryos at 1 μm intervals using the appropriate excitation wavelength (488 nm for GFP). A transmitted light image can be acquired simultaneously with each fluorescent image of GFP expression. For a Bio-Rad MRC 1024 we set laser power to 3% and photomultiplier gain to 1500 to maximize the detection of GFP. The developmental stage of the embryos can also be assessed at this time. In our experiments, embryos must attain a minimum age to be able to identify "mutant" from rescued embryos. The transmitted light image can be used to confirm the developmental stage of embryos.

2. We used an image scan size of 512×512 pixels. Increasing the image resolution results in a greater light exposure, which is unnecessary for the detection of our specific GFP. Control embryos from the same strain can be embedded in agarose, imaged by confocal microscopy, and filmed by four-dimensional microscopy to verify normal development and viability.

3.1.3. HPF and Freeze Substitution

1. Configure HPF specimen carriers to provide a 100- or 200-μm deep well. To allow access to the agarose-immobilized embryos, scrape the velap off the slide (which has already been imaged in a confocal microscope), and push the coverslip horizontally off the agarose pad using a single-edged razor blade. Cut out a small square of the agarose including the embedded embryos (maximum size is 2 mm diameter) and transfer this small pad to a bottom specimen carrier using the razor blade. The agarose pad can be pushed into a specimen carrier using a sharpened tooth pick that has been coated with 1-hexadecene to keep the tooth pick from adhering to the agarose pad (*see* **Note 23**). The specimen carrier can be filled with bacteria and 1-hexadecene.

2. HPF is preformed in a BAL-TEC HPM 010. As described by McDonald *(12)*, the filling of specimen carriers is the most important aspect of HPV. Specimen carriers must be full, without air bubbles or voids in order to reach 2100 Bar pressure. The actual freezing is initiated with the press of a button followed by a blast of liquid nitrogen. Within 2 to 3 s, the specimen holder must be removed and rapidly immersed in liquid nitrogen to separate the specimen carriers. The holders are split apart while under the liquid nitrogen, and all tools in contact with the specimen carriers are precooled in liquid nitrogen.

3. The use of a brass bottom and aluminum top (or vise versa) allows for quick identification of the bottom carrier, which is transferred, while under the LN_2, to a polypropylene freeze substitution vial filled with frozen 1% osmium tetroxide and 0.1% uranyl acetate in acetone. The vial is capped and transferred to an aluminum block also cooled in liquid nitrogen.

4. The empty holes in the aluminum block are filled with liquid nitrogen and the block is wrapped in aluminum foil and packed with crushed dry ice in a Styrofoam box taped to a a a rotary shaker. The box is shaken at 100 rpm for 3 to 4 d at −80°C (dry ice; *see* **Note 12**).

5. Shake vials on a microtiter plate shaker at 100 rpm for 2 to 3 d at −20°C.
6. Warm to 4°C overnight, transfer to room temperature, and rinse with three to four changes of dry acetone. Agarose blocks can usually be identified by the presence of the embryos.

3.1.4. Epoxy Infiltration and Polymerization for Agarose-Embedded Embryos

1. Transfer freeze substituted specimens into 20-mL scintillation vials (wash and oven dry) containing 30% Epon in acetone (EM grade) rotating on a rotary mixer for 4 hr to overnight at room temperature.
2. 50% and 75% Epon in acetone for 2 h each at room temperature.
3. Three changes of 100% Epon for 1 h each at 50 to 60°C.
4. Transfer the resin-infiltrated agarose pad to a Rain-X or Teflon coated slide with some fresh resin. Place 2 thicknesses of Parafilm on both ends of the slide as a spacer and place a second coated (Rain-X) slide over the Epon-infiltrated agarose.
5. Polymerize on a flat surface in a 60°C oven for 24 to 48 h.
6. Remove the resin from between the slides, rough up one side by scraping a razor blade across the surface, and mount/glue on a blank Epon block for sectioning (*see* **Note 24**).
7. Cut 0.5-μm sections and place on 3-APTS-coated cover slips. Stain thick sections with Richardson's stain. Cut thin and/or semi-thin sections after embryos are detected in thick sections by light microscopy.

3.1.5. TEM

Collect thin sections on formvar-coated slot grids to provide an unobstructed view of an entire section. Stain thin and ultrathin sections for 10 to 20 min in 1% aqueous uranyl acetate, followed by 3 min in Reynolds' lead before viewing in a TEM. Thinner sections may require longer staining times. Large montage images can be collected manually and "stitched together" in Photoshop (Adobe) or collected and montaged automatically with analySIS™ or similar software. When comparing different embryos within the same mount, it is helpful to make a map using a transmitted light image of the embedded group of embryos. Number the "mutant" and rescued embryos to keep track of higher magnification images. Because a TEM image may be a mirror image of the transmitted light image, flip the image in Photoshop with the same numbering scheme to be prepared for either orientation.

3.1.6. Overlaying Correlative Microscopic Images

LSCM fluorescent and transmitted light images are manipulated with Image J and Adobe Photoshop as described in **Subheading 3.2.6.** A single transmitted light image was chosen from the multiple focal planes acquired based on features of interest that were in focus. The fluorescent image was a brightest point

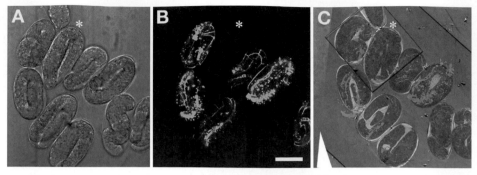

Fig. 1. The same group of *C. elegans* embryos imaged by laser scanning confocal microscopy (LSCM) and transmission electron microscopy (TEM). **(A)** A transmitted light image of live embryos acquired prior to high pressure freezing. **(B)** A brightest point projection of AJM-1::green fluorescent expression used to identify the genotype of individual embryos from a rescued transgenic strain. Scale bar, 20 μm. **(C)** TEM image of the same group of embryos after high-pressure freezing. A white asterisk (for orientation' purposes) marks the same position above a "mutant" embryo in **(A–C)**. The top left highlighted area of **(C)** is shown at higher magnification in **Fig.4A** (in color).

z projection (Image J) of all fluorescent images acquired. Embryos were alive and moving, so fluorescent image(s) may be blurred. The detection of GFP confirms the presence of the rescuing DNA and definitively identifies the genotype of embryos. **Figure 1** demonstrates the pre-embedding correlative method. **Figure 4A** (F-TEM image) integrates pre-embedding fluorescent information with a TEM image.

3.2. Postembedding Methods for Embryos and Adults

3.2.1. HPF and Freeze Substitution

1. *C. elegans* adults and/or embryos are loaded into specimen holders with bacteria off feeding plates and frozen as described in **Subheading 3.1.3.**
2. After freezing, samples are transferred to vials containing a substitution medium of 1% paraformaldehyde in 95% ethanol on a rotary shaker at 100 rpm in a Styrofoam box for 3-4 d at −80°C (dry ice).
3. Vials are transferred to a smaller Styrofoam box taped to a microtiter plate shaker, (rotating at 100 rpm) for 2 to 3 d in a −20°C freezer.

3.2.2. Infiltration and Polymerization With Acrylic Resin

1. After HPF and freeze substitution samples are infiltrated with 50% and 70% LR Gold monomer in absolute ethanol each for 8 to 12 h at −20°C.
2. 100% LR Gold for 12 to 24 h at −20°C.

3. Transfer samples from polypropylene specimen vials to glass scintillation vials in a 4°C cold room (*see* **Note 25**). Three changes in 100% LR-Gold + 0.5% Benzoin Methyl Ether (BME) (two x 8 hr changes and one overnight) at −20°C (*see* **Note 26**).
4. Just before UV polymerization, transfer samples to resin-filled gelatin capsules with precooled tweezers, add tops to gelatin capsules (oxygen will inhibit polymerization), UV polymerize for 15 to 20 h at −20°C.

3.2.3. Sectioning and Anti-GFP Immunolabeling

Thin sections (100 nm) are cut and picked up on formvar coated nickel finder EM grids. Immunogold labeling is conducted on drops on Parafilm in a hydrated chamber with care not to let the grids dry out during any of the immunolabeling steps. We made colloidal gold particles by the citrate reduction method *(13,14)*. Colloidal particles can be conjugated to a secondary antibody *(15,16)*. Because the antibody we conjugated is no longer available and commercial probes are readily available, we omit our gold reduction and conjugation methods until they can be confirmed using the currently available antibody. For immunogold labeling:

1. Block, 0.5% non-fat dry milk/ 0.02M PB for 15 min at room temperature.
2. Apply primary (anti-GFP) antibody diluted 1:1000 in 0.02M PB for 1 hr.
3. Rinse in 0.02 *M* PB 3X, then in 20 m*M* TBS 3X.
4. Apply secondary gold conjugate (Goat anti-Rabbit) diluted 1:50 in a mixture of 20 m*M* TBS containing 225 m*M* NaCl for 30 min.
5. Rinse in TBS, rinse in d H_2O, then mount for confocal microscopy.

3.2.4. Confocal Microscopy of Thick and Thin Sections

1. AJM-1::GFP is a brightly expressed GFP that can be observed in fluorescence equipped dissecting microscope. To examine whether a different GFP can be visualized and immunolabeled for TEM, examine the half-micron sections on 3-APTS-coated coverslips by LSCM before the thin sections. Half-micron sections on glass coverslips have five times the sample of a thin section, are flat, and therefore easier for acquiring BSL and fluorescent images. GFP fluorescence will be readily observed in these thicker sections, if they are present in thin sections. Half-micron sections also can be used to work out immunolabeling conditions using fluorescent secondary (detection) antibodies. Red secondary antibodies can be used with anti-GFP primary antibodies, allowing the comparison of immunolabeling to actual GFP fluorescence. GFP signal will be present throughout the entire section, whereas just the surface is exposed for antibody labeling, so secondary antibody fluorescence will be reduced compared with GFP fluorescence.
2. Examples of LSCM images of thick and thin sections of *C. elegans* worms cut from the same block are shown in **Figs. 2** and **3**. **Figure 2** is single focal plane, BSL image (**Fig. 2A**) and AJM-1::GFP image (**Fig. 2B**) of a thick section on a glass cover slip imaged with a 63X objective. **Figure 3** is a single focal plane, BSL image

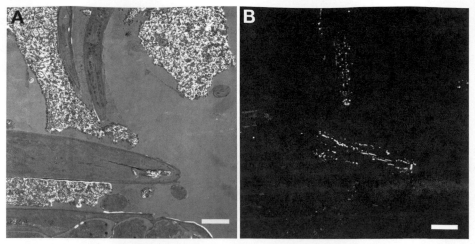

Fig. 2. Postembedding laser scanning confocal microscopy (LSCM) images of a 0.5-µm thick section adhered to a glass coverslip. (**A**) Back-scattered light (BSL) image visualized with a 63x oil objective in a confocal microscope. (**B**) AJM-1::green fluorescent protein (GFP) expression from the same thick section. BSL and GFP images are aligned with each other and acquired simultaneously. Thick sections are useful to confirm the presence of GFP and to optimize anti-GFP immunolabeling conditions by LSCM, before TEM. Bars = 20 µm.

(**Fig. 3A**) and AJM-1::GFP image (**Fig. 3B**) of a thin section on a coated nickel finder grid. These images demonstrate the reduction of GFP signal between thick and thin sections and the difference in BSL image quality when imaging a thick section on a flat surface (glass coverslip) vs a thin section finder grid, which is not flat. Although the z resolution of a LSCM is 500 nm, it may require more than one focal plane to acquire all the fluorescent signals from a thin section on an EM finder grid.

3. Thin sections are imaged by LSCM after immunogold labeling and before fixation and staining for TEM because heavy metal stains will quench fluorescence. For LSCM imaging, an EM finder grid is placed in 20 µL of water in the center of a slide with the section side up. Wet both sides of the finder grid to reduce trapping air bubbles between the grid and coverslip. Seal the edges of the coverslip with hot velap to keep the grid wet. BSL and fluorescent images can be obtained simultaneously with 488-nm excitation. We use a Bio-Rad 1024 LSCM with a 63X oil objective, 3% laser power, and collect 512 × 512 pixel images or 1014 × 768 pixel images. BSL photomultiplier gain is adjusted to obtain the best image of the surface of a thin section. Metal grid bars reflect light as seen in the upper and lower left of **Fig. 3A**. Saturation from metal grid bars is unavoidable when imaging nickel finder grids.

4. Thin sections on EM grids are not flat! Focus must be adjusted often to visualize the surface. The GFP signal is in focus when the BSL image is in focus, so the BSL

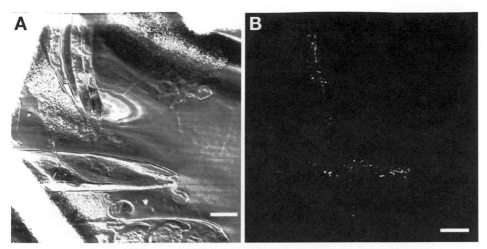

Fig. 3. Postembedding laser scanning confocal microscopy (LSCM) images of a 100-nm thin section. The same worms visualized in the thick section of **Fig. 2** are now visualized in this thin section. **(A).** A back-scattered light (BSL) image; **(B)** AJM-1::green fluorescence protein expression after high-pressure freezing and low-temperature embedding in methacrylate resin. BSL and fluorescent images are useful to visualize the same structures by LSCM and transmission electron microscopy (TEM) and to align (integrate) fluorescent information from the same thin section with TEM images. Bars = 20 μm.

image provides assurance that focus has been maintained when navigating across areas without fluorescent signal. Because the focal plane changes across a thin section, acquiring a stack of images at different focal planes is useful to acquire all the fluorescent information, especially at lower magnification, which covers a larger area. For the collection of GFP signal, avoid collecting saturated images; low-contrast images are preferable to high-contrast images, Because they can contain more information. Fluorescent image quality becomes increasingly important when images are enlarged 30 times and overlaid onto a TEM image, such as in the **Fig. 4B** inset image (top right).

3.2.5. TEM

After LSCM, methacrylate thin sections on Ni finder grids are fixed in 1% to 2% glutaraldehyde in PBS for 5 to 10 min, rinsed in dH$_2$O, dried, and stained in 0.5% aqueous uranyl acetate for 2 to 4 min and in Reynolds' lead citrate for 1 to 2 min before viewing in a Philips CM120 TEM operating at 60 to 100 kV.

3.2.6. Aligning LSCM and TEM Images From the Same Thin Section

LSCM images were prepared using ImageJ (an open-source image manipulation and analysis program available for free downloading at http://rsb.info.nih.

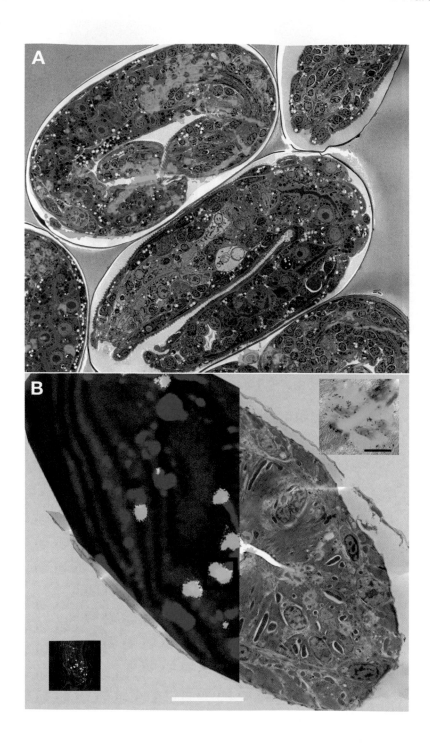

gov/ij/download.html, and Photoshop Elements (Adobe Software). Some of the light (LSCM) and EM images collected may be mirror images of each other; they can be corrected in Photoshop, command sequence Image_Rotate_Flip Horizontal. LSCM and TEM images are aligned starting with the TEM image, which usually contains more pixels; the LSCM BSL and GFP images are merged using ImageJ, Image_Color_RGB Merge command; the BSL image or image stack can be displayed in red and GFP expression in green. The BSL image also can be displayed in gray or blue; however, red provides good contrast to highlight GFP expression. Red and green image stacks are projected using ImageJ, command sequence Image_Stacks_Z project_Max intensity, with the resulting images saved as a TIFF file. This color projection or an image from a single focal plane is opened in Photoshop and pasted over a TEM image, ideally a montage composed of several higher magnification images stitched together to provide an image of the entire worm or cross section. TEM montages provide both low and intermediate magnification perspectives, which are useful for aligning with combined BSL and fluorescent images and for aligning higher magnification fluorescence with TEM images. The center image in **Fig. 4B** is a TEM montage of many images, roughly 4000 by 4000 pixels. A projected BSL (in red) and GFP (green) merged image is shown in the lower left. This image is enlarged and rotated (free transformed proportionally in Photoshop) to match the underlying TEM image. The BSL-GFP image overlies the left half of the TEM montage, whereas just the GFP signal overlies the entire TEM image.

Fig. 4. *(Opposite page)* **(A)** Shown is a pre-embedding fluorescence-integrated transmission electron microscopy (F-TEM) image, integrating green fluorescent protein fluorescence, as a green overlay on a TEM image. The TEM image is a montage of many TEM images stitched together. The center embryo without a green overlay is the same embryo marked with the white asterisks in **Fig. 1 A–C**. Field of view is 50 μm. **(B)** Shown is an example of a postembedding F-TEM image combined with anti-GFP immunogold labeling. A total of 30 separate TEM images are stitched together to form this TEM cross section of a *C. elegans* worm. This sample was HPF and freeze substituted in 1% paraformaldehyde in 95% ethanol and low temperature embedded in LR Gold. Lower left image combines a BSL image (red) and GFP expression (green) from this same thin section; actual size is 512×512 pixels. This combined BSL and GFP image is free transformed and overlaid on the left half of the TEM image. A combined BSL and fluorescent image is useful to align LSCM and TEM images. Once fluorescent information is aligned with the TEM image, it can be displayed as a green overlay, as seen in the right side of this image. White scale bar equals 5 μm. A F-TEM image of the boxed area near the center of the image is shown at higher magnification in the top right. This image integrates GFP expression from this same thin section and displays it as a green overlay on the TEM image. 25-nm colloidal gold particles label anti-GFP primary antibody. Nominal magnification, ×50,000. Black scale bar = 500 nm.

The BSL image is useful to find the same sample in the TEM and to align fluorescence with TEM images. The top right image in **Fig. 4B** is a F-TEM image acquired at 50,000X nominal magnification with GFP expression shown as a 30% transmittance green overlay. The higher magnification GFP overlay is aligned using the TEM montage for reference. The combined BSL and GFP images are used to accurately align LSCM and TEM images; a separate layer with just the fluorescence signal is usually the only layer displayed over the TEM image. The color overlay is usually at reduced opacity so as not to obscure underlying gold labeling or other TEM details.

4. Notes

1. Although coverslips of other sizes will work, a standard microscope slide is 25-mm wide. A 22-mm wide coverslip provide 1.5 mm of space on each side to seal the coverslip to the underlying slide before velap coats the edges and bottom of the slide. To position embryos in the center of a slide, we draw a small circle in the center on the back side of the slide. Embryos can be positioned in the center of the slide and agarose pad while working with a dissecting microscope. This black circle also is useful when covering the embryos with low melting temperature agarose and pushing down on the coverslip. The circle is wiped off with ethanol prior to confocal microscopy.

2. l-hexadecene is used to coat the top specimen carrier to promote release of carriers after freezing and as filler in the bottom carrier as described by McDonald *(12)* and Chapter 8 in this volume.

3. Most *C. elegans* worms are hermaphrodites, with an occasional male. Embryos are routinely collected by cutting open gravid hermaphrodites using a scalpel. Embryos and worms are transparent, so that mature worms containing embryos are readily identified in a dissecting microscope. Once embryos have been liberated from hermaphrodites in a watch glass filled with water, the optimal age or stage of embryo can be isolated with a hair glued to the end of a tooth pick. Embryos are roughly 20 μm by 50 μm in size and can be transferred to the agarose pad using a microcapillary pipet that has been heated and pulled to reduce the orifice size. Embryos are collected using the capillary action of the pipet, with a minimum amount of water and ejected in the center of the agarose pad using a mouth pipet.

4. The strength of agarose can be tested empirically by making a thin pad of agarose over a slide as described in **Subheading 3.1.1.** If the entire pad can be lifted off the slide without tearing, using a razor blade under one corner of the pad, it is strong enough to be used as a base for the correlative pad. We start with a 5% solution of agarose and dilute to between 4 and 5% with additional 0.1 *M* HEPES as needed to form the base pad.

5. Freeze substitution using a Styrofoam box has been described by McDonald *(12)*. We fit the aluminum block into a tight fitting piece of Styrofoam, which holds the block in place as the dry ice sublimes. We do not monitor the temperature during freeze substitution.

6. Any 3 mL or smaller disposable pipet with its own bulb should suffice. The use of disposable pipets for dispensing fixatives avoids contaminating more expensive pipets.

7. A single-edged razor blade is used to cut out a small piece of agarose containing the embryos to fit within the 2 mm specimen holder. The edge of a razor blade or a fine tipped weighing spatula can also be used to transfer to the specimen carrier.

8. Two standard 1 × 3-inch microscope slides are coated with Rain-X or Teflon release agent using a cotton tipped applicator or Kimwipe. Coat each slide 3X and buff clear with a Kimwipe. This coating prevents the epoxy resin from gluing the slides together.

9. Richardson's Stain *(17)*: Make: A. 1% methylene blue in 1% borax (w/v in dH$_2$0) and B. 1% Azure II in dH$_2$O. Mix equal volumes of A and B and apply to sections with a syringe equipped with a syringe filter to remove precipitates. Rinse coverslips gently after staining with distilled water.

10. A rotary mixer in a −20°C chest freezer provides good mixing while maintaining a low temperature during resin changes. To hold No. 00 gelatin capsules in a −20°C freezer for UV polymerization: we cut Eppendorf tubes in half and place the gelatin capsules inside the bottom half of the tubes. The sides of the gelatin capsules rise above the cut tubes so they can be capped with the other half of the gelatin capsule after filling with resin and sample. The Eppendorf tubes are held upright in a plastic test tube rack.

11. A cardboard box lined with aluminum foil with a hole cut out for the light source is used to distance the UV light source roughly 25 cm above the gelatin capsules. The technical data sheet provided with the resin also describes polymerization with a Thorn projector lamp (A1/209 FDX, 12V, 100W).

12. Finder grids are marked so that a specific location can be documented with a letter and number. This is helpful, but not necessary, especially if just one section is placed on a grid. In some TEMs, the outer perimeter of a grid may not be accessible, so placing a section in the center of a grid is important to assure access by TEM and will save time in a LSCM looking for the section.

13. LR Gold resin has very low viscosity and is formulated to penetrate biological tissue. Care should be taken to minimize skin contact, especially if there is a history of an allergy to methacrylates.

14. A stock solution of 1 *M* HEPES buffer is prepared by the addition of a solution of HEPES acid to a solution of HEPES base (sodium salt).

    13.01 g of HEPES sodium salt is added to 50 mL of dH$_2$O.

    23.8 g of HEPES acid is added to 100 mL of dH$_2$O.

    Add the acid solution to the base while stirring until pH = 7.3. Dilute to 0.1 *M*.

15. To prepare 100 mL of 20 m*M* Pb, add 0.052 g of sodium phosphate monobasic and 0.435 g of sodium phosphate dibasic heptahydrate to 100 mL of dH$_2$O pH to 7.4. (Use for block and for primary antibody dilutions).

16. For 1 mL of 0.5% blocking solution: add 0.005 g of nonfat dry milk to 1 mL of 20 m*M* phosphate buffer, warm to dissolve; spin for 3 min at 1500 rpm in an Eppendorf microfuge.

17. To prepare 100 mL of 20 mM Tris buffer, pH 8.2 at 25°C: add 0.142 g of Tris HCl and 0.134 g of Tris base to 100 mL of dH$_2$O, pH to 8.2.
18. To obtain 225 mM NaCl in 20 mM Tris; add 0.131 g of NaCl to 10 mL of 20 mM Tris buffer.
19. McDonald *(12)* has previously described the procedures to make various freeze substitution mixtures. Briefly, to prepare 25 mL of 1% osmium with 0.1% uranyl acetate, cool 24 mL of EM-grade acetone in a disposable 50 mL of polypropylene tube on crushed dry ice. If 0.5 grams of pure OsO$_4$ crystals are not consolidated in the bottom of the vial, freeze the unopened vial of solid osmium in liquid nitrogen. Osmium crystals will fall to the bottom of the ampule. Add 1 to 2 mL of the cold acetone to the ampule, mix, and add back to the 50-mL tube of acetone on dry ice. Repeat until all of the osmium is dissolved in the 25 mL of acetone. Add the UA in methanol (0.025g UA in 1 mL of methanol) to the acetone, keep cold on dry ice. Add 1 mL of freeze substitution mix (1% osmium tetroxide with 0.1% uranyl acetate in acetone) to each 2 mL substitution vial and freeze in liquid nitrogen.
20. For 10 mL of 20% paraformaldehyde: 2.0 g of paraformaldehyde is added to 8.5 mL of dH$_2$O; vortex, heat in a 60°C water bath for 5 min, add 0.2 mL of 0.1 M NaOH, and then add water to a total volume of 10 mL.
21. To coat 12-mm round coverslip with 4-APTS:
    Clean 12-mm round coverslips in detergent solution. Rinse 10x in dH$_2$O.
    Dehydrate using three changes in absolute ethanol.
    Coat slides in 2% solution of 3-APTS in dry acetone for several minutes.
    Rinse twice in dH$_2$O.
    Spread out coverslips on clean filter paper and dry at room temp.
22. The most difficult step is covering the embryos with the hot low-melting temperature agarose. This takes practice! This does not work with embryos that have been bleached, because they are too fragile. If visible water is removed the embryos usually do not move as the agarose spreads out over the pad. To obtain agarose of the right consistency, heat the low-melting temperature agarose to boiling in a glass test tube. The agarose is most easily dispensed by cutting 5 to 10 cm off the end of a yellow pipet tip and pre-heating the tip by rotating in the hot agarose. Fill the tip with 70 µL of hot agarose and then dispense the agarose in a line along the edge side of the slide, over the high strength agarose. The actual volume of agarose dispensed on the pad is less than 70 µL, as about half (or more) of the agarose stays in the tip. Use the 70 µL as a starting point and adjust as needed. This small volume cools rapidly, so the coverslip must be quickly placed over the low-melting temperature agarose and gently pressed down to spread a thin layer of agarose around the embryos. A microscope slide turned on edge can be used to apply pressure to the coverslip on either side of the embryos but not directly over the embryos. The top layer of agarose is thinner than the high melting temperature agarose used to make the base. If the mount is too thick, embryos will be beyond the focal depth of the confocal microscope, making it impossible to acquire a fluorescent image.
23. Loading tiny samples into specimen carriers becomes a battle with surface tension. Adhering carriers to the tops of Petri dishes with double stick tape holds carriers in place during the loading process. Double stick tape is available in two flavors,

"permanent" and a less aggressive "removable" (Scotch 667 from 3M). The less aggressive tape holds carriers without having to fight to remove them for loading in the freezing holder, just prior to freezing.

24. One side of the now polymerized agarose pad containing embryos is roughed up to remove residual Rain-X or Teflon release agent and to increase the surface area for adhesion to an Epoxy blank. The small sample piece can be attached to the blank using glue or additional Epoxy resin followed by polymerization in a 60°C oven. This orients the embryos parallel with the cutting plane to obtain a similar orientation in the TEM as the view already obtained by LSCM.

25. After freeze substitution in ethanol and paraformaldehyde worms, are usually still inside the bottom specimen carrier. Because the worms have not been exposed to osmium or tannic acid, they are translucent and can be difficult to see. To remove worms from the carriers, we use a dissecting microscope equipped with a light source to illuminate the inside of specimen holders. Ideally, we would remove worms from the holders while at −20°C. Because a walk-in freezer is not available, we use a 4°C cold room and dissecting microscope to remove samples from inside holders and transfer them to scintillation vials. With two pairs of cooled forceps, one holds the specimen cup and a second EM-grade (fine point) forceps scoops out the worms. Bacteria were used as a filler to help hold worms together; because of the light fixation, samples are only loosely held together and require gentle handling.

26. To remove resin from scintillation vials during resin changes without discarding the worms; we use a yellow pipet tip wedged on the tip of a disposable plastic pipet. The small orifice becomes plugged with larger pieces of sample and reduces the chance of discarding sample during resin changes.

Acknowledgments

We wish to acknowledge the assistance, advice, expertise, helpful discussions, and camaraderie of Jay Campbell of the White Laboratory, Ben August and Randall Massey of the Medical School EM Facility, and Dave Hoffman from the Zoology Department Instrumentation Shop. The editing assistance of Tina Tuskey was greatly appreciated. This work was supported by grants from the National Institutes of Health, (GM-058038) and from the National Science Foundation, (IBN-0712803).

References

1. Sims, P. A. and Hardin, J. D. (2005) Visualizing green (GFP) and red (DsRed) fluorescent proteins in thin sections with laser scanning confocal and transmission electron microscopy, in *Microscopy and Microanalysis 2005*, vol. 11. Cambridge University Press, Honolulu, pp. 6–7.

2. Sims, P., Albrecht, R., Pawley, J., Centonze, V., Deerinck, T., and Hardin, J. (2006) When light microscopy is not enough: Correlative light microscopy and electron microscopy, in *Handbook of Biological Confocal Microscopy*, 3rd ed. (Pawley, J., ed.), Springer, New York, pp. 846–860.

3. Koppen, M., Simske, J. S., Sims, P. A., et al. (2001) Cooperative regulation of AJM-1 controls junctional integrity in *Caenorhabditis elegans* epithelia. *Nat. Cell Biol.* **3,** 983–991.

4. Simske, J. S., Koppen, M., Sims, P., Hodgkin, J., Yonkof, A., and Hardin, J. (2003) The cell junction protein VAB-9 regulates adhesion and epidermal morphology in *C. elegans. Nat. Cell Biol.* **5,** 619–625.

5. Ward, W. W. (1998) *Green Fluorescent Proteins: Proteins, Applications and Protocols*, 1st ed., Wiley, New York.

6. Walther, P. and Ziegler, A. (2002) Freeze substitution of high-pressure frozen samples: the visibility of biological membranes is improved when the substitution medium contains water. *J. Microsc. (Oxford)* **208,** 3–10.

7. Pombo, A., Hollinshead, M., and Cook, P. R. (1999) Bridging the resolution gap: Imaging the same transcription factories in cryosections by light and electron microscopy. *J. Histochem. Cytochem.* **47,** 471–480.

8. Luby-Phelps, K., Ning, G., Fogerty, J., and Besharse, J. C. (2003) Visualization of identified GFP-expressing cells by light and electron microscopy. *J. Histochem. Cytochem.* **51,** 271–274.

9. Robinson, J. M., Takizawa, T., Pombo, A., and Cook, P. R. (2001) Correlative fluorescence and electron microscopy on ultrathin cryosections: bridging the resolution gap. *J. Histochem. Cytochem.* **49,** 803–808.

10. Kaur, R., Dikshit, K. L., and Raje, M. (2002) Optimization of immunogold labeling TEM: an ELISA-based method for evaluation of blocking agents for quantitative detection of antigen. *J. Histochem. Cytochem.* **50,** 863–873.

11. Paupard, M. C., Miller, A., Grant, B., Hirsh, D., and Hall, D. H. (2001) Immuno-EM localization of GFP-tagged yolk proteins in *C. elegans* using microwave fixation. *J. Histochem. Cytochem.* **49,** 949–956.

12. McDonald, K. (1999) High-pressure freezing for preservation of high resolution fine structure and antigenicity for immunolabeling, in *Electron Microscopy: Methods and Protocols*, 1st ed (Hajikbagheri, M. A. N., ed.), *Methods in Molecular Biology*, vol.117 (Walker, J., ed.), Humana Press, Totowa, NJ, pp. 77–97.

13. Slot, J. W. and Geuze, H. J. (1985) A new method of preparing gold probes for multiple-labeling cytochemistry. *Eur. J. Cell Biol.* **38,** 87–93.

14. Slot, J. W. and Geuze, H. J. (1981) Sizing of protein A-colloidal gold probes for immunoelectron microscopy. *J. Cell Biol.* **90,** 533–536.

15. Simmons, S. R., Sims, P. A., and Albrecht, R. M. (1997) Alpha IIb beta 3 redistribution triggered by receptor cross-linking. *Arteriosclerosis, Thrombosis & Vascular Biol.* **17,** 3311–3320.

16. De Roe, C., Courtoy, P. J., and Baudhuin, P. (1987) A model of protein-colloidal gold interactions. *J. Histochem. Cytochem.* **35,** 1191–1198.

17. Richardson, K. C., Jarrett, L., and Finke, E. H., (1960) Embedding in epoxy resins for ultrathin sectioning in electron microscopy. *Stain Technol.* **35,** 313.

15

Quantitative Immunoelectron Microscopy

Alternative Ways of Assessing Subcellular Patterns of Gold Labeling

Terry M. Mayhew

Summary

Using antibodies conjugated with colloidal gold particles, immunoelectron microscopy permits the high-resolution detection, localization, and quantification of one or more defined antigens in cellular compartments. These benefits reflect the properties of gold particles (they are electron dense, punctate, and available in different sizes) and the ability of transmission electron microscopy to resolve both particles and compartments. By relating gold marker to cellular fine structure and by taking into account the study design, three pertinent questions can be addressed. When studying a particular group of cells, we might ask: "What is the spatial distribution of gold particles between compartments within a group of cells?" and/or "Is the spatial distribution of gold particles within a group of cells random or due to preferential labeling of compartments?" When comparing two or more groups, a relevant question is: "Are there shifts in compartment labeling distributions in different groups of cells?" Recently, new ways of testing these basic questions have been developed. The efficiency and validity of all these methods rely on sampling, stereological, and statistical tools. Key processes include random selection of items at each sampling stage (specimen blocks, microscopical fields, etc.), stereological morphometry and/or unbiased counting, and statistical evaluation of a suitable null hypothesis (no difference in labeling between compartments or groups). This chapter reviews these new methods and illustrates their application with a consistent dataset.

Key Words: Transmission electron microscopy; colloidal gold; immunogold labeling; cellular compartments; organelles; membranes; filaments; frequency distributions; labeling density; relative labeling index.

1. Introduction

Immunoelectron microscopy using antibody probes conjugated with colloidal gold particles offers a high level of resolution for detecting, localizing, and

From: *Methods in Molecular Biology, vol. 369*
Electron Microscopy: Methods and Protocols, Second Edition
Edited by: J. Kuo © Humana Press Inc., Totowa, NJ

quantifying one or more defined antigens in subcellular compartments, including those antigens that are smaller than the resolution limit of the optical microscope. The detection and localization depend on the antigen-recognition specificity of the primary antibodies (*see* **Notes 1–4**) and the ability of transmission electron microscopy (TEM) to resolve different compartments and gold particles. The ability to quantify depends on the properties of the gold marker itself. In particular, colloidal gold particles are electron dense, punctuate, and available in different sizes. The precision and validity of quantification depend on the use of random sampling and estimation tools that provide unbiased estimates of the numbers of gold particles and sizes of compartments. This chapter summarizes recent developments aimed at achieving these objectives.

By relating gold particles to the fine structure of cells, the value of TEM in cell biology can be more fully exploited (*1,2*). By also taking into consideration the study design, three fundamental questions may be addressed:

1. What is the spatial distribution of gold particles between compartments within a group of cells?
2. Is the spatial distribution of gold particles within a group of cells random or is it due to preferential labeling of compartments?, and
3. Are observed patterns of compartment labeling the same or different in control and experimental groups of cells?

Recently, new ways of testing the alternative possibilities have been developed (*3–7*). All rely on random selection at each sampling stage (e.g., from cell pellets to specimen blocks to microscopical fields), unbiased counting or morphometry, and statistical evaluation of a null hypothesis (no difference between compartments or between cell groups).

In practice, question 1 is the easiest to answer. All that is required is the ability to count numbers of gold particles overlying each compartment of interest. The compartments may be membranes or organelles or cytoskeletal filaments/tubules (hereafter referred to simply as filaments) or any mixture of the three. The observed distribution of gold label provides a labeling frequency distribution and identifies compartments harboring large pools of antigen (*see* **Notes 1–4**). This simple gold-counting approach is particularly useful when gold marker is distributed across several compartments or when antigen localization varies with cell activity.

Finding an answer to question 2 is more demanding because the methods are ideally suited to analyzing gold label in one category of compartment (either organelles or membranes or filaments). However, it may be possible to modify them to analyse mixtures of some categories (notably, organelles and membranes). The methods also require information about compartment sizes. Two possible approaches exist, and both involve calculating a relative labeling index (RLI) for each compartment. This technique may be thought of as representing the degree to which a compartment is labeled in comparison with random (i.e., nonpreferential) labeling. For random labeling, the expected RLI = 1, but the expected RLI > 1 when there is preferential labeling. An efficient way of estimating RLI for a given compartment is to relate observed gold counts to the gold counts that would be expected if labeling was random. The latter can be created by a stereological device, viz, superimposing test lattices of points (for organelles) or lines (for membranes) and generating chance encounters with compartments. In fact, for each compartment, RLI_{comp} = observed gold particles/ expected gold particles *(3)*. Observed and expected distributions of gold particles may be compared using a two-sample χ^2 analysis for which the statistical degrees of freedom are determined by the number of compartments selected.

An alternative approach to addressing question 2 is to estimate RLI from a labeling density (LD), which relates the gold count to compartment size. Again, an efficient way of estimating the sizes of compartments is to relate gold counts to stereological test points (the numbers which hit organelle profiles) or test lines (the number of intersections with membrane traces or filament transections). In this way, a value of LD can be determined for each compartment and for the cell as a whole. For a given compartment, the ratio LD_{comp}/LD_{cell} provides an estimate of RLI_{comp} *(4)* and, again, observed (gold) and expected (point or intersection) counts can be compared by a two-sample chi-squared analysis.

Answering question 3 is very simple in practice *(5,7)*. No information about compartment size is needed and the magnification within and across groups need not be known or standardized. The method requires only that gold particles are counted and assigned to specified compartments (membranes or organelles or filaments or a mixture of all three). Distributions of raw gold counts in different groups are compared by contingency table analysis with the degrees of freedom now being determined by the numbers of compartments and numbers of study groups. Practical details of the implementation of these new methods are presented.

2. Materials

It is assumed that investigators have undertaken an appropriate immunocytochemical study using effective antibodies, gold, and cell/tissue preparation procedures for ultrathin transmission electron microscope (TEM) sectioning (*see*, for example, reviews by Griffiths *[1]* and Skepper *[8]*). The following provides detailed advice on how to sample such cells, abstract biologically relevant quantitative information from sections, and subject quantitative data to statistical testing.

3. Methods

3.1. Sectioning and Sampling

The aim of sectioning in TEM is to reveal the internal structure of a specimen (organ, tissue, cell culture, etc) at an acceptable level of resolution. In practice, optimal lateral resolution in the final image is achieved by cutting ultrathin (usually 40- to 90-nm thick) sections. This process has a number of consequences. First, it leads to loss of dimensional information. Second, thin slices provide an extremely small fraction of the total specimen and so must be sampled appropriately.

3.1.1. Loss of Dimensional Information

As a result of ultrathin sectioning, cellular compartments will not, in general, display their actual (three-dimensional) sizes or shapes on the plane represented by the cut surface of the section. Only objects smaller than the section thickness, totally contained within it and possessing sufficient opacity to be easily contrasted with their containing matrix, will be observed without loss of dimensionality. In all other circumstances, sectioning results in some loss of dimensional information. Thus, an organelle (which occupies a volume, cm^3) will appear on a section plane as a transection or profile that possesses a certain area (cm^2). A membrane (which presents as a surface, cm^2) will appear as a trace or boundary with a certain length (cm^1). A thin filament (which presents as a length, cm^1) will appear as one or more transections, the number (cm^0) of which is related mainly to filament length.

It is intuitively obvious, that the areas, lengths, or numbers observed on sections also will be determined by the position and orientation of section planes. Thus, although cutting a sphere will generate a circle, the area of that circular profile will vary according to whether the sphere was sectioned near its equator or its periphery. By the same token, a slice across a cube may appear on a section plane as a square, rectangle, or triangle. Also, depending on orientation, circles may be generated by slices through objects of different shapes (e.g., spheres, ellipsoids, cones, cylinders). The lesson here is that single sections through objects are not, in general, representative and, usually, they are misleading in terms of object size, shape, and number, which reinforces the need to take multiple samples that cover all locations and orientations within the specimen.

3.1.2. Importance of Random Sampling

TEM images (fields of view) on ultrathin sections account for a tiny fraction of the original specimen. For example, a single microscopical field on a 70-nm thick section might occupy an area of 25 μm^2 and represent a volume of only

1.75 μm³. If this field was sampled from a rat kidney (volume, say, 1.75 cm³), it would require a workload of 1×10^{12} fields to comprehensively cover the whole organ!

Given the gross disparity between specimen and field sizes and the impracticability of surveying the whole specimen, we need to be confident that the final sample of TEM images constitutes an unbiased selection of all parts of the specimen. In TEM, it is usual to sample specimens in a hierarchical fashion, and this process is termed multistage sampling. For example, each animal in a study group might provide an organ which, at stage I, is sampled to provide tissue blocks. At stage II, blocks are sampled by cutting sections and mounting them on support grids. At stage III, images are selected from the sections on those grids.

A chain is only as strong as its individual links. By analogy, the final set of images in multistage sampling is only as good as the process of selection adopted at each stage. This means that the selection of items (blocks, sections, images, etc.) should be randomized at every stage if the images are to represent an unbiased sample of the original specimen. By definition, random sampling provides an unbiased selection of items because it affords every part of the specimen the same chance of being selected and this is important regardless of the nature of the compartments (organelles, membranes, filaments) under investigation. Random sampling also can allow every orientation of the specimen the same chance of being selected. Indeed, the combination of random location and random orientation is necessary when dealing with compartments made up of membranes or filaments or mixtures of organelles, membranes and filaments *(9–13)*.

Random sampling can take different forms but, although these all share the property of unbiasedness, they are not equally efficient. In simple random sampling, the position (and, where necessary, the orientation) of every item is selected at random (e.g., by using the "lottery method," a table of random numbers or the random number generator of a personal computer). In contrast, with systematic uniform random (SUR) sampling, the position and orientation of the first item are selected at random, and a predetermined pattern decides the positions and orientations of other items *(6,10,13–15)*. This repeating pattern (the sampling interval) provides a more even coverage of the whole specimen and helps to explain why SUR sampling is usually more efficient than simple random sampling *(10)*.

3.2. Defining Cellular Compartments and Counting Gold Particles

3.2.1. Defining Compartments

Cellular compartments take the form of organelles (volumes), membranes (surfaces), or cytoskeletal filaments (linear features). A given compartment may be

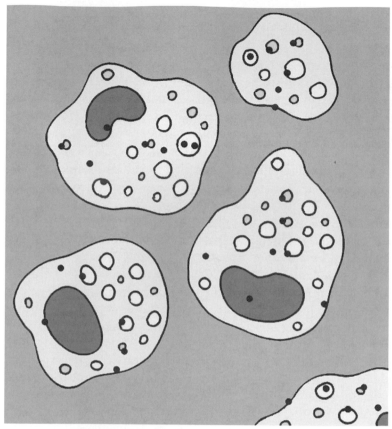

Fig. 1. Diagrammatic representation of sectioned cells within a cell pellet. Gold particles (black circles) are found on different subcellular structures associated with the nucleus, cytosol and intracytoplasmic granules. Are these gold particles preferentially located on any particular compartment? And, if so, which one(s)?

defined so as to be homogeneous (e.g., the sum total of all mitochondria constitutes the chondriome, an organelle compartment) or heterogeneous (e.g., early and late endosomes might be grouped together as another organelle compartment, the endosomal compartment; apical and basolateral membrane domains of a polarized epithelial cell might be grouped together as a membrane compartment, the plasma membrane). Having sampled randomly and selected optimal operating magnifications (the lowest sufficient to be able to identify each compartment), the next step is to define the compartments to be included in the analysis. The final choice will be governed largely by the aims of the study, prior experience and expectation. Thereafter, gold particles can be counted on the random sample of microscopical fields and assigned to the appropriate compartments (*see* **Fig. 1**). It is sensible to include in analyses not only the main ("labeled") compartments of principal interest but also other ("unlabeled" or "background-labeled") compartments not of individual interest (*see* **Note 4**). The latter can be lumped together for convenience as a composite compartment *(3)*.

It is prudent to aim for a compromise between the precision of intracellular localization (determined by the ability to identify several separate compartments) and noise (determined by variation of gold counts within a compartment). A reasonable approach is to limit the number of compartments to between 3 and 15. Having too few compartments is likely to compromise functional interpretation. Generally speaking, increasing the number of identified compartments will improve the precision of localization but may also increase the noise, especially for infrequent, small or poorly labeled compartments.

Comparing distributions in different groups involves calculating expected gold particles in each compartment and, for the resulting statistical testing to be exact (in terms of probability levels), no more than 20% of expected values should be less than 5 and no expected value should be less than 1. If these requirements are not satisfied, it would be expedient to reduce the number of compartments (by omission or conflation) or to count more gold particles.

3.2.2. Counting Gold Particles

Having chosen and identified interesting compartments, gold particles can be counted and each must be assigned to the compartment with which it is associated. Colloidal gold particles are available in various sizes (usually 5–15 nm) and this allows multiple-labeling experiments in which different antigens are localised simultaneously *(16)*. Although the smallest gold particles more closely resemble "points," larger particles may appear to straddle different compartments. Therefore, to count them rigorously, it is necessary to adopt an unbiased counting rule, the simplest of which is to count a particle as belonging to a given compartment if its centre lies on that compartment (*see* **Note 5**).

3.3. Relating Gold Particles to Compartments

Table 1 provides a synthetic dataset representing the numbers of gold particles counted on randomly sampled ultrathin sections of cells. The study design is one in which there are three groups: a control group and two experimental groups. The latter might differ in terms of the type, level, or duration of treatment. For consistency, and for descriptive convenience, the cells in each group have been partitioned into five compartments (1–5). This dataset will be used to illustrate how the three basic questions identified above can be tested.

3.3.1. Compartmental Frequency Distributions

The first of our trio of questions is: "What is the spatial distribution of gold particles between compartments within a group of cells?" Determining this

Table 1
Observed Numbers of Gold Particles Labeling the Same Set
of Defined Organelle Compartments (1–5) in Randomly Sampled Cells
From Different (Control and Experimentally Treated) Groups

Compartments	Control Group	Treated Group A	Treated Group B
1	41	34	65
2	7	6	6
3	29	36	14
4	38	70	40
5	86	107	94
Total on cell	201	253	219

distribution is relatively straightforward. All that is required is the ability to count numbers of gold particles overlying each identified compartment. In this context, compartments may be organelles or membranes or filaments or any mixture of the three. The observed distribution of gold label provides a labeling frequency distribution and indicates which compartments harbor the larger pools of antigen. Recently, we have shown that these distributions can be determined with minimal investment of effort involving counts of only 100 to 200 gold particles over one to two section grids *(6)*. This approach is especially useful when gold marker is located in several compartments or when antigen location changes with cell activity (or, more generally, with cell group).

Observed gold counts may be given as numerical frequencies (*see* **Table 1**) or converted into percentage-frequency distributions (*see* **Fig. 2** and Watt et al. *[17]*). Because the aim is merely to identify those compartments containing large pools of antigen, as many compartments as deemed appropriate (up to the recommended maximum of approx. 15) may be selected. Of course, although this secures better resolution in terms of where gold particles are located, selecting too many will lead to greater variability in gold particle counts in those compartments containing little antigen. In **Fig. 2**, the gold counts for the control group are presented as a percentage-frequency distribution and suggest that the highest proportions of label are found in compartments 1, 4, and 5.

3.3.2. Labeling Density

Although estimating compartmental frequency distributions provides a quick and simple screen as to where antigen resides, the observed labeling frequencies are "size-weighted," i.e., larger compartments occur, and are sampled, more often than smaller compartments. Therefore, even if two compartments contain the same concentration of antigen, and label equally efficiently *(1,18)*, more gold particles will be observed and counted on the larger compartment.

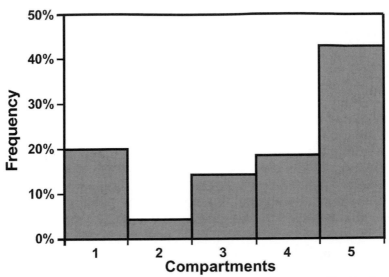

Fig. 2. Diagrammatic representation of the percentage-frequency distribution of gold particle label between five identified compartments in the control group of cells (*see* data in **Table 1**).

An alternative approach is to determine LD for each compartment (*see* **Notes 3** and **6**). This approach has tended to be applied only when, after a preliminary inspection, clear differences in compartment labeling are detected. However, failure to estimate LD values may lead to more subtle but still functionally significant, differences being overlooked.

LD values relate numbers of gold particles to the sizes of compartments. Therefore, two pieces of information are needed, viz. the number of gold particles associated with a compartment and some measure of compartment size. The latter can be obtained efficiently and unbiasedly by applying design-based stereological methods *(1,13,14,19)*. With these tools, lattices of test probes (points and lines) are superimposed randomly on sectional images and used to identify and count chance encounters with compartments. Unbiased estimation depends crucially on randomised sampling *(4,10,20)*. This is so because LD estimation depends on sampling compartments according to their relative sizes. Random sampling for both the position and orientation of section planes achieves these sampling requirements.

Depending on the nature of the compartment, LD values may be expressed in different ways. Usually, they are presented as numbers of gold particles per area (organelle profiles) or per length (membrane traces and filament transections) on the section plane or, less often, per volume of compartment *(1,18)*. LD values give an indication of local antigen concentrations (gold particles per μm, per μm² or per μm³). A compartment labeling preferentially might reasonably be expected to show a higher LD than one which is antigen-free, although it should be borne in mind that labeling efficiency may vary between compartments (*see* **Note 3**).

Before LD values can refer to the absolute sizes of compartments, the calibration of instrument magnification is required to convert areas and lengths on the magnified image into real dimensions on the scale of the specimen. As we shall see later, a stereological approach allows us to adopt more efficient estimators of LD that do not require knowledge of the magnification adopted or of the lattice constants used to convert test point counts into organelle profile areas or line intersections into membrane trace lengths.

3.3.3. Relative Labeling Index

If labeling efficiencies were identical in different compartments, compartmental LD values would provide relative measures of antigen concentrations (*see* **Note 3**). Unfortunately, labeling efficiencies often differ between compartments, and a useful alternative approach is to compare the LD of a given compartment with that of the entire cell. If compartments labelled randomly, we would predict that they would all share the same LD value (i.e., that of the cell as a whole). Therefore, the ratio between the LD of a compartment (LD_{comp}) and that of the entire cell (LD_{cell}) provides a measure of the degree to which the observed labeling of a compartment departs from a predicted (random) pattern. We have termed this measure the RLI and, as we shall see, it can be calculated from LD values or from raw gold counts and counts of chance encounters between stereological test probes (points, lines) and sectional images of compartments (profile areas of organelles, trace lengths of membranes or filaments).

3.4. Comparing Labeling Patterns

3.4.1. A Portfolio of Methods

Recently, a portfolio of different methods for rigorously comparing labeling patterns in different compartments and groups has been developed (*3–5,7*). All of these methods depend upon multistage random sampling (from specimens to fields), unbiased counting or morphometry, and statistical evaluation of a null hypothesis (no difference between compartments or groups) by χ^2 analysis with or without contingency table analysis. Statistical testing relies on comparing observed and expected distributions of gold particles and, in the case of comparing compartments within a group of cells, the expected distribution can be calculated by randomly superimposing lattices of stereological test probes (test points or lines).

3.4.2. Setting Up an Expected Random Distribution of Gold Particles Using Stereological Test Probes

A convenient way of constructing such a distribution is to take advantage of a fundamental stereological principle. For example, if all the compartments are volumes (i.e., organelles rather than membranes or filaments), test point probes distributed randomly in space hit compartments with probabilities determined solely by compartment volume. This is most simply appreciated by considering the extremes. Thus, if a cell is wholly occupied by compartment X, then random points hitting the cell will certainly hit X (i.e., the probability of hitting X is 1). In contrast, if a cell lacks ingredient X, then random points hitting the cell will certainly not hit X (i.e., the probability of hitting X is 0). By the same token, a compartment occupying 32% of cell volume has a probability of being hit equal to 0.32. We may take advantage of this principle by randomly superimposing a systematic array of test points on cell sections and recording the point totals that hit each compartment. The resulting distribution represents the expected spread of randomly positioned points and can be used to compare randomly-distributed gold particles with the observed distribution of gold particles.

In an exactly analogous way, expected distributions can be calculated for compartments that are surfaces (membranes) or lengths (filaments). These calculations are achieved via another basic stereological principle but, instead of applying test points, a lattice of test line probes is used. If such lines are randomly distributed and orientated in space, they will intersect membranes with probabilities determined by their surface areas and filaments with probabilities determined by their lengths. Now, the resulting distribution represents the spread of intersections for randomly positioned lines and can be used to compare expected and observed distributions of gold particles.

3.4.3. Comparing Different Compartments in the Same Group of Cells

In the context of our original trio of questions, the purpose here is to ask whether the observed distribution of gold particles between compartments within a given set of cells is random or non-random. In short, are some compartments preferentially labeled?

3.4.3.1. DIRECT ESTIMATES OF RLI

One way of addressing this question is to construct the random (expected) distribution and then test whether or not the observed gold particles are consistent with, or depart significantly from, this distribution. Imagine that this has been done for, say, the organelles in the control group of cells and produced the data summarised in **Table 2**. Because the total points on the cell (244) do

Table 2
Specimen Example of Analyses to Address the Question
"Are There Differences in the Observed Distributions of Gold Particles
Between Compartments in the Control Groups of Cells?" Deriving Relative
Labeling Index (RLI) From Observed and Expected Gold Particles

Compartments	Observed Golds	Test Points	Expected Golds	RLI (= obs/exp)	χ^2
1	41	13	10.709	$RLI_1 = \textbf{3.83}$	**85.68**
2	7	10	8.238	$RLI_2 = 0.85$	0.19
3	29	5	4.119	$RLI_3 = \textbf{7.04}$	**150.29**
4	38	56	46.131	$RLI_4 = 0.82$	1.43
5	86	160	131.803	$RLI_5 = 0.65$	15.92
Total on cell	201	244	201	$RLI_{cell} = 1.00$	253.52

For total χ^2 253.52 and 4 degrees of freedom (2–1 groups by 5–1 compartments), $p < 0.001$. The distribution of gold marker is not random and the highlighted values of RLI and χ^2 show that there is preferential labeling of compartments 1 and 3 (criteria: RLI > 1 and χ^2 makes substantial contribution to total of 253.52).

not coincide with the observed number of gold particles (201), we must now combine these totals to calculate the expected gold particles for each compartment. For example, the expected number of gold particles for compartment 1 is equal to $13 \times 201/244 = 10.709$. The benefit of doing this is that we now have a direct measure of the degree to which a particular compartment is labelled in comparison to random labeling. This RLI is calculated by dividing observed gold particles by expected gold particles. Therefore, for compartment 1, RLI = $41/10.709 = 3.83$ approximately, which implies that this compartment displays almost four times greater labeling than might be expected for a purely random deposition of gold particles. The corresponding partial χ^2 for any compartment is calculated from observed and expected gold counts as follows:

$$(\text{observed} - \text{expected})^2/\text{expected golds}$$

which, for compartment 1, amounts to $(41-10.709)^2/10.709=85.68$ approximately.

For the full dataset in **Table 2**, the total χ^2 is 253.52 for 4 degrees of freedom and $p < 0.001$. Therefore, the null hypothesis of no difference from random labeling must be rejected.

If observed and expected distributions are different, the criteria for deciding on preferential labeling of a compartment are twofold: first, the value of RLI must be greater than 1 and, second, the partial χ^2 must account for a significant proportion (say 10% or more) of total χ^2. On these grounds, the control cells display preferential labeling of compartments 1 and 3 (*see* **Table 2**). Although this seems to be inconsistent with the interpretation of **Fig. 2**, it is now clear that compartments 4 and 5 contain larger proportions of gold label primarily because they are larger compartments and not because they are preferentially labeled.

For statistical testing by χ^2 analysis, it is important to include a mix of labeled and unlabeled compartments, especially if the labeled compartments have similar RLI values *(3)*. The use of χ^2 analysis also imposes minimal conditions regarding the numbers of expected gold particles on individual compartments. For example, it is recommended that no more than 20%, and preferably none, of the compartments should have less than five expected gold particles *(5,7)*. This recommendation may influence the choice of compartments or the numbers of observed gold particles to be counted. For instance, if it is desirable to identify separately a rare or poorly labeled compartment, more effort will be required in counting gold particles associated with it. If the compartment is not of individual interest, it would be sensible simply to merge that compartment into some larger compartment such as "residuum" or "rest of cell."

3.4.3.2. RLI via LD

An alternative way of testing for nonrandom labeling of compartments within a group of cells is to compare the LD of each compartment with the LD of the cell as a whole. Conventionally, LD is expressed as the number of gold particles per area of organelle profiles or per length of membrane traces *(1,18,19, 21)*. However, an efficient shortcut is to use stereological principles to provide simplified estimators of LD. For instance, a convenient way of expressing the LD of an organelle compartment is to use the estimator

$$LD_{comp} = \text{number of gold particles on compartment/}$$
$$\text{number of test points on compartment.}$$

If all compartments within the cell were labeled randomly, then we would expect them all to display the *same* LD value. Therefore, the LD value for the cell as a whole (LD_{cell}) represents a very useful internal reference and, in fact, the RLI of a given compartment is given by the equation

$$RLI_{comp} = LD_{comp}/LD_{cell}.$$

Table 3 provides LD values for each compartment in the control-cell group. For example, the LD for compartment 1 is estimated as 41/13 = 3.154 golds per point, whereas that for the cell as a whole is estimated as 201/244 = 0.824 golds per point. Consequently, the RLI for compartment 1 is given by 3.154/ 0.824 = 3.83 with a partial χ^2 of 85.68. Again, it is clear that control cells have preferential labeling of compartments 1 and 3 (**Table 3**).

Since the methods based on analysis of RLI or LD *(3,4)* were introduced, they have been applied to localize a variety of antigens in diverse cells and tissues

Table 3
**Specimen Example of Analyses to Address the Question
"Are There Differences in the Observed Distributions of Gold Particles
Between Compartments in the Control Groups of Cells?" Deriving Relative
Labeling Index (RLI) From Labeling Density (LD)**

Compartments	Observed Golds	Test Points	LD (Golds/Point)	RLI $(= LD_{comp}/LD_{cell})$	χ^2
1	41	13	$LD_1 = 3.154$	**3.83**	**85.68**
2	7	10	$LD_2 = 0.700$	0.85	0.19
3	29	5	$LD_3 = 5.800$	**7.04**	**150.29**
4	38	56	$LD_4 = 0.679$	0.82	1.43
5	86	160	$LD_5 = 0.538$	0.65	15.92
Total on cell	201	244	$LD_{cell} = 0.824$	1.00	253.52

For total χ^2 253.52 and 4 degrees of freedom (2–1 groups by 5–1 compartments), $p < 0.001$. The distribution of gold marker is not random and the highlighted values of RLI and χ^2 show that there is preferential labeling of compartments 1 and 3 (criteria: RLI > 1 and chi-squared makes substantial contribution to total of 253.52).

(4,5,7,22–32). The methods deal effectively with between-compartment labeling differences when all compartments belong to the same category, for instance, they are all organelles or all membranes. Unfortunately, they are not ideal for dealing with situations in which antigens of interest are found in different categories of compartment or translocate from one to another, for example, from the cell membrane to an intracellular vesicular organelle (*see* **Note 7**).

3.4.4. Comparing the Same Set of Compartments in Different Groups of Cells

"Is the distribution of gold particles between compartments in a given set the same or different in different groups of cells?" To answer this question, the observed (rather than percentage) frequency distributions of raw gold counts in different groups of cells are compared directly by contingency table analysis (3,5,7). **Table 4** illustrates how this is achieved. For a particular compartment in a given group, the number of expected gold particles in a contingency table is calculated by multiplying the corresponding column sum by the corresponding row sum and then dividing by the grand row sum. For example, the expected gold particles on compartment 1 in treated group A is given by 253 × 140/673 = 52.63. With an observed gold count of 34, the partial χ^2 amounts to $(34 - 52.63)^2/52.63 = 6.59$ approximately.

The total χ^2 value for these cell groups is 29.29 and, for 8 degrees of freedom, $p < 0.001$. Therefore, the null hypothesis of no difference in distributions between groups must be rejected. Inspection of partial χ^2 values reveals that

Table 4
Specimen Example of Analyses to Address the Question
"Are There Differences in the Observed Distributions of Gold Particles
Between Control and Experimentally Treated Groups of Cells?"
Values for Each Group are Observed (Expected) Gold Particles

Compartments	Control Group	Treated Group A	Treated Group B	Row Sums	χ^2
1	41 (41.8)	**34 (52.6)**	**65 (45.6)**	140	0.02, **6.59, 8.30**
2	7 (5.7)	6 (7.1)	6 (6.2)	19	0.31, 0.18, 0.01
3	29 (23.6)	36 (29.7)	**14 (25.7)**	79	1.24, 1.34, **5.33**
4	38 (44.2)	**70 (55.6)**	40 (48.2)	148	0.87, **3.71**, 1.38
5	86 (85.7)	107 (107.9)	94 (93.4)	287	0.00, 0.01, 0.00
Column sums	201	253	219	673	29.29

For total χ^2 29.29 and 8 degrees of freedom (3–1 groups by 5–1 compartments), $p < 0.001$. The distributions are different and the highlighted values of particle number and χ^2 show that cells in treated group A have fewer-than-expected gold particles on compartment 1 and more-than-expected on compartment 4. Cells in treated group B have more-than-expected gold particles on compartment 1 and fewer-than-expected on compartment 3.

compartments 1, 3, and 4 in treated groups are the principal contributors to these differences. Cells in treated group A have fewer-than-expected gold particles on compartment 1 but more-than-expected particles on compartment 4. Treated group B shows more-than-expected gold particles on compartment 1 and fewer-than-expected particles on compartment 3 (*see* **Table 4**).

For this approach, the magnification need not be known or standardized between groups. For statistical evaluation by contingency table analysis, it is advisable that expected numbers of gold particles should not be smaller than 5 and, again, this number may influence the choice of compartments or numbers of sampled gold particles. It is also sensible to aim for similar column sums for total gold counts in each group of cells as statistical analysis may be distorted by large discrepancies between cell groups. Since this method was introduced, it has been used to follow shifts in antigen distributions in different groups of cells *(5,7,30,33)*.

A potential disadvantage of the between-group comparison of observed gold counts is that it may limit mechanistic interpretation of shifts in labeling patterns. For example, a shift of receptor labeling from the cell interior and towards the cell membrane might reflect an increase in the LD of the membrane (reflecting an increase in receptor concentration) or in the total amount of membrane (because of an increase in the surface area of cell membrane rather than a change in receptor concentration). In such cases, it may be better to supplement analysis by estimating labeling densities.

3.4.5. Other Recent Developments

Recent studies have examined the spatial patterns of gold label within particular compartments, notably cell membranes. One approach, computer simulation, aims to replicate the arrangement of gold particles on the actual (2D) membrane surface from the distribution of particles observed on the (1D) membrane traces generated by sectioning *(34)*. This approach has potential when 2D arrangements of gold particles cannot be viewed directly, e.g., by goniometry, use of membrane sheets or scanning electron microscopy *(35–37)*. As used to-date, the method has relied on relatively few simulation models (random, quadratic array and 'raft' cluster distributions) and is confined to membranes having extremely low curvature *(34)*. These authors have found that the 2D distribution of neural cell adhesion molecule on the plasma membrane of cultured hippocampal neurons best fits a 'raft' model.

Another approach takes advantage of the clustering of gold particles to define the location of cellular compartments that cannot be identified on purely morphological criteria *(38)*. Using this method, individual chromosomal domains were detected in HeLa cell nuclei by TEM *in situ* hybridization. Because the ability to distinguish labelled regions depends on differences in labeling intensity between those regions and their surroundings, the thresholding process used in this method works best when there is relatively low background labeling and when the interface between regions and surroundings is smooth rather than irregular *(38)*. At present, it is not clear how successful the method would be if it was necessary to resolve two or more compartments that share similar labeling densities.

An additional area of interest is the colocalization of different antigens (labeled with different sizes of gold marker) in membranes and other types of compartments. Methods based on correlation functions have been used to quantify colocalization of lipid raft markers in mast tumour cell membranes *(37)* and of nascent deoxyribonucleic acid with different nuclear proteins in HeLa cells *(39)*. It is worth noting that the LD and RLI methods described in this article could also be used to obtain indirect estimates of colocalization. For example, in a dual-labeling experiment using different sizes of gold particle to label two antigens, evidence for colocalization could be adduced if the distribution of gold label between compartments, or the labeling of a given compartment, was identical or very similar with both sizes of particle.

4. Notes

1. Even reproducible patterns of gold labeling may not represent the real distribution of antigen in the cell. Among the factors that confound interpretations are the specificity of the primary antibody and antibody-section interactions. When com-

partments or cells known to lack the antigen fail to label with immunogold, this strongly supports the presence of specificity. However, at present, there is no way of proving that all antibody molecules considered to bind specifically really do bind to the true epitope/antigen on the section. For these reasons, the term "label" is adopted when referring to reactions with counted gold particles (*3*) and the term "preferential labeling" is adopted to distinguish it from "specific labeling."

2. Indications of the specificity of labeling can be obtained by analyzing labeling distributions at different antibody dilutions. If a compartment contains an antigen recognized by the antibody, then labeling of this compartment should be more resistant to dilution because the affinity of the antibody is generally expected to be higher than non-specific adhesion. Recently, in ezrin-labeling experiments (*6*), we found that increasing antibody dilution was associated with lower frequencies of label over compartments, which were not expected to contain antigen, notably mitochondria. On the other hand, compartments expected to contain antigen showed increased frequencies with increasing dilution.

3. When gold labeling is specific, it will reflect the distribution of antibody binding sites that are accessible and, therefore, the distribution of antigen. In contrast to methods that report the density of antigen, labeling frequency distributions allow one to compare the total amounts of label in different compartments. The extent to which this pattern mirrors the true distribution of antigen is governed also by labeling efficiency, a term that refers to the mean number of gold particles per antigen in the section (*18*). On ultrathin sections, labeling efficiency varies between different compartments and influences the observed distribution of label. Much of the variation in labeling efficiency on cryosections depends on the degree of penetration into the section and, when this variation is minimised, labeling efficiencies are more uniform between compartments (*1*). Combining the present methods with more suitable embedding techniques may provide better assessment of the relative amounts of antigen in different subcellular compartments.

4. Labeled sections display both "specific" signal (apparent preferential labeling of compartments) as well as background "noise" (apparent labeling of compartments not expected to harbor appreciable amounts of antigen). Common reasons for this have been identified elsewhere (*1*). It is not unusual for raw gold counts (or gold labeling densities) to be "corrected" for background noise by subtracting golds from labeled compartments using a factor obtained from a compartment not expected to be labeled. Unfortunately, the choice of compartment and the threshold level for background labeling involve making rather subjective decisions. Moreover, it may not be possible to resolve different sources of background labeling, and it may not be safe to assume that background because of nonspecific adhesion, for example, affects all compartments to the same extent (for a practical example, *see* Mayhew et al. *[3]*) Consequently, there is no satisfactory (validated) procedure to correct for background and, without such a procedure, it seems inappropriate to try to correct it. Instead, it is sensible to keep background labeling extremely low by employing effective antibodies and gold preparation procedures.

With these precautions, the methods described here, based on null-hypothesis comparisons (of compartments within a cell type or between cell types), should retain their comparative worth.

5. Unfortunately, the simple rule of assigning a gold particle to a compartment if its centre lies on that compartment may not be universally applicable. For example, the ability to identify unambiguously the location of intramembrane or juxtamembrane antigens may be compromised by the angles at which membranes are tilted within the section. Membranes tilted beyond 20 to 26° to the incident electron beam appear as increasingly vague images and may, eventually, become unrecognizable *(40)*. Consequently, without additional investigation (e.g., by goniometry), it may not be possible to decide whether a given gold particle should be assigned to a membrane or, say, an organelle interior. Even on a clear membrane image (sectioned orthogonally; no tilt angle), it may be necessary to define it as being labelled if a gold particle lies on it or at a specified distance from it (e.g., within one unit of particle diameter).

6. A variant on the LD method is to calculate a labeling index *(41,42)*. This method is the product of organelle LD and volume fraction (the proportion of cell or cytoplasmic volume occupied by that organelle compartment). Given random positioning of sections, volume fraction is equivalent to a real fraction (the proportion of total cell or cytoplasmic profile area occupied by the compartment). Consequently, labeling index merely changes the reference space from the specified organelle to a larger compartment (the cell or its cytoplasm) and expresses gold density per micrometer squared of that larger reference space. Therefore, it conveys no more useful comparative information than organelle LD. In fact, because the number of gold particles on a given compartment will not correlate so well with the size of the larger reference space, labeling index is likely to be a less efficient estimator than organelle LD.

7. Where a given antigen is found in both organelle and membrane compartments, possibilities exist for treating membranes as volumes rather than surfaces *(43)* or membrane-associated label as belonging to a parent organelle. However, these approaches do not have general utility and have not been subjected to validation studies. The former method treats a membrane as a volume by counting gold particles that fall on the membrane trace and within a fixed distance from it. Unfortunately, this approach is susceptible to errors resulting from membrane tilting and loss (*see* **Note 5**). A possible advance would be to restrict gold counts to those membranes sectioned orthogonally. Membrane traces so generated are clear but represent only a fraction of the total membrane trace which is present on the section. Fortunately, it is possible to compensate for this loss by applying correction factors determined by goniometry or estimated from theoretical considerations *(40)*. An alternative, and superior, approach would be to sample orthogonally sectioned membranes in order to estimate their labeling densities. The labeling densities so obtained may then be used conveniently to calculate the relative labeling indices of those membrane compartments (*see* text and **Table 3**).

Acknowledgments

It has been my good fortune over many years to enjoy fruitful interactions with colleagues who are not only expert stereologists and cell biologists but also fine teachers and friends. We have learned a lot together in a spirit of friendly competition accompanied by much laughter.

References

1. Griffiths, G. (1993) *Fine Structure Immunocytochemistry.* Springer Verlag, Heidelberg.
2. Koster, A. J. and Klumperman, J. (2003) Electron microscopy in cell biology: integrating structure and function. *Nature Rev. Mol. Cell Biol.* **4,** Suppl., SS6–SS10.
3. Mayhew, T. M., Lucocq, J. M., and Griffiths, G. (2002) Relative labelling index: a novel stereological approach to test for non-random immunogold labelling of organelles and membranes on transmission electron microscopy thin sections. *J. Microsc. (Oxford)* **205,** 153–164.
4. Mayhew, T., Griffiths, G., Habermann, A., Lucocq, J., Emre, N., and Webster, P. (2003) A simpler way of comparing the labelling densities of cellular compartments illustrated using data from VPARP and LAMP-1 immunogold labelling experiments. *Histochem. Cell Biol.* **119,** 333–341.
5. Mayhew, T. M., Griffiths, G., and Lucocq, J. M. (2004) Applications of an efficient method for comparing immunogold labelling patterns in the same sets of compartments in different groups of cells. *Histochem. Cell Biol.* **122,** 171–177.
6. Lucocq, J. M., Habermann, A., Watt, S., Backer, J. M., Mayhew, T. M., and Griffiths, G. (2004) A rapid method for assessing the distribution of gold labeling on thin sections. *J. Histochem. Cytochem.* **52,** 991–1000.
7. Mayhew, T. M. and Desoye, G. (2004) A simple method for comparing immunogold distributions in two or more experimental groups illustrated using GLUT1 labelling of isolated trophoblast cells. *Placenta* **25,** 580–584.
8. Skepper, J. N. (2000) Immunocytochemical strategies for electron microscopy: choice or compromise. *J. Microsc. (Oxford)* **199,** 1–36.
9. Baddeley, A. J., Gundersen, H. J. G., and Cruz-Orive, L. M. (1986) Estimation of surface area from vertical sections. *J. Microsc. (Oxford)* **142,** 259–276.
10. Gundersen, H. J. G. and Jensen, E. B. (1987) The efficiency of systematic sampling in stereology and its prediction. *J. Microsc. (Oxford)* **147,** 229–263.
11. Mattfeldt, T., Mall, G., Gharehbaghi, H., and Möller, P. (1990) Estimation of surface area and length with the orientator. *J. Microsc. (Oxford)* **159,** 301–317.
12. Nyengaard, J. R. and Gundersen, H. J. G. (1992) The isector: a simple and direct method for generating isotropic, uniform random sections from small specimens. *J. Microsc. (Oxford)* **165,** 427–431.
13. Howard, C. V. and Reed, M. G. (1998) *Unbiased Stereology. Three-Dimensional Measurement in Microscopy.* Bios Scientific, Oxford.
14. Mayhew, T. M. (1991) The new stereological methods for interpreting functional morphology from slices of cells and organs. *Exp. Physiol.* **76,** 639–665.

15. Nyengaard, J. R. (1999) Stereologic methods and their application in kidney research. *J. Am. Soc. Nephrol.* **10**, 1100–1123.

16. Slot, J. W. and Geuze, H. J. (1985) A new method for preparing gold probes for multiple-labeling cytochemistry. *Eur. J. Cell Biol.* **38**, 87–93.

17. Watt, S. A., Kular, G., Fleming, I. N., Downes, C. P., and Lucocq, J. M. (2002) Subcellular localization of phosphatidylinositol 4,5-bisphosphate using the pleckstrin homology domain of phospholipase Cδ_1. *Biochem. J.* **363**, 657–666.

18. Lucocq, J. (1992) Quantitation of gold labeling and estimation of labeling efficiency with a stereological counting method. *J. Histochem. Cytochem.* **40**, 1929–1936.

19. Lucocq, J. (1994) Quantitation of gold labelling and antigens in immunolabelled ultrathin sections. *J. Anat.* **184**, 1–13.

20. Gundersen, H. J. G., Jensen, E. B. V., Kiêu, K., and Nielsen, J. (1999) The efficiency of systematic sampling in stereology: reconsidered. *J. Microsc. (Oxford)* **193**, 199–211.

21. Lucocq, J., Manifava, M., Bi, K., Roth, M. G., and Ktistakis, N. T. (2001) Immunolocalisation of phospholipase D1 on tubular vesicular membranes of endocytotic and secretory origin. *Eur. J. Cell Biol.* **79**, 508–520.

22. Ochs, M., Johnen, G., Müller, K.-M., et al. (2002) Intracellular and intraalveolar localization of surfactant protein A (SP-A) in the parenchymal region of the human lung. *Am. J. Respir. Cell Mol. Biol.* **26**, 91–98.

23. Cernadas, M., Sugita, M., van der Wel, N., et al. (2003) Lysosomal localization of murine CD1d mediated by AP-3 is necessary for NK T cell development. *J. Immunol.* **171**, 4149–4155.

24. Mironov, A., Latawiec, D., Wille, H., et al. (2003) Cytosolic prion protein in neurons. *J. Neurosci.* **23**, 7183–7193.

25. Bennett, P. M., Baines, A. J., LeComte, M-C., Maggs, A. M., and Pinder, J. C. (2004) Not just a plasma membrane protein: in cardiac muscle cells alpha-II spectrin also shows a close association with myofibrils. *J. Muscle Res. Cell Motil.* **25**, 119–126.

26. Kweon, H. S., Beznoussenko, G. V., Micaroni, M., et al. (2004) Golgi enzymes are enriched in perforated zones of Golgi cisternae but are depleted in COPI vesicles. *Mol. Biol. Cell* **15**, 4710–4724.

27. Mazzone, M., Baldassarre, M., Beznoussenko, G., et al. (2004) Intracellular processing and activation of membrane type 1 matrix metalloprotease depends on its partitioning into lipid domains. *J. Cell Sci.* **117**, 6275–6287.

28. Zhang, L.-H., McManus, D. P., Sunderland, P., Lu, X.-M., Ye, J.-J., Loukas, A., and Jones, M. K. (2005) The cellular distribution and stage-specific expression of two dynein light chains from the human blood fluke *Schistosoma japonicum*. *Int. J. Biochem. Cell Biol.* **37**, 1511–1524.

29. Young, F. M., Thomson, C., Metcalf, J. S., Lucocq, J. M., and Codd, G. A. (2005) Immunogold localization of microcystins in cryosectioned cells of *Microcystis*. *J. Struct. Biol.* **151**, 208–214.

30. Potolicchio, I., Chitta, S., Xu, X., et al. (2005) Conformational variation of surface class II MHC proteins during myeloid dendritic cell differentiation accompanies structural changes in lysosomal MIIC. *J. Immunol.* **175**, 4935–4947.

31. Fehrenbach, H., Tews, S., Fehrenbach, A., Ochs, M., Wittwer, T., Wahlers, T., and Richter, J. (2005) Improved lung preservation relates to an increase in tubular myelin-associated surfactant protein A. *Resp. Res.* **6,** 1–12.
32. Schmiedl, A., Ochs, M., Mühlfeld, C., Johnen, G., and Brasch, F. (2005) Distribution of surfactant proteins in type II pneumocytes of newborn, 14-day old, and adult rats: an immunoelectron microscopic and stereological study. *Histochem. Cell Biol.* **124,** 465–476.
33. Santambrogio, L., Potolicchio, I., Fessler, S. P., Wong, S.-H., Raposo, G., and Strominger, J. L. (2005) Involvement of caspase-cleaved and intact adaptor protein 1 complex in endosomal remodeling in maturing dendritic cells. *Nature Immunol.* **6,** 1020–1028.
34. Nikonenko, A. G., Nikonenko, I. R., and Skibo, G. G. (2000) Computer simulation approach to the quantification of immunogold labelling on plasma membrane of cultured neurons. *J. Neurosci. Meth.* **96,** 11–17.
35. Prior, I. A., Muncke, C., Parton, R. G., and Hancock, J. F. (2003) Direct visualization of Ras proteins in spatially distinct cell surface microdomains. *J. Cell Biol.* **160,** 165–170.
36. Meredith, D. O., Owen, G. Rh., Ap Gwynn, I., and Richards, R. G. (2004) Variation in cell-substratum adhesion in relation to cell cycle phases. *Exp. Cell Res.* **293,** 58–67.
37. Wilson, B. S., Steinberg, S. L., Liederman, L., et al. (2004) Markers for detergent-resistant lipid rafts occupy distinct and dynamic domains in native membranes. *Mol. Biol. Cell* **15,** 2580–2592.
38. Schöfer, C., Janáček, J., Weipoltshammer, K., Pourani, J., and Hozák, P. (2004) Mapping of cellular compartments based on ultrastructural immunogold labeling. *J. Struct. Biol.* **147,** 128–135.
39. Philimonenko, A. A., Janáček, J., and Hozák, P. (2000) Statistical evaluation of colocalization patterns in immunogold labelling experiments. *J. Struct. Biol.* **132,** 201–210.
40. Mayhew, T. M. and Reith, A. (1988) Practical ways to correct cytomembrane surface densities for the loss of membrane images that results from oblique sectioning, in *Stereology and Morphometry in Electron Microscopy. Problems and Solutions* (Reith, A. and Mayhew, T. M., eds.), Hemisphere Publ. Co., New York/ Washington/ Philadelphia/ London, pp. 99–110.
41. Nemali, M. R., Usuda, N., Reddy, M. K., et al. (1988) Comparison of constitutive and inducible levels of expression of peroxisomal β-oxidation and catalase genes in liver and extrahepatic tissues of rat. *Cancer Res.* **48,** 5316–5324.
42. Ogiwara, N., Usuda, N., Yamada, M., Johkura, K., Kametani, K., and Nakazawa, A. (1999) Quantification of protein A-gold staining for peroxisomal enzymes by confocal laser scanning microscopy. *J. Histochem. Cytochem.* **47,** 1343–1349.
43. Slot, J. W., Geuze, H. J., Gigengack, S., James, D. E., and Lienhard, G. E. (1991) Translocation of the glucose transporter GLUT4 in cardiac myocytes. *Proc. Natl. Acad. Sci. USA* **88,** 7815–7819.

16

Electron Crystallography of Membrane Proteins

Hui-Ting Chou, James E. Evans, and Henning Stahlberg

Summary

Electron crystallography studies the structure of two-dimensional crystals of membrane proteins or other crystalline arrays. This method has been used to determine the atomic structures of six membrane proteins and tubulin, as well as several other structures at a slightly lower resolution, where secondary structure motifs could be identified. To preserve the high-resolution structure of 2D crystals, the meticulous sample preparation for electron crystallography is of outmost importance. Charge-induced specimen drift and lack of specimen flatness can severely affect the resolution of images for tilted samples. However, sample preparations that sandwich the two-dimensional crystals between symmetrical carbon films reduce the charge-induced specimen drift, and the flatness of the preparations can be optimized by the choice of the grid material and the preparation protocol. Data collection in the cryoelectron microscope using either the imaging or the electron diffraction mode has to be performed after low-dose procedures. Spot scanning further reduces the charge-induced specimen drift.

Key Words: 2D membrane protein crystals; back-injection; sandwich method; spot-scanning; carbon flatness; low-dose; cryoelectron microscopy.

1. Introduction

Structural biology of membrane proteins is of central importance for cellular biology and for the development of new drugs. Membrane proteins represent the majority of today's drug targets in pharmaceutical research. Nevertheless, our databases contain less than 50 unique folds of membrane proteins as compared with several thousand for soluble proteins. Electron crystallography studies the structure of membrane proteins in a two-dimensional crystalline arrangement in a phospholipid bilayer membrane. The atomic models of tubulin (*1*) and of the membrane proteins BR (*2*), LHCII (*3*), AQP1 (*4,5*), the C-ring of Na$^+$ ATPases (*6*), nAChR (*7*), and AQP0 (*8*) were determined by electron crystallography. Several other membrane proteins classified as transporters, ion pumps, receptors

From: *Methods in Molecular Biology, vol. 369*
Electron Microscopy: Methods and Protocols, Second Edition
Edited by: J. Kuo © Humana Press Inc., Totowa, NJ

and membrane bound enzymes have been studied at slightly lower resolution allowing the localization of secondary structure motifs such as transmembrane helices, and are likely to produce atomic models in the near future (e.g., Hirai et al. *[9]*).

Electron crystallography can be used to study membrane protein structures at resolutions of 3 Å or better (e.g., *[8,10,11]*), demonstrating the value of this approach. Electron crystallography represents an alternative method for structure determination, when fragile membrane protein complexes cannot be grown into three-dimensional (3D) protein crystals for X-ray diffraction or are not available in sufficient amounts for NMR measurements. Membrane-inserting proteins that undergo conformational changes between the soluble and the membrane-inserted state are likely to be best studied by electron crystallography. Two-dimensional (2D) membrane crystals are frequently grown easier than 3D crystals of membrane proteins and offer to the membrane proteins a more native environment than most 3D crystal forms. Membrane crystals also are advantageous for the structure determination of co-crystals, when preformed crystals are to be incubated with protein binding partners.

Technological advancements like the availability of coherent intermediate voltage electron sources and helium-cooled and stable sample stages *(12)* allow the recording of high-resolution data of biological macromolecules. Improvement in CCD camera sizes and recording bit depth allow efficient data recording of electron diffraction pattern *(13)*. In addition, recent advancements in sample preparation with the sandwich-back-injection technique using non-wrinkled carbon films *(14–16)*, and the spot scanning data collection method *(17)* strongly reduce the resolution-limiting charge effect during data acquisition of tilted samples.

2D membrane protein crystallization usually requires detergent-solubilized and purified protein typically at a concentration of 1 mg/mL. Crystallization can be achieved by various methods *(18–22)*. Although the approach that has so far yielded the best-ordered crystals involves slow and controlled detergent dialysis in the presence of added phospholipids *(19)*, there is no one method or condition that is optimized for all proteins.

1.1. Sample Preparation

Sample preparation is as important as the operation of the electron microscope for the recording of high-resolution data of 2D crystal samples. The 2D crystalline arrangement of the sample allows efficient extraction of the structure's signal from extremely noisy images by computer processing. For the choice of the sample preparation, preference is therefore given to methods that perfectly preserve the high-resolution order of the 2D crystals, while the contrast of the preparation is of minor importance.

Cryoelectron microscopy (cryo-EM) grids of 2D crystal samples can be prepared using holey or continuous carbon film support grids. However, holey carbon film grids embed the samples in viscous (liquid) vitrified ice, which in most cases provides inferior electrical conductivity, physical stability, and sample flatness than a continuous carbon film support *(23)*. Grids for cryo-EM imaging of 2D crystals therefore usually employ continuous carbon film support. The following protocol describes the preparation of such grids.

2. Materials

1. Carbon evaporator (should have an oil-free high vacuum).
2. Mica sheets.
3. Petri dishes.
4. Anticapillary self-closing tweezers.
5. Molybdenum transmission electron microscopy (TEM) grids (e.g., 300 mesh Mo grids from Pacific Grid-Tech, TX; see also *[16]*).
6. Humid chamber (home-made, constructed from Petri dishes).
7. Filter paper (Whatman, #1).
8. Liquid nitrogen in a Styrofoam cup, covered with aluminum foil to reduce boiling, for freezing of the sugar embedded grids.
9. For plunge freezing: A plunge freezer, if possible with a closed humid chamber (e.g., the Vitrobot, *see* www.vitrobot.com; or a homebuilt device, *see* **Fig. 1**).
10. For the sandwich method: 4-mm diameter platinum wire loop (available for example from Ted Pella, Inc., Redding, CA).

3. Methods

3.1. Plunge Freezing

2D crystals can be adsorbed to glow-discharged carbon-coated grids, blotted and plunge frozen in ethane slush, cooled by liquid nitrogen (*see* **Fig. 1**), similar to the preparation of cryo-EM grids with the holey carbon film method.

1. Evaporate carbon onto freshly cleaved mica (*see* **Note 1**).
2. Place mica with evaporated carbon into a humid chamber over night, to increase ease in floating the carbon off the mica.
3. Float carbon off the mica onto the surface of a buffer solution. This piece of carbon should be slightly larger than a TEM grid.
4. Pickup the carbon with a TEM grid (*see* **Note 2**).
5. Turn the tweezers with the grid upside down.
6. Remove a part of the liquid on the grid with a pipet.
7. Add 1 to 3 µL of 2D crystal solution (through the grid bars; *see* **Note 3**).
8. Place the tweezers into a plunger (Guillotine).
9. Blot the grid with a filter paper and plunge-freeze in ethane slush (*see* **Note 4**).
10. Transfer the grid into a cryo-holder and image with minimal-dose cryo-EM techniques.

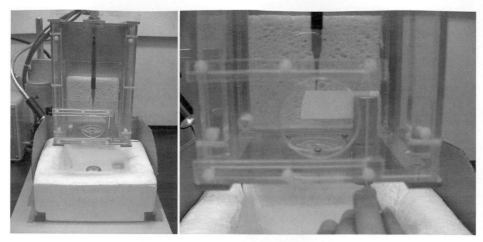

Fig. 1. The plunge-freezing device with a humid chamber. The grid is held by a tweezers in the center of the device (left). The sponge behind the tweezers can be soaked in warm water to increase the air humidity. Samples can be added through the sliding door in front of the tweezers. Blotting for elongated periods of time can be performed with the closed chamber from the outside by rotating the lever manually so that the filter paper blots the sample (right). The tweezers and the tweezers' holding arm is slim, so that plunging of the sample into the LN_2-cooled ethane slush below the chamber can be done without prior removing of the blotting paper.

3.2. Sugar Embedding

Drying of membrane protein crystals in the presence of sugars such as tannic acid *(3)*, trehalose *(1,4,24)*, or glucose *(10)* can preserve the intact ultrastructure of the proteins, while eliminating the need for quick-freezing.

1. Evaporate carbon onto freshly cleaved mica (*see* **Note 1**).
2. Place mica with evaporated carbon into humid chamber over night, to increase ease in floating the carbon off the mica.
3. Float carbon off the mica onto the surface of a buffer solution.
4. Pickup the carbon with a TEM grid (*see* **Notes 2** and **5**).
5. Place the grid with the carbon film facing up onto three drops of sugar solution (*see* **Fig. 2B** and also **Notes 6** and **7**).
6. Turn the tweezers with the grid upside down. The carbon film is now on the lower side of a hanging drop of sugar under the grid.
7. Remove a part of the liquid on the grid with a pipet (from the top through the grid bars).
8. Add 1 to 3 μL of 2D crystal solution (1 mg/mL) through the grid bars (*see* **Note 3**).
9. Place grid in tweezers for 60 s into a humid chamber, which can be constructed from three plastic Petri dishes (design by Dr. T. Braun, *see* **Fig. 2D**).

Fig. 2. Sugar embedding of two-dimensional crystals requires a piece of carbon the size of a grid. This carbon can be cut from mica with evaporated carbon (**A**). The carbon is floated onto a water surface (**B**) and placed onto three drops of sugar containing buffer solution (**C**). After adding the crystal solution, the grid is allowed to rest for a few minutes in a humid atmosphere that can be created using Petri dishes (**D**). The edge of the cover of the right Petri dish was broken off over a stretch of 2 cm, forming a hole where the tip of the tweezers can enter. The bottom of the right Petri dish is covered with a wet filter paper, which increases air humidity when the Petri dish is closed.

10. With a pipet, take excess solution off the grid, leaving approx. 1 µL on the grid.
11. Turn the tweezers again upside down, and place the grid flat onto two layers of filter paper with the carbon film facing up.
12. After sufficient blotting time (e.g., 20 s), and potential freezing in liquid nitrogen, observe the grid in the TEM (*see* **Note 7**).

3.3. The Sandwich Method

The sandwich sample preparation method embeds the 2D crystals between two layers of carbon film. This preparation offers two conductive surfaces around the sample, which may help reduce specimen charging and presents a symmetrical charge situation to the electron beam. It may thereby increase the image stability under the electron beam. Another advantage of this method arises when it is used in conjunction with a staining solution, since it can provide a more even staining of the two sample surfaces, even though the second carbon film adds noise to the image (e.g., Golas et al. *[25]*). The carbon sandwich can be made by placing both carbon films on the same surface of the TEM grid, resulting in the order grid → carbon → sample → carbon *(25,26)*. Alternatively, the two carbon films can cover both sides of the TEM grid, as described in *(15)*, resulting in the order carbon → grid/sample → carbon. The latter is presented here:

Perform **steps 1** through **8** as in **Subheading 3.2.** for sugar embedding.

9. Float a second piece of carbon film onto buffer solution (*see* **Note 8**).
10. Lift the carbon film up with a 4 mm diameter platinum wire loop and place it onto the grid from the top side, so that the grid and 2D crystals are sandwiched between the two carbon films (*see* **Fig. 3**).

Fig. 3. Transfer of the second carbon film onto the grid to form the carbon film sandwich.

11. Blot the sandwich construction from the edge of the grid, using a piece of torn filter paper.
12. Plunge the grid into liquid nitrogen and transfer into the cryo-EM.

3.4. Electron Microscope Operation

Data collection on the TEM from 2D crystals can be performed by recording images or electron diffraction pattern. Although the Fourier transformations of the images contain amplitudes *and* phases of the sample structure, electron diffraction pattern allow a more reliable determination of the amplitudes but lack the phase information.

3.4.1. Recording of Images

Images are preferably recorded on photographic film and subsequently digitized with a high-resolution scanner, because the current image processing methods benefit from large and coherently connected single crystal images, and the current generation of CCD cameras does not yet offer for higher acceleration voltages (200 kV and higher) a better transfer function at high resolution than the conventional Kodak SO 163 film *(27)*.

Recording of images of 2D crystals requires operation of the electron microscope under low-dose conditions, in which the microscope offers so-called "search," "focus," and "photo" positions. Chose a second condenser aperture of 100 μm or smaller and an objective aperture of 100 μm or larger. Align the TEM in the "photo" position, then setup the "focus" position, and finally the "search" position. Cycle through the modes exclusively in the order "search" → "focus" → "photo", to prevent hysteresis effects that may make you loose the alignment.

1. Search position: The "search" position uses the defocused electron diffraction mode. Suitable 2D crystal samples are identified at very low illumination intensity, which is best done by using the strong contrast of the shadow image obtained when operating the instrument in over-focused diffraction mode. Use an electron diffraction camera length of 1 m or longer. Set the second condenser lens to strong overfocus to spread the illumination over a large sample area. Set the intermediate lens to overfocus to create a strongly contrasted "shadow image" of the sample. Use a combination of image shift and beam tilt to align the center of the "search" position with the "photo position."

2. Focus position: The "focus" and the "photo" position use the image mode of the electron microscope. In the "focus" position the instrument is operated under identical lens settings as used for the "photo" position, but with additional beam- and image-shift, so that focusing the objective lens can be done adjacent to the sample location of interest. Align first the "photo" position, then copy those settings to the "focus" position, and finally add sufficient beam- and image-shift so that the beam in the "focus" position does not illuminate the sample area of the "photo" position. At 50 kx magnification, this usually corresponds to 3.5-μm beam deflection.

3. Photo position: Images are recorded with the settings of the "photo" position (e.g., 50 kx magnification, 0.5 seconds exposure time on a FEG instrument). Recorded images are modulated by the microscope's contrast transfer function, which can be corrected computationally. To reduce the number of Thon rings, try to work at lower defocus values, if the crystal quality is good enough. When images of tilted samples are recorded, a resolution loss in the direction perpendicular to the tilt axis can frequently be observed, resulting from charge-induced specimen drift and vibration of the sample. Images recorded on photographic film are best inspected by optical diffraction *(28)* before digitization.

4. Spot-scanning: When recording images of highly tilted samples, the resolution perpendicular to the tilt axis can be severely affected by charge-induced specimen drift during the acquisition: The illuminated sample area can accumulate electrical charge during the exposure, which gradually deflects electrons into the direction perpendicular to the tilt axis. This effect can be minimized by the sandwich sample preparation method but also by using the so-called *spot-canning* illumination method *(17)*: The electron beam on the sample is concentrated into an area of 50 to 100 nm diameter, and step-wise scanned over the sample area, whereas the entire hexagonally arranged spot-scan pattern is recorded onto the same photographic film *(see* **Fig. 4** *[29]*). Each *spot-scanning* spot has an exposure time of for example 50 ms (when using a FEG instrument), whereas the camera shutter remains open for a few minutes to record the entire *spot-scanning* pattern on one photographic film. Attention has to be paid to adjusting the illumination conditions so that a sufficiently small illumination opening angle is maintained also in the *spot-scanning* mode.

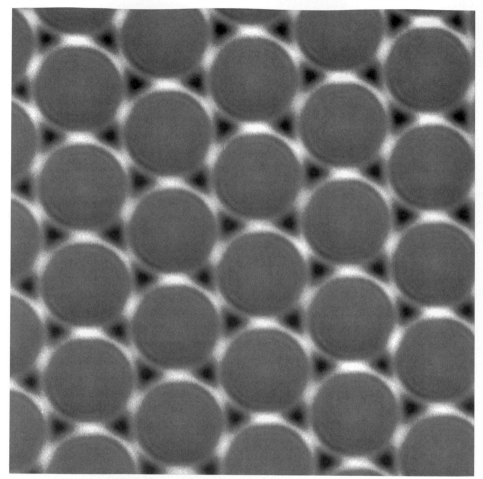

Fig. 4. A micrograph recorded by spot scanning of AQP2 2D crystals *(37)*. The approx. 100 nm diameter spots cover in a hexagonal pattern the micrograph, which in total spans an area of approx. 0.6 μm² (dimensions on the sample level). The finer AQP2 2D crystal lattice is not recognizable in the unprocessed image. Such an image can be computer processed to mask the nonexposed dark triangles and the double-exposed bright contact edges between the spot-scan spots, and replace these regions with the average gray value.

3.4.2. Recording of Electron Diffraction Pattern

Electron diffraction pattern are not affected by the contrast transfer function, nor suffer from specimen vibration or drift, and have only very limited dependence on specimen charging during the exposure. Although electron diffraction usually requires well-ordered 2D crystal samples of more than a micrometer in diameter, its recording can be a time-efficient way to obtain a high-resolution amplitude dataset from tilted samples *(30)*. Electron diffraction pattern are best recorded on a digital CCD camera, where the superior dynamic range (typically 12 to 16 bit) allows capturing intense low-resolution and weak high-resolution diffraction spots in a single exposure at good signal-to-noise ratio *(13)*. Depending on the unit cell spacing of the 2D crystal and the expected resolution in a diffraction pattern, a CCD camera of 2048×2048 or higher pixel number is required, to sufficiently resolve individual diffraction spots in a diffraction pattern *(13)*.

To record diffraction patterns, all positions of the low-dose system are setup in diffraction mode. Chose a small second condenser aperture of 10 or 20 μm in diameter, and remove the objective aperture.

1. Search position: The "search" position is set-up as described previously, using the "shadow" image of the overfocused diffraction mode with an over-focused second condenser lens and overfocused intermediate lens.

2. Focus position: The use of this position is optional and may not be required, if the microscope is sufficiently stable, because the electron diffraction focus does not depend on the sample position or sample height. The "focus" position is a copy of the "photo" position and therefore also uses the electron diffraction mode. Focusing is performed *on* the sample. The "focus" mode does not use additional beam or image shift. The alignments of the "photo" and the "focus" mode only differ in the illumination system: A highly excited first condenser lens in the "focus" position is used to dim the sample illumination to the lowest possible intensity. The second (and/or third) condenser lens is then adjusted to again achieve parallel illumination onto the sample. Focusing the diffraction pattern with the intermediate lens can then be done by focusing the *direct* (zero-order) beam on the screen. The focusing settings from the "focus" position are only valid for the "photo" position, if both "focus" and "photo" positions are using the identical parallel illumination conditions onto the sample, which can be verified by checking if the objective aperture border appears sharp in both modes. (This requires the objective aperture to be aligned to the correct height, which is the back-focal plane of the objective lens. Don't forget to remove the objective aperture from the beam for recording the diffraction pattern).

3. Photo position: In the "photo" position, the diffraction pattern of the crystal sample is recorded, using a long exposure time (e.g., 30 s or longer). This position has to provide an alignment with parallel illumination of the sample. Recording of electron diffraction pattern can be done with or without a "selected area diffraction" aperture. If you chose to limit the electrons that contribute to the diffraction pattern using a selected area diffraction aperture, you can illuminate a sample area somewhat larger than the diffracting crystal, allowing a more coherent and homogeneous illumination of the crystal.

The dose for recording one image should be in the order of 500 electrons/nm^2 at liquid nitrogen temperature but can be as high as 2000 electrons/nm^2 at liquid helium temperature, when information in the resolution range beyond 4 Å is to be recorded. Recent observations of changes in electrical conductivity, viscosity and density of vitreous ice at liquid helium temperatures *(31,32)* might not apply to phospholipid bilayers and/or sugar-embedded samples, which would explain why for 2D crystal samples the usage of helium-cooled instruments has proven strongly beneficial *(33)*.

4. Notes

1. The flatness of the supporting carbon film is of high importance when attempting to record images at high resolution. It is especially important when images of tilted 2D crystal samples are to be recorded because small tilt-angle variations will strongly affect the resolution perpendicular to the tilt axis *(14)*. Uneven or rough carbon film can perturb the specimen flatness, which also can be severely affected by so-called cryo-crinkling of the carbon film *(15,34)* because of different thermal expansion coefficients between the sample and the support grid. Carbon film usually is prepared by carbon evaporation onto freshly cleaved mica, which should be performed at a vacuum better than 5×10^{-6} mbar. Care should be taken that only carbon films from evaporation processes without sparking are used. Carbon films prepared in this way will be smoother on the mica-facing side than on the carbon source-facing side *(35)*. Therefore, the 2D crystal samples should be adsorbed onto the side of the carbon film, which previously was facing the mica.

2. Carbon film should be floated onto the darker, less shiny side of the TEM support grid, which will attach better to the carbon film and result in smoother films *(16)*. Cryocrinkling can be reduced significantly by using grid materials with thermal expansion coefficients similar to that of the sample: Cryocrinkling is strongest with copper, less with titanium, and best with molybdenum grids *(14,36)*. The costly molybdenum grids can be reused after ultrasound cleaning in ethanol. Carbon film will rupture less often when using grids with thicker metal bars, but the visible sample area on tilted grids will be smaller, and the amount of cryocrinkling will increase *(16)*. For high-resolution imaging, 300 mesh molybdenum grids are recommended.

3. The spreading of 2D crystals onto the carbon film can be facilitated by the addition of small amounts of detergents. Bacitracin (0.25 mg/mL) in the sugar solution can, for example, be used as wetting agent to help increase spreading *(37)*. In addition, a pipet can be used to take up a part of the sample/sugar solution at the edge of the grid, and readmit it to the center of the grid several times, to physically increase spreading of the crystals onto the carbon film. Spreading also can improve if longer adsorption times are allowed. This requires the availability of a humidity chamber to prevent sample evaporation during the adsorption time of a few minutes.

4. The sample solution does not evaporate as fast on a closed carbon film as it would do on holey carbon film, allowing longer blotting times. Although the electrical charging of ice contamination on the grid is more difficult to control with this method, vitrification by plunge freezing of 2D crystals on a continuous carbon

film may be preferred over sugar embedding, when a good contrast of undisturbed protein surface structures is required.

5. This method is best done by holding the grid with a self-closing inverted anti-capillary tweezers.

6. The grid is placed for a few seconds onto three drops of buffer solution containing 1 to 7% (w/v) of sugar (e.g., trehalose, glucose, tannic acid), in order to replace the water under the carbon film with the sugar solution (*see* **Fig. 2C**). To maintain an intact carbon film, attention has to be paid to avoid wetting the carbon film on the upper surface.

7. Many membrane protein 2D crystals support complete drying in glucose, without loss of the high-resolution order. A grid with a glucose embedded sample can usually be dried extensively, and even be loaded with a cryosample holder into the vacuum of the TEM, while still at room temperature. The grid quality can then be assessed at room temperature in the TEM, which, on most microscopes, is faster than handling a cryogrid at low temperatures. Only after verifying that the grid shows a suitable sample density and glucose thickness, the cryo-EM holder is filled with liquid nitrogen to cool the sample. (A different procedure may be required for Helium cooled instruments, which usually are cooled before sample insertion.) This sample preparation results in cryogrids that are completely free of ice contamination. Trehalose-embedded membrane proteins will in most cases still require the presence of traces of water. A grid prepared with trehalose should therefore only be blotted for approx 20 s (depending on air humidity), and then frozen by manual plunging into liquid nitrogen. Because trehalose prevents ice crystal formation, quick-freezing in ethane slush is not necessary. The grid can then be mounted into a pre-cooled cryo sample holder and transferred into the TEM.

8. The second piece of carbon film should be slightly smaller than the grid. To yield a symmetrical carbon film sandwich, this second carbon film should be from the same evaporation process as the first. The symmetrical carbon is essential to reduce the beam-induced specimen drift (Y. Fujiyoshi, personal communication, 2004, and Glaeser and Downing *[33]*).

References

1. Nogales, E., Wolf, S. G., and Downing, K. H. (1998) Structure of the alpha beta tubulin dimer by electron crystallography. *Nature (London)* **391,** 199–203.
2. Henderson, R., Baldwin, J. M., Ceska, T. A., Zemlin, F., Beckmann, E., and Downing, K. H. (1990) Model for the structure of Bacteriorhodopsin based on high-resolution electron cryo-microscopy. *J. Mol. Biol.* **213,** 899–929.
3. Kühlbrandt, W., Wang, D. N., and Fujiyoshi, Y. (1994) Atomic model of plant light-harvesting complex by electron crystallography. *Nature (London)* **367,** 614–621.
4. Murata, K., Mitsuoka, K., Hirai, T., et al. (2000) Structural determinants of water permeation through aquaporin-1. *Nature (London)* **407,** 599–605.
5. Ren, G., Reddy, V. S., Cheng, A., Melnyk, P., and Mitra, A. K. (2001) Visualization of a water-selective pore by electron crystallography in vitreous ice. *Proc. Natl. Acad. Sci. USA* **98,** 1398–1403.

6. Vonck, J., von Nidda, T. K., Meier, T., et al. (2002) Molecular architecture of the undecameric rotor of a bacterial Na+-ATP synthase. *J. Mol. Biol.* **321**, 307–316.

7. Miyazawa, A., Fujiyoshi, Y., and Unwin, N. (2003) Structure and gating mechanism of the acetylcholine receptor pore. *Nature (London)* **424**, 949–955.

8. Gonen, T., Sliz, P., Kistler, J., Cheng, Y., and Walz, T. (2004) Aquaporin-0 membrane junctions reveal the structure of a closed water pore. *Nature (London)* **429**, 193–197.

9. Hirai, T., Heymann, J. A., Shi, D., Sarker, R., Maloney, P. C., and Subramaniam, S. (2002) Three-dimensional structure of a bacterial oxalate transporter. *Nat. Struct. Biol.* **9**, 597–600.

10. Grigorieff, N., Ceska, T. A., Downing, K. H., Baldwin, J. M., and Henderson, R. (1996) Electron-crystallographic refinement of the structure of bacteriorhodopsin. *J. Mol. Biol.* **259**, 393–421.

11. Mitsuoka, K., Hirai, T., Murata, K., Miyazawa, A., Kidera, A., Kimura, Y., and Fujiyoshi, Y. (1999) The structure of bacteriorhodopsin at 3.0 Å resolution based on electron crystallography: implication of the charge distribution. *J. Mol. Biol.* **286**, 861–882.

12. Fujiyoshi, Y., Mizusaki, T., Morikawa, K., et al. (1991) Development of a superfluid helium stage for high-resolution electron microscopy. *Ultramicroscopy* **38**, 241–251.

13. Downing, K. H. and Li, H. (2001) Accurate recording and measurement of electron diffraction data in structural and difference Fourier studies of proteins. *Microsc. Microanal.* **7**, 407–417.

14. Glaeser, R. M. (1992) Specimen flatness of thin crystalline arrays: influence of the substrate. *Ultramicroscopy* **46**, 33–43.

15. Gyobu, N., Tani, K., Hiroaki, Y., Kamegawa, A., Mitsuoka, K., and Fujiyoshi, Y. (2004) Improved specimen preparation for cryo-electron microscopy using a symmetric carbon sandwich technique. *J. Struct. Biol.* **146**, 325–333.

16. Vonck, J. (2000) Parameters affecting specimen flatness of two-dimensional crystals for electron crystallography. *Ultramicroscopy* **85**, 123–129.

17. Downing, K. H. (1991) Spot-scan imaging in transmission electron microscopy. *Science* **251**, 53–59.

18. Remigy, H. W., Caujolle-Bert, D., Suda, K., Schenk, A., Chami, M., and Engel, A. (2003) Membrane protein reconstitution and crystallization by controlled dilution. *FEBS Lett.* **555**, 160–169.

19. Jap, B. K., Zulauf, M., Scheybani, T., Hefti, A., Baumeister, W., Aebi, U., and Engel, A. (1992) 2D crystallization: from art to science. *Ultramicroscopy* **46**, 45–84.

20. Levy, D., Chami, M., and Rigaud, J. L. (2001) Two-dimensional crystallization of membrane proteins: the lipid layer strategy. *FEBS Lett.* **504**, 187–193.

21. Kühlbrandt, W. (1992) Two-dimensional crystallization of membrane proteins. *Q. Rev. Biophys.* **25**, 1–49.

22. Hasler, L., Heymann, J. B., Engel, A., Kistler, J., and Walz, T. (1998) 2D crystallization of membrane proteins: rationales and examples. *J. Struct. Biol.* **121**, 162–171.

23. Henderson, R. (1992) Image contrast in high-resolution electron microscopy of biological macromolecules: TMV in ice. *Ultramicroscopy* **46**, 1–18.

24. Kimura, Y., Vassylyev, D. G., Miyazawa, A., et al. (1997) Surface of bacterio-rhodopsin revealed by high-resolution electron crystallography. *Nature (London)* **389**, 206–211.

25. Golas, M. M., Sander, B., Will, C. L., Lührmann, R., and Stark, H. (2003) Molecular architecture of the multiprotein splicing factor SF3b. *Science* **300**, 980–984.

26. Golas, M. M., Sander, B., Will, C. L., Lührmann, R., and Stark, H. (2005) Major conformational change in the complex SF3b upon integration into the spliceo-somal U11/U12 di-snRNP as revealed by electron cryomicroscopy. *Mol. Cell.* **17**, 869–883.

27. Sander, B., Golas, M. M., and Stark, H. (2005) Advantages of CCD detectors for de novo three-dimensional structure determination in single-particle electron microscopy. *J. Struct. Biol.* **151**, 92–105.

28. Aebi, U., Smith, P. R., Dubochet, J., Henry, C., and Kellenberger, E. (1973) A study of the structure of the T-layer of *Bacillus brevis*. *J. Supramol. Struct.* **1**, 498–522.

29. Schenk, A. D., Werten, P. J., Scheuring, S., et al. (2005) The 4.5 Å structure of human AQP2. *J. Mol. Biol.* **350**, 278–289.

30. Walz, T. and Grigorieff, N. (1998) Electron crystallography of two-dimensional crystals of membrane proteins. *J. Struct. Biol.* **121**, 142–161.

31. Iancu, C. V., Wright, E. R., Heymann, J. B., and Jensen, G. J. (2006) A comparison of liquid nitrogen and liquid helium as potential cryogens for electron cryotomography. *J. Struct. Biol.* **153**, 231–240.

32. Comolli, L. R. and Downing, K. H. (2005) Dose tolerance at helium and nitrogen temperatures for whole cell electron tomography. *J. Struct. Biol.* **152**, 149–156.

33. Fujiyoshi, Y. (1998) The structural study of membrane proteins by electron crystallography. *Adv. Biophys.* **35**, 25–80.

34. Glaeser, R. M. and Downing, K. H. (2004) Specimen charging on thin films with one conducting layer: Discussion of physical principles. *Microsc. Microanal.* **10**, 790–796.

35. Butt, H.-J., Wang, D. N., Hansma, P. K., and Kühlbrandt, W. (1991) Effect of surface roughness of carbon support films on high-resolution electron diffraction of two-dimensional protein crystals. *Ultramicroscopy* **36**, 307–318.

36. Booy, F. P. and Pawley, J. B. (1993) Cryo-crinkling: what happens to carbon films on copper grids at low temperature. *Ultramicroscopy* **48**, 273–280.

37. Mindell, J. A., Maduke, M., Miller, C., and Grigorieff, N. (2001) Projection structure of a ClC-type chloride channel at 6.5 Å resolution. *Nature (London)* **409**, 219–223.

17

Cryoelectron Microscopy of Icosahedral Virus Particles

Wen Jiang and Wah Chiu

Summary

With the rapid progresses in both instrumentation and computing, it is increasingly straightforward and routine to determine the structures of icosahedral viruses to subnanometer resolutions (6–10 Å) by cryoelectron microscopy and image reconstruction. In this resolution range, secondary structure elements of protein subunits can be clearly discerned. Combining the three-dimensional density map and bioinformatics of the protein components, the folds of the virus capsid shell proteins can be derived. This chapter will describe the experimental and computational procedures that lead to subnanometer resolution structural determinations of icosahedral virus particles. In addition, we will describe how to extract useful structural information from the three-dimensional maps.

Key Words: Cryo-EM; cryoelectron microscopy; icosahedral virus; 3D reconstruction; subnanometer resolution; secondary structure elements; structural fitting

1. Introduction

Icosahedral virus particles were among the first biological specimen for which three-dimensional (3D) molecular structures have been solved using electron microscopy, image processing, and 3D reconstruction *(1,2)*. Because of their large size, high symmetry, and availability in large quantities, icosahedral virus particles frequently have been studied structurally by single particle cryoelectron microscopy (cryo-EM). These studies, for example, the hepatitis B virus *(3,4)*, herpes simplex type-1 capsid *(5)*, and rice dwarf virus *(6)*, have played and are continuing to play key roles in the development of cryo-EM technologies and in pushing for increasingly higher resolutions, accuracy, and throughput.

In recent years, we have seen significant progress in cryo-EM studies of icosahedral virus structures toward subnanometer resolutions (6–10 Å; Chiu et al. *[7]*). At these resolutions, it is possible to discern the protein boundaries at most of the regions and allow the dissection of the assembly mechanisms of multiple components in the large virus capsid. Secondary structure elements (alpha helices and beta sheets) can also be discerned. The visualization of these

From: *Methods in Molecular Biology, vol. 369*
Electron Microscopy: Methods and Protocols, Second Edition
Edited by: J. Kuo © Humana Press Inc., Totowa, NJ

structural features not only validates the structures but also allows searching homologue protein fold using the spatial information of (alpha helices and beta sheets; Jiang et al. *[8]*). By integrating bioinformatics, homology modeling, the subnanometer resolution structures, and the identified secondary structure elements, atomic models can be built for the protein subunits *(9)*. This progress was made possible by advances in all aspects of cryo-EM study: instrumentation, image acquisition, image processing/3D reconstruction software, computing resource, and structural analysis and interpretation tools. The following sections will describe each of these steps and discuss the protocols in the order that parallels the pipeline for a typical cryo-EM study.

2. Materials

There are many different types of equipment needed for successful cryo-EM imaging and 3D reconstruction of icosahedral viruses. Examples of the major equipments and their suppliers used in our Center are listed:

1. Quantifoil® R2/2 grids (Quantifoil Micro Tools GmbH, Germany).
2. Emitech glow discharger (http://www.emitech.co.uk).
3. Vitrobot™ (http://www.vitrobot.com) or equivalent freeze plunger (http://www.gatan.com/holders/cryoplunge.html).
4. Gatan or Oxford cryo-holder and cryo-transfer station (http://www.gatan.com).
5. JEOL or FEI 200-300kV electron microscopes with LaB_6 or field emission gun with cryo-stage and low dose kit.
6. Gatan or Tietz 4K × 4K CCD camera for direct digital recording.
7. Kodak SO163 photographic films and darkroom.
8. Nikon CoolScan 9000ED scanner or equivalent scanner.
9. Tweezers, liquid N_2 tank, and other small accessories for cryo-EM imaging.
10. Computer cluster for image processing (10–20 CPUs of current Intel Xeon or AMD Athlon/Opetron running at 1–3 GHz with 1–2 GB memory per CPU).
11. Graphical workstation for image and 3D map visualization.
12. Image processing software: EMAN *(10)* and SAVR *(11)* (http://ncmi.bcm.edu/software). There are other software packages suitable for this task. In this article, we will refer to the programs in the freely available EMAN or SAVR software packages if not otherwise specified.
13. 3D Visualization software: Amira™ (http://www.amiravis.com) or Chimera (http://www.cgl.ucsf.edu/chimera).

3. Methods

3.1. Data Collection

3.1.1. Specimen Preparation

The virus particles should be purified to homogeneity, usually by gradient centrifugation followed by thorough dialysis. The final buffer solutions should keep the virus particles stable. In general, the preferred buffer has <300 m*M* salt. Detergent and sucrose/glycerol should be avoided. Buffers with a smaller pH dependence on temperature, such as phosphate-buffered saline instead of Tris buffer, are preferred to minimize the pH change from the approx 200° temperature decrease during sample freezing *(12)*. The optimal concentration of each virus sample is dependent on the sample and buffer, but a good starting point is at approx 10^{12} particles/mL or approx 1 mg/mL in general. At this concentration, a total sample volume of 50 to 200 µL generally is adequate for collecting a sufficient number of images for a subnanometer resolution 3D reconstruction.

3.1.2. Sample Freezing

The goal is to preserve the virus particles in a thin layer of vitreous ice on a holey grid. Flash freezing of the sample is required to prevent formation of ice crystals that can damage the virus particles *(13,14)*.

1. Holey grids: 400 mesh copper grids coated with a thin layer of carbon film with 2- to 4-µm diameter holes. Commercial Quantifoil grids or home-made holey grids *(15)* can be used. The grids should be thoroughly cleaned using organic solvents, such as acetone, to remove residual plastics that can cause charging and interfere with image quality. The grids also typically are glow discharged before use to clean the grid surface and to make the carbon film surface more hydrophilic so that the aqueous biological sample will stick to the grid better. The optimal strength and duration of the glow discharging varies with the grids and the samples, and needs to be determined for each study. A reasonable starting condition is approx 20 s, with a current of 20 mA using an Emitech glow discharger. Note that gloves should be worn when handling the grids. The grids should be handled very gently to avoid bending or other deformations.
2. Plunge freezing: A homemade gravity-driven plunger has been used successfully *(16)*. However, a more elaborate plunge freezer, the Vitrobot, with environmental chamber and computer control capability is also commercially available. When using these freezing apparatuses, a sharp-tipped tweezer with a holey grid clamped at the edge is held by a vertically movable rod. A small volume of sample solution (3–5 µL) is pipetted on the surface of the grid and excess solution is then blotted away for a few seconds using filter paper (*see* **Note 1**). After blotting, the grid is plunged into a receiving well with ethane slush cooled by liquid N_2. Rapid plunging, driven either by gravity or pneumatic pressure, is essential to ensure a fast cooling rate to form a vitreous ice layer embedding the particles. A slower cooling rate will result in the formation of ice crystals which may damage the specimen. The frozen grid is placed into a transfer grid box for storage in a liquid N_2 dewar or tank for later imaging in the microscope.

3.1.3. Imaging

Acquisition of high-quality images of the virus particles is the critical step to ensure a 3D reconstruction towards subnanometer resolutions. As the result of instrument advances, this task can be routinely performed using modern electron microscopes.

1. Microscope: A 200- to 300-kV microscope with LaB_6 or field emission gun can be used. The microscope should have liquid nitrogen or helium cooled stage, cryotransfer system, anticontamination blades and low dose imaging kit. Microscopes with these accessories are available from many vendors, with the majority of installations from JEOL and FEI.
2. Cryotransfer of grid into the microscope column: A side entry Gatan or Oxford cryoholder typically is used. This type of holder has a dewar on the far end that can be filled with liquid N_2 to cool the holder tip (where the grid is located) via thermal coupling. The holder is inserted into a pre-cooled cryotransfer station where the holder tip is semi-immersed within a liquid nitrogen bath. The dewar of the cryoholder is filled with liquid nitrogen. When the tip of the cryoholder reaches −170°C, the frozen grid is transferred from its storage slot of the grid box to the grid-seat at the holder tip using a pre-cooled tweezer. A small clip or screw ring secures the grid firmly in place. With the grid shutter closed, the cryoholder is then retracted from the cryotransfer station and rapidly inserted into the pre-pumped airlock of the microscope column (*see* **Note 2**). After cryotransfer of a grid into the column, a short period of wait time (several minutes) is generally needed to allow the full recovery of the vacuum in the column and mechanical stability of the specimen stage before proceeding to data collection.
3. Image acquisition: The microscope needs to be aligned well to acquire high resolution images. Astigmatism should be corrected to a negligible amount by minimizing either the ellipticity of the power spectra at a small under-focus when using a CCD camera or by judging the image contrast at different focuses using a TV monitor or the microscope fluorescent screen. Smaller condenser apertures and suitable spot sizes should be used to increase the spatial coherence of the beam while still providing efficient beam brightness. A small objective aperture is used to maximize the image contrast but should be large enough not to cut off high frequency information.

   Low-dose imaging is used to minimize radiation damage of the specimen by the electron beam. A low magnification (5000X) or defocused diffraction mode and low-intensity beam is used for the search mode to survey the grid and identify areas suitable for imaging. A focused beam at the same magnification as used in the final imaging is used to adjust the defocus using an area adjacent to the area of interest. It is critical to record the image only when there is no apparent stage and specimen drift. For studies aiming at subnanometer resolutions, typically 4000 to 8000 particles are collected which can usually be accomplished with a few imaging sessions.

   The aforementioned search/focus/imaging cycle for the whole grid typically is performed manually by the user. However, semi or fully automated imaging is becoming available on the new generation microscopes using software packages such as JAMES *(17)*, Leginon *(18)*, AutoEM *(19)*, or AutoEMation *(20)*.

4. Recording medium: Traditionally, photographic films such as Kodak SO163 are the recording medium for cryo-EM images. The exposed negatives are developed for 12 minutes using full strength developer solution at 20°C and subsequently fixed. Alternatively, high-resolution CCD cameras with 4k × 4k pixels, such as the Gatan UltraScan™ 4000 and Tietz TemCam F415, have become available for 200- to 300-kV electron microscopes in recent years and have been shown to be capable of recording virus particle images that yield subnanometer resolution 3D reconstructions *(17,21)*. Using a CCD provides several advantages over the use of photographic film for medium resolution reconstructions because: direct digitization eliminates the need for film development and subsequent digitization, water contamination in the electron microscope column is reduced due to the absence of photographic film, and image contrast is improved *(17)*. CCD recording also simplifies the automated image collection protocol.

5. Imaging conditions: 50,000 to 60,000 magnifications typically are used with film as a recording medium, when targeting at subnanometer resolutions. However, higher microscope magnifications (60,000–80,000) are needed for CCD imaging due to the modulation transfer function characteristics of the current 4k CCD *(17)*. Underfocuses in the range of 1 to 3 μm are good choices to minimize the signal decay at high resolutions but still maintain adequate image contrast for easy particle selection and orientation determination. The final image is usually taken using a total dose of <20 e/$Å^2$ on the sample typically in a 1 second exposure.

6. Digitization of images recorded on photographic films: Currently, the Nikon CoolScan 9000ED scanner (*see* **Note 3**) can be used to digitize the film at 6.35 μm/pixel for 3D reconstructions at subnanometer resolution *(22)*. The drawback of the Nikon scanner is that only approx 90% of film width can be scanned and requires a modified film holder to clamp the film into the proper position. The quality of the scans needs to be checked periodically to ensure that the scanner works at top performance. The scanning parameters should also be adjusted so that the histograms of the scanned images will adequately sample the whole dynamic range (16 bit) of the scanner.

7. Digital sampling: Based on Shannon information theory, at least 2X spatial sampling is required to preserve the structural details at the intended resolutions. In practice, 3X sampling is generally needed. When targeting at subnanometer resolution (6–10 Å) 3D reconstructions, images should be sampled at 2 to 3.5 Å/pixel to satisfy this requirement. The practical approach is to use the above-suggested magnification for imaging and the scanning step sizes, and then average the data at 1.5 or 2X.

3.2. Preprocessing

Before actual image processing to determine the orientation of all particles, the following preprocessing tasks should be completed.

3.2.1. Image Screening

Cryo-EM images (*see* **Fig. 1**) are rarely 100% suitable for data processing for 3D structure determination. Many factors, such as ice contamination and

Fig. 1. A typical low-dose image of ice-embedded rice dwarf virus particles imaged at 2.1 μm underfocus on a JEM4000 electron microscope equipped with a Gatan liquid nitrogen cryoholder.

specimen drift, will render many images unusable. Visual inspection using the graphical programs *boxer* or *eman*, can be used to discard the images with too few particles, too much contamination or obvious charging/drift. The power spectra of the digital images should be inspected (using *ctfit*) to discard those without Thon rings at the targeted resolutions, those with obviously elongated Thon rings due to astigmatism, and those exhibiting anisotropic Thon rings caused by significant charging or drift (*see* **Fig. 2**).

3.2.2. Particle Selection

Individual good particles would be selected for the later image processing. Both manual and automated particle selection are available; however, in practice, combined manual screening post automatic selection is the best practical approach to ensure high quality selection.

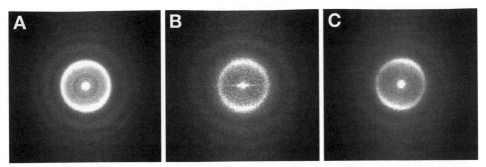

Fig. 2. Examples of power spectra of incoherent averages of virus particles with different image qualities. (**A**) good; (**B**) astigmatic; (**C**) drift.

1. Pre-filtering of images: because the purpose of particle selection is to mark the center position of selected particles, the fully sampled image is not necessary. In fact, prefiltered images by averaging and low-pass filtering can speed up particle selection by enhancing the visibility of the particles *(23)* and shrinking the number of pixels for rapid computation turnaround. Appropriate averaging and filtering could be performed using the *proc2d* program so that the shrunken particle diameter is about 80–100 pixels. The coordinates of the particles in the original fully sampled image are easily computed by multiplying the amount of averaging.

2. Automated selection: Many automated particle selection programs are available with a broad range of applicable sample types, selection accuracy and speed *(24)*. Our preferred method for the spherical virus particle selection is the *ethan* program (http://www.cs.helsinki.fi/group/bimcom/spade) using the ring filter algorithm *(25)*. It is fast (a few seconds per micrograph), nearly parameter free and has good selection accuracy (*see* **Fig. 3A**). The *ethan.py* program runs *ethan* and automatically handles the format conversions from/to EMAN conventions.

3. Manual screening: *ethan* also would select the particles on the carbon film area between the holes, nonparticle contaminations, or particles that are too crowded. Manual screening with the graphic program *boxer* could be used to discard those undesirable particles (*see* **Fig. 3B**).

3.2.3. CTF Parameter Determination

The images taken on an electron microscope are modulated by the contrast transfer function (CTF; *see* **Fig. 4A**) of the electron optical system and the experimental envelope functions *(26–28)*. Because the modulations vary for different micrographs, it is critical to accurately determine the parameters for the CTF and envelope function for each micrographs or CCD frames. The determined parameters would be used in the image refinement/3D reconstruction steps.

Defocus is the primary parameter, which varies in each micrograph. It will dictate the frequency ranges in which phase flipping of the Fourier transform complex values must be made. In addition, the experimental B factor *(29)* needs to be corrected for the Fourier signal amplitudes during the 3D reconstruction.

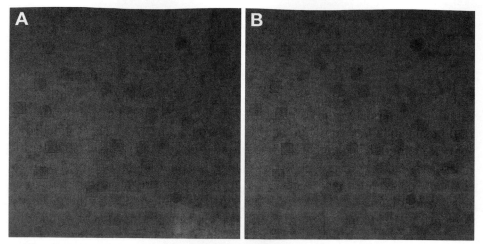

Fig. 3. Examples of particle selection. **(A)** automatically selected particles. **(B)** selected particles after manual screening.

The 2D power spectra of all particles within the same micrograph are incoherently averaged to generate the 2D power spectrum of each micrograph (using *proc2d fftavg* option or *ctfit*). Because the average power spectrum of a good image with minimal drift and astigmatism is essentially isotropic, the 2D power spectrum can be further rotationally averaged to generate a 1D power spectrum curve (using *ctfit*). The CTF fitting process in general is then performed with this 1D curve. It is essentially a curve fitting problem by adjusting the CTF parameters to match the predicted power spectra to the experimental power spectra.

Using *ctfit*, the noise background of the power spectra should first be fitted to touch the minima. The defocus and experimental B factor are then fitted by adjusting the corresponding slider controls in the GUI. A good fit will have the peak/minima positions aligned and the peak heights matched (*see* **Fig. 4B** and also **Note 4**).

Once the CTF parameters are fitted, the corresponding CTF and associated parameters should be recorded in the image headers of each particle. When the EMAN *refine* program is used for the image refinement, the particles should also be CTF phase corrected by flipping the phases in the frequency ranges where CTF amplitude is negative. Both operations can be performed using the program *applyctf* or *ctfit*.

3.3. 2D Image Alignment and 3D Reconstruction

The 3D reconstruction of icosahedral viruses was first introduced by Crowther et al. *(1,2)*. There are numerous software packages such as Spider *(30)*, IMAGIC *(31)*, Frealign *(32)*, PFT *(33)*, XMIPP *(34)*, IMIRS *(35)*, SAVR *(11)*, and EMAN

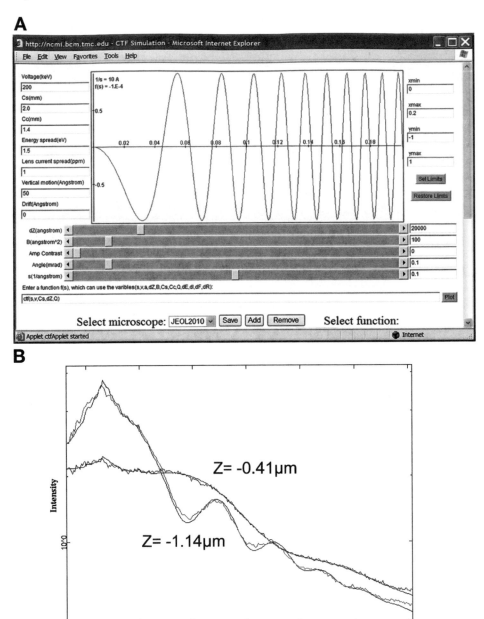

Fig. 4. Contrast transfer function. (**A**) Snapshot of the online CTF simulation (http://
ncmi.bcm.edu/homes/wen/ctf). (**B**) Examples of good fits of contrast transfer function.
Shown are two micrographs of rice dwarf virus imaged on a JEM4000 electron microscope with underfocuses of 0.41 μm and 1.14 μm, respectively.

(10) that can be used for these image processing steps. SAVR is a subset of programs within EMAN that has been developed based on Crowther's algorithm for icosahedral particle reconstruction. This method specifically takes advantage of the high symmetries of icosahedral viruses and is very efficient in computing. It is particularly useful for initial model building. Nevertheless, EMAN is a generic single particle image processing software package that also supports other symmetries in addition to performing icosahedral reconstruction. Adopting the generic single particle package, EMAN, offers additional benefits in that the same software can be used to determine structures of the non-icosahedral components in an icosahedral particle *(36)*.

3.3.1. Build Initial Model

Several approaches can be used to generate the initial model. The first method is to use the *starticos* program that will find the images with best five-, three-, and twofold symmetry and build a low-resolution initial model from these images. The second method is simply using a synthetic icosahedron-shaped model that is roughly similar in size to the virus under investigation. The later step in EMAN refinement cycles are generally powerful enough to converge to the proper structure. The third method is to use the *buildim.py* in SAVR that relies on the self common-line method and Fourier-Bessel synthesis method.

3.3.2. 2D Alignment

2D alignment refers to the task that determines the five parameters for each particle image: three for rotational angles and two for particle center positions in 2D. Euler angles generally are used to represent the three rotational parameters (*see* **Note 5**).

In EMAN, many different 2D alignment algorithms are available, ranging from projection matching, common-lines, to principal component analysis. The projection matching method is the most mature and recommended approach for this step. In this method, an exhaustive collection of projections of the 3D model are first generated using the *project3d* program to cover the complete orientational asymmetric unit. Each particle is aligned to all the projections and the orientation of the best matched projection is assigned to the particle. This step is also often referred to as the classification step because each projection from the model serves as a reference to group the particles with similar orientations. This step is performed using the *classesbymra* program. Many similarity scoring functions, such as dot product, phase difference, integrated Fourier ring correlation, and linear least square difference, etc., are available. In general, the linear least square difference in combination with matched filter (option *dfilt*) is a good choice for the scoring function. After the orientations of all particles have been determined, the particles with the same orientations (i.e., assigned to the same best matched projection) are averaged to generate a class average with improved signals over individual particles using the *classalignall* program.

Note that the 2D alignment by *classesbymra* simply ranks the similarity of each particle compared with all projection images. It does not decide which subsets of particles are "bad" or "good." The class averaging step by *classalignall* makes these decisions and excludes those "bad" particles in the final class averages. This step is performed by evaluating the similarities among the particles within the same class, building the similarity histogram, and discarding the particles with least similarities to other particles by using a user-selectable threshold (typically 1 sigma or less above mean specified using the *keep* option). Additionally, the particles with vastly different orientation assignments in different refinement iterations can be detected and removed using the program *ptcltrace* with *zap* option.

3.3.3. 3D Reconstruction

Once the orientations of each particle is determined and class averages are generated, a 3D reconstruction procedure will be performed to coherently merge the 2D class averages into a single 3D density map at the highest possible resolution allowed by the data. In EMAN, the *make3d* (*see* **Note 6**) program implementing the direct Fourier inversion method is the default reconstruction program. In this step, it is also possible to further remove the bad particle images by discarding the class averages that don't agree well with the reconstruction (using *hard* option).

3.3.4. CTF Correction

In EMAN, CTF correction is performed at two separate stages. After CTF determination, the 2D particle raw images are immediately phase-corrected using the graphic program *ctfit* or the command-line program *applyctf* before orientation determination. These two programs also record the CTF parameters in the image header of each particle for later CTF amplitude correction (*see* **Note 7**). The amplitude correction (using the *refine* program *ctfcw* option) is performed at the class averaging stage by using a Wiener filter and relative SNR weighting of contributions from different particles *(10)*. A satisfactory CTF correction is critically dependent on accurate determination of the CTF and associated parameters, both defocus and amplitude decay factor, of the raw particle images. A useful visual criterion of proper CTF compensation is the absence of apparent black rings just outside of the particle images in the class averages.

3.3.5. Automated Image Refinement

The aforementioned tasks (2D alignment, class averaging, CTF correction, and 3D reconstruction) and the iteration of these tasks are automated by the

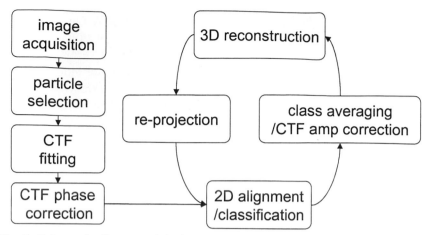

Fig. 5. Schematic diagram of the image processing for icosahedral particles using EMAN.

program *refine*. In the working directory, the phase-corrected raw particles should be pooled into a single image file named *start.hed/start.img* (*see* **Note 8**) with the CTF parameters embedded in the image headers of each particle. An initial model named *threed.0a.mrc* should also be placed in this directory. The structure factor file used for CTF determination should also be present (*see* **Note 4**). Then the *refine* command can be started to perform the iterative 3D map reprojection, orientation determination using projection matching, class averaging, CTF correction, and 3D reconstruction until convergence (*see* **Fig. 5**). A sample *refine* command is given here, which can be a good starting point to tailor to different data sets:

refine < number of iterations > sym=icos ang= < angular stepsize in degree for reprojection > mask= < mask radius in pixel > ctfcw= < structural factor filename > classkeep=1 classiter=3 dfilt refine proc= < number of CPU >

3.4. Structural Analysis

Once the 3D reconstruction refinement has converged, further structural analysis is necessary to segment out the individual subunits and examine their characteristic structural features.

3.4.1. Resolution Determination

The resolution of the 3D reconstruction is, in general, determined using the Fourier shell correlation (FSC) between two reconstructions that are generated using the half datasets *(37)*. Different researchers all are in agreement about the usage of FSC, differences among them exist regarding how it is performed, for instance, what stage the datasets should be split, what type of masking is legitimate, and what threshold to use to read the resolution from the FSC curve *(38,39)*. In EMAN, the resolution evaluation is performed using program *eotest*, which splits each of the classes into halves and performs class averaging and 3D reconstruction by treating the two sets of classes as independent datasets. Before computing the FSC curve, the two 3D reconstructions were masked optionally using an adaptive auto-masking procedure (the *proc3d automask2* option) to remove the background noises without using fixed mask geometry to avoid artificially inflated correlation from the masks. From the FSC curve, we use the 0.5 criterion (the resolution where the FSC value falls to 0.5) as the reported resolution for the final 3D reconstruction. We feel that the resolution number determined this way matches best to the visible structural features especially when the secondary structural elements are resolved in the subnanometer resolution range (*see* **Note 9**).

3.4.2. Segmentation, Visualization, and Animation

Visualization and segmentation of the 3D density map are essential tasks to structural interpretation. Both commercial graphical softwares such as *IRIX Explorer*, *Amira*, and open source softwares such as *Chimera* and *pymol* can be used. In our experience, *Amira* currently has the best overall features, covering all needs of visualization and segmentation and is relatively easy to learn and use. Animations can also be easily created using *Amira*. Alternatively, *Chimera* is an excellent open source alternative especially for hybrid visualization of density maps and PDB atomic models. Preliminary animation capabilities in *Chimera* are also available as a plugin (*emanimator*) provided by EMAN.

The visualization and segmentation of large virus density maps at subnanometer resolutions is very demanding and requires a decent desktop computer or workstation with a good graphical card. A desktop computer equipped with >2GHz CPU, >1 GB memory, and a Nvidia GeForce 4 or later graphical card will be a good configuration.

For an icosahedral virus structure at subnanometer resolutions, typical segmentation tasks are to cut out the asymmetric unit and each of the composing subunits in this asymmetry unit. Subunit boundaries should be resolved at the majority of places except for areas where intimate interactions exist among neighboring subunits. These individual subunits were generally visualized as isosurface views with different colors in isolation or in the context of whole capsid. The whole capsids were also commonly visualized as isosurface views along its five-, three- or twofold symmetry axes, viewed from either outside or inside, often radially colored (*see* **Fig. 6**).

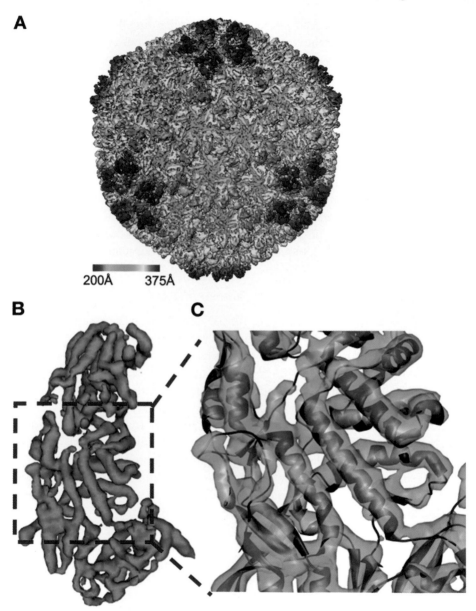

Fig. 6. Example of rice dwarf virus map at subnanometer resolution. (**A**) The whole capsid with radial coloring. (**B**) A single subunit of the inner shell protein P3b. (**C**) Zoomed view of the carapace domain of P3b with the superimposition of subsequently X-ray crystallographically determined model.

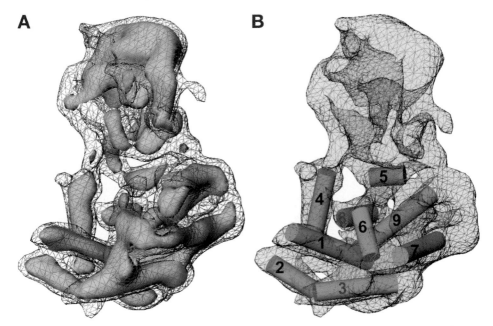

Fig. 7. Secondary structural elements identified in rice dwarf virus outer shell protein P8. (**A**) The average monomer subunit; (**B**) the identified 9 helices in the lower domain.

3.4.3. Identification of Secondary Structure Elements

At subnanometer resolutions, the alpha helices would appear as rod-like shapes of 5 to 6 Å diameter and varying lengths. The sheets are flat, smooth densities with varying curvatures. The helices and sheets can be visually identified by interactively examining the graphical display or through a more objective process using the programs *helixhunter* and *ssehunter* provided in EMAN *(8)*. These processes have been used to correctly identify the majority of the SSEs in many 3D reconstructions at subnanometer resolutions *(6,40,41)*. The identified SSEs encode rich information. The relative positions and orientations of these SSEs can uniquely represent the protein fold and can be used to search for homologue structures in the PDB. This procedure has been successfully employed to identify homologous protein folds for several virus capsid proteins and have lead to a substantially deeper understanding of the assembly principles of these virus capsids (*see* **Fig. 7,** and also **refs. *6,17,40,41***). In addition, direct visualization of secondary structure features is a good indicator whether a 3D reconstruction reaches truly subnanometer resolutions.

4. Notes

1. The blotting is a critical step that will ultimately decide if the resulting grid is usable. Insufficient blotting will result in too thick ice and over-blotting will leave a bare grid. Since the blotting task is to remove >99.9% (3–5 µL) of the solution and leave <0.1% (~1 nL) of the sample on the grid, it is extremely sensitive to the sample viscosity, grid surface hydrophobicity, environmental humidity, filter paper wetness, contact between the filter paper and grid, and the duration of the blotting. One should be prepared for the potential difficulties in this step, especially for sample buffers including glycerol/sucrose, detergent or high concentration of salt (1 M). Currently, there is no method that can guarantee the success of every blotting/freezing. Practice and experience through systematic trial-and-error is the only solution to get around the difficulties.
2. The cryotransfer process should be practiced very carefully and be performed rapidly and smoothly; it is essential to ensure minimal exposure to the room moisture and to avoid ice-contamination and excessive temperature increase of the grid.
3. The Nikon scanner records the transmittance rather than the optical density directly. A log transform is needed to convert the image values so that the values are linearly proportional to optical density for later image processing. Different image processing packages have different assumptions about the scanned image's contrast: some, such as EMAN, expect that the pixel values for the particles are larger than the values of background pixels, whereas others expect the opposite. The image contrast of CCD images and Z/I SCAI scanner images need to be inverted before proceeding to image processing using EMAN.
4. The structure factor curve of the imaged specimen is needed for a good fit of the CTF parameters. X-ray solution scattering experiments usually are used to acquire these curves (29). Alternatively, simultaneous fitting of multiple micrographs could be used (http://ncmi.bcm.edu/homes/stevel/EMAN/doc/faq.html).
5. There are many different conventions of Euler angles used in different software packages. The user must be very careful to convert the parameters properly if different software packages are used for one dataset. In EMAN, conversions among most of the Euler angle conventions and other representations of rotation (spin axis and quaternion) are supported.
6. *make3d* needs to hold four copies of the 3D reconstruction in memory. For large reconstructions, a 64bit computer with 4+ GB memory might be required for a reconstruction map larger than 600^3 pixels.
7. Currently, all particles in the same micrograph are treated to have the same CTF modulation and astigmatism is ignored by discarding apparently astigmatic micrographs.
8. The 2D particle images are typically in IMAGIC format or the special EMAN LST format. The LST format is a text file with links to the actual binary image files. It is especially useful when there are too many particles that cause the total IMAGIC format file size to exceed 2 GB, which is not supported properly on many computer systems. Multiple image files in IMAGIC format can be consolidated into a single LST file using the *lstcat.py* program.

9. The resolution number judged from FSC using any threshold criterion is just a rough estimate of the map quality, and it is somewhat sensitive to the exact masking and other operations before the curve is calculated. It is often true that different areas of the structure have different qualities mostly due to conformational variability. This type of quality variation is poorly represented by a single global resolution number. It is not an absolutely reliable quality ranking criterion for judging the quality or the correctness of the map.

Acknowledgments

We thank Joanita Jakana and Z. Hong Zhou for the RDV images used in the figures. The research has been supported by NIH grants P41RR02250, R01GM 070557, R01AI38469, the Agouron Institute and the Robert Welch Foundation.

References

1. Crowther, R. A., Amos, L. A., Finch, J. T., De Rosier, D. J., and Klug, A. (1970) Three dimensional reconstructions of spherical viruses by fourier synthesis from electron micrographs. *Nature* **226,** 421–425.
2. Crowther, R. A. (1971) Procedures for three-dimensional reconstruction of spherical viruses by fourier synthesis from electron micrographs. *Phil. Trans. Roy. Soc. Lond. B.* **261,** 221–230.
3. Bottcher, B., Wynne, S. A., and Crowther, R. A. (1997) Determination of the fold of the core protein of hepatitis B virus by electron cryomicroscopy. *Nature* **386,** 88–91.
4. Conway, J. F., Cheng, N., Zlotnick, A., Wingfield, P. T., Stahl, S. J., and Steven, A. C. (1997) Visualization of a 4-helix bundle in the hepatitis B virus capsid by cryo-electron microscopy. *Nature* **386,** 91–94.
5. Zhou, Z. H., Dougherty, M., Jakana, J., He, J., Rixon, F. J., and Chiu, W. (2000) Seeing the herpesvirus capsid at 8.5 Å. *Science* **288,** 877–880.
6. Zhou, Z. H., Baker, M. L., Jiang, W., Dougherty, M., Jakana, J., Dong, G., Lu, G., and Chiu, W. (2001) Electron cryomicroscopy and bioinformatics suggest protein fold models for rice dwarf virus. *Nat. Struct. Biol.* **8,** 868–873.
7. Chiu, W., Baker, M. L., Jiang, W., Dougherty, M., and Schmid, M. F. (2005) Electron cryomicroscopy of biological machines at subnanometer resolution. *Structure* **13,** 363–372.
8. Jiang, W., Baker, M. L., Ludtke, S. J., and Chiu, W. (2001) Bridging the information gap: computational tools for intermediate resolution structure interpretation. *J. Mol. Biol.* **308,** 1033–1044.
9. Chiu, W., Baker, M. L., Jiang, W., and Zhou, Z. H. (2002) Deriving folds of macromolecular complexes through electron cryomicroscopy and bioinformatics approaches. *Curr. Opin. Struct. Biol.* **12,** 263–269.
10. Ludtke, S. J., Baldwin, P. R., and Chiu, W. (1999) EMAN: semiautomated software for high-resolution single-particle reconstructions. *J. Struct. Biol.* **128,** 82–97.

11. Jiang, W., Li, Z., Zhang, Z., Booth, C. R., Baker, M. L., and Chiu, W. (2001) Semi-automated icosahedral particle reconstruction at sub-nanometer resolution. *J. Struct. Biol.* **136,** 214–225.

12. Good, N. E., Winget, G. D., Winter, W., Connolly, T. N., Izawa, S., and Singh, R. M. (1966) Hydrogen ion buffers for biological research. *Biochemistry* **5,** 467–477.

13. Adrian, M., Dubochet, J., Lepault, J., and McDowall, A. W. (1984) Cryo-electron microscopy of viruses. *Nature* **308,** 32–36.

14. Dubochet, J., Adrian, M., Chang, J. J., et al. (1988) Cryo-electron microscopy of vitrified specimens. *Q. Rev. Biophys.* **21,** 129–228.

15. Fukami, A. and Adachi, K. (1965) A new method of preparation of a self-perforated micro plastic grid and its application. *J. Electron Microsc.* **14,** 112–118.

16. Jeng, T. W., Talmon, Y., and Chiu, W. (1988) Containment system for the preparation of vitrified-hydrated virus specimens. *J. Electron Microsc. Tech.* **8,** 343–348.

17. Booth, C. R., Jiang, W., Baker, M. L., Zhou, Z. H., Ludtke, S. J., and Chiu, W. (2004) A 9 Å single particle reconstruction from CCD captured images on a 200 kV electron cryomicroscope. *J. Struct. Biol.* **147,** 116–127.

18. Suloway, C., Pulokas, J., Fellmann, D., et al. (2005) Automated molecular microscopy: the new Leginon system. *J. Struct. Biol.* **151,** 41–60.

19. Zhang, P., Beatty, A., Milne, J. L., and Subramaniam, S. (2001) Automated data collection with a Tecnai 12 electron microscope: applications for molecular imaging by cryomicroscopy. *J. Struct. Biol.* **135,** 251–261.

20. Lei, J. and Frank, J. (2005) Automated acquisition of cryo-electron micrographs for single particle reconstruction on an FEI Tecnai electron microscope. *J. Struct. Biol.* **150,** 69–80.

21. Saban, S. D., Nepomuceno, R. R., Gritton, L. D., Nemerow, G. R., and Stewart, P. L. (2005) CryoEM structure at 9 Å resolution of an adenovirus vector targeted to hematopoietic cells. *J. Mol. Biol.* **349,** 526–537.

22. Ludtke, S. J., Chen, D. H., Song, J. L., Chuang, D. T., and Chiu, W. (2004) Seeing GroEL at 6 Å resolution by single particle electron cryomicroscopy. *Structure* **12,** 1129–1136.

23. Jiang, W., Baker, M. L., Wu, Q., Bajaj, C., and Chiu, W. (2003) Applications of a bilateral denoising filter in biological electron microscopy. *J. Struct. Biol.* **144,** 114–122.

24. Zhu, Y., Carragher, B., Glaeser, R. M., et al. (2004) Automatic particle selection: results of a comparative study. *J. Struct. Biol.* **145,** 3–14.

25. Kivioja, T., Ravantti, J., Verkhovsky, A., Ukkonen, E., and Bamford, D. (2000) Local average intensity-based method for identifying spherical particles in electron micrographs. *J. Struct. Biol.* **131,** 126–134.

26. Thon, F. (1971) Phase contrast electron microscopy, in *Electron Microscopy in Material Sciences* (Valdre, U., ed.), Academic Press, New York, pp. 571–625.

27. Erickson, H. P. and Klug, A. (1971) Measurement and compensation of de-focusing and aberrations by Fourier processing of electron micrographs. *Phil. Trans. Roy. Soc. Lond. B.* **261,** 105–118.

28. Hanszen, K. J. (1967) New knowledge on resolution and contrast in the electron microscope image. *Naturwissenschaften* **54,** 125–133.
29. Saad, A., Ludtke, S. J., Jakana, J., Rixon, F. J., Tsuruta, H., and Chiu, W. (2001) Fourier amplitude decay of electron cryomicroscopic images of single particles and effects on structure determination. *J. Struct. Biol.* **133,** 32–42.
30. Frank, J., Radermacher, M., Penczek, P., et al. (1996) SPIDER and WEB: processing and visualization of images in 3D electron microscopy and related fields. *J. Struct. Biol.* **116,** 190–199.
31. van Heel, M., Harauz, G., Orlova, E. V., Schmidt, R., and Schatz, M. (1996) A new generation of the IMAGIC image processing system. *J. Struct. Biol.* **116,** 17–24.
32. Grigorieff, N. (1998) Three-dimensional structure of bovine NADH:ubiquinone oxidoreductase (complex I) at 22 Å in ice. *J. Mol. Biol.* **277,** 1033–1046.
33. Baker, T. S. and Cheng, R. H. (1996) A model-based approach for determining orientations of biological macromolecules imaged by cryoelectron microscopy. *J. Struct. Biol.* **116,** 120–130.
34. Sorzano, C. O., Marabini, R., Velazquez-Muriel, J., et al. (2004) XMIPP: a new generation of an open-source image processing package for electron microscopy. *J. Struct. Biol.* **148,** 194–204.
35. Liang, Y., Ke, E. Y., and Zhou, Z. H. (2002) IMIRS: a high-resolution 3D reconstruction package integrated with a relational image database. *J. Struct. Biol.* **137,** 292–304.
36. Jiang, W., Chang, J., Jakana, J., Weigele, P., King, J., and Chiu, W. (2006) Structure of Epsilon15 phage reveals organization of genome and DNA packaging/injection apparatus. *Nature* **439,** 612–616.
37. Harauz, G. and van Heel, M. (1986) Exact filters for general geometry three dimensional reconstruction. *Optik* **73,** 146–156.
38. van Heel, M. and Schatz, M. (2005) Fourier shell correlation threshold criteria. *J. Struct. Biol.* **151,** 250–262.
39. Rosenthal, P. B. and Henderson, R. (2003) Optimal determination of particle orientation, absolute hand, and contrast loss in single-particle electron cryomicroscopy. *J. Mol. Biol.* **333,** 721–745.
40. Baker, M. L., Jiang, W., Bowman, B. R., et al. (2003) Architecture of the herpes simplex virus major capsid protein derived from structural bioinformatics. *J. Mol. Biol.* **331,** 447–456.
41. Jiang, W., Li, Z., Zhang, Z., Baker, M. L., Prevelige, P. E., Jr., and Chiu, W. (2003) Coat protein fold and maturation transition of bacteriophage P22 seen at subnanometer resolutions. *Nat. Struct. Biol.* **10,** 131–135.

18

Three-Dimensional Reconstruction of Chromosomes Using Electron Tomography

Peter Engelhardt

Summary

The evolvement of preparative methods in structural studies has always been as important as the development of sophisticated equipment. Software development is also a significant part for three-dimensional (3D) structural studies using electron tomography methods (ETMs). Advanced computing makes amenable procedures that relatively recently were only visionary, such as the 3D reconstruction of chromosomes with ETM. Morphological guidelines and beauty are occasionally the only standard for a method to be acceptable in the realms of preparative as well as software development. Bulk isolation of metaphase chromosomes using acetic acid is such an apparent accomplishment in preparative methods. Our ETM with maximum entropy and, more so, the ongoing development toward fully automatic alignments, are contributions in the software line. Furthermore, whole mounting of chromosomes on holey-carbon grids makes it possible to use even yesterday's 80-kV transmission electron microscope with a standard goniometer to collect tilt series. These advances in preparing whole-mount metaphase chromosomes enable laboratories that do not have access to a medium- or high-voltage transmission electron microscope to study complex structures like chromosomes in 3D using today's desktop computers.

Key Words: Whole-mounted chromosomes; electron microscopy; 3D reconstruction; stereo pairs; electron tomography; maximum entropy; tilt-series; automatic alignment.

1. Introduction

Electron tomography *(1)* methods (ETMs), that is, transmission electron microscopy (TEM) tomography, have brought powerful approaches to explore the molecular machinery of cells in three dimensions (3D). 3D reconstructions are a more genuine presentation of structures, as compared with the examination

From: *Methods in Molecular Biology, vol. 369*
Electron Microscopy: Methods and Protocols, Second Edition
Edited by: J. Kuo © Humana Press Inc., Totowa, NJ

of ordinary TEM images that actually present two-dimensional (2D) projections of complex 3D configurations. This does not only concern whole mounts but also sections. An easy solution to extract 3D information from 2D pictures and simultaneously adequately improve the resolution has been to use stereo pairs of the structures under study. Photogrammetry has extensively used stereo pairs with the parallax method to extract 3D data to produce accurate geomorphic cartographic maps. In theory, the exactness of measuring with the parallax method exceeds the resolution of the original stereo photographs.

ETM is far superior compared with the static limitations of stereo pairs. Translucent reconstruction can be viewed from all directions in stereo at high resolution. The reconstruction can also be sectioned virtually to thinner sections that cannot be accomplished with ultramicrotomes. Accurate measurements and analyses can be performed in any direction. Tomograms are possible to accomplish in days or even hours today as a result of the increase in computer power and the availability of programs that can automatically collect digital tilt series, like the free UCSF-tomography software *(2)*.

In the study of the structure of complex molecular machines like chromosomes, the availability of remarkable modern TEM is not enough because the preparative methods to preserve the 3D morphology are essential *(3)*. Accordingly, the development of preparative methods is as important as development of the equipment.

Actually, the whole-mount methods that we have developed do not necessarily require medium (200–400 kV) or high-voltage (1 MV) TEM, as we have found that even an 80-kV conventional TEM provided with an ordinary goniometer is suitable. Previously, tomography calculations, that is, the 3D reconstruction, required supercomputer capacity that today, to a certain extent, may be replaced by desktop computers or clusters to which tomography programs are available, for example, IMOD *(4)*.

2. Materials

2.1. TEM

Medium (200–400 kV) or high-voltage (1 MV or more) TEM equipped with a eucentric goniometer specimen holder and high-performance CCD cameras ultimately are preferred. However, the minimum requirement is an 80-kV TEM with a goniometer specimen holder, tilting range from 0° to ± 60° *(see* **Note 1***)*. This is possible because of a new mounting method using special holey-carbon grids *(see* **Subheading 2.4., item 18***)* that we introduce for whole mounts of chromosomes and, in addition, for whole mounts of cell cytoskeletons, as described in Chapter 19 *(5)*, that will be clearly visible even with 80-kV TEM *(see* **Note 2***)*.

2.2. Computer Requirements

Modern desktop computers (Mac OSX or PC) and clusters appear sufficient to reconstruct small- or medium-sized volumes (e.g., $100 \times 100 \times 200$ voxels) with e.g., IMOD *(4)* that basically use the back-projection method for tomography processing. However, more advanced procedures like the maximum-entropy method (MEM *[3]*) require greater computer capacity (*see* **Note 3**).

2.3. Software

IMOD *(4)* is one of those freely available electron tomography program kits that cover different versions to many platforms (*see* **Note 4**). Our JPEGANIM and MEM (MaxEntropy) axial tomography program kit package that produces the final tomogram using the MEM procedure is freely available and basically easy to use due to automation (*see* **Note 5**). JPEGANIM, that is, the manual alignment steps of the tilt series, is restricted to SGI (Silicon Graphics Inc.) machines *(1)*. However, our MEM is available for different platforms. We have automatic alignment programs, one that relies on fiducial markers *(6)* and another that does not need them *(7)*. Our team is regularly developing more sufficient methods, that is, most accurate fully automatic alignment algorithms that are independent of fiducial markers (*see* **Note 6**), for example, second-generation, "trifocal alignment" *(8)*, and third-generation, "alignment without correspondence" *(9)*. More details of these newly developed automatic alignment procedures will be found in an up-to-date review *(10)*.

2.4. Other Equipment

1. Critical-point dryer (CPD) apparatus (Bal-Tec or other suppliers).
2. CPD holders for grids, commercial or self-made (*see* **Fig. 1**).
3. Containers with liquid CO_2 (preferable only with trace amounts 2 ppm water for CPD).
4. Desktop centrifuge, for example, 50-mL tubes ($260–2580g$; Heraeus Biofuge).
5. 50-mL sterile test tubes.
6. Grid microchamber, self-made (*see* **Fig. 2**).
7. Desktop centrifuge for Eppendorf tubes.
8. Eppendorf microcentrifuge tubes: 1.5 mL, 0.5 mL.
9. Hematocrit tubes (*see* **Fig. 2**).
10. Parafilm.
11. Depression-making device for Parafilm (self-made, *see* **Fig. 3**).
12. Incubation chambers with 96 wells (*see* **Fig. 3**).
13. Sterile filter units, 0.22 μm (Millipore).
14. Latex or rubber gloves (washed if powdered).
15. Curved forceps, preferably anticapillary for grids.
16. Straight forceps for grids (same requirements as above).

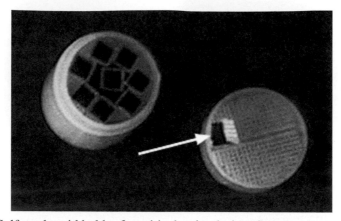

Fig. 1. Self-made grid holder for critical-point drying. Square tubes (approx 3-mm diagonal-fitting electron microscopy grids) are cut and filled tightly into a plastic tissue holder with perforated bottom and lid. Square tube places must somehow be labeled, or their position mapped for recognition. To the removable lid, a square hole (size of the tube) is partially cut (to make a hatch, arrow). The grids are inserted to the grid electron microscopy holder (in methanol) into the square tubes (the same kind of preparations can be piled cross-wise in the same tube). After critical-point drying, each tube is separately emptied by turning it upside down after rotating the lid so that the square hole comes in place, in the tube to be unloaded. We have found that this self-made grid holder is much easier to load (and unload) than commercial ones and more protective to grid films. For example, the film of large single-hole slot grids are better preserved than in other holders, which might be attributable to the fact that fluids are forced to flow through the tubes, i.e., along grid surface to cause a minimum of pressure on the grid film.

17. Ni grids or Au grids, with parallel bars (50×100, 100×200 mesh) or honeycomb or square 50–150 mesh, coated with polystyrene film *(3)*.
18. Holey-carbon-Au grids, MultiA, 200×400 mesh (Quantifoil Micro Tools, GmbH) (*see* **Note 7**).
19. Petri dishes (15-cm diameter).
20. Adjustable shaking table (IKA KS 260).
21. Adjustable heating device (100°C) for 1.5-mL Eppendorf tubes (Grant QBT2).
22. Vacuum device (Savant Speed Vac Concentrator).
23. Flatbed scanner with device for scanning negatives.
24. High-resolution (4000 dpi) scanner for EM film plates (6.5×9 cm) (Nikon Super Coolscan 8000 ED).

2.5. Stock Solutions and Chemicals

Double-distilled water (DDW) is used for water solutions and filtered DDW (FDDW), using 0.22-µm filter units (Millipore), for small amounts of dilution immunogold conjugates and similar solutions that cannot be filtered.

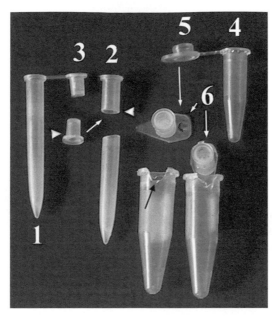

Fig. 2. The assembly of the grid microchamber is nearly self-explained with the numbers and arrows. A plastic hematocrit tube (1) is cut into a smaller tube, arrowhead in (2), and the top of the plug (3) is cut off (arrowhead) to be forced (arrow) to the required depth to function as a bottom for the grid. Suitable-sized chambers (50–200 µL) are thus obtained. In a swing-out desktop centrifuge a 0.5-mL Eppendorf tube (4) will serve as an adapter to the grid microchambers for 1.5-mL Eppendorf tubes. However, without a swing-out centrifuge the caps (5) of the 0.5-mL tubes (4) will serve as adapters to fit the microchambers (6) that are placed (arrow) into a 1.5-mL Eppendorf tube with u-form cut (black arrow). This substitute will put the microchamber in a 45° angle to cause an even pelleting of the sample on the grid, as with the swing-out centrifuge.

Fig. 3. Method to make Parafilm with depressions (*see* **Note 16**). Briefly, two 96-well chambers are pressed together. The Parafilm, with the paper cover, is placed between the chambers. Note, the upper chamber is round-bottomed and has frames removed.

1. Acetic acid (AA).
2. 5 *M* NaCl, stock solution.
3. 1 *M* HEPES [*N*-(2-hydroxyethyl)piperazine-*N'*-(2-ethanesulfonic acid), Sigma] buffer, pH 7.4.
4. 1 *M* MgCl$_2$, stock solution.
5. 1 *M* CaCl$_2$, stock solution.
6. 2 *M* KCl, stock solution.
7. 75 m*M* KCl (*see* **Note 8**).
8. 10% Sodium azide (NaN$_3$), stock solution.
9. HM buffer: 10 m*M* HEPES buffer, pH 7.4, 5 m*M* MgCl$_2$, with 0.02% NaN$_3$.
10. HC buffer: 10 m*M* HEPES buffer, pH 7.4, 5 m*M* CaCl$_2$, with 0.02% NaN$_3$.
11. Immunogold conjugates: 5-, 10-, and 15-nm sizes (Sigma; *see* **Note 9**).
12. DNase I (Sigma, DN-25).
13. RNase A (Boehringer).
14. DAPI (4',6-diamidino-2-phenylindole; Sigma).
15. Glycerol, fluorescence free (Merck).
16. DTT (dithiotreitol, Sigma): 200 m*M* in FDDW, divided into aliquotes (e.g., 0.5 mL), stored at −20°C.
17. ME (2-mercaptoethanol, Sigma).
18. DAPI-*MM* (mounting medium): 90% glycerol, 0.25–0.5 μg/mL DAPI, in HM or HC, with antifading substances, such as 10–20 m*M* DTT or ME.
19. Methanol (*see* **Note 10**).
20. 1% Uranyl acetate (UA), in DDW, filtered (0.22 μm, Millipore), stock solution, (stored at +4°C, in e.g., 50-mL Falcon tube wrapped in aluminum foil). Used as a stock solution for the UA staining solution: 0.001 to 0.002% UA in 100% methanol.
21. 1,2-dichloroethane or chloroform (*see* **Note 11**).
22. 0.5% polystyrene, e.g., pieces of Petri dish (Nalge, Nunc Inc.), dissolved in 1,2-dichloroethane (Sigma-Aldrich, cat. no. 154784; *see* **Note 11**).
23. Tert-Butanol, water free (Riedel-de Haen, cat. no. 33067).
24. Glutaraldehyde (GA), 25% stock solution in glass vial (Electron Microscopy Sciences, PA, cat. no. 16200).
25. 0.2 *M* Sodium cacodylate (SC) stock solution buffer, pH 7.4.
26. 1% Osmium tetraoxide (OT) in 0.1 *M* SC.
27. 1% Tannic acid (Mallinckrodt Inc.) in 0.1 *M* SC and HM buffers (1:1).

2.6. Cell Cultures for Chromosome Isolation

1. Standard outfits for HeLa cell culturing, *see* Chapter 19 *(5)*.
2. Colcemid (KaryoMAX, Gibco, cat. no. 0920).
3. Thymidine (Sigma, cat. no. T1895).

3. Methods

Working solutions are preferably filtered trough 0.22-μm filters directly before use (*see* **Note 12**).

3D Reconstruction of Chromosomes

3.1. Cell Cultures for Chromosome Isolation

1. After refreshing HeLa cell (ATCC CCL-2) culture, the cells are plated for growing to four 600-mL Nunclon Surface cell culture bottles in logarithmic growth phase.
2. When the confluence of the cells has reached to 50–60%, the cells are synchronized with the addition of 2 mM thymidine in the cultivation medium and incubated for 16 to 22 h.
3. The growth medium, with thymidine, is washed three times with a fresh medium without thymidine and replaced with fresh medium including a mitotic inhibitor, e.g., 0.06 to 0.1 µg/mL Colcemid, with ordinary 10% fetal bovine serum and incubated 8 to 12 h.
4. Cells in mitosis are released, naturally or aided by gentle shaking, into the growth medium. Mitotic cells are collected from the medium by centrifugation (*see* **Subheading 3.2.1.**).

3.2. Whole Mounts and Sections

The methods described are basically intended to obtain whole mounts of metaphase chromosomes from cultured cell lines, for example, HeLa cells (*see* **Subheading 3.1.**) that are used as example source material. The methods described for isolating chromosomes in bulk will work with many cell cultures and tissues. If metaphase chromosomes are the goal, the amount of cells at metaphase stage must be significant or otherwise the amount obtained might be difficult to handle. Yet, in some instances, chromosomes are observed in section preparations. It is possible to obtain reconstruction also from sections, as has been explained in more detail *(3)*. The protocols presented here are, however, restricted to prepare whole mounts from isolated chromosomes and their 3D reconstruction.

3.2.1. Chromosome Isolation for Whole Mounts

1. Collecting mitotic cells: The supernatant (after 8–12 h of incubation with Colcemid) of the growth medium (after gently shaking) is put into 50-mL Falcon tubes (in sterile conditions) and collected at 260g for 15 to 20 min (desktop centrifuge, with swing-out rotator with adapters for 50-mL tubes). When placing the supernatant (cell free and sterile) back to the growth bottles, it is possible to collect another harvest of metaphase cells after an additional 8 to 12 h of growth. However, some chromosomes might appear too condensed for some purposes, but the second harvest usually contains more material than the first, also including a fraction of less condensed chromosomes.
2. Hypotonic treatment: The cell pellets are carefully dispersed in filtered 75 mM KCl (0.22-µm filter, *see* **Note 8**), kept at room temperature or at 37°C for 5–10 min, and the hypotonic-treated swollen cells are then collected gently by centrifugation, 260g for 5 to 10 min. The total time of hypotonic treatment is 15 to 20 min, including centrifugation. From this step forward various ways to prepare chromosomes are possible, e.g., direct spreading on a hypo phase *(3)* or, using more physiological

methods, *see* Chapter 19 *(5)*. However, the acetic acid (60% AA) method of isolating chromosomes will be described in this chapter. The method promises an almost foolproof way to get morphologically perfect chromosomes. As a criterion, the chromosome coiling appears clearly in phase-contrast microscopy (especially in long chromosomes, less in very condensed small ones). Notably, for unknown reasons, chromosome coiling usually is not found in physiologically isolated chromosomes *(3)*. Nevertheless, although the method seems physiologically harsh, it works effectively with various source materials, and no aggregated chromosomes or severe contamination of other cellular fragments are noticed.

3. Chromosome isolation with 60% AA method: The loose pellet after hypotonic is rapidly and thoroughly dispersed with 60% AA with approx. 5 to 10 times the amount of 60% AA to the pellet volume. Centrifuged at 400 to 2600*g* for 5 to 10 min. The pellet will excessively shrink in volume and appear in some cases indistinct depending on the bulk of cells used. Washing the pellet is repeated, for example, twice with 60% AA 5 to 10 times of the pellet volume. If many tubes have been used initially, all can usually be collected to one tube, such as to a 1.5-mL Eppendorf tube. In the final wash, the remaining pellet is diluted with a volume of 60% AA two to three times the volume of the chromosome pellet to obtain a concentrated chromosome sample. The sample contains beside metaphase chromosomes approx. 10% to 30% prophase and interphase nuclei depending on how the cell culture has responded to the synchronization, the mitotic inhibitor and to other conditions. Casually, because of the mixed composition, other stages might be useful for 3D reconstruction in addition to metaphase chromosomes. The chromosomes in 60% AA are stored at +4°C or at −20°C, and they seem to be preserved for long periods, at least months.

3.2.2. Application of Fiducial Markers on the Grids

Ni or Au grids (with 100–150 mesh parallel bar or 50–100 mesh with a honeycomb pattern) are coated with polystyrene film according to standard methods for preparing grid coatings (*see* **Notes 7** and **11**). For the coating of grids with fiducial gold markers, immunogold conjugates are diluted with FDDW 1:10, e.g., Au conjugate 5 µL to 45 µL of FDDW (also, different sizes can be used as a mixture). The grids used are either polystyrene film-coated Ni grids or holey-carbon-Au grids that are dipped on both sides to the immunogold dilution drop on Parafilm using a curved forceps and directly immersed in 100% ethanol in 1.5-mL Eppendorf tubes and air dried (AD). The ethanol precipitates the gold on the grid. Repeat if necessary to get the right amount and size of fiducial markers needed.

3.2.3. Mounting Chromosomes on Grids

1. Before the chromosome sample is applied to the grid microchamber (100–150 µL; *see* **Fig. 2**), the chamber is filled to the rim with 60% AA and a grid with (or without) fiducial Au markers is immersed with the film coating side pointing up (*see*

Note 13). By dipping a few times the likely air bubbles will pass away and the grid will settle to the bottom. An amount of 60% AA corresponding to the sample applied is removed. The amount applied is usually 5 to 10 μL of the chromosome sample (*see* **Subheading 3.2.1.**).

2. After the sample has been introduced, the preparation is ready for the centrifuge. The sample is spun through a cushion of 60% AA, that is, chromosomes will settle appropriately on the grid. Centrifuge for approx 1 to 5 min at 720*g* in an Eppendorf desktop centrifuge (*see* **Note 14**). After centrifugation the grid microchamber, detached from the adapter (*see* **Fig. 1**), is filled with 60% AA to a convex surface. The grid microchamber is flipped upside down (e.g., held by forceps), and the grid will usually detach by gently rocking or if not then after a slight touch on the periphery of the grid with a needle and the grid will float on the hanging drop, where it can be picked with the curved anticapillary (and antimagnetic, for Ni grids) forceps.

Different procedures are then followed:

3. For more procedures, *see* **Subheading 3.3.**
4. For CPD or tert-butanol drying (TBD), *see* **Subheading 3.5.**
5. For Cryo-EM methods, *see* **Subheading 3.6.**
6. For Immunoelectron tomography (IET), *see* Chapter 19.

3.3. Further Treatments: Boiling in 50% AA, Salt Extraction, and Enzyme Digestion

Comments: The advantage is that the chromosomes are attached on the grid and treatments are easily performed by moving the grids to different solutions in drops, cf. immunogold labeling, (*see* Chapter 19 *[5]*).

1. Boiling in 50% AA (*see* **Note 15**): Continuing from **Subheading 3.2.3., step 2**, the chromosome preparation on the grid is held with the forceps and boiled in 50% AA (30–60 s) in a 1.5-mL Eppendorf tubes placed in an adjustable heater.
2. Proceed to CPD or TBD (*see* **Subheading 3.5.**), or cryoelectron microscopy (*see* **Subheading 3.6.**) or continue with the steps below.
3. Enzyme and fixation treatments are accomplished in a large, with filter paper moistened. Petri dish provided with a Parafilm with drop-sized depressions (*[3] see* **Fig. 4** and also **Note 16**) on an adjustable shaking (set to minimum) table.
4. Washing: Grids are brought from 60% AA to salt solutions at neutral pH (*see* **Note 17**). Grid preparations, (**Subheading 3.2.3., step 2**) are washed in several drops (SD), that is, four to five drops for 1–10 min/drop of HM or HC buffer on the Parafilm (*see* **Fig. 4**).
5. Note that in every step, the grids are submerged in the drops.
6. Nuclease digestion steps with DNase 1 (10–100 μg/mL) and RNase A (10–100 μg/mL) preferably accomplished in HM or 2 *M* NaCl in HM buffer for 10 min to several hours. The digestion is monitored with fluorescence microscopy using DAPI (0.25–0.5 μg/mL) in the digestion solution (*see* **Note 17**).

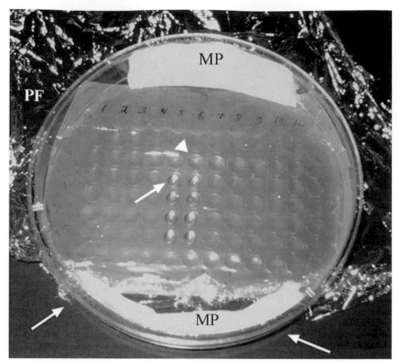

Fig. 4. Parafilm with "dimples" (arrowhead) holding e.g., 30- to 50-µL drops (arrow) in a large (15 cm) Petri dish with moistened papers (MP) (*see* **Note 16**). As seen in the picture, a thin plastic foil (PF) has been stretched between the lid (wrapped around the dish, as seen in the lower part of the picture, two arrows) and the base to keep the chamber tight and moistened for longer incubation times, e.g. enzymes digestions and immuno labeling, *see* Chapter 19 *(5)*.

7. After the treatments the grids are washed in HM (or HC) buffer in SD. After washing steps continue to **Subheading 3.5.**
8. For immunogold IET labeling procedures, *see* Chapter 19.

3.4. Fixation

Comments: 60% AA-isolated chromosomes are "fixed" by 60% AA acting as a precipitator; thus, minor differences are detected using GA especially when DNA is present. However, more drastic changes appear if OT and tannic acid (TA) are used *(3)*. When DNA is present the chromosomes are easily overstained for examination with conventional 80 to 120 kV TEM. Thus only diluted (0.001–0.002%) UA in methanol is used as a stain (*see* **Note 18**). However, after nuclease digestion (*see* **Subheading 3.3.**), the DNA-depleted chromosome, i.e. chromosome scaffolds, is rather electron transparent with UA staining in methanol. We have noticed this appearance earlier *(3)* and when the chromosome scaffold is immunogold labeled with scaffolding proteins (*see* Chapter 19). Thus, the chromosome scaffold is preferably fixed with GA, OS, and TA including UA in methanol to reveal more details as we do with whole-mounted cells in IET (*see* Chapter 19).

Fixation procedures are as follows:

1. No postfixation, proceed to **Subheading 3.5.**
2. Rapid TA method, e.g. for chromosome scaffold (**Subheading 3.3.**)
   a. Washing in HM buffer, SD 1 to 5 min/drop.
   b. Fixing, for 2 to 5 min, in mixture i:ii (1:1): (i) 1% GA in HM buffer, pH 7.4; (ii) 0.5% OT in 0.1 M SC buffer, pH 7.4.
   c. Washing in HM buffer SD 1 to 5 min/drop.
   d. 0.1% to 1% TA in 0.1 M NC/HM buffer (1:1) for 5 min.
   e. Washing in HM buffer SD (*see* **Subheading 3.4., step 2a**).
3. Controls: Leaving out systematically OT, and/or TA and/or post staining with UA (0.001–0.002%) in methanol (*see* **Subheading 3.5.**).

3.5. 3D Preservation Methods

To preserve the 3D appearance of the chromosomes, the preparations must be dried, preferably with a critical-point drying (CPD) apparatus using liquid CO_2 (water-free, *see* **Subheading 2.4., step 3**). However, if a CPD apparatus is not available, TBD principally produces analogous results *(3)*.

3.5.1. CPD of Whole-Mounted Chromosome Grid Preparations

Grid preparations are dehydrated with methanol, placed in a series of 1.5-mL Eppendorf microcentrifuge tubes, holding the grid samples with the curved anticapillary (antimagnetic, for Ni grids) forceps:

1. Perform a stepwise dehydration in 30%, 50%, 75% methanol for 10 to 30 s/step.
2. Stain for 30 s with 0.001–0.002% UA in 100% methanol.
3. Wash three times in 100% methanol, for approx 10 s/step.
4. Grid samples are collected in methanol in CPD holders for grids (*see* **Fig. 1**).
5. CPD is performed from methanol, e.g., following instructions for the CPD apparatus in use. General principles and procedures are found in Engelhardt *(3)*.

3.5.2. TBD (Tert-butanol-Dried) Preparations

1. Grids after the same procedures as in **step 1** (*see* **Subheading 3.5.1.**).
2. After the steps in 1–3 (*see* **Subheading 3.5.1.**), the preparations are brought to a second series of 1.5-mL Eppendorf microcentrifuge tubes containing methanol and tert-butanol (1:1) for 30 s.
3. Complete a stepwise series of three to four tubes with 100% tert-butanol for 10 s/step.

4. The last step is placed in a separate tube with approx 0.5 mL of 100% tert-butanol (preferably changing the solution several times). Bring the grid to be attached to the wall of the tube by capillary force, letting the solution to settle to the bottom, and withdraw the final drops of tert-butanol with a micropipet before closing the tube, which is cooled quickly in a cold place (less than 25.5°C or immersed shortly in an ice bath) to solidify the tert-butanol.

5. Grids in the tubes are brought to a vacuum device (Savant Speed Vac Concentrator, with centrifugation off). The grids in the tubes (with lids open) drop usually rather quickly (in less than 5–10 min) to the bottom of the tubes when the vacuum is turned on. After additional 10 min, the grids are ready for investigation using TEM.

3.6. Cryoelectron Microscopy (Cryo-EM)

The cryo-EM method gives one the advantage of examining fully hydrated samples in vitrified stage, that is, close to native state. A modern, computer-controlled TEM is needed with cryoequipment, for example, a goniometer stage including a cryoholder cooled with liquid nitrogen (N_2) and including accessories for preparing the vitrified samples. Furthermore, an automatic program for cryo-EM to collect the tilt series is freely available *(2)*. The disadvantage is that in the visualization of the images, collected in a tilt series, the signal-to-noise ratio tends to be poor because of the unstained specimens in the vitrified ice. However, we have noticed that our MEM procedure significantly improves the 3D reconstruction, and a very noise-free reconstruction can be achieved compared with the ordinary weighted back projection (WBP) method *(see* **Fig. 5**). Because of the advantage of using MEM in cryo-EM, a short description of the preparative steps we used is presented. There is no need to describe in detail the cryo-EM procedures, as these are standard methods.

1. Preparations, mounted on holey-carbon-Au grids, from **Subheading 3.2.3.,** and following the steps in **Subheading 3.3.,** including **step 1** (boiling in 50% AA) or not, following the procedures described in **step 2 (CPD or TBD)** including **steps 3–7** (nuclease digestions) or not.

2. The grids are washed in HM or HC with reduced amount of HEPES buffer (≤5 m*M*) in SD.

   Then, follow the standard procedures for cryo-EM preparations, that is, blotting with filter paper, and quickly dropping them in a small container (inserted in liquid N_2) with liquid ethane or propane (or a mixture of them at 3:1 that prevents solidification of the freeze-transferring cryogens). The grids are stored in liquid N_2 until examination in cryo-TEM of the chromosomes, using an automatic program for collecting the tilt series, for example, UCSF tomography software *(2)*.

3.7. 3D Reconstruction of Chromosomes

It is relevant that every step in the methods of isolation and data processing be performed for securing sufficient reconstructions. Thus, the steps following the

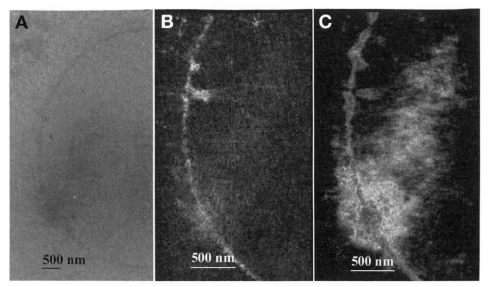

Fig. 5. 3D reconstructions of 60% AA chromosomes (unstained) using cryoEM with 5° increments (0° to ± 60°), collecting manually the tilt series on film (SO163, Kodak) with "minimal dose focusing" at 120 kV with TECNAI 12 TEM. The tilt series was manually aligned using fiducial markers. The raw data were binned (4X) and the chromosome was reconstructed with WBP and MEM. The images shown, (**B**) and (**C**), are from animation series (increment 10°) interpolated (3X) with FUNCS (*see* **Subheading 3.10.**). (**A**) Example of an image of the tilt series at 0° tilt, where the chromatin material is dimly seen within the rim in a hole of the holey-carbon-Au grid. (**B**) 3D reconstruction with WBP. (**C**) 3D reconstruction using MEM. The improvement with MEM is clearly noticeable.

collecting of the tilt series, their alignment, and the electron tomography methods (ETM) must all be carefully accomplished *(3)*.

3.7.1. Manual vs Automatic Recording of Tilt Series

1. For axial tomography the first step is collecting a tilt series. Manual collecting of tilt series sounds time consuming, but manual recording may be more secure than the automatic procedure. In the manual procedure, the pictures usually are collected on photography plates (SO163, Kodak) that are convenient for covering much larger areas than CCD cameras. An additional advantage is that photographic negatives can be collected at a much lower magnification, as compensating with scanning at high resolution is possible with film plates.
2. For whole-mounted chromosome samples, our prevailing rule is to collect a tilt series with 3° increments from 0° to ± 60°, which results in 41 pictures and can be manually achieved after some practice within an hour or so. An increment of 5°

gives less satisfactory results, and increments smaller than 3° are, in practice, difficult, producing too many pictures to handle. In addition, the risk of beam damage is possible, which easily ruins the whole set. The maximum magnification usable is approx 50 to 60,000×, without severe radiation risk. The beam is kept at low intensity, for example, the same as for film exposure (1 s). With the eucentric gonimeter accurately aligned, the focusing and specimen movements are small. At high tilts, some adjustments may be necessary.

3. In cryo-EM, automation is usually necessary because the beam radiation is too destructive for collecting tilt series at high magnification or with small increments. As mentioned, an automatic program for the purpose is freely available *(2)*. However, if the TEM is equipped with a "minimal-beam-focusing device," cryo-EM tilt series are possible to collect manually without severe beam damage at least with 5° increments (*see* **Fig. 5**) or even with 3°, according to our experience.

3.7.2. Digitalization of Negatives

As mentioned previously, if the tilt series is collected on film plates (SO163, Kodak) they have to be digitalized, that is, converted so that they can be used by the data procedures. The digitalization is performed by scanning procedures that actually might take longer than the collecting of tilt series. For this procedure, ordinary flatbed scanners for low resolution with a device for negatives (resolution 1000–1200 dpi) are available, and so are special film scanners for EM plates. Recently, reasonably priced high-resolution flatbed scanners with devices for film have become available. Flatbed scanners usually include programs that can automatically scan a film batch of eight films. The high-resolution film scanners usually scan only one film plate at a time (4000 dpi). Thus, to scan a set of 41 film plates takes approx 4 to 6 h.

3.7.3. Automatic Recording Using CCD Camera

The real advantage with automatic recording tilt series is that series with much smaller increments are possible to collect within reasonable time. As mentioned, such a program is freely available *(2)*. The more pictures, that is, projections, can be collected in a tilt series, the better the resolution will be in tomograms. An additional advantage is that the beam damage is reduced, which also contributes to better reconstructions.

3.7.4. Alignment of Tilt Series

The accuracy of the alignment procedure is also of great importance as the amount of pictures in a tilt series increase. The accuracy of the alignment must increase in proportion to the theoretical increase in resolution and because the number of pictures is getting larger the procedure must also become automatic. All the steps seem to converge now to carry through high accuracy with automation to accomplish high-resolution reconstructions.

3.7.4.1. MANUAL ALIGNMENT USING MARKERS

In practice, if a device for high-performance automatic alignment procedures is not available, we depend on manual alignment using markers. The accuracy needed is usually sufficient as long as the number of projections, that is, pictures, in the tilt series does not exceed 41 (3° increments), using some 5 to 10 markers (that can be picked reasonably fast). The collection of tilt series with 2 to 3 times more pictures must be automated and aligned with high accuracy, as manual alignment is practically impossible.

3.7.4.2. AUTOMATIC ALIGNMENT USING MARKERS

When there are a lot of markers, automatic alignment is of great help. To introduce many fiducial markers is, in practice, not advisable as they will cover and interfere with the structures under study. With immunogold labeling (*see* Chapter 19 *[5]*), different-sized gold labels are introduced that point to relevant structures and proteins. However, the gold markers easily cluster and overlap and these gold markers are impossible or very difficult to pick manually. In these cases automatic marker picking is useful and works quickly *(6)*.

3.7.4.3. AUTOMATIC ALIGNMENT WITHOUT MARKERS

Naturally, as already mentioned, the automatic alignment procedures will soon be the method that will replace all the aforementioned procedures because, according to the laboratory results *(7–10)*, they perform excellently. As mentioned, we already maintain different test variations.

3.8. ETM

The alignment program (e.g., our JPEGANIM subprogram ALIGN) produces accordingly *(3)* from the accurately aligned tilt series (*see* **Subheading 3.7.4.**) a subset of images, that is, Radon transforms (covering twice the area) of the selected object from the original images, for example, part of or whole chromosome. Alignment procedures possess (e.g., JPEGANIM) options for binning images in a tilt series to reconstruct any structures, big or small, with necessary details. The aligned set can be visualized as a movie to test the precision of the midpoint, smoothness, and viewing of details, for example, magnified and visualized also in stereo using JPEGANIM. In addition data info-file ("backproj. inp," with corrected tilt angles, volume size, low-pass filter parameter, and output file name) is also produced by the program. The set of aligned images and the "backproj.inp" file are all that is needed to produce a 3D reconstruction with methods to follow.

3.8.1. WBP Method (Backproject)

In our WBP program: the command (Backproject) is user friendly *(3)* because an ordinary default value, e.g., for the low pass-filter 0.5, will do. WBP is practical because it is a rapid way to produce a reconstruction for checking. However, the reconstruction is usually rather noisy and a high threshold, that is, alpha values (*see* **Subheading 3.9.**) for visualization, is needed, and worse, in depth (i.e., along the electron beam, z-axis) the structures are stretched because of the missing data wedges between 60° and 90°. However, if the reconstruction with WBP does not emerge too bad or noisy, we can be confident of getting an accurate reconstruction using MEM.

3.8.2. MEM

The MEM (our MaxEntropy) is definitely the reconstruction procedure we prefer *(3)*. The noise of the reconstruction is strikingly diminished, with alpha threshold values at 0–10 (*see* **Subheading 3.9.**). MEM also shows phenomenal improvements in cryo-EM axial tomography (*see* **Fig. 5**). Stretching in depth (*see* **Subheading 3.8.1.**) is also notably reduced, for instance, when comparing gold particles of WBP with MEM. The procedure is also transparent to the user, with the only necessary action the adding of a regularization parameter, i.e. the Lagrange coefficient (lambda that with according to our MEM will be, e.g., 1.0e–3, 5.0e–3 or 1.0e–2) in the same line after the low-pass filter (0.5) in the backproj. inp file. An accurate reconstruction is achieved with a small lambda value and, consequently, the procedure will be more calculation intensive. If minor stripes appear that distort the reconstruction, a higher value is chosen.

3.9. Visualization of 3D Reconstruction Volumes

Accurate visualization is of essential importance to resolve details in the 3D configuration of chromosomes and in estimating differences in procedures used. The only correct way to study a 3D reconstruction in detail is to use a true 3D approach, that is, stereo evaluation. I have noticed that many have difficulty in stereoviewing and thus grasping its true value. Many useful and free programs are available. BOB *(11)* shows translucent volumes that are regulated by alpha blending. BOB is our standard, easy to use, and a fast way for routine viewing (SGI) of small volumes or selected details of large volumes or a whole volume using binning and stereo mode (*see* **Fig. 6**), and cf. also figures in *(3)*. A newer version for BOB, AnimaBob *(12)*, is available in many platforms. Other popular programs are Chimera *(13)* and VMD *(14)*. VMD is accomplished in SITUS *(15)*, i.e., docking of PDB molecules with tomograms. Some are powerful but appear quite complicated, such as Vis5D *(16)*. Promising next-generation exceptional technologies for an easier display of 3D volumes at higher resolution are already in sight *(17)*.

3.10. Stereo Animations from 3D Reconstruction Volumes

To viewing volumes freely in all directions at high resolution in a fast, continuous, and smooth manner, one needs to use a high-performance computer.

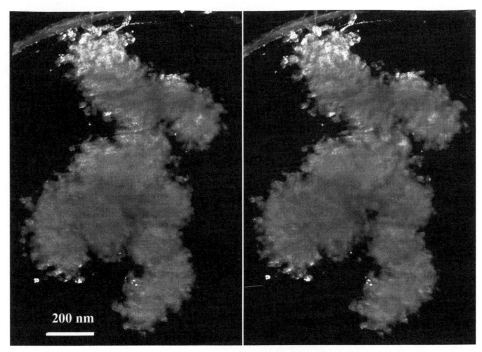

Fig. 6. 60% AA-isolated HeLa chromosome on a holey-carbon-Au grid (prepared as in **Subheading 3.5.1.**). Manually aligned with JPEGANIM (using 15-nm fiducial Au markers). The tilt series was collected manually, increment 3° (0° to ± 60°) at 80 kV with Jeol 1200 EX, on film plates (SO163, Kodak). The tilt series was automatically registered using "trifocal alignment" method *(8)* and the data processed further as in **Fig. 5**. The rim of the hole is visible as an arch on the top of the chromosome. The chromosome coiling emerges in stereo.

Usually, blurring occurs during movement or changing the views and/or the procedure becomes too slow for efficient examination of details. An effective solution to this problem is to make accurate high-resolution animations along the tilt-axis (or any other axis) with increments of 1 to 10°. For this purpose, we prefer to use our volume-visualization program, FUNCS *(3)*. However, many of the aforementioned (*see* **Subheading 3.9.**) programs also seem fitted for producing movies. FUNCS has many advantages, including different-colored lights from many directions, and it includes interpolation of volumes. Different colors for any of the 256 gray scales can be used. The animations produced with FUNCS are viewed in stereo mode with JPEGANIM that can magnify, move to different areas, and measure details with high accuracy. Blurring is eliminated in the animations because all the frames of the movie are already in the memory of the computer that runs the movie smoothly, fast or stepwise for detailed inspections. Most complex scenarios should be analyzed in stereo mode in detail as transparent animations with high accuracy approaching the presumed real 3D organization. However, much of a scenario is lost with magnifications as the displays are too small. Before we have an access to the next-generation 3D displays *(17)*, a solution to look at big scenarios with less restriction is CAVE, i.e., computer-aided virtual environment. Such facilities have been available already for some time *(3)*.

3.11. Hard Copies, Solid 3D Models

Another and rather simple solution that fits some reconstructions is to make solid 3D hardcopies of the volumes in analogy with photographic prints. These technologies also have been available *(3)*. A true 3D hard copy preferably transparent and in different colors according to different voxel densities might partly substitute and refine sufficiently a virtual reconstruction in many details.

4. Notes

1. An eucentric goniometer is preferable because object movements from the center and the focusing adjustments will be small when collecting tilt series. Reconstruction programs usually correct the scaling, that is, differences in magnification, but if not, then an eucentric goniometer is necessary.
2. Higher voltages (100–200 kV or more) are preferable for adequate transparency and better resolution. However, it might be a relief for those laboratories that do not have access to a high-voltage TEM to know that a conventional 80 kV will operate satisfactorily with samples mounted on holey-carbon-Au grids (*see* **Subheading 2.4., item 18**).
3. Big volumes (e.g., 500 × 500 × 1000 voxels or more) and MEM *(3)* need more computing capacity. These calculations need the best possible high-performance computers available preferably with parallel processing setups or clusters that will complete the work within reasonable time, overnight (o/n) or within few days.
4. The axial tomography kits are as the word indicates a large set of many different (sub) programs many of which can be independently used and are useful in many other ways, e.g., converting different formats. With the free IMOD as an example, reconstructions from serial sections can be obtained and many other useful methods can be used, such as semiautomatic alignments and dual-axial tomography, which compensate the missing information to produce potentially better tomograms.
5. Our axial tomography program kit also contains a large set of different (sub)programs. The program kit is freely available (to the academic community and other non-commercial institutions). The problem is that, because we are constantly improving our methods, there is only partial documentation available for using the programs. We are planning to set up a website where to download the programs, including instructions. However, as it appears *(3)*, the whole package is very easy to use, and the best and fastest way to learn to use our tomography program kit is to visit us for a few days.

6. Our tomography team has produced recently new and sufficient automatic alignment algorithms that are independent of markers but they work as Matlab versions; thus, accurate alignment procedures take long to complete. Naturally, after the Matlab versions have been fully tested, faster algorithms will be produced that complete an alignment o/n or in hours depending on the platform.

7. We recently found that the holey-carbon-Au grids, MultiA 200 × 400 mesh (Quantifoil Micro Tools, GmbH) are extraordinarily useful and amazingly firm. We found them casually suitable to grow cells on (*see* Engelhardt et al. *[5]*). Presumably, the strength of the holey-carbon grids (though only covered by a very thin fragile carbon film) is caused by the perforations that can withstand the treatment procedures. A further advantage is that, with ordinary 80-kV TEM, details can be observed that usually are only detected with high voltages such as 100 and 200 kV. However, we detected some variations in different batches because, in a new batch obtained, the film of some grids tended to peel off. In the first initial batch, this was not the case. Such shortcomings can be fixed by using some adherent. To generally prevent film from peeling off, the grids may first be treated with an adhesive solution such as polystyrene in benzene. We also have found that Rubber Cement (Ross Chemical Company), diluted in chloroform (1:100–1000) effectively secures plastic (polystyrene) films on grids that we have earlier used for mounting chromosomes *(3)*, as polystyrene withstands acetic acid. Note that carbon-film-Au grids are resistant to any chemicals.

8. 75 mM KCl solution for the hypotonic treatment is freshly made and filtered (like all solutions) through 0.22-µm filters (Millipore). If a stock solution is made, it must be kept sterile and additionally before use filtered through 0.22-µm filters (Millipore).

9. Immunogold particles, even old and expired ones, will serve as fiducial markers for manual alignment of the tilt series. Commercial products are preferred, as they appear more round, i.e. more accurate as markers, compared to self-made. Immunogold fits, as it does not aggregate as fast as pure colloidal gold. A mixture of different sizes is useful, because the smallest size seen will be best for alignments. However, all depends on the resolution and with small magnifications only the big sizes will be clear enough to be useful.

10. Methanol has a smaller MW than ethanol and presumably reacts faster in the dehydration steps. Methanol is more suitable for CPD because it is not so hygroscopic as ethanol. The trace amount of water bound is 0.05% for methanol and for absolute ethanol 0.5%.

11. When considering solutions for making film coatings to EM grids: 1,2-dichloroethane is recommended because it contains less trace amounts of water than chloroform. Grid film coatings with almost no holes can be made with 1,2-dichloroethane.

12. We keep most used working solutions in syringes fitted with 0.22-µm filter (Millipore) from which droplets are easy to dispense.

13. In holey-carbon-Au Multi A grids (**Subheading 2.4.**), the coating side is darker and oriented, according to the manufacturer (Quantifoil Micro Tools, GmbH), in the grid box toward the centre from both sides of the box.

14. Centrifugation is an advisable method to preserve the 3D configuration of chromosomes, as other more direct applications according to our experience appear more destructive, such as stretching and other deformations of chromosomes are prominent.
15. Boiling in 50% AA will additionally purify chromosomes primarily from histones and other cellular components and fragments that may attach to chromosomes in the 60% AA procedure. The chromosomes appear sharper after the treatment.
16. The depressions on Parafilm are obtained with two incubation chambers with 96 wells (one with round bottom and frames removed) and inserting the Parafilm between the chambers (the round-bottomed laying on the side of the Parafilm with the cover paper on) and pressing them together (see **Fig. 3**). The Parafilm with the "dimples" is put in a large Petri dish with moistened (e.g., filter) paper. To effectively prevent any drying of the drops (e.g. o/n incubations) thin plastic foil is stretched between the Petri dish and its lid (see **Fig. 4**).
17. At neutral pH, the primary requirement to keep chromosomes stable is the presence of divalent cations (e.g. 1–5 mM Ca^{2+} or Mg^{2+}) or polycations such as polyamines spermine (0.0125 mM) and spermidine (0.05 mM), see Chapter 19 . As we have experienced, monovalent salts only cause aggregation of chromosomes and nuclei in isolation procedures at neutral pH.
18. Transparency is primary importance in ETM. If the chromosomes are impenetrable to the electron beam, a successful 3D reconstruction is not obtained. However, without UA staining chromosomes appear somewhat diffuse.

Acknowledgments

For electron microscopy services, I thank Fang Zhao at the Haartman Institute of the University of Helsinki. For cryo-EM facilities and technical help, I thank Sarah Butcher, Pasi Laurinmäki, and Benita Koli in Electron Microscopy Unit at Institute of Biotechnology, University of Helsinki. For helping to collect the electron tomography tilt series, I thank Janne Ruokolainen and Antti Nykänen from Helsinki University of Technology (HUT). I thank Sami Brandt for the automatic alignment of tilt series and Vibhor Kumar, both from HUT, for image processing, and Jari Meriläinen for HeLa cell cultures. For linguistic and editorial help, I thank Kalevi Pusa.

References

1. Frank, J. (ed.) (1992) *Electron Tomography: Three-Dimensional Imaging With the Transmission Electron Microscopy.* Plenum Press, New York.
2. Zheng, S. Q., Braunfeld, B. M., Sedat, W. J., and Agard, A. D. (2004) An improved strategy for automated electron microscopic tomography. *J. Struct. Biol.* **147,** 10–22.
3. Engelhardt, P. (2000) Electron tomography of chromosome structure, in *Encyclopedia of Analytical Chemistry* vol. 6 (Meyers, R. A., ed.), John Wiley & Sons Ltd, Chichester, pp. 4948–4984. Available at: http://www.lce.hut.fi/~engelhar/. Accessed May 29, 2006.

4. IMOD (2005). Available at: URL: http://bio3d.colorado.edu/imod/. Accessed May 29, 2006.
5. Engelhardt, P., Meriläinen, J., Zhao, F., Uchiyama, S., Fukui, K., and Lehto, V.-P. (2006) Whole-mount immunoelectron tomography of chromosomes and cells, in *Electron Microscopy Methods and Protocols*, 2nd ed (Kuo, J., ed.), Humana, Totowa, NJ, pp. 387–405.
6. Brandt, S., Heikkonen, J., and Engelhardt, P. (2001) Multiphase method for automatic alignment of transmission electron microscope images using markers. *J. Struct. Biol.* **133**, 10–22.
7. Brandt, S., Heikkonen, J., and Engelhardt, P. (2001) Automatic alignment of transmission electron microscope tilt-series without fiducial markers. *J. Struct. Biol.* **136**, 201–213.
8. Brandt, S. S. and Ziese, U. (2006) Automatic TEM image alignment by trifocal geometry. *J. Microsc. (Oxford)* **222**, 1–4.
9. Brandt, S. S. and Kolehmainen, V. (2004) Motion without correspondence from tomographic projections by Bayesian Inversion Theory, in *Proceedings of the IEEE Computer Society Conference on Computer Vision and Pattern Recognition (CVPR 2004)* Vol. 1, Washington DC, pp. 582–587.
10. Brandt, S. S. (2006) Markerless alignment in electron tomography in *Electron Tomography*, 2nd ed. (Frank, J., ed.), Springer, New York, pp. 187–215.
11. BOB (2005) Available at: http://www.ahpcrc.org/software/bob/. Accessed May 30, 2006.
12. AnimaBob (2005) Available at: http://www.borg.umn.edu/~grant/AnimaBob/. Accessed May 30, 2006.
13. Chimera (2005) Available at: http://www.cgl.ucsf.edu/chimera/. Accessed May 30, 2006.
14. VMD (2005) Available at: http://www.ks.uiuc.edu/Research/vmd/. Accessed May 30, 2006.
15. SITUS (2005) Available at: http://situs.biomachina.org/. Accessed May 30, 2006.
16. Vis5D (2005) Available at: http://www.ssec.wisc.edu/~billh/vis5d.html. Accessed May 30, 2006.
17. Sullivan, A. (2005) 3-Deep. IEEE Spectrum, INT. April 22-27.

19

Whole-Mount Immunoelectron Tomography of Chromosomes and Cells

Peter Engelhardt, Jari Meriläinen, Fang Zhao,
Susumu Uchiyama, Kiichi Fukui, and Veli-Pekka Lehto

Summary

Standard immunogold-labeling methods in transmission electron microscopy (TEM) are unable to locate immunogold particles in the depth direction. This inability does not only concern bulky whole mounts, but also sections. A partial solution to the problem is stereo inspection. However, three-dimensional reconstruction of immunogold-labeled structures, that is, immuno-electron tomography (IET), is a correct solution for this inconsistency. Striking improvement in resolution is achieved: the 1.4-nm immunogold particles are shown in IET that are not detected in the original tilt series. IET is not restricted to laboratories with advanced medium- or high-voltage TEM and super-computing facilities; the methods we have developed for whole-mounted chromosomes and also for whole-mounted cytoskeleton of fibroblasts work remarkably well with ordinary 80-kV TEMs equipped with a goniometer to collect tilt series for IET on film. In addition, free programs are available to produce three-dimensional reconstructions even without high-performance computers. These improvements make it possible to many laboratories without modern facilities to perform IET reconstruction with standard TEM apparatus.

Key Words: Immunoelectron tomography; immunogold-labeling; whole-mounted chromosomes; whole-mounted cells; transmission electron microscopy; electron tomography; 3D reconstruction; maximum entropy; stereo pairs.

1. Introduction

The superimposing of structures in two-dimensional (2D) transmission electron microscopy (TEM) images not only makes it difficult to understand the three-dimensional (3D) configuration of complex molecular machines like chromosomes, but it likewise affects immunogold labeling, because it is impossible, particularly in depth along the z-axis, that is, in the direction of the electron

From: *Methods in Molecular Biology, vol. 369*
Electron Microscopy: Methods and Protocols, Second Edition
Edited by: J. Kuo © Humana Press Inc., Totowa, NJ

beam, to locate the position of the labeling. This total insufficiency concerns whole mounts as well as thin sections. A partial solution is to use stereo pairs, as mentioned in Chapter 18 *(1)*. The most appropriate solution for TEM immunogold-labeling is the electron tomography method (ETM), i.e., immunoelectron tomography (IET). With IET, different-sized gold markers are used to label different antigens in the 3D reconstruction, making the immunolabeling essential for contributing to understanding the 3D complexity. We have developed IET methods not only to study chromosomes and chromosome scaffold elements but also to label components of the cytoskeleton using whole-mounted fibroblast cells as an example. To demonstrate the power of our IET method, we have shown that 1.4-nm gold markers are clearly visible in the 3D reconstruction but are usually not detectable in the original 2D images of tilt series *(2)*. Furthermore, as mentioned in Chapter 18, we have developed whole-mount preparative methods that do not only suit modern high-performance TEM instruments equipped with modern devices but also work for ordinary 80-kV TEM provided with a standard goniometer. This versatility is convenient for laboratories that do not have access to high-voltage TEM facilities.

In Chapter 18 *(1)*, we used the almost "foolproof" acetic acid isolation method that preserves the higher-order chromosome structure, including chromosome coiling, which also is retained in DNA-depleted chromosomes, i.e., the chromosome scaffold. Chromosome coiling usually is not prominent with "physiological" isolation methods *(2)*. For IET, we will use chromosomes that have been isolated with less harsh methods, i.e. purification of chromosomes using polyamine (PA), as such chromosomes are fitted for explicit analyses of chromosomal proteins *(3)*. We use here PA-isolated chromosomes and the 60% acetic acid-isolated chromosomes as described in Chapter 18 to compare the preservation of scaffolding proteins with immunogold-labeling methods. The procedures for PA chromosome isolation methods will be described using human metaphase chromosomes. Furthermore, our preparation methods for whole-mounted cells on holey-carbon grids fit remarkably well for IET analysis of cytoskeleton proteins of fibroblasts. These whole-mount methods also work well with 80-kV TEM as they do for whole-mount chromosomes (*see* Chapter 18). A protocol that describes how to produce 3.5-nm immunogold conjugates must be included because we have noticed that commercial 3-nm immunogold conjugates are not only costly but also, even worse, not regularly available. Essentially, we prefer to use small immunogold conjugates: 1.4-nm together with our self-made 3.5-nm conjugates to acquire accurate and densely immunogold-labeled IET reconstructions. The methods of 3D reconstruction steps for IET do not need to be repeated, because they are analogous to ETM described in the Chapter 18.

2. Materials
1. Double-distilled water (DDW) is used for all water solutions.
2. Filtered DDW (FDDW), 0.22-μm filter (Millipore) for small amounts of dilutions that cannot be filtered.

2.1. Isolation of PA Chromosomes
2.1.1. Major Equipment
1. Avanti HP30 centrifuge (Beckman-Coulter) or equivalent.
2. Swinging bucket rotor (SW24.38 for Avanti HP or SW28 for XL series; centrifuge tube capacity: 38.5 mL) and angle rotor (JA20; centrifuge tube capacity: 50 mL).
3. Gradient former with a piston (Bio-Rad).
4. Gradient Mate (Biocomp, Canada).
5. Dounce homogenizer with a loose type (B type) pestle (Wheaton).
6. Centrifuge tubes (50 mL for the angle rotor and 38.5 mL for the swinging bucket rotor).
7. Adjustable shaking table (IKA KS 260).

2.1.2. Chemicals and Solutions
1. Appropriate culture medium for each cell line.
2. Digitonin (Sigma) or Empigen (30% solution; Calbiochem).
3. 89% Percoll solution (Amersham).
4. Polyamine (PA) buffer: 15 mM Tris-HCl, pH 7.4, 2 mM ethylene diamine tetraacetic acid (EDTA), 80 mM KCl, 20 mM NaCl, 0.5 mM ethylenebis(oxyethylenenitrilo)tetraacetic acid (EGTA), 0.2 mM spermine and 0.5 mM spermidine, and 0.1 mM phenylmethyl sulfonyl fluoride (PMSF).
5. Hypotonic (HP) buffer: 7.5 mM Tris-HCl, pH 7.4, 40 mM KCl, 1 mM EDTA, 0.1 mM spermine, 0.25 mM spermidine, 1% thiodiglycol, and 0.1 mM PMSF.
6. Lysis (LS) buffer: 15 mM Tris-HCl, pH 7.4, 80 mM KCl, 2 mM EDTA, 0.2 mM spermine, 0.5 mM spermidine, 1% thiodiglycol, 0.1 mM PMSF and 0.05% Empigen.
7. Storage (ST) buffer: 3.75 mM Tris-HCl pH 7.4, 20 mM KCl, 0.5 mM EDTA, 0.05 mM spermine, 0.125 mM spermidine, 1% thiodiglycol and 0.1 mM PMSF.
8. Isolation (IS) buffer: 5 mM Tris-HCl pH 7.4, 20 mM KCl, 20 mM EDTA, 0.25 mM spermidine, 1% thiodiglycol and 0.1 mM PMSF.

2.2. Producing 3.5-nm Immunogold Conjugates
1. Beckman L8-70M ultracentrifuge, SW 41 rotor, 13.2-mL tube.
2. 1% (w/v) Hydrogen tetrachloroaurate (III) hydrate, chloroauric acid (Fluka, cat. no. 50790) stock solution in FDDW.
3. 1% Trisodium citrate in DDW, stock solution.
4. 1% Tannic acid (TA; Mallinckrodt Inc.) in DDW, stock solution.
5. 250 mM Potassium carbonate (Merck) in DDW, stock solution.
6. 1% Polyethylene glycol (PEG) 20 000 (Fluka, cat. no. 81275) in FDDW, stock solution.

7. 5 M NaCl in DDW, stock solution.
8. 10% NaCl in DDW, stock solution.
9. pH 9-Buffer solution: 5 mM NaCl, in 40 mM Tris-HCl buffer, pH 9.0.
10. Anti-mouse IgG, developed in goat, F(ab')$_2$ fragment of affinity-isolated antibody (Sigma, cat. no. M0659).
11. Anti-rabbit IgG, developed in goat, whole molecule of affinity-purified isolated antibody (Sigma, cat. no. R4880).
12. Glycerol.

2.3. Cell Cultures and Reagents for IET

1. HeLa cells (ATCC CCL-2, RCB0007, RCB0191) cultures and lymphocytes K562 (RCB0027), Ball-1 (RCB02561).
2. The chicken embryo heart fibroblast (CEHF) cells *(4)*.
3. Standard outfit for cell culturing *(4)*.
4. Colcemid (KaryoMAX, Gibco, cat. no. 0920).
5. Thymidine (Sigma, cat. no. T1895).
6. Mouse monoclonal antibody (J. Meriläinen, unpublished results) to FAP52 *(5)*.
7. Rabbit polyclonal antivinculin *(6)*.
8. Mouse monoclonal: anti-Topo IIα (Topogen Inc.).
9. Rabbit polyclonal: anti-hCAP-E part of the condensin complex *(2)*.
10. GAM: 1.4-nm gold-conjugated IgG (attached to affinity-purified Fab' fragment, raised in goat, against mouse IgG, whole molecule, Nanoprobes, cat. no. 2002).
11. GAR: 1.4-nm gold-conjugated IgG (attached to affinity-purified Fab' fragment raised in goat, against rabbit IgG, whole molecule, Nanoprobes, cat. no. 2004).
12. GAM: 3.5-nm gold-conjugated F(ab')$_2$ (*see* **Subheading 2.2., step 10**, and **Subheading 3.2.**).
13. GAR: 3.5-nm gold-conjugated IgG (*see* **Subheading 2.2., step 11**, and **Subheading 3.2.**).
14. TXHS + FG solution: 0.05% Triton X-100, 30 mM HEPES, pH 7.4, 0.1 to 0.5 M NaCl, 20 mM KCl, 5 mM MgCl$_2$, 20 mM glycine, 0.5% fish gelatin (FG), 0.02% NaN$_3$.
15. Washing solution: TXHS (no FG).
16. Fish gelatin (Sigma, cat. no. G7765).
17. Glutaraldehyde (GA), 25% stock solution in glass vial (Electron Microscopy Sciences, PA, cat. no. 16200).
18. 0.2 M Sodium cacodylate (SC) stock solution buffer, pH 7.4.
19. HM buffer: 10 mM HEPES buffer, pH 7.4, 5 mM MgCl$_2$, with 0.02% NaN$_3$.
20. HC buffer: 10 mM HEPES buffer, pH 7,4, 5 mM CaCl$_2$, with 0.02% NaN$_3$.
21. 1% Osmium tetraoxide (OT) in 0.1 M SC.
22. 1% TA (Mallinckrodt Inc.) in 0.1 M SC and HM buffers (1:1).

2.4. Grids for IET of Whole-Mounts of Chromosomes and Cells

Holey-Carbon-Au grids, Multi A, 100 × 400 mesh (Quantifoil Micro Tools, GmbH).

3. Methods

Working solutions are preferably filtered through 0.22-μm filter directly before use (*see* **Note 12** in Chapter 18).

3.1. Isolation of PA Chromosomes

Mass isolation of metaphase chromosomes is useful for morphological studies combined with biochemical analyses. In particular, comparisons of their fine structures revealed by electron microscopy and identification of their constituent proteins by mass spectrometry allow the structural roles of each component to be deduced. Although several isolation procedures have been developed to date *(7,8)*, it should be noted that the higher-order structure and components of chromosomes largely depend on the isolation method and degree of purification *(3,9,10)*. Chromosomes are large complexes composed of nucleic acids and proteins, and they usually require Mg^{2+} ions to preserve their condensed structure at physiological pH by reducing the repulsion between the highly negatively charged nucleic acids. However, Mg^{2+} ions also activate several proteases and nucleases at the same time, thereby causing problems for biochemical analyses *(8)*. Therefore, polyamines usually are used to stabilize the condensed structure. Meanwhile, chromosome structure is known to be maintained at acidic pH or in ethanol/acetic acid (3:1) solution. Chemical fixation of chromosomes by glutaraldehyde or formaldehyde also allows the condensed structure to be preserved. These solution conditions mainly have been used for genetic studies. However, under these nonphysiological conditions, the interactions of proteins with other proteins or nucleic acids differ from those under physiological conditions and, therefore, these solutions are not suitable for biochemical analyses, especially protein studies. In general, chemical fixation causes almost irreversible crosslinking of proteins with each other, which largely interferes with further biochemical analyses.

Isolation procedures for chromosomes can roughly be divided into two steps, namely isolation of purified chromosomes and further purification. Purification of chromosomes is a trade-off relationship because an increase in purification step leads to decreased yields. An optimal combination of initial isolation and further purification procedures would provide samples suitable for different purposes. In the next subheading, chromosome isolation and purification procedures using PA buffers are presented. Protein components and structural differences of the isolated chromosomes are also described.

3.1.1. Isolation of PA Chromosomes by Multistep Centrifugation

The first step of metaphase chromosome isolation is completed by either multistep centrifugation or glycerol density gradient centrifugation in the presence of PAs *(3,8,10)*, which both produce chromosomes with similar protein components. Although large amounts of PA chromosomes can be prepared by these methods, the isolated chromosomes contain large numbers but small amounts of chromosome-coating proteins, such as mitochondrial and endoplasmic reticulum proteins *(3)*. Furthermore, cytoskeletal proteins, such as actin and keratin, are rather strongly attached to PA chromosomes, and further purification is necessary for biochemical or structural analyses. However, PA chromosomes are sufficient for light-microscopic studies, such as protein localization analyses after immunochemical labeling. PA chromosomes can be stored at 4°C for 1 to 2 wk or at −20°C in the presence of glycerol for a longer period, and subsequently used for further purification by sucrose density gradient centrifugation or Percoll density gradient centrifugation.

1. For 4000 mL of synchronized cells (*see* **Notes 1–3**), harvest the cells by centrifugation at 1500*g* for 10 min.
2. Resuspend the cells in 100 mL of 75 m*M* KCl and transfer the solution into two 50-mL tubes. Perform hypotonic treatment for 15 min at room temperature (RT), then gently mix the cells and perform a further hypotonic treatment for 15 min at RT, followed by centrifugation at 730*g* at 4°C for 10 min. Resuspend the cells in 80 mL of ice-cold PA buffer containing 0.1% Digitonin or 0.05% Empigen.
3. Take 40 mL of the solution, and disrupt the cell membranes on ice using a 40-mL capacity Dounce homogenizer with a loose type (B type) pestle. Repeat the same procedure with another 40 mL of solution (*see* **Note 4**).
4. Mix the solutions and add PA buffer to a final volume of 90 mL. Transfer 15 mL aliquots of the solution into six 50-mL tubes and centrifuge at 190*g* at 4°C for 3 min. Approximately 10 mL of chromosome containing supernatant is usually recovered (*see* **Note 5**).
5. Add buffer with PA to the pellets up to a final volume of 15 mL after mixing and centrifuge at 190 *g* at 4°C for 3 min. Collect the supernatant.
6. Mix solutions from step 4 and step 5 (total of 120 mL) and gently overlay the mixture onto four 50-mL tubes each containing a 10-mL cushion of 70% glycerol in PA buffer and centrifuge at 1750*g* at 4°C for 20 min. Most of the chromosomes are recovered as a white band at the boundary region between the supernatant and the cushion (*see* **Note 6**).
7. Discard the supernatant (30 mL), collect the chromosomes and store them at −20°C (*see* **Notes 7–9**). Chromosome morphology is maintained for at least 1 month under these conditions.

3.1.2. Isolation of PA Chromosomes by Glycerol Density Gradient Centrifugation

The glycerol density gradient centrifugation method was developed by Laemmli and his collaborators *(7,11)* and is rather complicated compared with the multistep centrifugation method. However, in this method, chromosomes are treated in a dispersed state in solution until the final concentration step, thereby avoiding chromosome aggregation.

This protocol is for a 400-mL culture:

1. Prepare four exponential glycerol density gradients (5–70% in LS buffer; *see* **Note 10**) in separate centrifuge tubes (4X 38.5 mL), and keep them at 4°C in a cold room or on ice (*see* **Note 11**).
2. For 400 mL of synchronized cells, harvest the cells by centrifugation at 1500*g* at RT for 10 min.
3. Gently, but completely, wash the pellet with 20 mL of HP buffer and centrifuge at 1200*g* at RT for 5 min. Transfer the pellet of cells to a 50-mL tube and repeat the cell washing twice. Centrifuge the suspension at 1200*g* for 5 min.
4. Resuspend the pellet in 20 mL of ice-cold LS buffer. From this step on, all the procedures are conducted at 4°C.
5. Disrupt the cell membranes using a 40-mL capacity Dounce homogenizer with a loose type (B type) pestle.
6. Transfer separate aliquots of the solution onto the four exponential glycerol density gradients, and centrifuge at 200*g* for 5 min using a SW24.38 swinging bucket rotor and then at 700*g* for 15 min.
7. Collect 3-mL fractions from the top to the bottom of the gradient. Check the fractions containing chromosomes for their authentic morphology by optical microscopy. Layer the fractions containing chromosomes onto a 5-mL cushion of 70% glycerol in ST buffer and centrifuge at 3000*g* for 15 min. Chromosomes are recovered as a slightly white band at the boundary region between the supernatant and the cushion.
8. Collect the chromosomes and store at −20°C.

3.1.3. Further Purification of PA Chromosomes by Sucrose Density Gradient Centrifugation

This purification removes proteins weakly adhered to chromosomes, such as several chromosome peripheral proteins and chromosome coating proteins (*see* **Note 5**).

1. Prepare 38.5 mL of a linear (20–60% w/v) sucrose density gradient in PA buffer using a Gradient Mate.
2. Discard a small amount (1–2 mL) of the gradient solution from the top of the tube and layer 1–2 mL of PA chromosomes onto the gradient solution in PA buffer containing 0.1% digitonin. Centrifuge at 2500*g* for 15 min in a JS-24.38 swing rotor.
3. Combine the fractions containing chromosomes with native morphology. Dilute the sucrose solution with PA buffer and collect the purified chromosomes by centrifugation at 1000*g* for 10 min.

3.1.4. Further Purification of PA Chromosomes by Percoll Density Gradient Centrifugation

This method is the most effective technique for removing proteins adhering to chromosomes during isolation procedures (*see* **Notes 9, 12, 13**). However, the amount of chromosomes reduces to one-third to one-fifth of the starting material after the purification. All these procedures are carried out at 4°C.

1. Mix 10 mL of chromosomes in 70% glycerol with 10 mL of 89% Percoll solution.
2. Gently homogenize the solution using a 40-mL capacity Dounce homogenizer with a loose type (B type) pestle (10 strokes).
3. Add 15 mL of 89% Percoll solution and homogenize the solution gently and thoroughly (30–40 strokes). Transfer the solution to a 50-mL capacity polycarbonate tube and centrifuge at 45,440g for 30 min using an angle rotor.
4. Collect the band located closest to the bottom among the two visible broad bands. This band is positioned at one-fifth of the length of the tube from the bottom of the tube.
5. Dilute the collected solution with two volumes of IS buffer and centrifuge at 3000g for 15 min.
6. Wash the precipitated chromosomes with ST buffer.
7. Collect the chromosomes as a precipitate or recover on a 70% glycerol cushion as described in **Subheading 3.1.3.**

PA chromosomes are appropriate for IET purposes just as for biochemical analyses (*see* **Note 14**).

3.2. Production of Gold-labeled Secondary Antibodies

3.2.1. Preparation of a 3.5-nm Colloid Gold by the TA Reducing Method

Please also refer to Lucocq (*12*) and Slot and Geuze (*13*).

1. 1% Hydrogen tetrachloroaurate gold solution in FDDW.
2. Dilute 1 mL of 1% gold solution into 79 mL of FDDW in an Erlenmeyer flask (*see* **Note 15**).
3. Add 1% trisodium citrate 4 mL, 1% TA 5 mL, 25 mM KCO$_3$ 5 mL, FDDW 6 mL to make total 20 mL reducing solution in another Erlenmeyer flask.
4. Heat both solutions to 60°C.
5. Add reducing solution to gold solution while it is being vigorously stirred. The color of the mixture turns into purple within seconds.
6. Heat the mixture gently until boiling, keep for 10 min (*see* **Note 16**).
7. Cool it down on ice, keep in ice overnight (o/n) for use in gold conjugation.

3.2.2. Determination of the Amount of Minimal Stabilizing Protein by Electrolyte Aggregation

1. Adjusting pH of gold solution: to 10 mL of gold solution, add 250 mM KCO$_3$ until pH is 9.0, which ensures that the antibodies are negatively charged. About 30 µL/ mL gold solution is needed (*see* **Note 17**).
2. Dilute antibody in a series of Eppendorf tubes: from 0 to 10 µg, 1 µg per step in 100 µL of FDDW.
3. Add 100 µL of pH 9-buffer solution (*see* **Subheading 2.2., step 9**) to the antibody.
4. Add 500 µL of pH-adjusted gold solution to the tube, mix thoroughly.
5. Add 100 µL of 10% NaCl and mix together. Leave the tubes in RT, overnight for gold particles smaller than 5 nm. The electrolyte-induced gold aggregation and the

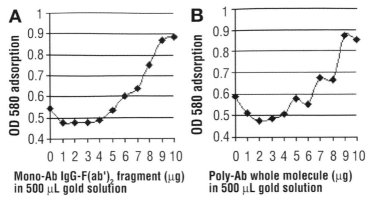

Fig. 1. (A) Antimouse IgG, F(ab')$_2$ fragment. **(B)** Antirabbit IgG (whole molecule).

prevention of aggregation by protein, i.e. stabilization, can be observed visually or spectrophotometrically. Color change from purple to red to light-blue indicates aggregation; unchanged color indicates stabilization. More precisely, OD$_{580}$ absorption is used to find the aggregation point (*see* **Note 18**).

6. Choose the first tube where the aggregation is prevented as the value for calculation. For example, in **Fig. 1,** the tube with 9 μg of antibody in 500 μL of gold solution is the first tube without aggregation, i.e., 18 μg of antibody per 1 mL of gold solution. Take this as the minimal stabilizing antibody concentration.

3.2.3. Antibody-Colloidal-Gold Conjugation

1. Add 250 mM KCO$_3$ to 10 mL fresh colloidal gold solutions, adding drop-wise while measure the pH until pH reaches 9.0. 20 to 30 μL 250 mM KCO$_3$ is needed.
2. Take certain amount of antibody to a new tube (the amount is determined in **Subheading 3.2.2., step 6**), taking from the same batch of antibody used for determining minimal stabilization amount (*see* **Fig. 1** and also **Note 18**). Add pH 9-buffer solution to the antibody tube until total volume is 100 μL.
3. Dialyzing antibody against borate is optional and usually not necessary.
4. Add antibody drop-wise to the pH-adjusted gold solution while shaking, wait for 2 min.
5. Add 500 μL of 1% PEG 20000 (final 0.05%) to gold-antibody mixture, wait for 10 min.

3.2.4. Purifying Antibody-Gold Conjugates

This step removes unconjugated free antibodies, gold particles and aggregates. Both pelleting and gradient centrifugation can be used but the latter gives better results.

1. Glycerol step gradient is made in three steps, each 2.5 mL (from the bottom of the tube): with 70%, 35%, 17.5% glycerol, using the stock solutions pH 9-buffer

solution (*see* **Subheading 2.2., step 9**) and 1% PEG 20000 to get final 0.05% PEG 20000 concentration and pH 9 for each gradient.

2. Add 5 mL of antibody-gold mixture (in 3% glycerol) very slowly, to the top of the tube.
3. Centrifugation: Beckman SW41 rotor, tube 13.2 mL, 100 000 g at 4°C, 90 min for whole antibody, 150 min for F(ab')$_2$ fragment.
4. Collect the thin red band, which is relocated, to the lower middle of the tube. If the bands are not sharply defined after the suggested time, continue centrifugation until the bands get sharp.

3.3. Cell Culturing

3.3.1. Cell Culture Synchronization for Chromosome Isolation

For synchronization of the HeLa cell cultures for chromosome isolation, *see* Chapter 18.

3.3.2. Primary Chicken Embryo Heart Fibroblast (CEHF) Cell Line Culture

The CEHF cells are purified and grown with standard cell culture purification and growth protocols *(4)*.

3.4. Immunogold Labeling of Whole-Mounted Chromosomes and Cells for IET

3.4.1. Precoating of Grids With Fiducial Gold Markers

Application of markers (*see* **Note 19**, and also Chapter 18). Immunogold conjugates, 10 or 15 nm, are diluted with FDDW, 1:10 (e.g. 5 μL to 45 μL FDDW). A Holey-Carbon-Au (HCAu), 100 × 400 mesh, grid is dipped or touched on both sides to the immunogold dilution drop on Parafilm using the curved forceps. The grid is then immersed in 100% ethanol (including the tip of the forceps to secure the sterile conditions when growing cells on the grids (*see* **Subheading 3.4.3.**) in a 1.5-mL Eppendorf tube. The ethanol precipitates the gold on the grid. The procedure is repeated if necessary to get the needed amount of fiducial markers.

3.4.2. Whole-Mount Chromosomes on Grid

1. For 60% acetic acid (AA)-isolated chromosomes, the procedure has been described previously (*see* **Subheading 3.2.3.** in Chapter 18).
2. For PA chromosomes (*see* **Subheading 3.1.**), their application on HCAu grids (precoated as mentioned previously, with fiducial markers or not) is similar, but instead of 60% AA, a cushion of HM or HC buffer is used.
3. A PA chromosome sample (in 60% glycerol, kept at −20°C) of approx 20 to 30 μL is centrifuged in the grid microchamber (*see* **Fig. 2** and also **Subheading 3.2.3.** of Chapter 18) through a cushion of 60 μL of HM or HC buffer.

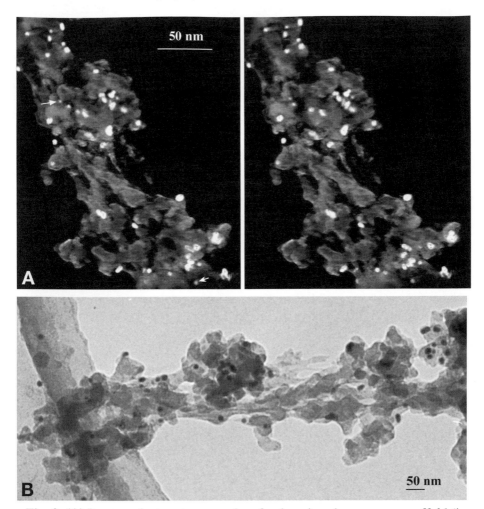

Fig. 2. (A) Immunoelectron tomography of polyamine chromosome scaffold (i.e., nuclease digested, *see* **Subheading 3.4.2.**), with immunogold labeling (*see* **Subheading 3.4.6.**) with 3.5-nm gold particles to scaffold proteins Topo IIα and 1.4 nm to hCAP-E (*see* **Subheading 2.3.**). After immunolabeling, the chromosome scaffold was postfixed as described (*see* **Subheading 3.4.6.**). Several 3.5-nm gold particles appear in the depth dimension together, with some 1.4-nm gold particles (small arrows) showing the advantage of using immunoelectron tomography and stereo viewing. (B) Region of the scaffold from the tilt series. The tilt series of the chromosome scaffold mounted on holey-carbon-Au grid was collected at 80 kV in Jeol 1200 EX TEM, 3° increment (0° to ± 60°), on film plates (SO163, Kodak). The tilt series was automatically registered with 'Trifocal Alignment' method (8). The raw data in **A** (above) were binned (2X) and the chromosome scaffold was reconstructed using the maximum-entropy method, The stereo pair shown is from animation series interpolated (3X) with FUNCS (*see* **Subheading 3.10.** in Chapter 18).

4. Subsequently, followed by nuclease digestion (*see* **Fig. 2**) or not (*see* **Subheading 3.3., steps 3–7** in Chapter 18 and *see* **Subheading 3.4.6,** for IET).

3.4.3. Growing Cells on a Grid

1. For IET the CEHF cells are grown on the fiducial-marked grids (above) with 2 to 10×10^5 cells/mL (with approx four to six grids in a small cell culture dish). The medium is changed after each 24 h. The culture on the grids is after 2 to 3 d at 75% confluence that we consider advantageous for IET.
2. Grids are washed in the cell culture dishes with Dulbecco's PBS (DPBS) without bovine serum albumin, two to three times for some 3 to5 min/wash.
3. Follow the steps for pretreatment (*see* **Subheading 3.4.4.**) or prefixation (*see* **Subheading 3.4.5.**).

3.4.4. Pretreatment with TXHS solution

We have observed that a brief pretreatment with TXHS + FG solution increases substantially immunolabeling of cytoskeleton elements compared with the earlier method with a brief prefixation (*see* **Subheading 3.4.5.**). We recommend both methods to be used in parallel for comparison as delicate structures might be preserved better or differently with the prefixation compared to the pretreatment method (*see* **Fig. 3**).

1. Put grids in drops of the TXHS + FG solution for approx 2 min.
2. Wash briefly in two to three drops (1–3 min/drop) of DPBS.
3. Follow the steps in **Subheading 3.4.5.**

3.4.5. Prefixation Method

The procedures are carried out on the parafilm in a Petri-dish with drop-sized depressions moistened with filter paper (*see* **Fig. 4, Subheading 3.3,** Chapter 18).

1. Prefixing in 0.125% glutaraldehyde (15 min), in DPBS, for cells, for chromosomes in HM or HC buffer.
2. Wash in DPBS for cells and in HM or HC buffer for chromosomes, several drops (SD): three or four drops, 1–10 min/drop.

3.4.6. Immunogold Labeling for IET

1. Blocking solution: incubate o/n in TXHS + FG, 1 to 2 h at RT and/or at +4°C.
2. Incubation in primary antibody mixture: (a) Rabbit polyclonal antibody for protein 1 (e.g., $1{:}40$–$1{:}10^3$); (b) mouse monoclonal antibody for protein 2 (e.g., $1{:}25$–$1{:}10^2$) Diluted in TXHS + FG and incubated, 1 to 2 h at RT and/or o/n at +4°C.
3. Wash with TXHS, in SD.
4. Incubation in secondary gold-conjugated antibody mixtures, examples: Mixture: GAR (goat anti-rabbit) 3.5-nm gold conjugates (1:40–100 dilution), self-made (*see* **Subheading 3.2.** and also **Note 20**) with GAM (goat anti-mouse) 1.4-nm gold conjugates (1:25–40 dilution) diluted with TXHS + FG.

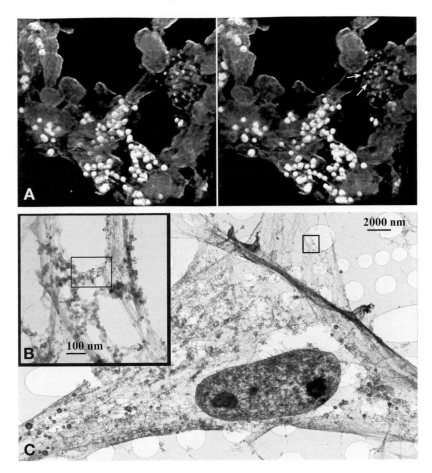

Fig. 3. (A) Stereo pair of immunoelectron tomography of pretreated chicken fibro-blast cells with TXHS + FG solution (*see* **Subheading 3.4.4.**) showing the difference of immunolabeling with prefixation (cf. **Fig. 4**). With this method, intracytoplasmic staining is very prominent. The cells underwent immunogold labeling (*see* **Subheading 3.4.6.**) for vinculin using 3.5-nm and for FAP52 using 1.4-nm gold conjugates (*see* **Subheading 2.3.**) and postfixed as described (*see* **Subheading 3.4.6.**). Vinculin appears widely distributed; however small delicate clusters of 1.4-nm FAP52 labels also are clearly distinguished (arrows). **(B)** Detail of the region from which the immu-noelectron tomography was collected (framed). **(C)** The advantage of growing whole cells on a holey-carbon-Au grid is apparent, e.g. note that also the cell nucleus is clearly transparent at 80 kV. The region of insertion in **(B)** is framed in **(C)**. Tilt series was collected with 3° increment (0° to ± 60°) at 80 kV, Jeol 1200 EX, on film plates (SO163, Kodak). The tilt series was automatically registered with the 'Trifocal Align-ment' method *(8)*. The raw data in (A) were not binned and the object was reconstructed with maximum-entropy method, The stereo pair shown is from animation series inter-polated (3X) with FUNCS (*see* **Subheading 3.10.** in Chapter 18).

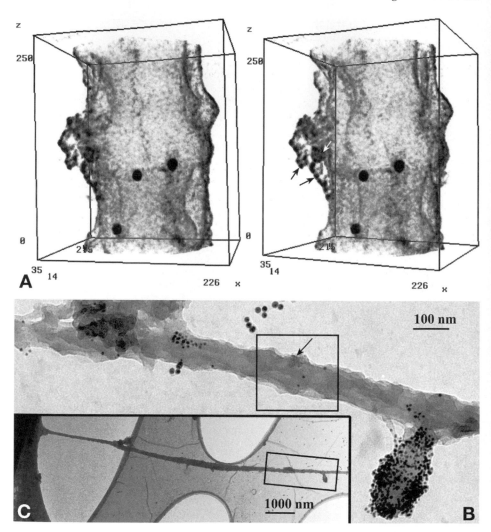

Fig. 4. (A) IET stereo pair of prefixed (*see* **Subheading 3.4.5.**) chicken fibroblast cell (cf. **Fig. 3**). The cell periphery appears less penetrable to immunogold labeling (*see* **Subheading 3.4.6.**), as the label appears prominent on the surface even in thin filopodia protrusions. Vinculin is shown using 3.5-nm and FAP52 using 1.4-nm immunogold conjugates (as in **Fig. 3**). The cells were postfixed after immunolabeling as described (*see* **Subheading 3.4.6.**). A delicate knotty fiber-like cluster that sticks out from the filopodia surface is densely labeled by 1.4-nm gold (black arrows on the fiber-like structure, a white arrow points to a knotty cluster). **(B)** An image of the tilt series of the filopodia: the immunoelectron tomography region is framed with an arrow pointing to the faint cluster, showing that only with immunoelectron tomography the 1.4-nm gold markers are recognizable. On the right side along the filopodia, a heavily, 3.5-nm-gold vinculin stained protrusion is visible and another smaller 3.5-nm cluster

5. Incubation: 1–2 h at RT and o/n at 4°C, preferably on an adjustable shaking table, set to minimum (*see* **Subheading 2.1.1., item 7** and also **Note 21**).
6. Wash in TXHS, SD.
7. Wash in DPBS for cells and for chromosomes in HM or HC buffer, SD.
8. Wash in 0.1 *M* sodium cacodylate (SC), SD.
9. Postfixing: 1.25% GA in 0.1 M SC, 30 min–1 h or o/n (for chromosomes in HM, or HC buffer), add osmium tetraoxide (OT) (from e.g. 1% stock solution) in 0.1 *M* SC buffer, pH 7.4, to 0.3% OT in SC, 1 to 5 min, RT. Control: no OT.
10. Wash in HM or HC buffer, SD.
11. TA treatment: (a) 1% TA in 0.1 M SC and HM (1:1), 5 min; (b) Control: no TA.
12. Wash in HM, SD.
13. Dehydrate in 30, 50, 75, 100% methanol, in a series of Eppendorf 1.5-mL tubes.
14. Stain in 0.001% to 0.002% uranyl acetate (UA) in 100% methanol (30 s).
15. Wash in a series of Eppendorf 1.5-mL tubes 3–4 with 100% methanol.
16. Critical point drying or other 3D-preserving methods (*see* **Subheading 3.5.** in Chapter 18).
17. Controls: In IET, control (preimmune) serum was used or primary serum Abs were omitted and samples were postfixed only in GA (with or without TA), with no UA before drying step.

3.5. 3D Reconstruction Steps for IET and Visualization

The steps are equivalent to ETM described (*see* **Subheading 3.7.** in Chapter 18).

4. Notes

1. To avoid disruption of the chromosome structure by shearing forces during the isolation procedures, cut the pipet tips to create a larger drainage hole.
2. The formation of metaphase chromosomes occurs within a short period during the cell cycle and thus only a small percentage of cultured cells are usually at metaphase. The cell cycle can be arrested at metaphase with inhibitors of microtubule formation, such as Colcemid. In general, 50% to 60% of cells become synchronized at metaphase after treatment with 0.02 µg/mL Colcemid for 13 to 16 h, whereas at the best, a synchronization of more than 90% has been reported (*14*). At the initial stage of the experiment, the synchronization of the cells should be checked by counting the numbers of mitotic and interphase cells. In cases where the synchronization decreases a few months after the beginning of a culture, restarting the cell culture from a frozen stock is recommended.

Fig. 4. (*Continued*) on the left side is seen. The larger, 15-nm gold particles in the background are fiducial markers (*see* **Subheading 3.4.1.**). The insertion **(C)** in (B) shows the filopodia protruding from a cell at low magnification. The tilt series was collected with 3° increment (0° to ± 60°) at 80 kV, Jeol 1200 EX EM, on film plates (SO163, Kodak). The raw data presented in **A** (above) were processed as for **Fig. 3A**, except SNAPSHOT stereo pictures (inverted) of the volume shown with BOB (*see* **Subheadings 3.8.2.** and **3.9.** in Chapter 18) has been used to show the volume (no interpolation).

3. Cell cycle synchronization does not have a large effect on the components of metaphase chromosomes, as evaluated by 2D electrophoresis *(10)*.

4. For HeLa S3 cells, complete disruption requires 20 strokes. However, the number of strokes necessary for cell membrane disruption depends on the cell type and culture conditions. Complete disruption of the cell membranes after the homogenization procedure should be checked by microscopic observation of chromosomes stained with DAPI. Metaphase chromosomes are released and dispersed into the buffer after disruption of the cell membranes.

5. Leave as much of the sediment fraction as possible, which avoids contamination of the nuclei.

6. Chromosomes at boundary region between the supernatant and the cushion can be seen only when they are large in amount. In case of Piston Fractionator, lighting up the tube from the bottom direction helps the chromosome band to be visualized. Before storage at −20°C, washing the chromosomes three times by dilution in PA buffer followed by centrifugation at 1750*g* for 10 min removes many of the contaminating proteins.

7. In general, chromosomes are very sticky toward one another, and therefore concentrating them by centrifugation requires special attention. Once chromosomes form visible aggregates, their dispersion is almost impossible. Therefore, high speed centrifugation over 1000*g* should be avoided in the absence of glycerol cushions. Glycerol cushions are very useful for preventing the formation of chromosome aggregates and make it possible to collect chromosomes at 3000*g*. It should also be noted that a low glycerol concentration results in freezing during storage at −20°C and, therefore, most of the supernatant needs to be removed prior to storage in order to avoid dilution of the glycerol.

8. At a synchronization rate of 50%, 400 ml of a cell culture at 0.7×10^6 cells/mL typically produces approximately 5×10^9 PA chromosomes using both the multistep centrifugation and glycerol density gradient centrifugation methods, from which 500 µg of chromosome proteins can be extracted with acetic acid.

9. In the case of PA chromosomes, the protein components vary among cell lines, since the chromosome coating or fibrous proteins that adhere to PA chromosomes during isolation generally differ among cell lines. Purification using Percoll density gradient centrifugation (*see* **Subheading 3.1.4.**) eliminates these proteins and allows chromosomes predominantly composed of essential chromosome structural proteins to be obtained.

10. Linear density and exponential density gradients have different gradient profiles. A linear density gradient can be formed using an open-top Gradient Maker or Gradient Mate, while an exponential density gradient can be prepared using a Gradient Maker equipped with a piston.

11. As pointed out by Gasser and Laemmli *(11)*, cytoskeletal proteins tend to bind strongly to chromosomes and facilitate chromosome aggregate formation (chromosome cluster), when cell collection and subsequent procedures prior to cell membrane disruption are conducted at 4°C. However, the amounts and types of cytoskeletal proteins that adhere to isolated chromosomes depend on the cell type

used. PA chromosomes from lymphocytes contain a smaller amount of cytoskeletal proteins.

12. At metaphase, the nuclear envelope has already broken down and chromosomes are surrounded by cytoskeletal proteins in the cells. Most of the cytoskeletal proteins that adhere to the chromosomes can be removed by repeated washing with Percoll using a Dounce homogenizer. Percoll is composed of small particles that form a density gradient during centrifugation. The purified chromosomes are recovered after the centrifugation as a visible band at the position where their buoyant force becomes zero.

13. Percoll consists of silica particles with a tightly bound layer of polyvinylpyrrolidone (PVP). The diameter and density of these particles are 15 to 30 nm and 1.130 g/mL, respectively. Percoll seriously interferes with electrophoretic separation, and therefore its complete removal is necessary for biochemical studies.

14. Chromosome proteins can be extracted efficiently using acetic acid as follows. Add two volumes of glacial acetic acid and a one-tenth volume of 1 M MgCl$_2$ to the isolated chromosomes resuspended in PA buffer. After rapid mixing for 1 h at 4°C, centrifuge at 15,000g for 10 min at 4°C and collect the supernatant, which contains chromosomal proteins in a soluble state. Extract proteins from the pellet once more in a solution of 33 mM MgCl$_2$ and 66% glacial acetic acid at 4°C as in the first extraction. Combine the two supernatants, and completely dialyze the solution against 2% acetic acid. Dispense aliquots, and subject them to lyophilization to obtain the chromosomal proteins as a powder. This powder is highly soluble in various solvents for electrophoresis in gels containing 8 M urea.

15. Hydrogen tetrachloroaurate has a different molecular weight due to its different hydrate status: with 2H$_2$O, 4H$_2$O, or more, adjust the amount of gold accordingly to get 1% concentration.

16. For gold of 3.5-nm size, the colloidal particle forms almost immediately. For bigger sizes, a longer time is needed, for example, up to 1 h for a 100-nm size particle.

17. 0.25 mM KCO$_3$ is used for adjusting the pH of the colloidal gold (the KCO$_3$ concentration here is 10 times higher than that used in colloid formation). The amount of KCO$_3$ needed has to be determined empirically by measuring the pH. Use precision pH paper instead of a pH meter because the colloidal gold can impair the function of the electrode. (If pH meter must be used, the gold solution has to be prestabilized by BSA or PEG 20000, and the prestabilized solution is only for pH measuring, it cannot be used for antibody conjugating). Adjusting the pH of the gold solution is very important, if pH of the gold colloid made is lower than 7.0; thus, the antibody molecules are positively charged in this environment and no stabilizing amount of the antibody can be found. Even worse, colloidal gold particles cannot be conjugated to antibody under this condition.

18. Two spectrophotometer-based methods are available: 400- to 700-nm wavelength scan and single OD$_{580}$ measuring. The latter is better for this experiment and described here. Measure OD of all tubes at 580 nm with reference set by water. Tubes with raised absorption are the stabilized ones. Only one curve is needed for all 11 tubes and the first stabilized tube can be picked up at a glance (*see* **Fig. 1**). The

minimal amount in these two examples is 8 and 9 μg per 500 μL, corresponding to 16 and 18 μg of antibody in 1 mL of gold solution.

19. Introducing fiducial gold markers is preferred if manual or automatic procedure with markers is used (*see* Chapter 18). Fiducial markers are usually helpful as the immunogold markers used in labeling are usually not useful for alignment, as they are too small or hidden or otherwise overlap in the tilt series. For alignment, fiducial markers must be clearly seen (e.g., 10 or 15 nm) and are applied, i.e. precipitated e.g., with ethanol on the grid. Simultaneously the grids are also disinfected with ethanol to fit the sterile conditions necessary in culturing cells on grids.

20. Naturally, any other sizes and mixtures also are functional. However, we now prefer to use a combination of the smallest sizes of immunogold conjugates available (1.4 nm and 3 nm or self-made 3.5 nm), as the penetration, labeling range and precision are then optimal for high resolution IET.

21. All the subsequent steps are preferably performed on an adjustable shaking (set to minimum) table.

Acknowledgments

We thank Professor Ismo Virtanen, Department of Anatomy, University of Helsinki, for a gift of vinculin antibody and Lauri Ylimaa, Haartman Institute of University of Helsinki, for technical assistance in preparing the figures. For grants, we thank The Finnish Academy of Science and Helsinki University Central Hospital (EVO). We also thank Special Coordination Funds of the Ministry of Education, Culture, Sports, Science and Technology, Japan. For help with the text, we thank Kalevi Pusa.

References

1. Engelhardt, P. (2006) Three-dimensional reconstruction of chromosomes using electron tomography, in *Electron Microscopy Methods and Protocols,* 2nd ed (Kuo, J., ed.), Humana, Totowa, NJ, pp. 365–385.
2. Engelhardt, P. (2000) Electron tomography of chromosome structure, in *Encyclopedia of Analytical Chemistry* (Meyers, R. A., ed.), vol. 6, John Wiley & Sons Ltd, Chichester, pp. 4948–4984. Available at: http://www.lce.hut.fi/~engelhar/. Accessed May 30, 2006.
3. Uchiyama, S., Kobayashi, S., Takata, H., et al. (2005) Proteome analysis of human metaphase chromosomes. *J. Biol. Chem.* **280,** 16994–17004.
4. Stacey, G. (1998) Chick embryo fibroblast preparation, in *Cell & Tissue Culture: Laboratory Procedures* (Doyle, A., Griffiths, J. B., and Newell, D. G., eds.), John Wiley and Sons Ltd., England, pp. 3E:1.1–3E:1.3.
5. Meriläinen, J., Lehto, V. P., and Wasenius, V. M. (1997) FAP52, a novel, SH3 domain-containing focal adhesion protein. *J. Biol. Chem.* **272,** 23278–23284.
6. Lehto, V. P. and Virtanen, I. (1985) Vinculin in cultured bovine lens-forming cells. *Cell Differ.* **16,** 153–160.

7. Lewis, C. and Laemmli, U. (1982) Higher order metaphase chromosome structure: evidence for metalloprotein interactions. *Cell* **29**, 171–181.
8. Spector, D., Goldman, R., and Leinwand, L. (1998) Chromosome isolation for biochemical and morphological analysis, in *Cells: A Laboratory Manual* (Janssen, K., ed.), vol. 1: *Culture and Biochemical Analysis of Cells*, Cold Springer Harbor Laboratory Press, New York, pp. 49.1–49.12.
9. Sone, T., Iwano, M., Kobayashi, S., et al. (2002) Changes in chromosomal surface structure by different isolation conditions. *Arch. Histol. Cytol.* **65**, 445–455.
10. Uchiyama, S., Kobayashi, S., Takata, H., et al. (2004) Protein composition of human metaphase chromosomes analyzed by two-dimensional electrophoreses. *Cytogenet. Genome Res.* **107**, 49–55.
11. Gasser, S. and Laemmli, U. (1987) Improved methods for the isolation of individual and clustered mitotic chromosomes. *Exp. Cell Res.* **173**, 85–98.
12. Lucocq, J. (1992) Particulate markers for immunoelectron Microscopy, in *Fine Structure Immunocytochemistry* (Griffiths, G., ed.), Springer-Verlag, Berlin, pp. 279–305.
13. Slot, J. W. and Geuze, H. J. (1985) A new method for preparing gold probes for multiple-labelling cytochemistry. *Eur. J. Cell Biol.* **38**, 87–93.
14. Gassmann, R., Henzing, A. J., and Earnshaw, W. C. (2004) Novel components of human mitotic chromosomes identified by proteomic analysis of the chromosome scaffold fraction. *Chromosoma* **113**, 385–397.

20

Three-Dimensional Cryotransmission Electron Microscopy of Cells and Organelles

Michael Marko and Chyong-Ere Hsieh

Summary

Cryoelectron microscopy of frozen-hydrated specimens is currently the only available technique for determining the "native" three-dimensional ultrastructure of individual examples of organelles and cells. Two techniques are available, stereo pair imaging and electron tomography, the latter providing full three-dimensional information about the specimen. A resolution of 4 to 10 nm can currently be obtained with cryotomography. We describe specimen preparation by means of plunge-freezing, which is straightforward and rapid compared with conventional EM techniques. We detail the considerations and preparation needed for successful cryotomography. Frozen-hydrated specimens are very radiation-sensitive and have low contrast because they lack heavy metal stains. The total electron dose that can be applied without damage to the specimen at a given resolution must be estimated, and this dose is fractionated among the images in the tilt series. The desired resolution determines the number and magnification of the images in the tilt series, as well as the objective lens defocus used for phase contrast imaging. The combination of the desired resolution and the maximum number of images into which a given dose can be fractionated sets an upper limit on specimen thickness. Because of these constraints, careful choice of imaging conditions, use of a sensitive CCD camera system, and microscope automation, are important requirements for conducting cryoelectron tomography.

Key Words: Plunge-freezing; cryotransfer; electron imaging; stereo pairs; electron tomography; macromolecules.

1. Introduction

This chapter deals with specimens prepared in the "native" state by rapid plunge freezing (*1*), without the use of chemical fixatives or subsequent dehydration and staining. Once frozen, specimens are imaged in the electron microscope (EM) at liquid nitrogen temperature. With this technique, specimens up

From: *Methods in Molecular Biology, vol. 369*
Electron Microscopy: Methods and Protocols, Second Edition
Edited by: J. Kuo © Humana Press Inc., Totowa, NJ

to approx 1 μm in thickness can be frozen with the water in the "vitreous" (noncrystalline) state. The formation of ice crystals that disrupt the ultrastructure is thereby avoided. Unlike high-pressure freezing, as described in Chapters 8 and 9 in this volume, plunge-freezing requires little effort and expense to achieve specimen vitrification. Three-dimensional (3D) information can be obtained from these specimens by recording stereo pairs in the EM and, in some cases, may be sufficient. However, electron tomography (*[2]*, *see* also Chapters 18 and 19 in this volume), yields more complete 3D information and usually is preferred. Cryoelectron tomography of plunge-frozen specimens is a well-established technique. Just in the last 5 yr, structures studied include (1) macromolecules such as the nuclear pore complex *(3,4)*, viruses *(5–7)*, the protein-like ATPase, VAT *(8)*, ribosomal subunits *(9)*, and immunoglobin *(10,11)*; (2) isolated organelles such as mitochondria *(12,13)*, axonemes *(14,15)*, and triad junctions *(16)*; (3) small cells such as *Dictyostelium (17)*, *Pyrodictium (18)*, and *Spiroplasma melliferum (19)*, and others *(20,21)*. For reviews, see Koster through McIntosh *(22–30)*.

Of particular interest is correlation between the high-resolution 3D structure of macromolecules reconstructed using the "single-particle" approach (*[31]*, *see* also Chapter 17 in this volume), and the same macromolecules, as they exist *in situ*, as revealed by electron tomography of cells and organelles *(32–36)*. The single-particle approach usually involves imaging of thin, frozen suspensions of isolated macromolecular complexes. The reconstruction represents an average of thousands of nominally identical particles, making use of symmetry when it exists, and the resolution can approach atomic dimensions. On the other hand, an electron tomographic reconstruction is computed from a tilt series of projections from a single cell or organelle. Such a reconstruction can provide information about location and orientation of macromolecular complexes in their native cellular environment, and can potentially reveal their individual variation.

Because all projections are recorded from an individual cell or organelle, the resolution attainable with electron tomography of frozen-hydrated specimens is limited by electron irradiation damage to approx 3 nm. The application of symmetry, and of averaging when there are multiple examples of a given macromolecule within the tomogram, may improve the resolution. Electron irradiation is a less-severe problem with the single-particle approach, because the electron dose is effectively distributed over a large number of identical macromolecules.

The approach described here is restricted to cells and organelles small enough to be plunge-frozen. For larger cells, or cells in tissue, frozen-hydrated sections are required (*see* Chapter 9). We have considerable experience in applying electron tomography to frozen-hydrated tissue sections *(37,38)*, but the methodology is challenging and far from routine, so it is not included here.

In this chapter, we concentrate on the first two steps in 3D cryoelectron microscopy of cells and organelles: preparing the specimen and acquiring the tilt series at the EM. We leave alignment and reconstruction of tomograms to Chapters 18 and 19, and only briefly discuss visualization of the 3D data. We hope that by understanding the capabilities and limitations of cryo-EM of frozen-hydrated specimens, and the reader can evaluate applications and design appropriate experiments.

2. Materials

2.1. Specimen Preparation

1. Quantifoil grids, R2/1 or R3.5/1 (Quantifoil Microtools, Jena, Germany; www. quantifoil.com; *see* **Note 1**).
2. "Plasma cleaner" glow-discharge device (e.g,, Harrick Plasma, Ithaca, NY; www. harricksci.com; *see* **Note 2**).
3. 10- or 20-nm Colloidal gold solution (Sigma, St. Louis, MO; *see* **Note 3**).
4. Carbon evaporator and carbon rods.
5. Plunging machine, home-made or purchased (*see* **Fig. 1** and also **Note 4**).
6. Filter paper (Fisher Scientific, Whatman No. 1).
7. Ethane (Class 2.1, lecture bottle, e.g., Advanced Gas Technologies, Palm, PA).
8. Liquid nitrogen.
9. Cryo grid storage boxes (TGS technologies, Cranberry Township, PA; www.tgs technologies.net; Gatan, Warrington, PA; www.gatan.com; or homemade as shown in **Fig. 2**).

2.2. Microscopy

Items marked with (*) are needed for automated tomography but not for stereopair imaging.

1. "Low-dose imaging" capability with the TEM.
2. Provision for control of the TEM by a general-purpose computer.*
3. Goniometer that can be computer-controlled.*
4. Cryotransfer specimen holder (*see* **Fig. 2**).
5. Additional/alternate anticontaminator for cryo-EM work, as recommended by TEM manufacturer.
6. Sensitive TV-rate camera.*
7. CCD camera with a CCD element size large enough to accommodate a thick, sensitive phosphor scintillator (*see* **Subheading 3.5.**).*
8. Faraday cage for measuring electron dose (*see* **Note 5**).

2.3. Software

1. Tilt series acquisition software for the TEM (*see* **Note 6**).
2. Alignment and reconstruction software (*see* **Note 7**).
3. Visualization software (*see* **Note 8**).

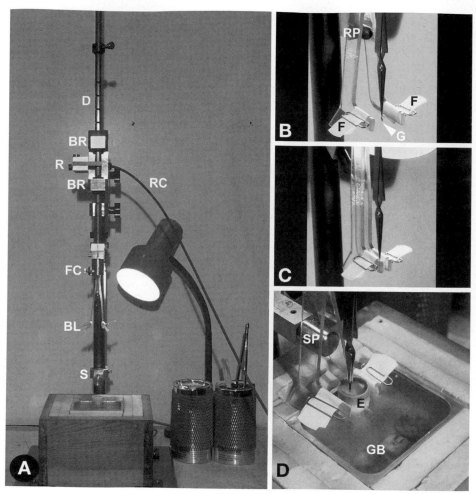

Fig. 1. Plunge-freezing machine built at Albany. (**A**) Overall view: The plunging rod runs in bearings (BR) and a forceps is clamped (FC) at the bottom end. The height from which the plunge starts is determined by engaging the release (R) in one of the detents (D). The plunge is initiated by a foot pedal connected to the release cable (RC). The plunge stop (S) ends the plunge. The two Dewars on the right are for holding the plastic tubes carrying the grid boxes, and for pre-cooling the necessary tools (*see* (D)). (**B**) The plastic L-shaped holders for the strips of filter paper (F) are held apart with the release pin (RP) while a drop of the suspension is applied to the grid (G). (**C**) When the release pin is removed, the grid is blotted from both sides. (**D**) After a manually timed interval of blotting, the cable release is activated and the grid plunges into the liquid ethane. The filter paper is moved out of the way by the spreader pin (SP) just before plunging. The forceps are then unclamped from the plunging rod and the grid is quickly transferred into the liquid nitrogen well that surrounds the ethane (E). There, the grid is loaded into a storage box (GB), which is also shown in **Fig. 2**.

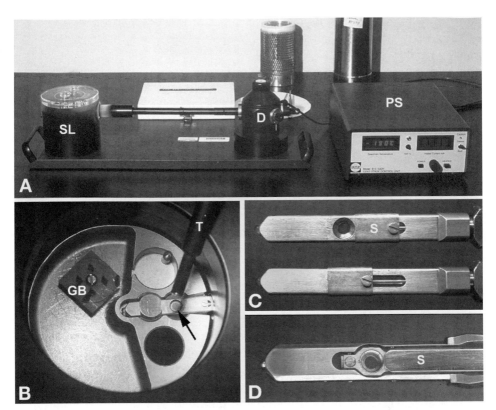

Fig. 2. Cryotransfer specimen holder. (**A**) Workstation for loading the grid into a Gatan 626 cryotransfer holder. Both the specimen loading chamber (SL) and the holder Dewar (D) are filled with liquid nitrogen. The power supply (PS) should indicate −180°C or lower. The two Dewars in the rear are for replenishing the liquid nitrogen, and for holding the plastic tubes carrying the grid boxes. (**B**) Detail of the specimen loading chamber, shown without liquid nitrogen for clarity. The specimen is removed from the storage box (GB) and placed into the specimen holder as indicated by the arrow. A hold-down ring is inserted using tool (T). (**C**) Detail of the tip of a Gatan 626 cryotransfer holder with shutter (S) open and closed. The shutter is closed after the grid is loaded, and not opened again until the holder is in the TEM and the vacuum and temperature are stable. With the 626 design, the entire tip is liquid nitrogen cooled. (**D**) Detail of the tip of a Gatan CT3500RT cryotransfer holder. In this design, only the specimen grid area and the shutter (S) are liquid nitrogen cooled. This specimen holder is a modification of the standard CT3500, and allows in-plane rotation of the grid to enable selection of the tomographic tilt axis relative to the specimen feature of interest. More importantly, it allows collection of cryotomographic tilt series from two axes, which improves the depth resolution as shown in **Fig. 3**.

3. Methods

3.1. Planning the Experiment

3.1.1. Specimen Thickness

The first consideration in planning an experiment is the size of the specimen. It should not be larger than 1 μm in thickness; otherwise, vitrification during plunge-freezing cannot be assured. More importantly, the specimen must be thin enough to image optimally in the EM. Imaging of frozen-hydrated specimens is by phase contrast from elastically scattered electrons, and as the specimen thickness increases, the increased proportion of inelastically scattered electrons adds blur to the image as the result of chromatic aberration in the objective lens. Thus, zero-loss energy filtering, which can remove the inelastically scattered electrons, is of considerable benefit *(39,40)*. However, even with zero-loss energy filtering, the upper limit of specimen thickness for good phase-contrast imaging is approx 1.5 times the inelastic mean-free path. Thus, based on our unpublished data, and extrapolating from Egerton *(41)*, the maximum specimen thickness is approx 600 nm at 400 KeV, 500 nm at 300 keV, 400 nm at 200 keV, and 250 nm at 100 keV. However, note that the effective thickness of the specimen doubles at 60° tilt and triples at 70° tilt. Thus, the quality of the high-tilt projection images in a tomographic tilt series will be seriously degraded using the maximum thickness stated previously.

3.1.2. Stereo Pair or Tomography?

The second consideration is whether tomography is necessary, or if a stereo pair will suffice. A small number of stereo pairs can be created from a tilt series with a coarse tilt interval, such as 10°. This may suffice to determine relative positions of structural elements, provided that the specimen is not excessively crowded with overlapping features. Use of stereo pairs may be a good strategy for initial investigations. Since they can be obtained manually, without external computer control of the TEM or goniometer, stereo imaging may be the only option for laboratories not equipped for low-dose automated tomographic tilt series acquisition.

3.1.3. Estimation of Tomographic Resolution

In planning a tomography experiment, one must first determine whether sufficient resolution is possible with a specimen of given thickness. A quick estimate can be made using the formula known as the "Crowther criterion," $d = \pi D/N$ *(42)*. The resolution in nm (d) depends on the size—usually thickness—of the specimen in nm (D) and the number of tilt images that can be recorded (N). The actual 3D resolution will be lower, however, because it is not possible to tilt the specimen to 90° because of the occlusion of the specimen holder or the

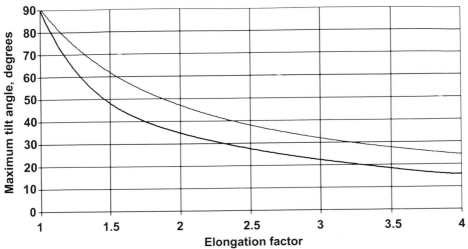

Fig. 3. Effect of maximum tilt angle on elongation along z-axis. The lower (dark) line represents single-axis tilting, while the upper (light) line represents conical tilting (very similar to dual-axis tilting). Data from Crowther et al. *(42)*. Although the tomographic resolution in depth approaches that in the *x-y* plane when a high maximum tilt angle is used, with the slab geometry of typical specimens the effective specimen thickness rapidly becomes excessive. A practical limit may be 70°, where the specimen is already three times thicker than at 0°. Assuming a maximum tilt of 60°, which is typical for many transmission electron microscopes, note that the depth resolution with dual-axis tilting is as good as with a single-axis tilting to approx 73°.

specimen grid. The resulting "missing wedge" of information degrades the resolution in z by an "enlongation factor" *(43)*. The elongation factor (*see* **Fig. 3**) can be reduced by the use of double-tilt data collection *(44,45)*. A further reduction in resolution occurs because typical specimens have slab rather than cylindrical geometry, thus the specimen thickness increases with tilt. As explained in **Subheading 3.6.**, there is a limit to the number of projection images that can be recorded, so reduction of the specimen thickness may be required to achieve a given resolution. Nonetheless, despite slab geometry and the missing wedge, experience has shown that visibility of features (at least in slices of the tomogram taken in the x-y plane) is often close to, and sometimes better than, that which would be expected at the Crowther resolution.

3.2. Preparation of Specimen Grids

1. Plasma clean Quantifoil grids for 30 s before the procedures to follow are conducted.
2. If a carbon film is desired (usually for very small specimens), evaporate approx 10 nm of carbon on freshly cleaved mica, strip off on distilled water, collect on the

grids, and let dry. Continuous carbon films are also used when cells are grown directly on EM grids *(21)*.

3. If the gold particles are not to be mixed with the specimen before plunge freezing, apply 5 μL of gold solution to the support film side of the grid (*see* **Note 9**).
4. After about 5 min, blot by touching the edge of the grid with filter paper. Let dry and store for later use.

3.3. Plunge Freezing

1. For the first use of a particular specimen, determine the appropriate concentration of the specimen by applying 5 μL of the suspension to one side of an ordinary Formvar-carbon coated grid and blot it. Then apply 5 μL of 0.5% aqueous uranyl acetate negative stain to the specimen, wait approx 20 s and blot. Examine in the TEM. The concentration on the cryogrid may, however, be less that seen on these test grids.
2. If gold particles are to be mixed with the specimen, centrifuge and resuspend them in the same medium or buffer used for the specimen. If the color of the resuspended gold solution is clear rather than the original pink color, add 0.2% PEG-20 to the buffer before resuspending the gold.
3. Pour nitrogen into the cooling chamber of the plunging machine (*see* **Fig. 1**).
4. Place a plastic pipet on the end of the tubing connected to the ethane cylinder and place the pipet in the ethane chamber. Open the tank valve to fill the chamber. This step must be performed in an area with good ventilation and far from any source of flame or electrical sparks. The ethane will look cloudy when it has reached the proper temperature (approx −186°C).
5. If gold particles are to be mixed with the specimen, mix 10% to 20% of the resuspended gold solution with the specimen just before plunging.
6. Load a freshly plasma-cleaned grid on the forceps, and add 5 μL of the specimen to the support film side.
7. Blot for 1 to 2 s. The thickness of the ice layer is largely determined by the blotting time but also by the size of the cell or organelle and the sample concentration. The appropriate blotting time for a particular sample can only be determined by experience after examining the specimen in the cryo-TEM. If the ice is too thin, the specimen may be distorted by flattening, which will be evident in tomographic cross-sections. If it is too thick, the resolution will be unnecessarily reduced. When working with thin (approx 100 nm) ice layers, it may be necessary to do the plunging in a humidity chamber, or in a cold room, to avoid evaporation.
8. Release the holding pin and plunge the grid.
9. Transfer the grid under liquid nitrogen to a grid storage box. Transfer the grid box to a small plastic cup, such as a water bottle cap, so that it can be moved to a large Dewar.
10. In the Dewar, place the grid box in a 50 mL plastic centrifuge tube that has a cap in which holes were made to admit liquid nitrogen and for attachment of a fishline.
11. Submerge the plastic tube in the liquid nitrogen refrigerator. Note that in the above steps the grids always remain under liquid nitrogen.

3.4. Cryotransfer Into the TEM

1. Cool down the TEM anticontaminator(s). The temperature of the anticontaminator should be at least 10° colder than that of the specimen.
2. Set the goniometer to the maximum tilt (often −60°), to prevent spillage of liquid nitrogen when the specimen holder is placed in the airlock.
3. With the cryotransfer holder in the workstation, pour liquid nitrogen into both the workstation specimen loading chamber and the specimen holder Dewar (*see* **Fig. 2**). Wait until the holder tip temperature stabilizes at −180°C or below, and bubbling in the workstation chamber has stopped. Do not perform this step too early because frosting at the specimen holder o-ring will occur over time.
4. Precool the necessary tools, and transfer a grid storage box into the workstation specimen loading chamber. This step can be performed by transferring the grid box under liquid nitrogen to a small cup, then submerging the cup in the workstation.
5. Load the specimen into the holder according to the manufacturer's instructions.
6. Close the shutter of the specimen holder to cover the grid.
7. Place a cover on the TEM console to protect it against liquid nitrogen spilled from the workstation when the holder is removed, then move the workstation to the TEM console, as close to the goniometer as possible.
8. Ensure that dry nitrogen is audibly flowing out of the specimen airlock.
9. Quickly and carefully remove the holder from the workstation and insert it into the airlock.
10. Prepump the airlock several minutes longer than usual to sublimate most of the frost that had formed on the holder tip.
11. The specimen holder Dewar will be close to vertical when it is inserted into the airlock because the goniometer was pretilted. When opening the airlock into the column, hold the Dewar vertical while tilting the goniometer back to zero degrees.
12. Top up the specimen holder Dewar as soon as the holder is in the column.
13. After inserting the holder into the column, wait until the holder temperature stabilizes and the column vacuum is again as good as it was with the holder out (or at least 2×10^{-7} Torr). In any case, the shutter should not be opened earlier than 20 min after the transfer.
14. Move the stage to the center position, open the shutter, and preview the grid, as in **Subheading 3.5.3.**
15. Keep the shutter closed when the specimen is not being examined.

3.5. Imaging in the TEM

3.5.1. Choice of Pixel Size or Magnification

When a CCD camera is used, it is necessary to choose an EM magnification that gives an appropriate image pixel size on the specimen scale. Although according to the Nyquist theorem a pixel size of 2 nm should suffice for 4-nm resolution, we find that oversampling gives better results. Thus, if the desired resolution for the image were 4 nm, a pixel size of 1 nm would be appropriate. The resolution of the camera itself also must be considered. For 300- or 400-kV

TEMs, a thick phosphor scintillator is needed for high sensitivity. However, CCD cameras employing 1:1 fiberoptic coupling may have reduced resolution when a thick phosphor scintillator is used. Electrons may be scattered in the phosphor such that a given electron could be detected by one or another adjacent CCD elements. In other words, the point-spread function of a thick, sensitive phosphor may be too broad for full CCD resolution. We find that a thick phosphor suitable for cryotomography requires a CCD element size of about 30 μm at 300 kV or 50 μm for 400 kV. Depending on the camera, the CCD element size may be between 14 and 30 μm. Thus, it may be necessary to bin the camera (combine adjacent pixels) to create an effective CCD element size that is large enough to accommodate the point-spread function. This, of course, will reduce the number of pixels in the image, and consequently increase the image pixel size on the specimen scale.

3.5.2. Choice of Optimum Underfocus Value

Frozen-hydrated specimens have inherently low contrast and are very sensitive to radiation. For electron tomography, it is important to maximize the signal-to-noise ratio with respect to the resolution requirements of a given experiment. Frozen-hydrated specimens lack heavy elements that scatter strongly and to high angles; thus, they have little amplitude or "aperture" contrast. Instead, these specimens are imaged using strong underfocus of the objective lens, which generates phase contrast from elastically scattered electrons. The transfer of information from the specimen to the image follows the contrast-transfer function (CTF; *see* **Fig. 4**). The CTF has the form of a damped sine wave, and represents the transfer of information in Fourier space as a function of spatial frequency (k in **Fig. 4**). Starting at low frequencies, the CTF rises to a maximum and falls to zero; then, at higher spatial frequencies it rises again but with opposite contrast. The result is that intensity in the image is dependent on feature size (d in **Fig. 4**), as determined by the CTF.

Because resolution is limited by the geometry of tilt-series collection (*see* **Subheading 3.1.3.**), it is a common convention to set the defocus so that the CTF first zero falls past the spatial frequency corresponding to the maximum expected resolution. The resulting tomogram can be low-pass filtered to eliminate information past the first zero because this information typically only adds noise to the images. A better practice is to place the CTF first *maximum* at the highest resolution to be expected. This step will optimize the transfer of information at the highest expected tomographic resolution *(47)*.

The CTF is determined primarily by the accelerating voltage, objective underfocus, and objective lens parameters. Freeware is available that easily and interactively plots the CTF for any underfocus setting with any TEM. One such application is "CTF Explorer" (M. V. Sidorov; http://click.to/ctfexplorer). As an

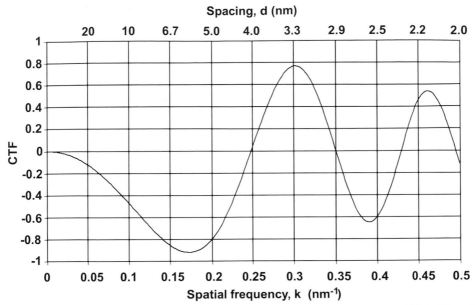

Fig. 4. Phase-contrast transfer function of our JEOL JEM4000FX at 400 keV and 10-μm underfocus. In this case, no information is transferred from the specimen to the image for spacings of 4 nm, and features with smaller spacings have reversed contrast. More importantly, spacings of 5.7 nm are transferred with maximum intensity. Cryotomograms of typical specimens have a resolution in the range of 4 to 8 nm. The signal-to-noise ratio in the tilt images can be optimized by choosing an underfocus setting that places the contrast transfer function first maximum at the best possible expected tomographic resolution (from d = πD/N).

example, for a typical tomographic resolution of 5 nm, an underfocus value of 3.3 μm might be used with a 100 keV TEM, 5.0 μm at 200 keV, 6.3 μm at 300 keV, or 7.6 μm at 400 keV.

3.5.3. Previewing Grids

The first look at a new grid should be in the low-magnification mode, so that many grid squares can be seen at once on the viewing screen. The relative intensity of the grid squares will, as experience is gained, indicate which grid squares have ice of optimal thickness. Note that at low magnification (100 to 1000x), an electron flux on the specimen of less than 0.01 $e^-/Å^2/s$, and as low as 0.001 $e^-/Å^2/s$, can be used. At selected grid squares, manually tilt the goniometer through the full tilt range to ensure that a full tomographic tilt series can be recorded. Make a note of the stage coordinates of suitable grid squares, so that they can be recalled without the need to return to the low-magnification mode.

The grid may be sitting at a slightly different height each time it is loaded, or it may be slightly bent. Before a tilt series is collected, the eucentric height must be set. This should be done at not less than one-tenth the magnification to be used for recording the tilt series. The most convenient way to set the eucentric height is to set the objective lens to the standard focus value, and then focus the specimen with the mechanical z-height control, making use of the image wobbler and the gold particles (*see* **Note 10**). This step also ensures that the magnification calibration will be correct. If a sensitive TV-rate camera is available, this operation can be performed quickly and without significant irradiation damage to the specimen.

If an energy filter is used, the thickness of the ice layer can be roughly estimated by noting the mean pixel intensity in the area of interest with the slit in (zero-loss electrons, I_0) and out (both zero-loss and inelastically-scattered electrons, I_{tot}). The thickness (t) is approximately equal to $\ln(I_{tot}/I_0)T$ where T is the thickness in nm of one inelastic mean-free path. For ice-embedded biological material, the value of T is approximately 170 nm at 100 keV, 270 nm at 200 keV, 330 nm at 300 keV, and 380 nm at 400 keV, extrapolated *(41)* from our unpublished experimental data.

3.5.4. Tilt Axis Determination

The direction of the specimen stage tilt axis relative to the x-y image coordinates must be known, both for tomography and for stereo pair viewing. Stereo pairs must be displayed with the tilt axis vertical, and a vertical tilt axis is also the convention for tomographic reconstruction. During alignment of the tomographic tilt series, the precise direction of the tilt axis is determined, and this operation is simplified if the approximate direction of the tilt axis is known *a priori*.

With side-entry goniometers, it is straightforward to approximately determine the tilt axis direction because one of the specimen translation knobs (or motor drives) moves the specimen along the axis of the specimen holder rod. Note that with some TEMs, the direction of the tilt axis may change with magnification, thus the below procedure must be repeated for each magnification to be used.

1. Record an image.
2. Shift the specimen by about half the image field, using only the control that moves the specimen along the specimen holder rod axis.
3. Record another image.
4. Overlay the two images, or place then next to each other with the first image on the left.
5. Draw a line from a specimen feature in the first image to the same feature in the second image.

6. The line will be parallel to the tilt axis, and the images should be rotated accordingly before further processing.

3.5.5. Recording Stereo Pairs

At normal reading distance, the average angular separation between the eyes is 12 to 14°. Thus, stereo pairs in the electron microscope are often recorded at tilt angles of plus and minus 6 to 7°. Using these angles, high-magnification stereo pairs of thin specimens, or lower magnification stereo pairs of thicker specimens, can be comfortably viewed. To maintain the same stereo depth effect with a given specimen thickness, the tilt angle is increased as the magnification is reduced. Note that this procedure does not ensure that the depth dimension is represented at the same scale as the in-plane dimensions. However, if the tilt angle and the magnification are known, the actual vertical separation between two points in the specimen can be measured from the stereo pair (*see* **Subheading 3.7.**).

TEM automation is not needed for recording stereo pairs, but it is helpful to use a sensitive TV camera for checking the centering of the object after tilting, and for focusing. Because high underfocus is used, focusing with the image wobbler at a ten-times reduced magnification is adequate, and the electron dose for focusing is thereby reduced by a factor of 100.

3.6. Collecting a Tomographic Tilt Series

3.6.1. Choice of Total Electron Dose

The thickness of the ice layer depends on the size of the cell or organelle (unless frozen-hydrated sections are used). As explained in **Subheading 3.1.3.**, for a given specimen thickness, resolution can only be increased by collecting more tilt images. Thus, as many tilt images as possible must be recorded without damaging the specimen by excessive electron irradiation.

The cumulative dose that can be applied to the specimen is resolution dependent; the higher the desired resolution, the less the total permissible irradiation *(46)*. Radiation sensitivity varies with the specimen type and also is influenced by the ice thickness (thicker ice may tolerate a higher total dose). The total permissible electron dose increases with the accelerating voltage, since the interaction of the beam with the specimen decreases (the mean-free path for scattering increases). This by itself is not a benefit, however, because electrons that do not interact with the specimen convey no information. At 400 kV, we have found that the intensity of information at 8 nm is reduced 50% after 25 e$^-$/Å^2, although there is still useful information after 100 e$^-$/Å^2 *(47)*. As a general guideline, the total dose should be no greater than 100 e$^-$/Å^2, and should be kept to 30 e$^-$/Å^2 for tomograms with desired resolution approaching 3 nm. Because it may not be straightforward to assess radiation damage directly in the tomogram, the best practice is to use the lowest possible dose, assessed as described in **Subheading 3.6.2.**

3.6.2. Choice of Electron Dose Per Tilt Image

As explained in **Note 5**, the incident electron dose usually is measured by recording an image at an empty hole in the grid, using the CCD camera. The same illumination conditions are then used for recording the tilt series images. In theory, the total permissible electron dose can be fractionated into as many projection tilt images as needed for the resolution desired *(48)*. However, in practice, the residual noise of the camera systems limits dose fractionation. The practical sensitivity of the camera system should be measured under the imaging conditions (specimen thickness, magnification, and underfocus) to be used for a particular experiment, and the lowest dose that allows loss-less dose fractionation should be determined. The minimum dose per image determined in this way will dictate the resolution that can be obtained with a given experimental protocol. In the example that follows, a pixel size of 1 nm on the specimen scale is used. Sample results are shown in **Fig. 5**.

1. Prepare a frozen-hydrated test specimen that has a helical or crystalline repeat. The specimen should be mixed with colloidal gold as described in **Subheading 3.3.**
2. Record ten images of the specimen at 0.1 e$^-$/Å^2, and one image at 1 e$^-$/Å^2 (*see* **Note 11**).
3. Move to another area of the specimen.
4. Record 10 images of the specimen at 1 e$^-$/Å^2, and one image at 10 e$^-$/Å^2.
5. For each of the sets of 10 images, align the images so that the gold particles are in the same positions. Alignment can in principle be done by cross correlation, but with very noisy images, experience has shown that the use of gold markers is more reliable and accurate.
6. Sum the ten aligned images, do a Fourier transform, and display the power spectrum.
7. Compare the power spectrum of the summed low-dose images with that of the single high-dose image (the same total electron dose will be represented in both).
8. Acquire a line profile through the layer lines or spots and compare the "peak heights". If the peak heights are nearly the same, the camera is not limiting dose fractionation.

Loss-less fractionation down to 0.1 e$^-$/Å^2 per image, for a 1 nm pixel size relative to the specimen, should be possible with most camera systems, and this is adequate for most purposes. The experiment can be repeated with higher or lower dose as needed to characterize the camera. If loss-less fractionation to 1 e$^-$/Å^2 is not possible, the camera system is probably not suitable for cryo-tomography work.

Once the minimum dose for loss-less fractionation is determined, note the mean (gain-corrected) pixel intensity of the low-dose image. When recording

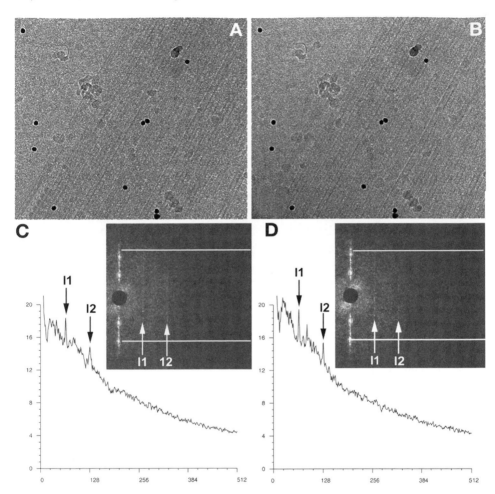

Fig. 5. Evaluation of imaging sensitivity. (**A**) Image of a frozen-hydratred sea urchin axoneme (demembranated sperm tail) taken with an electron dose of 0.86 e⁻/Å². The central sheath of the axoneme has helical spacings at 8 and 16 nm, which produce layer lines l1 and l2 shown in the power spectrum in (**C**). (**B**) The sum of ten images recorded at 0.086 e⁻/Å². The corresponding power spectrum is shown in (**D**). To obtain the plots in (C) and (D), the images were rotated so that the layer lines in the power spectra were vertical. Pixels were summed vertically between the two white lines shown in the power spectra, and then plotted. The power spectrum is symmetrical, so only the right side is shown. Note that the intensity obtained from summing 10 images at the lower dose is not less than that of a single image at 10 times the dose. In fact, there is a slight improvement, due to averaging of noise. The experiment shows that, when recording a tilt series, a dose higher than 0.08 e⁻/Å² is not necessary. Any frozen-hydrated specimen having a repeating structure with a spacing in the size range close to the expected tomographic resolution can be used.

a tilt series, the mean pixel value should not fall below this value (*see* **Subheading 3.6.3.**).

3.6.3. Choice of Protocol for Tilt Series Collection

In acquiring a tilt series, an image is recorded, the goniometer is tilted by a defined increment, and another image is recorded, continuing until the full tilt range is covered. The object of interest has to be kept centered in the image, and the images must all have the proper focus. For frozen-hydrated specimens, this is accomplished at low electron dose by means of TEM automation. The classical method *(22,49–51)* uses cross correlation to automatically center the object and refocus the image after each tilt. These operations are performed at a location on the specimen that is along the stage tilt axis and adjacent to the object of interest. Thus, in principle, no irradiation of the selected specimen area is required except that needed to record the tilt images. This basic method has been refined with new concepts such as precalibration *(52)* and prediction *(53,54)*, and these are incorporated in some of the software listed in **Note 6**. In any case, there is nearly always some "overhead" dose required to initially locate the object of interest, and sometimes to recover a tilt series if there is an interruption. The overhead can be reduced to a minimum if real-time observation of object of interest is done only with a sensitive TV-rate camera, and at one-tenth the magnification used to record the tilt images.

Plunge-frozen specimens typically have "slab" geometry, and as mentioned in **Subheading 3.1.3.**, the effective thickness increases with tilt. In principle, if a uniform tilt interval is used (e.g., 1 or 2°), information from high-tilt images will be under-represented in the tomogram. To overcome this problem, a graduated tilt increment was proposed by Saxton et al. *(55)*. Most software for tomographic tilt series acquisition includes an option for the Saxton tilt scheme, although some laboratories find that it makes no significant difference in the results.

The intensity of any voxel in the electron tomogram should have a simple relationship to the mass of the specimen at the corresponding 3D location. This will roughly be the case if the tilt series is collected so that the exposure time and illumination from the electron source stays constant as the specimen is tilted. However, as the specimen is tilted to higher angles, the image intensity will decrease due to the increased effective specimen thickness, and camera sensitivity will limit the information that can be recorded. Therefore, the best practice is to adjust the illumination so that the minimum mean pixel value, as determined in **Subheading 3.6., step 2**, is obtained at the highest tilt, with an exposure time of not more than approx 2 s (to avoid effects of possible drift). During the course of the tilt series, the exposure time is progressively lowered or raised to keep the mean pixel value approximately constant. Again, most software for tomographic tilt series acquisition includes an option for this.

3.6.4. Image Intensity Scaling and Correction

With the aforementioned scheme, the electron dose varies with tilt, thus the intensity of each image should be rescaled to simulate the case of a constant electron dose. If images are recorded on film instead of on a CCD camera, the log of the image should be taken when scanning the film. Note that, because imaging is by phase-contrast, the spatial frequencies in the specimen are not transferred uniformly, but according to the CTF. Thus, to obtain a clearer relationship between the voxel intensity in the reconstruction and the corresponding mass in the specimen, a correction for the CTF would also be needed. In many studies, the mass-intensity relationship is of lesser importance than the shapes and spatial arrangement of features, and not all investigators make these corrections.

3.7. Alignment, Reconstruction, and Visualization

These topics are covered in Chapters 18 and 19 and, in general, the methods described there apply equally well to tomography of frozen-hydrated specimens. In the case of alignment, however, our experience has shown that markerless schemes, which are based on some form of cross correlation, usually are not effective because the low-dose tilt images are very noisy. The presence of markers is in any case very useful for following the specimen as it is imaged during the tilt series.

In some cases, tomography is not needed and stereo pairs suffice. We have written software ("Stereocon") to make 3D measurements and contour-based models by interactive tracing during stereoscopic viewing *(56)*. Stereocon also can be used to stereoscopically view and analyze 3D windows from tomographic volumes. This software differs from that used in scanning electron microscopy because TEM images are projections through the depth of the specimen, not views of an opaque surface.

The vertical distance between two points in the specimen can be measured manually from a stereo pair. This method is especially convenient for measurement of the section thickness, without resorting to tomography *(57)*.

1. Orient the stereo pair with the tilt axis vertical.
2. Measure the horizontal component of the distance between the two points in question on each micrograph. This measurement should be in units relative to the specimen (e.g., nm).
3. Subtract the two horizontal distances. This difference is known as the parallax (P).
4. The vertical distance (t) equals the parallax (P) divided by two times the sine of the tilt half-angle (α): $t = P/(2\sin\alpha)$. The half angle is used because a stereo pair is commonly recorded by tilting to equal increments both directions from zero.
5. The 3D distance between the two points can then be determined by a further trigonometric step.

4. Notes

1. Quantifoil grids are available in various mesh sizes and materials. They are pre-coated with a carbon film that has a regular array of holes. For tomography of organelles and small cells, 200-mesh grids with a Quantifoil hole size of 2 or 3.5 μm are suitable. The regular array of holes is very useful for tracking and locating specimens on the grid. Before availability of Quantifoil grids, homemade "lacy" grids were used, which are now commercially available (Product no. 01890, Ted Pella, Inc., Redding, CA; www.tedpella.com).

2. Some carbon evaporators have a plasma cleaning accessory. A sophisticated, powerful plasma cleaner such as this is not necessary for use in materials science.

3. Store the stock colloidal gold solution at 4°C.

4. The plunging machine shown in **Fig. 1** *(58)* is quite adequate for the work described here but does require some experience to obtain good results. The object is to obtain suspension layers that are as thin as possible, without flattening the cell or organelle. Two important parameters, blotting time and humidity in the vicinity of the grid, are controlled manually and may vary from day to day. Two commercial plunging machines offer automatic blotting and humidity control, the Gatan Cryoplunge, (Pleasanton, CA; www.gatan.com) and the FEI Vitobot (AIM Company BV, Brunssum, The Netherlands; www.vitrobot.com). The latter offers computer control of drop application, blotting, plunging, and humidity. In addition, Leica (Leica Microsystems, Vienna, Austria; www.em-preparation.com) offers the EM CPC, and EMS (Hatfield, PA; www.emsdiasum.com) offers the EMS-002. These two are less-specialized, multipurpose instruments that can also be used for other types of cryopreparation.

5. A Faraday cage is used to measure the current in the entire beam. Some room-temperature specimen holders have a built-in Faraday cage, and an accessory is available for most TEMs. The beam current at the Faraday cage is measured with an electrometer. The beam current in amps is converted to electrons per second by multiplying by 6.25×10^{18}. To calibrate a CCD camera, adjust the diameter of the illuminated area to be slightly smaller than the CCD image area. The total number of electrons for a one-second exposure is divided by the sum of the gain-corrected CCD counts in the illuminated area, yielding the number of CCD counts per primary electron. For low-dose cryotomography, this value should be at least 5. If no Faraday cage is available, the TEM's readout of screen current density can be calibrated with photographic film, making use of the film manufacturer's sensitivity data for the accelerating voltages employed. When using this method, be sure that the entire diameter of the beam is recorded on the film.

6. The classic approach to TEM automation for low-dose tomographic tilt series acquisition *(49–51)* is available from TVIPS (Gauting, Germany; www.tvips.com) and can be installed on any modern TEM that has a computer-controllable goniometer. There have been a number of implementations of tomography software for FEI Tecnai TEMs. In addition to FEI's own software, freeware is available from the Boulder Laboratory for 3-D Electron Microscopy of Cells (SerialEM *[53]*; http://bio3d.colorado.edu/), the Max-Planck Institute for Biochemistry (TOM *[59]*; http://

www.biochem.mpg.de/baumeister/), as well as from Utrecht University (www.bio.uu. nl/mcb/3dem/Electron_tomography.html). The Boulder software also is available for JEOL TEMs, and JEOL offers its own tomography software.

7. Several laboratories have written their own software for alignment and tomographic reconstruction, often with specialized features. A popular example is IMOD, developed by David Mastronarde of the Boulder Laboratory (*[60]*; http://bio3d.colorado. edu/). TOM *(59)* also has this capability. Our laboratory has recently developed a graphical user interface called SPIRE, which has a subsection specifically for tomography. This application uses our SPIDER software system *(61)* and has the advantage that the user can easily customize the image processing software and the graphical user interface to fill special needs, using the extensive repertory of image-processing operations available in SPIDER. *See* www.wadsworth.org/spider_doc/.

8. As with alignment and reconstruction software, several laboratories have written their own software for visualization. The first step in visualization of electron tomograms is simply viewing slices in arbitrary directions and making animations of slice walk-throughs. The multipurpose image-processing freeware "ImageJ" (http:// rsb.info.nih.gov/nih-image) is convenient to use for this purpose. The next step may be manual segmentation of structures using traced contours, followed by surface rendering. Both steps can be accomplished with IMOD (see previous). More sophisticated methods include computationally based segmentation, with both surface and volume rendering. Several laboratories use IRIS Explorer for this (NAG, Oxford, UK; www.nag.com). The most commonly used visualization software, which is suitable for many steps in visualization of tomograms, is Amira (Mercury Computer Syetems, San Diego, CA; www.tgs.com).

9. The gold particles will accumulate on the edges of the Quantifoil holes, and can be used for alignment when the area of interest is near the edge of the hole. Some of the gold particles will also be carried into the specimen. If a carbon film is used, the gold particles will be evenly-distributed. When doing this first time, the distribution of the gold should be checked in the TEM. Note that at least 4, but preferably 8 to 12, gold particles should be present within the area to be imaged for tomography.

10. If the TEM has not previously been used for tomography, it may be necessary to carefully align the stage mechanically. This step usually is performed by a service engineer. If the TEM has a computer-controllable stage, the focus variation over the entire tilt range should be no more than approx 20 μm, and the image shift no more than 5 μm. In any case, it will be very worthwhile to spend time "tuning up" the goniometer for tomography use. Once the goniometer is optimally adjusted, the focus setting for eucentricity should be noted. The eucentric focus setting should be stored either in the microscope's computer or an external computer so that it can easily be precisely reset.

11. Use the same exposure time (preferably 1 second) for all images. A simple way to reduce the dose is to slightly mis-align the gun tilt for the low-dose images and restore it to the normal value for the high dose images. Remember that the incident dose can only be measured in an empty area of the specimen.

Acknowledgments

We thank Drs. Joachim Frank and Carmen Mannella for critical reading of this contribution. This work was supported by NIH/NCRR Biomedical Research Technology Program Grant RR01219 (P.I. J. Frank), which funds the Wadsworth Center's Resource for Visualization of Biological Complexity.

References

1. Dubochet, J., Adrian, M., Chang, J. J., et al. (1988) Cryo-electron microscopy of vitrified specimens. *Quart. Rev. Biophys.* **21,** 129–228.
2. Frank, J. (ed.) (2006) *Electron Tomography of Cells and Tissue.* Springer, New York.
3. Stoffler, D., Feja, B., Fahrenkrog, J., Walz, J., Typke, D., and Aebi, U. (2003) Cryo-electron tomography provides novel insights into nuclear pore architecture: implications for nucleocytoplasmic transport. *J. Mol. Biol.* **328,** 119–130.
4. Beck, M., Förster, F., Ecke, M., et al. (2004) Nuclear pore complex structure and dynamics revealed by cryoelectron tomography. *Science* **306,** 1387–1390.
5. Grünewald, K., Desai, P., Winkler, D. C., et al. (2003) Three-dimensional structure of herpes simplex virus from cryo-electron tomography. *Science* **302,** 1396–1398.
6. Förster, F., Medalia, O., Zauberman, N., Baumeister, W., and Fass, D. (2005) Retrovirus envelope protein complex structure in situ studied by cryo-electron tomography. *PNAS USA* **102,** 4729-4734.
7. Cyrklaff, M., Risco, C., Fernandez, J. J., et al. (2005) Cryo-electron tomography of vaccinia virus. *Proc. Natl. Acad. Sci. USA* **102,** 2772–2777.
8. Rockel, B., Jakana, J., Chiu, W., and Baumeister, W. (2002) Electron cryo-microscopy of VAT, the archaeal p97/CDC48 homologue from *Thermoplasma acidophilum. J. Mol. Biol.* **317,** 673–681.
9. Zhao, Q. Ofverstedt, L. G., Skoglund, U., and Isaksson, L. A. (2004) Morphological variation of individual *Escherichia coli* 30S ribosomal subunits in vitro and in situ, as revealed by cryo-electron tomography. *Exp. Cell Res.* **297,** 495–507.
10. Sandin, S., Ofverstedt, L. G., Wikstrom, A. C., Wrange, O., and Skoglund, U. (2004) Structure and flexibility of individual immunoglobulin G molecules in solution. *Structure* **12,** 409-415.
11. Bongini, L. Fanell,i D., Piazza, F., De los Rios, P., Sandin, S., and Skoglund, U. (2004) Freezing immunoglobulins to see them move. *Proc. Natl. Acad. Sci. USA* **101,** 6466–6471.
12. Nicastro, D., Frangakis, A. S., Typke, D., and Baumeister, W. (2000) Cryoelectron tomography of Neurospora mitochondria. *J. Struct. Biol.* **129,** 48–56.
13. Mannella, C. A., Pfeiffer, D. R., Bradshaw, P. C., et al. (2001) Topology of the mitochondrial inner membrane: dynamics and bioenergetic implications. *IUBMB Life* **52,** 93–100.
14. McEwen, B. F., Marko, M., Hsieh, C.-E., and Mannella, C. (2002) Use of frozen-hydrated axonemes to assess imaging parameters and resolution limits in cryo-electron tomography. *J. Struct. Biol.* **138,** 47–57.

15. Nicastro, D., McIntosh, J. R., and Baumeister, W. (2005) 3-D structure of eukaryotic flagella in a quiescent state revealed by cryo-electron tomography. *Proc. Natl. Acad. Sci. USA* **102**, 15889–15894.

16. Wagenknecht, T., Hsieh, C.-E., Rath, B., Fleischer, S., and Marko, M. (2002) Electron tomography of frozen-hydrated isolated triad junctions. *Biophys. J.* **83**, 2491–2501.

17. Jonkman, T., Kohler, J., Medalia, O., et al. (2002) Dynamic organization of the actin system in the motile cells of Dictyostelium. *J. Muscle Res. Cell Motility* **23**, 639–649.

18. Nickell, S., Hegerl, R., Baumeister, W., and Rachel, R. (2003) *Pyrodictium cannulae* enter the periplasmic space but do not enter the cytoplasm, as revealed by cryo-electron tomography. *J. Struct. Biol.* **141**, 34–42.

19. Kürner, J., Frangakis, A. S., and Baumeister, W. (2005) Cryo-electron tomography reveals the cytoskeletal structure of *Spiroplasma melliferum*. *Science* **307**, 436–438.

20. Kürner, J., Medalia, O., Linaroudis, A. A., and Baumeister, W. (2004) New insights into the structural organization of eukaryotic and prokaryotic ctyoskeletons using cryo-electron tomography. *Exp. Cell Res.* **301**, 38–42.

21. Medalia, O., Weber, I., Frangakis, A. S., Nicastro, D., Gerisch, G., and Baumeister, W. (2002) Macromolecular architecture in eukaryotic cells visualized by cryo-electron tomography. *Science* **289**, 1209–1213.

22. Koster, A. J., Grimm, R., Typke, D., et al. (1997) Perspectives of molecular and cellular electron tomography. *J. Struct. Biol.* **120**, 276–308.

23. McIntosh, J. R. (2001) Electron microscopy of cells: a new beginning for a new century. *J. Cell Biol.* **153**, F25–F32.

24. Frank, J., Wagenknecht, T., McEwen, B. F., Marko, M., Hsieh, C.-E., and Mannella, C. A. (2002) Three-dimensional imaging of biological complexity. *J. Struct. Biol.* **138**, 85–91.

25. Plitzko, J. M., Frangakis, A. S., Foerster, F., Gross, A., and Baumeister, W. (2002) In vivo veritas: electron cryotomography of intact cells with molecular resolution. *Trends Biotechnol.* **20**, 40–44.

26. Steven, A. C. and Aebi, U. (2003) The next ice age: cryo-electron tomography of intact cells. *Trends Cell Biol.* **13**, 107–110.

27. Frangakis, A. S. and Förster, F. (2004) Computational exploration of structural information from cryo-electron tomograms. *Current Opinion Struct. Biol.* **14**, 325–331.

28. Baumeister, W. (2004) Mapping molecular landscapes inside cells. *Biol. Chem.* **385**, 865–872.

29. Subramaniam, S. and Milne, J. L. S. (2004) Three-dimensional electron microscopy at molecular resolution. *Ann. Rev. Biophys. Biomol. Struct.* **33**, 141–155.

30. McIntosh, J. R., Nicastro, D., and Mastronarde, D. (2005) New views of cells in 3D: an introduction to electron tomography. *Trends Cell Biol.* **15**, 43–51.

31. Frank, J. (ed.) (2006) *Three-dimensional Electron Microscopy of Macromolecular Assemblies—Visualization of Biological Macromolecules in Their Native State,* Oxford University Press, New York.

32. Böhm, J., Frangakis, A. S., Hegerl, R., Nickell, S., Typke, D., and Baumeister, W. (2000) Toward detecting and identifying macromolecules in a cellular context: template matching applied to electron tomograms. *Proc. Natl. Acad. Sci. USA* **97,** 14245–14250.

33. Böhm, J., Lambert, O., Frangakis, A. S., et al. (2001) FhuA-mediated phage genome transfer into liposomes: a cryo-electron tomography study. *Curr. Biol.* **11,** 1168–1175.

34. Frangakis, A. S., Böhm, J., Förster, F., et al. (2002) Identification of macromolecular complexes in cryoelectron tomograms of phantom cells. *Proc. Natl. Acad. Sci. USA* **99,** 14153–14158.

35. Grünewald, K., Medallia, O., Gross, A., Steven, A., and Baumeister, W. (2002) Prospects of electron cryotomography to visualize macromolecular complexes inside cellular compartments: implications of crowding. *Biophys. Chem.* **100,** 577–591.

36. Rath, B. K., Hegerl, R., Leith, A., Shaikh, T. R., Wagenknecht, T., and Frank, J. (2003) Fast 3D motif search of EM density maps using a locally normalized cross-correlation function. *J. Struct. Biol.* **144,** 95–103.

37. Hsieh, C.-E., Marko, M., Frank, J., and Mannella, C. A. (2002) Electron tomographic analysis of frozen-hydrated tissue sections. *J. Struct. Biol.* **138,** 63–73.

38. Hsieh, C.-E., Leith, A., Mannella, C. A., Frank, J., and Marko, M. (2006) Towards high-resolution three-dimensional imaging of native mammalian tissue: Electron tomography of frozen-hydrated rat liver sections. *J. Struct. Biol.* **153,** 1–13.

39. Grimm, R., Typke, D., and Baumeister, W. (1998) Improving image quality by zero-loss energy filtering: quantitative assessment by means of image cross-correlation. *J. Microsc. (Oxford)* **190,** 339–349.

40. McEwen, B. F., Marko, M., Hsieh, C.-E., and Mannella, C. (2002) Use of frozen-hydrated axonemes to assess imaging parameters and resolution limits in cryo-electron tomography. *J. Struct. Biol.* **138,** 47–57.

41. Egerton, R. F. (1996) *Electron Energy-Loss Spectoscopy in the Electron Microscope,* 2nd ed. Plenum, New York.

42. Crowther, R. A., DeRosier, D. J., and Klug, A. (1970) The reconstruction of a three-dimensional structure from its projections and its application to electron microscopy. *Proc. R. Soc. Lond. (A)* **317,** 319–340.

43. Radermacher, M. and Hoppe, W. (1980) Properties of three-dimensionally reconstructed objects from projections by conical tilting compared to single axis tilting. *Proc. 7th Eur. Congr. Electron Microsc. Den Haag* **1,** 132–133.

44. Penczek, P., Marko, M., Buttle, K., and Frank, J. (1995) Double-tilt electron tomography. *Ultramicroscopy* **60,** 393–410.

45. Mastronarde, D. N. (1997) Dual-axis tomography: an approach with alignment methods that preserve resolution. *J. Struct. Biol.* **120,** 343–352.

46. Jeng, T. W. and Chiu, W. (1984) Quantitative assessment of radiation damage in a thin protein crystal. *J. Microsc. (Oxford)* **136,** 35–44.

47. McEwen, B. F., Marko, M., Hsieh, C.-E., and Mannella, C. (2002) Use of frozen-hydrated axonemes to assess imaging parameters and resolution limits in cryo-electron tomography. *J. Struct. Biol.* **138,** 47–57.

48. McEwen, B. F., Downing, K. H., and Glaeser, R. M. (1995) The relevance of dose-fractionation in tomography of radiation-sensitive specimens. *Ultramicroscopy* **60**, 357–373.

49. Dierksen, K., Typke, D., Hegerl, R., and Baumeister, W. (1993) Towards automatic electron tomography. II. Implementation of autofocus and low-dose procedures. *Ultramicroscopy* **49**, 109–120.

50. Dierksen, K., Typke, D., Hegerl, R., Walz, J., Sackmann, E., and Baumeister, W. (1995) Three-dimensional structure of lipid vesicles embedded in vitreous ice and investigated by automated electron tomography. *Biophys. J.* **68**, 1416–1422.

51. Rath, B. K., Marko, M., Radermacher, M., and Frank, J. (1997) Low-dose automated electron tomography: a recent implementation. *J. Struct. Biol.* **120**, 210–218.

52. Ziese, U., Janssen, A. H., Murk, J. L., et al. (2002) Automated high-throughput electron tomography by pre-calibration of image shifts. *J. Microsc. (Oxford)* **205**, 187–200.

53. Mastronarde, D. N. (2005) Automated electron microscope tomography using robust prediction of specimen movements. *J. Struct. Biol.* **152**, 36–51.

54. Zheng, Q. S., Braunfeld, M. B., Sedat, J. W., and Agard, D. A. (2004) An improved strategy for automated electron microscopic tomography. *J. Struct. Biol.* **147**, 91–101.

55. Saxton, W. O., Baumeister, W., and Hahn, M. (1984) Three-dimensional reconstruction of imperfect two-dimensional crystals. *Ultramicroscopy* **13**, 57–70.

56. Marko, M. and Leith, A. (1996) Stereocon—three-dimensional reconstruction from stereoscopic contouring. *J. Struct. Biol.* **116**, 93–98.

57. Hama, K. (1973) High voltage electron microscopy, in *Advances in Biological Electron Microscopy* (Koehler, J. K., ed.), Springer, New York, pp. 275–296.

58. Frank, J., Penczek, P., Agrawal, R. K., Grassucci, R. A., and Heagle, A. B. (2000) Three-dimensional cryoelectron microscopy of ribosomes. *Meth. Enzymol.* **317**, 276–291.

59. Nickell, S., Förster, F., Alexandros, L., et al. (2005) TOM software toolbox: acquisition and analysis for electron tomorgrapy. *J. Struct. Biol.* **149**, 227–234.

60. Kremer, J. R., Mastronarde, D. N., and McIntosh, J. R. (1996) Computer visualization of three-dimensional image data using IMOD. *J. Struct. Biol.* **116**, 71–76.

61. Frank, J., Radermacher, M., Penczek, P., et al. (1996) SPIDER and WEB: processing and visualization of images in 3D electron microscopy and related fields. *J. Struct. Biol.* **116**, 190–199.

21

Electron Energy-Loss Spectroscopy as a Tool for Elemental Analysis in Biological Specimens

Nadine Kapp, Daniel Studer, Peter Gehr, and Marianne Geiser

Summary

A transmission electron microscope (TEM) accessory, the energy filter, enables the establishment of a method for elemental microanalysis, the electron energy-loss spectroscopy (EELS). In conventional TEM, unscattered, elastic, and inelastic scattered electrons contribute to image information. Energy-filtering TEM (EFTEM) allows elemental analysis at the ultrastructural level by using selected inelastic scattered electrons. EELS is an excellent method for elemental microanalysis and nanoanalysis with good sensitivity and accuracy. However, it is a complex method whose potential is seldom completely exploited, especially for biological specimens. In addition to spectral analysis, parallel-EELS, we present two different imaging techniques in this chapter, namely electron spectroscopic imaging (ESI) and image-EELS. We aim to introduce these techniques in this chapter with the elemental microanalysis of titanium. Ultrafine, 22-nm titanium dioxide particles are used in an inhalation study in rats to investigate the distribution of nanoparticles in lung tissue.

Key Words: Electron energy-loss spectroscopy; EELS; energy-filter transmission electron microscope; EFTEM; electron spectroscopic imaging; ESI; lung, nanoparticles; titanium.

1. Introduction

Analytical transmission electron microscopy (TEM) techniques allow the location of chemical elements in solids. The potential of electron energy loss spectroscopy (EELS) was recognized some time ago (**1**,**2**). These methods are used routinely in material sciences. The application of these techniques in biological specimens is important because they allow one to discriminate cellular compartments and to identify intracellular components of different chemical elements at the ultrastructural level. However, EELS rarely has been applied

From: *Methods in Molecular Biology, vol. 369*
Electron Microscopy: Methods and Protocols, Second Edition
Edited by: J. Kuo © Humana Press Inc., Totowa, NJ

in biological specimens, mainly as the result of difficulties in the preparation of very thin tissue sections *(3)*. So far, of special interest for elemental micro-analysis in biological specimen have been calcium *(4,5)*, iron *(6)*, lanthanum tracer *(7)*, as well as titanium *(8)*. In 1994, researchers reported a resolution of 10 nm and detection limits of a few atoms for elements such as calcium, phosphorus and iron using a field emission scanning transmission electron microscope (STEM) under optimal conditions *(9)*. In 2003, Leapman *(10)* detected single atoms of calcium and iron in biological molecules and in 2004, he reported the three-dimensional distribution of phosphorus in biological samples by energy-filtered electron tomography *(11)*. In our study we investigated the distribution of inhaled and deposited ultrafine titanium dioxide (TiO_2) particles in lung tissue *(12,13)*.

Electron energy-loss spectroscopy is an analytical technique as well as an imaging method in a TEM or STEM. It is based on inelastic scattering events of the primary electrons of the beam with atoms in the specimen. In contrast to elastic scattering, the inelastic scattering causes an element-specific energy-loss of the primary electrons of the beam (*see* **Fig. 1**).

In conventional TEM unscattered, elastic and inelastic scattered electrons contribute to image information. Image phase contrast is enhanced by excluding electrons scattered through angles that are larger than the objective aperture. These are mostly elastic-scattered electrons. The remaining elastic electrons, as well as the inelastic and unscattered ones, form the final image. Therefore, the resulting image suffers from the chromatic aberration generated by inelastic scattered electrons. Because inelastic scattered electrons carry an energy loss specific for a chemical element in the specimen, these electrons can be used in an energy-filtering TEM (EFTEM) to locate the corresponding elements at the ultrastructural level.

The TEM is equipped with an energy spectrometer, which offers three methods for elemental analysis based on the measurement of the energy loss of the beam electrons that interacted with the specimen. In the following, three methods for elemental analysis using EELS are described.

The first, parallel-EELS, is a method for fast spectrum acquisition of a small region in the specimen. The intensity is recorded simultaneously in all channels of the energy loss spectrum. The most intense signal of an EEL spectrum is the zero-loss peak, whereas the specific ionization edges are of least intensity. The low-loss spectrum reflects beam electron interaction with loosely bound conduction and valence bound electrons. Chromophores and dyes, which usually are used for histochemical staining of light microscopic sections, induce low energy-loss signals in the spectrum. Hence, the specific dyes also can be used for analytical electron microscopy using EELS *(14)*. The energy-loss spectrum contains small ionization edges on a strong plural-scattering background. The elemental

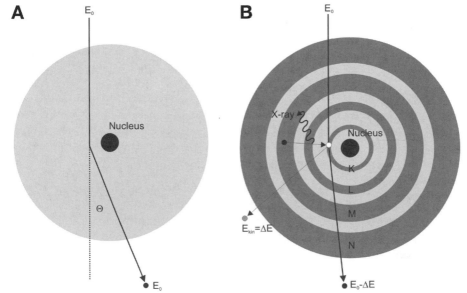

Fig. 1. Schematic drawing of interactions of beam electrons with electrons or nuclei of atoms. **(A)** During elastic scattering at the positively-charged nuclei of the atoms, electrons of the beam with the primary energy E_0 are scattered for large scattering angles Θ without loosing any measurable energy. **(B)** Inelastically scattered electrons of the beam scatter at small angles only, but loose energy and thus change their wavelength ($E_0-\Delta E$). This change depends on the chemical element the interaction has occurred with. The electrons of the atom get ionized at this interaction process. The free orbit becomes occupied by electrons with higher energies. Due to this process characteristic X-ray radiation is emitted, which is also element specific. K, L, M, N indicate energy levels according to Bohr.

composition data of a specimen can be extracted from the specific energy-loss-ionization edges *(15)*. Another feature is the so-called energy-loss near edge structure (ELNES) that is associated with each ionization edge in the EEL spectrum, which provides information about chemical bonds and electronic structure of nanomaterials *(16)*.

The second method, electron spectroscopic imaging (ESI), provides information about the nature, the spatial distribution, as well as the concentration of chemical elements within the specimen *(17)*. In the zero-loss mode of ESI, only electrons of an energy-loss of a few eV (depending on the slit width) contribute to the image formation. Because of the energy-selecting window, most inelastic scattered electrons are excluded, and chromatic aberration is minimized. Zero-loss filtering allows imaging at an improved contrast compared to conventional TEM *(18)*.

The principal of elemental mapping is that by selecting scattered electrons at a particular element-specific energy loss edge for imaging, a two-dimensional distribution map of the respective element can be obtained. However, becuse unspecific and multiple scattering events also occur, there is considerable background noise *(19)*.

The third method, Image-EELS, was first proposed by Jeanguillaume et al. *(20)*. An image stack is recorded over a defined energy-loss range with a given energy step between each image of the series. Spectra are obtained from any cluster of interest of the image by plotting a curve, energy loss versus grey level, through the image stack, resulting in the corresponding EEL spectrum *(21)*. This technique allows the detection of very small objects.

The aim of the chapter is to provide an overview about the methods. A protocol for the qualitative analysis of ultrafine titanium dioxide particles within ultrathin sections of heavy metal-stained lung tissue is described using ESI, parallel-EELS and image-EELS in a LEO 912 EFTEM with an in-column energy filter.

2. Materials

1. TiO_2 particles generated with a spark generator (GFG 1000, Palas Karlsruhe, Germany).
2. Young male WKY/NCrl BR rats (Charles River, Sulzfeld, Germany).
3. Tissue fixation: 2.5% glutaraldehyde in 0.03 *M* potassium phosphate buffer (511 mOsm), 1% osmium tetroxide in 0.1 *M* sodium cacodylate-HCl (350 mOsm), 0.5% uranyl acetate in 0.05 *M* sodium hydride maleate-NaOH (100 mOsm).
4. Ethanol and propylene dioxide for tissue dehydration.
5. Epon for tissue embedding.
6. Ultra-microtome (LKB, Bromma, Sweden).
7. 600-Mesh hexagonal EM copper grids.
8. TEM equipped with an energy filter/spectrometer (EFTEM). We used a LEO 912 EFTEM (Zeiss, Oberkochen, Germany) with the in-column Omega Filter in our studies. This microscope is equipped with a lanthanum hexaboride (LaB_6) cathode that has an energy spread of approx 2eV (*see* **Note 1** and also Williams and Carter *[22]*).
9. Computer, CCD camera, and software for digital image acquisition and processing. We used the Proscan software "ProDia" version 2.3, a slowscan CCD camera with 1024×1024 pixel resolution and a standard personal computer (*see* **Note 2**).

3. Methods

Three methods for elemental analysis are introduced, which provide high resolution and sensitivity: (1) Images taken at a defined energy loss (ESI), (2) spectra recorded over a small morphological area on the ultrathin tissue section (parallel-EELS), and (3) a mixture of the two methods called image-EELS, the simultaneous recording of multiple electron energy-loss spectra from a series of electron spectroscopic images *(23)*. For elemental identification, the $L_{2,3}$-edge of titanium was used *(see* **Note 3** and also Starosud et al. *[24]*).

3.1. Sample Preparation

1. Aerosols of ultrafine TiO_2 particles were generated with a spark generator (GFG 1000, Palas Karlsruhe, Germany) in a pure argon plus 0.1% oxygen stream *(25)*. The anesthetized rats (n = 5) were placed in a plethysmograph box and inhaled the aerosol for one hour via an endotracheal tube by negative-pressure ventilation, as described previously *(25)*.
2. Lungs of rats were fixed by intravascular triple perfusion of buffered 2.5% glutaraldehyde, 1% osmium tetroxide and 0.5% uranyl acetate applied in sequence *(see* **Note 4** and also Im Hof *[26]*).
3. The lung tissue is systematically sampled, dehydrated in a graded series of ethanol, and embedded in epoxy resin.
4. Ultrathin sections (<50 nm) were cut with an ultramicrotome (LKB, Sweden) *(see* **Note 5**).
5. Ultrathin sections of lung tissue were transferred onto uncoated 600-mesh hexagonal copper grids for elemental microanalysis of ultrafine TiO_2 particles *(see* **Note 6**).

3.2. General Adjustment of the Microscope and the Software (see Note 7)

1. Operate the microscope at a voltage of 120 kV.
2. Adjust the zero-loss peak and focus the spectrometer to achieve good energy resolution.
3. Do the beam alignment and the adjustment of all apertures, as well as the correction of objective astigmatism, as described in the manual of the microscope.
4. Use the index point in the middle of the small screen as reference point for beam alignment.
5. Before starting any measurements, use the software tool for background and flat-field correction of the CCD camera.

3.3. ESI

3.3.1 Adjustment of the Microscope

1. Switch to the spectrum mode in order to adjust the spectrum. Place the spectrum caustic to the index point of the small screen. This step usually has been done already during the general adjustment of the microscope. However, always check whether it has shifted.
2. Insert the slit aperture into the beam path. Be sure not to cut off the zero-loss of the spectrum (a dark spot in the center of your image would appear after returning to image mode). The slit width should be as narrow as possible *(see* **Note 8**).
3. Change to the image mode and shift the energy loss from zero to about 60-80 eV energy loss using the δE steps. A "dark field image" will appear, an image with reversed contrast (the structures appear bright and the background dark), because

ESI uses solely inelastic scattered electrons with an energy loss *(27)*. Adjust the brightness to have sufficient illumination and focus the structure for analysis in the energy loss.

4. Set an adequate condensor aperture for illumination manually and adjust it to the index point. The aperture should be as small as possible, but large enough not to disturb the image acquisition by the CCD camera, and it depends on the magnification used for analysis.

3.3.2. Adjustment of the Software

1. Start the ESI mode.
2. Set the "three window method" as default for background subtraction (*see* **Note 9**).
3. Select the chemical element of interest in the periodic table of elements appearing on the computer screen. For some elements, several energy loss edges can be chosen, e.g., for Ti: L_2 with its maximum at 464eV, or L_3 with its maximum at 469eV energy loss (*see* **Note 10**). The measurement will be done according to the standard operating procedures: the energy loss value at which images are taken are given as default values for each element (*see* **Note 11**).
4. Set the dwell time for image acquisition (*see* **Note 12**) and start the measurement. Three sequential images are automatically acquired, the element specific signal (*see* **Fig. 2E**) and the two background images (*see* **Fig. 2C, D**). The background windows can also be manually set to improve the signal to noise ratio *(28)*. The last image is calculated and reflects the net distribution of the selected element, which appears in white (*see* **Fig. 2F**; and also **Note 13**).
5. Save the data.

3.4. Parallel-EELS

For all following adjustments be sure to have carefully performed for the general microscope adjustment! Before doing any measurement, the calibration of the spectrum magnification is needed:

1. Choose a spectrum magnification, set energy loss ΔE to 0eV energy loss at the microscope and set the spectrum caustic to the center of the small illumination screen.

Fig. 2. *(Opposite page)* Electron spectroscopic imaging. **(A)** Illustration of background extrapolation using the three-window method. Three images are made, two below the element-specific edge to extrapolate a background image (ΔE_{W1} and ΔE_{W2}) and one image within the maximum (ΔE_{max}) of the element-specific signal. The dE reflects the energy selecting window width; $S(\Delta E_{W1})$, $S(\Delta_{W2})$, $S(\Delta W_{max})$ reflect the signal intensity and the energy loss values set for ΔE_{W1}, ΔE_{W2}, and ΔE_{max} respectively. S_N reflects the net signal of the element after background subtraction. Ti indicates the edge of titanium. **(B)** Zero-loss transmission electron micrograph of an unstained macrophage incubated with ultrafine titanium dioxide particles, found as large aggregates. **(C–F)** ES image series of the same region as in **(B)**. **(C–D)** Background images at 390

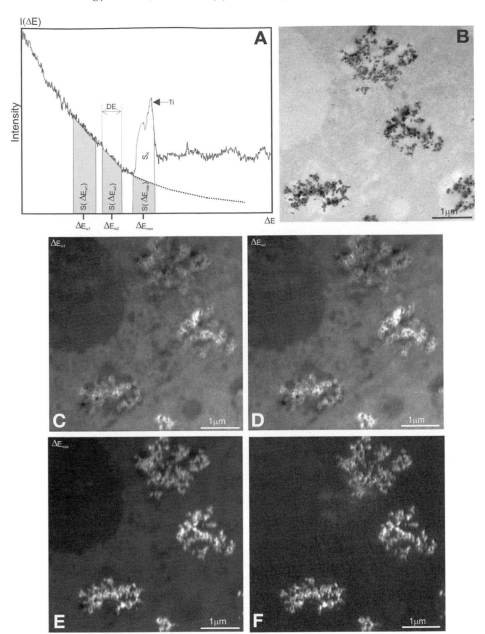

Fig. 2. *(Continued)* eV (ΔE_{W1}) (**C**) and 440 eV (ΔE_{W2}) (**D**) energy loss. (**E**) Image taken at the maximum of the titanium specific signal at 464 eV energy loss (ΔW_{max}). (**F**) Net titanium signal calculated by subtracting the extrapolated background image from the titanium specific signal. The obtained image reflects the titanium distribution, is displayed as white pixels.

2. Set the dwell time in the software to 50 ms and start the spectrum acquisition. The spectrum, including the zero-loss peak, is displayed on the computer screen.

3. Stop the spectrum acquisition. The position of the zero-loss peak reflects the start point.

4. Start the spectrum acquisition again and shift the spectrum for 30 eV. The new position of the zero-loss peak reflects the endpoint of measurement. The range between starting point and end point of the zero-loss peak reflects the energy loss value shifted (here: 3 eV; *see* **Fig. 3**). Repeat for every spectrum magnification.

3.4.1. Adjustment at the Microscope

1. Use a magnification higher than ×25,000 in the image mode. Shift the structure you want to analyze close to the index point within the small illumination screen (*see* **Note 14**).

2. Adjust the spectrometer entrance aperture to the center of the image screen in TEM mode.

3. Set the spot size as small as possible, since the zero-loss signal has a very high intensity (*see* **Note 15**).

4. Be sure that the structure you want to measure overlaps with the spectrum you set (*see* **Note 16**).

5. Change to spot mode, diffraction mode and spectrum mode: a very slim spectrum is displayed on the image screen.

6. Shift the zero-loss peak to the center of the small illumination screen (*see* **Note 17**).

3.4.2. Adjustment of the Software

1. Start the parallel-EELS mode.

2. Choose an adequate spectrum magnification for the energy range you like to measure. For the analysis of ultrafine TiO_2 particles (<100 nm), a convenient spectrum magnification was 80×. The energy loss interval ranged from 430 eV to 580 eV energy loss (*see* **Fig. 4B–D**), and, hence, the Ti-$L_{2/3}$ edge (452 eV energy loss) as well as the O-K edge (532 eV energy loss) can be measured in one spectrum (*see* **Note 18**).

3. Choose a 50-ms acquisition time to set the zero-loss peak for calibration.

4. Start the spectrum acquisition: The image of the spectrum is displayed on the computer screen. If the illumination intensity is correctly set, the zero-loss spectrum appears as a sharp peak in the spectrum window.

Fig. 3. (*Opposite page*) Parallel electron energy-loss spectroscopy spectrum calibration at a spectrum magnification of 40× defining the energy loss range displayed. (**A**) Spectrum of the zero-loss peak. The location of the zero-loss is set to be 0 eV, which defines the starting point. Energy loss range at the X-axis, intensity at the Y-axis. (**B** and **C**) The corresponding image of the spectrum in (A) and (D), respectively. (**D**) Spectrum of the zero-loss peak shifted for 30 eV. The location of the zero-loss peak after shifting defines the end point. Energy loss range at the X-axis, intensity at the y-axis.

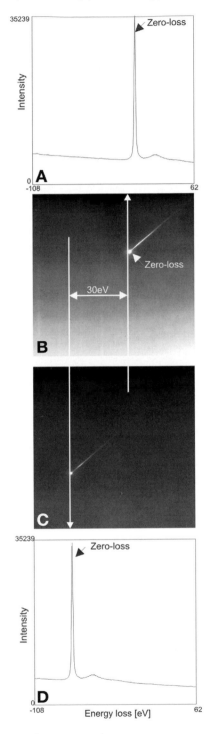

5. Stop the spectrum acquisition.
6. Set the maximum of the zero-loss peak to be 0 eV energy loss on the x-axis.
7. To measure the spectrum of the elements of interest, e.g., Ti and O, shift the energy loss δE close to the edge of interest (for a Ti of approx 430 eV). Adjust the spot size to 80 nm and the dwell time up to 500 ms for signal improvement. If the specimen contains titanium, a sharp edge is displayed at 452 eV energy loss (*see* **Fig. 4B–D**).
8. Save the data.

3.5. Image-EELS

3.5.1. Adjustment of the Microscope

The adjustment of the microscope is the same as for ESI, since an image-EELS series reflects a stack of ES images. Adjust the slit width as narrow as possible to obtain best energy resolution (*see* **Note 8**).

3.5.2. Adjustment of the Software

1. Start the image–EELS mode.
2. Choose an energy loss range for the measurement, e.g., from 400–510 eV energy loss including the energy loss edge of interest (here Ti at 453 eV).
3. Set the step width between individual ES images. Use 10-eV increments for a first screen and 2 eV for a spectrum with good energy resolution (*see* **Note 19**).
4. Set the dwell time for image acquisition to 1 sec.
5. Start the image acquisition (*see* **Notes 20** and **21**). The image acquisition stops automatically.
6. Save the data.
7. Create regions of interests on the first image of the stack. The intensity of the energy loss signal is plotted through the image stack and the spectrum is displayed in the spectrum window. The spectrum of any point on the image can be displayed by shifting the region of interest over the image (*see* **Fig. 5** and also **Note 22**).

Fig. 4. (*Opposite page*) Parallel electron energy-loss spectroscopy (EELS) spectra acquisition. (**A**) energy-filtering transmission electron microscopy image taken at 0 eV of a particle profile found in lung tissue, resolving a small particle cluster. Bar = 50 nm. (**B**) The parallel-EELS-spectrum for the particle shown in (A) at a spectrum magnification of 40×, the energy-loss ranges from 264 eV to 610 eV. The edges for carbon at 280 eV, titanium at 456 eV and for oxygen at 532 eV are shown. A detailed spectral analysis at higher spectrum magnifications of the small box is shown in (C) and (D). (**C**) The parallel-EELS spectrum for the particle shown in (**A**) at a spectrum magnification of 80×, the energy-loss ranges from 427eV to 607eV. The edges for titanium at 456eV and for oxygen at 532eV are shown. (**D**) Parallel-EELS-spectrum for the particle shown in (A) at a spectrum magnification of 100×, the energy-loss ranges from 436 eV to 577 eV. The edges for titanium at 456 eV and for oxygen at 532 eV are shown.

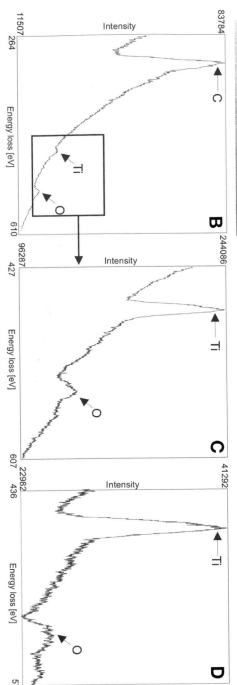

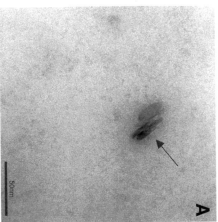

4. Notes

1. At present, there are two types of EFTEM on the market. Those from Zeiss and JEOL are equipped with in-column energy filters. The microscopes from other manufacturers contain post-column Gatan Image Filters (GIF). These two types of spectrometers have been built for different purpose. The GIF has been designed for energy spectrometry as its primary function. The in-column filter has been mainly developed for energy-filtered images but spectra can also be obtained.

2. Because there are computational image calculations in the different applications of EELS, it is necessary that the EFTEM is equipped with digital image acquisition technology.

   The software strongly depends on the type of microscope. For EFTEM with the post-column GIF, the software "digital micrograph" from Gatan (Pleasanton, CA) is used. For systems using in-column filters, there are different software solutions for elemental microanalysis: the "ESI vision" from Soft Imaging Systems (Muenster, Germany), "Pro Dia" from Proscan (Scheuring, Germany), and also the aforementioned Gatan software can be used.

3. Before starting an experiment, determine the location of the edge within the energy loss spectrum. In our purpose for identification of titanium we chose the sharp $L_{2,3}$ edge, which is the best to use for titanium. The shape of an edge depends on the element and the chemical state of that element. For elements with low atomic number, L-edges will not be available and K-edges must be used. For heavier elements, it is the M or higher edges that contain nice sharp features ideal for EFTEM. The energy-range containing useful signals will depend on the amount of the elements present, the chemical state of the element, and the instrumentation used to obtain the data. We achieved good results using elements with energy-loss energies beyond the carbon K edge and up to 1000 eV energy loss, since there was sufficient intensity for measuring.

4. If the edge of the element to analyze is close to the edge of the heavy metal used for tissue fixation and staining, elemental analysis of the structure may be difficult, because the signal from the structure may be overlapped by the signal of the heavy metal. In our experiment, the use of heavy metals for tissue fixation is inevitable to preserve all components of the inner lung surface. However, poststaining ultrathin

---

Fig. 5. *(Opposite page)* Image- electron energy-loss spectroscopy (EELS) analysis of TiO_2 particle aggregates in a cell culture of macrophages. (**A**) Stack of images taken for analysis. The first image of the series was taken at 400 eV energy loss, the last one at 510 eV. In each series different step sizes were taken: 10 eV, 5 eV, and 2 eV. Note that the image contrast is optimized to visualize structural details of the particle profile. Two regions of interest were selected for elemental analysis. The x- and y-axes of the coordinate system represent the plane image, the z-axis represents the energy loss. Bar = 200 nm. (**B**) The corresponding image-EELS spectrum of the region as selected in (A) within the particle aggregate shows the specific detection of the titanium edge at 456 eV. The step size for recording of the image stack was 10 eV. (**C**) The corresponding image-EELS spectrum of the region as selected in (A) within the particle

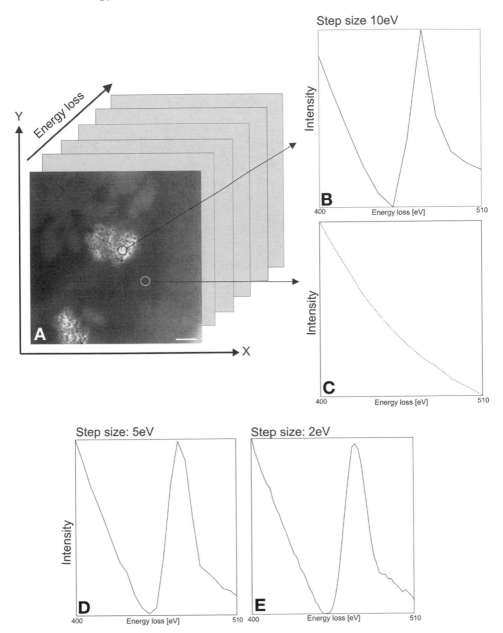

Fig. 5. *(Continued)* aggregate shows the background signal. The step size for recording of the image stack was 10 eV. (**D** and **E**) The corresponding image-EEL spectrum of the region as selected in (A) within the aggregate of particles shows the specific detection of the edge of titanium at 456 eV. The step size for recording of the image stack were 5 eV (D) and 2 eV (E), respectively.

sections with heavy metals for contrast enhancement is no longer necessary because energy filtering enables the generation of images with an enhanced contrast.

5. For our purpose we cut the ultrathin sections as thin as possible (less than 50 nm), to preclude multiple scattering events, that can cause false results due to the sum of multiple energy losses *(29)*. For the same reason, we used TEM grids without any film-coating in order to minimize thickness effects. The optimum section thickness depends very much on the microscope used (accelerating voltage, electron source). Working at 300 kV permits thicker sections to be analyzed. However the microscope we used for our analysis runs at a high voltage up to 120 kV.

6. Spectral analysis requires high energy doses, which may destroy the specimen. The higher the magnification and the energy loss are to measure, the higher is the electron dose needed for analysis. We obtained good results using TEM grids with small meshes (600- or 700-mesh), for stabilization of the ultrathin sections. However, for most biological systems larger meshes are required to ensure sufficient fields of view.

7. Adjustment of the microscope in general: (a) Use a test specimen for adjustment of the microscope. Beam-sensitive specimens are not adequate for the adjustment procedure. (b) What to do after having started the high voltage and the filament but there is no beam visible?

   Increase the emission current when values are less than 1 µA.
   Use higher beam intensity that are adequate for the magnification used.
   Reduce the magnification.
   Set the energy loss SE = 0.
   Remove the beam blanker from the beam path.
   Remove all apertures from the beam path.

   (c) Use the possibilities of the spectrometer for imaging: zero-loss filtering or structure-sensitive imaging for contrast enhancement. Insert the slit aperture into the beam path and set the spectrometer! The use of the energy filter has several advantages for imaging:

   Contrast enhancement.
   Minimization of chromatic aberration.
   Better resolution (when objective aperture is removed).

8. Set the slit as narrow as possible. The thinner the slit width, the better the energy resolution.

9. Use the three-window method for background subtraction, which improves the signal to noise ratio *(18,30)*. This will give you better results, as compared with those obtained with the two-window method or the jump-ratio method.

10. The actual edge energies depend not only on the element but also on the chemical state of the element. For example, attempts to analysis pure metals or metal oxides would require different choices of energy-loss for the energy-selecting slit to obtain the best results. Any change in the chemical state of the element results in a small shift for up to a few eV.

11. Background values can also be set manually in order to achieve better results. Be sure to set the background windows correctly. Incorrect settings may result in ele-

mental signal cut off, eventually leading to cause false negative artifacts *(17)*. Errors in the background estimation make it difficult to use the signal reliably when the signal is small compared to the background *(31)*.

12. The acquisition time should be optimized to obtain the best signal-to-noise ratio without damaging the sample. We achieved good results using 500 ms for the ESI acquisition in our specimen.

13. With increasing magnification and higher energy loss energy is needed for analysis. This means the longer dwell times and/or a larger illumination angle are required. Be aware of beam damage of the specimen and of the scintillator!

14. Use high magnifications (minimal 25,000×) to be sure that the spot is focused on the structure to analyze.

15. Be aware of burning damage on the screen and on the scintillator of the CCD camera! Work fast and do not leave the spectrum on the camera or the image screen after finishing an acquisition.

16. Always verify that the electron beam in spot mode hits the structure you like to analyze. Any drift of the specimen may cause false negative result.

17. Always check the zero-loss peak calibration (*see* **Note 21**).

18. The spectrum magnification correlating with the energy loss range is displayed on the screen: a low spectrum magnification means a large energy loss range, a high spectrum magnification means a small energy loss range (*see* **Fig. 4**).

19. For the initial screening purpose, use larger eV steps; for fine spectral analysis at high-energy resolution on the image, use small eV steps (*see* **Fig. 5**).

20. Be sure that the specimen does not drift during the image acquisition. Drift may cause false-negative results because the spectrum is obtained by linear plotting through the image stack. In the case of drift, use drift-correction tools of the software (if available).

21. Energy shift should be considered. Unpredictable energy drift occurs from changes in the energy of the beam electrons and from spectrometer instabilities. The energy scale of the spectrum in each pixel of the Image-EELS series is calibrated by matching a feature of known energy-loss to a feature in the spectrum. If no feature can be found then no energy shift is assumed *(32)*.

22. Regions of interest should surround the structures to measure as close as possible in order to obtain only the signal from the structure of interest and not from the surrounding areas.

Acknowledgments

This work was supported by the Swiss National Science Foundation and the Swiss Agency for the Environment, Forest and Landscape.

References

1. Egerton, R. F. (1982) Electron energy loss analysis in biology. *Electron Microsc.* **1,** 151–158.

2. Jeanguillaume, C. (1987) Electron energy loss spectroscopy and biology. *Scanning Microsc.* **1,** 437–450.

3. Roomans, G. M., Wroblewski, J., and Wroblewski, R. (1988) Elemental micro-analysis of biological specimens. *Scanning Microsc.* **2,** 937–946.
4. Pezzati, R., Bossi, M., Podini, P., Meldolesi, J., and Grohovaz, F. (1997) High-resolution calcium mapping of the endoplasmatic reticulum-golgi-exocytic membrane system. *Mol. Biol. Cell* **8,** 1501–1512.
5. Bordat, C., Bouet, O., and Cournot, G. (1998) Calcium distribution in high-pressure frozen bone cells by electron energy loss spectroscopy and electron spectroscopic imaging. *Histochem. Cell Biol.* **109,** 167–174.
6. Bordat, C., Sich, M. S., Réty, F., Bouet, O., Cournot, G., Cuénod, C. A., and Clément, O. (2000) Distribution of iron oxide nanoparticles in rat lymph nodes studied using electron energy loss spectroscopy (EELS) and electron spectroscopic imaging (ESI). *J. Magnetic Res. Imaging* **12,** 505–509.
7. Fehrenbach, H., Schmiedl, A., Brasch, F., and Richter, J. (1994) Evaluation of lanthanide tracer methods in the study of mammalian pulmonary parenchyma and cardiac muscle by electron energy-loss spectroscopy. *J. Microsc. (Oxford)* **174,** 207–223.
8. Stearns, R. C., Paulauskis, J. D., and Godleski, J. J. (2001) Endocytosis of ultrafine particles by A549 cells. *Am. J. Respir. Cell Mol. Biol.* **24,** 108–115.
9. Leapman, R. D., Sun, S. Q., Hunt, J. A., and Andrews, S. B. (1994) Biological electron energy loss spectroscopy in the field-emission scanning transmission electron microscope. *Scanning Microsc. Suppl.* **8,** 245–258.
10. Leapman, R. D. (2003) Detecting single atoms of calcium and iron in biological structures by electron energy-loss spectrum-imaging. *J. Microsc. (Oxford)* **210,** 5–15.
11. Leapman, R. D., Kocsis, E., Zhang, G., Talbot, T. L., and Laquerriere, P. (2004) Three-dimensional distribution of elements in biological samples by energy-filtered electron tomography. *Ultramicroscopy* **100,** 115–125.
12. Kapp, N., Kreyling, W., Schulz, H., Im Hof, V., Gehr, P., Semmler, M., and Geiser, M. (2004) Electron energy loss spectroscopy for analysis of inhaled ultrafine particles in rat lungs. *Microsc. Res. Tech.* **63,** 298–305.
13. Geiser, M., Rothen-Rutishauser, B., Kapp, N., et al. (2005) Ultrafine particles cross cellular membranes by non-phagocytic mechanisms in lungs and in cultured cells. *Environ. Health Perspect.* **113,** 1155–1160.
14. Barfels, M. M. G., Jiang, X., Heng, X. J., Arsenault, A. L., and Ottensmeyer F. P. (1998) Low energy loss electron microscopy of chromophores. *Micron* **29,** 97–104.
15. Williams, D. B. and Carter, C. B. (eds.) (1996) *Transmission Electron Microscopy: A Textbook for Material Scientists. IV Spectrometry.* Plenum Press, New York.
16. Brydson, R. (1991) Interpretation of near-edge structure in the electron energy-loss spectrum. *EMSA Bull.* **21,** 57–67.
17. Körtje, K. H. (1994) Image-EELS: simultaneous recording of multiple electron energy-loss spectra from series of electron spectroscopic images. *J. Microsc. (Oxford)* **174,** 149–159.
18. Egerton, R. F. (ed.) (1986) *Electron Energy-loss Spectroscopy in the Electron Microscope.* Plenum Press, New York.

19. Reimer, L., Zepke, U., Moesch, J., Schulze-Hillert, S., Ross-Messemer, M., Probst, W., and Weimer, E. (eds.) (1992) *EEL Spectroscopy. A Reference Handbook of Standard Data for Identification and Interpretation of electron energy loss spectra and for generation of electron spectroscopic images.* Carl Zeiss, Oberkochen.
20. Jeanguillaume, C., Trebbia, P., and Colliex, C. (1978). About the use of EELS for chemical mapping of thin foils with high spatial resolution. *Ultramicroscopy* **3,** 137–142.
21. Lavergne, J. L., Foa, C., Bongrand, P., Seux, D., and Martin, J. M. (1994) Application of recording and processing of energy-filtered image sequences for elemental mapping of biological specimens: Image-spectrum. *J. Microsc. (Oxford)* **174,** 195–206.
22. Williams, D. B. and Carter, C. B. (eds.) (1996) *Transmission Electron Microscopy: A Textbook for Material Scientists. I. Basic.* Plenum Press, New York.
23. Gelsema, E. S., Beckers, A. L., Sorber, C. W., and de Bruijn, W. C. (1992) Correspondence analysis for quantification in electron energy loss spectroscopy and imaging. *Methods Inf. Med.* **31,** 29–35.
24. Starosud, A., Bazett-Jones, D. P., and Langford, C. H. (1997) Energy filtered transmission electron microscopy (EFTEM) in the characterization of supported TiO_2 photocatalysts. *Chem. Commun.* **5,** 443–444.
25. Kreyling, W. G., Semmler, M., Erbe, F., Mayer, P., Takenaka, S., and Schulz, H. (2002) Translocation of ultrafine insoluble iridium particles from lung epithelium to extrapulmonary organs is size dependent but very low. *J. Toxicol. Environ. Health* **65,** 1513–1530.
26. Im Hof, V., Scheuch, G., Geiser, M., Gebhart, G., Gehr, P., and Heyder, J. (1989) Techniques for the determination of particle deposition in lungs of hamsters. *J. Aerosol. Med.* **2,** 247–259.
27. Ottensmeyer, F. P. (1984) Electron Spectroscopic Imaging: Parallel energy filtering and microanalysis in the fixed-beam electron microscope. *J. Ultrastruct. Res.* **88,** 121–134.
28. Ottensmeyer, F. P. and Andrews, J. W. (1980) High-resolution microanalysis of biological spsecimens by electron energy-loss spectroscopy and by electron spectroscopic imaging. *J. Ultrastruct. Res.* **72,** 336–348.
29. Ottensmeyer, F. P. (1982) Scattered electrons in microscopy and microanalysis. *Science* **215,** 461–466.
30. Colliex, C. (1986) Electron energy-loss spectroscopy analysis and imaging of biological specimen. *Ann. NY Acad. Sci.* **483,** 311–325.
31. Leapman, R. D. and Newbury, D. E. (1993) Trace element analysis at nanometer spatial resolution by paralle-detection electron energy-loss spectroscopy. *Anal. Chem.* **65,** 2409–2414.
32. Hunt, J. A. and Williams, D. B. (1991) Energy-loss spectrum-imaging. *Ultramicroscopy* **38,** 47–73.

22

Conventional Specimen Preparation Techniques for Scanning Electron Microscopy of Biological Specimens

John J. Bozzola

Summary

This chapter covers conventional methods for preparing biological specimens for examination in the scanning electron microscope (SEM). Techniques for handling cells grown in liquid culture, as well as on substrates such as culture dishes, slide culture chambers or agar, are discussed. These methods may be used to process most cultured organisms as well as whole botanical and zoological specimens.

Key Words: Scanning electron microscopy; biological specimen preparation; conventional SEM preparatory techniques.

1. Introduction

Specimens for scanning electron microscopy (SEM) examination may vary from individual cells grown in culture to solid tissues to entire organisms measuring several centimeters in size. As was the case with specimen preparation for transmission electron microscopy (TEM), the basic steps for SEM preparation are very similar: fixing it in buffered aldehyde, postfixing it in osmium tetroxide, dehydrating it in ethanol, drying it, mounting it on a specimen stub, coating it with a heavy metal, and examining it in the SEM. Detailed considerations for the choice of buffer and type of fixatives are discussed in Chapters 1 and 2. In the present chapter, only the most commonly used reagents and procedures used to prepare specimens for SEM examination are discussed.

The major advantage of the SEM is that it permits one to study the morphology and surface detail of solid material. It literally permits an "in-depth" study of specimens with great relief because of the tremendous depth of field available to the operator. SEM studies typically investigate the external features of a specimen, in contrast to TEM studies, where intracellular exploration is the

From: *Methods in Molecular Biology, vol. 369*
Electron Microscopy: Methods and Protocols, Second Edition
Edited by: J. Kuo © Humana Press Inc., Totowa, NJ

major focus. However, the SEM can be used to probe internal cellular detail; if one removes the overlying material, perhaps by fracturing, cutting, or tearing into the specimen. An understanding of the operational principle of the conventional SEM is useful because the quality of images obtained will be affected by how specimens are prepared for microscopic examination.

The SEM is unlike most conventional imaging systems, including light microscopes and the TEM, because it does not contain image-forming lenses. The lenses in the SEM do not form images but act as a set of three condenser lenses, in a demagnification series, to focus an extremely small spot, or probe, of electrons on a solid specimen. When the high-energy, accelerated electrons strike the specimen surface, they generate different forms of energy, or signals. These signals include secondary electrons, backscattered electrons, X-rays, light, and heat. With the appropriate detector, these signals can be captured, yielding different sorts of information. For example, secondary electrons give information related to topography (the three-dimensional image), backscattered electrons give information primarily about differences in atomic number, X-ray signals reveal the types of elements present, whereas light and heat may reveal information about the compositional nature of the specimen area being probed.

The SEM probe is not static but is scanned rapidly over the specimen much like the scanning of the electron beam that takes place in a television monitor or cathode ray tube (CRT). When the SEM probe strikes the specimen, each point that is impacted will yield a certain number of secondary electrons, depending primarily on specimen topography. This information is displayed on a viewing monitor, or CRT, so the number of secondary electrons from the specimen is displayed in terms of brightness and contrast. Areas of the specimen that yield many secondary electrons will appear bright, whereas areas that yield fewer secondary electrons will appear proportionally darker on the CRT.

The quality of the image displayed on the CRT depends on the quality of the signal, or the overall yield of secondary electrons from the specimen. Stronger signals are the result of generating and collecting higher numbers of secondary electrons. The elements that make up the surface of the specimen are the source of the secondary electrons. Elements with a high atomic number yield a high number of secondary electrons and, ultimately, a higher quality image. Unfortunately, biological systems are composed of lighter elements (e.g., carbon, hydrogen, oxygen, nitrogen) and yield a poor signal in the conventional SEM. To overcome this obstacle, biological specimens are coated with a thin layer of a high atomic numbered element such as gold, palladium, platinum, or osmium. After fixation, drying, and coating with a heavy metal, specimens are ready for study in the SEM. For greater coverage of specimen preparation and SEM operation, several reference books are recommended *(1–3)*.

2. Materials

1. Critical point dryer.
2. Sputter coater or vacuum evaporator.
3. Specimen collection supplies (tubes, pipettes, dishes, containers).
4. Buffers. A wide variety of buffer systems are available; however, the most commonly used buffer for SEM preparation is the phosphate buffer of Sorenson. Additional buffers are described in Chapters 1 and 2 on specimen preparation for the TEM. Sorenson's phosphate buffer consists of two parts: 0.2 M monobasic sodium phosphate, NaH_2PO_4, and 0.2 M dibasic sodium phosphate, Na_2HPO_4. The pH is adjusted by mixing the two components as shown in **Table 2**, in Chapter 1.
5. Fixatives (*see* Chapters 1 and 2 on specimen preparation for TEM). The usual fixative regime for SEM specimen preparation consists of 2% to 2.5% glutaraldehyde in phosphate buffer followed by postfixation in 1% osmium tetroxide, with or without phosphate buffer. It is especially important that one be aware of the dangers associated with using these chemicals.
6. Holders or chambers to hold small specimens during processing. The reusable, microporous chambers, sized at 12×12 mm, available through various microscopy suppliers (Electron Microscopy Sciences, cat. no. 70188; Structure Probe Inc., cat. no. 13215), are particularly useful to safely store very small specimens. Some are available in pore sizes ranging from 30- to 200 μm.
7. Poly-L-lysine-coated slides (commercial or self-prepared). Clean several conventional slides with soap and water, rinse in distilled water, and dry the slides in a dust-free environment. Prepare a solution containing 1 mg of poly-L-lysine (70,000–150,000 MW) per milliliter of distilled water. Flood or dip the slides in the poly-L-lysine solution for 10 min at room temperature. Rinse in distilled water, allow to air dry, and store until needed. Slides coated with poly-L-lysine are available from vendors of electron microscopy supplies.
8. Microfiltration apparatus consisting of micropore filter, holder, 10-mL syringe.
9. 1% Aqueous carbohydrazide (w/v).
10. Hexamethyldisilazane.
11. Mounting supplies (sticky tabs, carbon tape, conductive cement, china cement).
12. SEM specimen stubs.
13. Desiccated or dry compartment to store SEM specimen stubs.
14. Ethanol dehydration series (*see* Chapters 1 and 2 on specimen preparation for TEM).

3. Methods

3.1. Preparation of Cell Suspensions (see Note 1)

Cells that are grown as suspensions in culture liquids must be deposited onto a substrate of some sort. This step is accomplished either by allowing the cells to settle and adhere onto the substrate (*see* **Fig. 1**) or by trapping them on a microporous filter substrate.

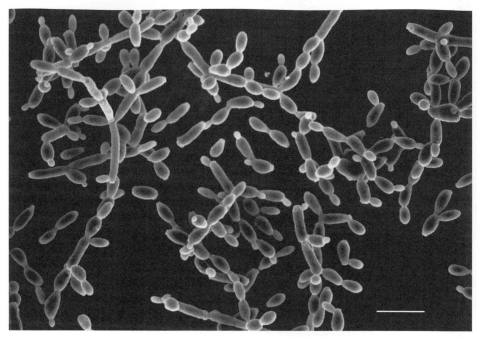

Fig. 1. *Candida albicans* yeast cells attached to the surface of a poly-L-lysine coated microscope slide. Specimen was prepared as described in **Subheading 3.1.1.** Bar = 10 μm.

3.1.1. Adherence of Cells to Poly-L-Lysine-Coated Slides

1. To adhere cells to a substrate, transfer several milliliters of culture to a microscope slide coated with poly-L-lysine and allow the cells to settle at room temperature for 1 hr. If cells do not adhere adequately using the poly-L-lysine procedure, then the organosilane *(4)* procedure should be used. Protect the slide from evaporation (*see* **Note 2**).
2. Tip the slide to drain the culture liquid, leaving some cells adhering to the surface of the slide.
3. Use a pipet to very gently flow 2.5% glutaraldehyde in phosphate buffer over the slide. Take care not to dislodge the attached cells. Allow the slide to stand undisturbed for 1 h at room temperature.
4. Drain the glutaraldehyde from the slide and rinse three times in phosphate buffer, each for 5 min.
5. Replace the distilled water with 1% osmium tetroxide in either phosphate buffer or distilled water. Allow slide to stand for 1 h at room temperature.
6. Rinse in distilled water and dehydrate the specimen using a graded ethanol series consisting of 25%, 50%, 75%, and two x 100% each for 10 min (*see* **Note 3**).
7. Transfer the slide, still submerged in absolute ethanol, to the critical point drying apparatus (*see* **Subheading 3.5.2.**) to complete the drying of the specimen.

8. Affix the slide to a specimen stub (*see* **Subheading 3.6.**) and apply a heavy metal coating using either a sputter coater or vacuum evaporator (*see* **Subheading 3.7.1– 3.7.2.**).

3.1.2. Cells Deposited on a Microporous Filter

1. Transfer several mL of culture into the barrel of a 10-mL syringe that contains an attached micropore filtration apparatus (*see* **Note 4**).
2. Place the plunger into the syringe barrel and apply pressure to the syringe to start the flow of liquid. If the flow stops immediately, too many cells were loaded into the syringe and a higher dilution is needed.
3. Remove the syringe barrel and associated plunger from the filter apparatus (*see* **Note 5**) and dispose of any unneeded cell culture.
4. Take up several ml of glutaraldehyde fixative into the syringe, place syringe onto the filter apparatus and press plunger gently to flow fixative onto the cells. Allow to stand at room temperature for 15 min.
5. Remove syringe from filter, discard unused fixative, and draw buffer into the syringe.
6. Rinse cells with buffer by attaching syringe to filter and pressing plunger.
7. Apply 1% osmium fixative using the syringe delivery method described in **steps 5** and **6**.
8. Discard unused osmium solution and rinse in distilled water using the syringe.
9. Dehydrate using a graded ethanol series applied onto the filter through the syringe.
10. After the cells have been rinsed in absolute ethanol, remove the filter holder and transfer the entire holder, including filter, into the critical point dryer.
11. Critical point dry the membrane, remove from filter apparatus, mount onto a specimen stub using double-stick tab and coat with heavy metal.

3.2. Preparation of Cells Grown on Substrates

3.2.1. Cultured Cells on the Surface of Petri Dishes

Many cultured cells will grow onto the surfaces of Petri dishes (*see* **Fig. 2 A–C**), glass microscope slides or cover glasses, plastic flasks, or Permanox slide chambers. Attached cells are processed as follows (*5*):

1. Decant (or gently aspirate) the culture medium and replace with 2.5% glutaraldehyde in 0.1 *M* phosphate buffer (warmed to the same temperature as the culture). Fix at room temperature for 30 min.
2. Pour off fixative solution and rinse three times with phosphate buffer, each for 5 min.
3. Postfix for 1 h at room temperature in 1% osmium tetroxide in distilled water.
4. Rinse three times in distilled water, 5 min each.
5. Incubate in freshly prepared 1% carbohydrazide in distilled water for 15 to 30 min.
6. Rinse five times with distilled water over the course of a 15-min period.
7. Incubate in 1% osmium tetroxide in distilled water for 30 min.

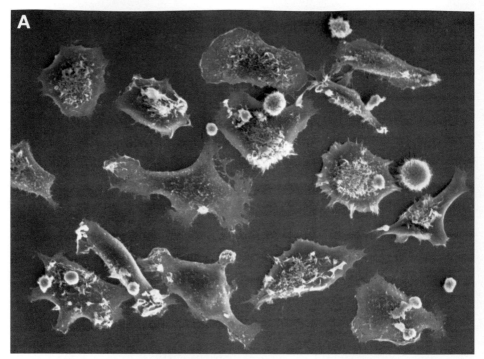

Fig. 2. (A) Monolayer of mammalian cells cultured on surface of Petri dish. Cells were prepared as described in **Subheading 3.3.1.** Bar = 25 μm. (Courtesy of William Kournikakis).

8. Rinse five times with distilled water over the course of a 15-min period.
9. Taking care not to let the specimen dry out, use wire cutters or clippers to cut the plastic dishes or flasks into pieces that will fit in the critical point dryer (*see* **Note 6**).
10. Place trimmed pieces, specimen side up, in another Petri dish.
11. Follow **steps 6–8** in **Subheading 3.1.1.**

3.2.2. Cultured Cells on an Agar Surface

If the cells are growing on an agar surface (*see* **Figs. 3** and **4**), excise 1 × 1 1-cm pieces of agar containing the cells, removing as much underlying agar as possible.

1. Place excised agar pieces in a Petri dish or other vessel containing 2.5% glutaraldehyde in phosphate buffer to submerge the cells (*see* **Note 7**). Keep the cell layer uppermost.
2. After 30 min fixation at room temperature, rinse the specimen in three changes of phosphate buffer for 5 min each.
3. Post-fix the cells 1 h at room temperature in 1% osmium tetroxide, phosphate buffered or in distilled water.

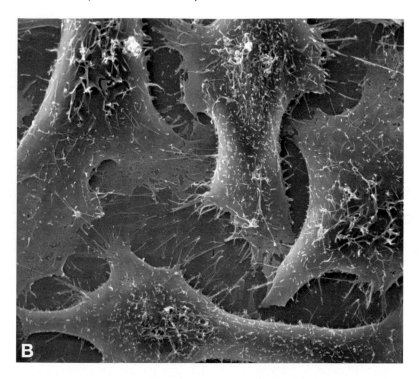

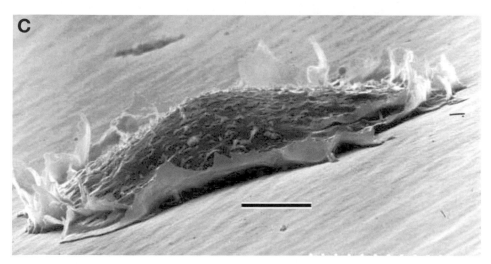

Fig. 2. (B) Overhead view of mammalian cells cultured on plastic surface. Cells were prepared as in (A). Bar = 10 μm. (Courtesy of William Kournikakis). **(C)** Low angle view of mammalian cell cultured on surface of Petri dish. Cell was prepared as described in **Subheading 3.2.1.** Bar = 5 μm. (Courtesy of Carol Heckman).

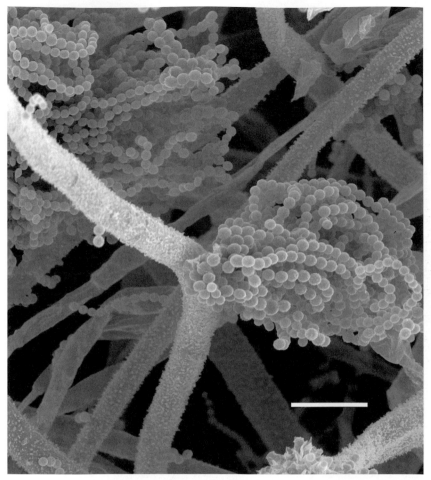

Fig. 3. *Aspergillus* fungi grown on agar medium. Specimen was prepared as described in **Subheading 3.6.2.** Bars = 20 μm. (Courtesy of Saara Mansouri).

4. Rinse three times in distilled water for 5 min each. After the last rinse, transfer the fragile specimens into holders, such as the microporous holders described in **Section 2**, for subsequent steps.
5. Follow **steps 6–8** in **Subheading 3.1.1.**

3.3. Preparation of Tissues and Large Pieces of Biological Material

3.3.1. Normal Specimens

Some organisms (insects, small plants) may be examined in their entirety using the following procedure (*see* **Note 8**):

Fig. 4. *Aspergillus* fungi grown on agar medium. Specimen was prepared as described in **Subheading 3.6.2.** Bars = 20 μm. (Courtesy of Saara Mansouri).

1. Clean the surface of the organism using either mechanical means or chemicals (*see* **Note 9**) because surface debris may obscure surface features.
2. Transfer the specimen into a container of 2.5% glutaraldehyde in phosphate buffer so it is submerged. Fixation is normally conducted for 1 to 3 h at room temperature; however, specimens may be stored in this fixative for several weeks (*see* **Note 10**).
3. Rinse specimen in three changes of phosphate buffer, 5 min each.
4. Post-fix for 1 to 3 h at room temperature in 1% osmium tetroxide.
5. Rinse three times in distilled water, 5 min each.
6. Follow **steps 6–8** in **Subheading 3.1.1.**

3.3.2. Specimens Larger Than Several Centimeters

With larger specimens, it is necessary to excise pieces, which are then processed in the following way:

1. Clean the surface of the organism to remove any debris (*see* **Note 9**).
2. With a scalpel, or fresh razor blade, excise a piece of tissue from the organism (*see* **Note 11**) and transfer it to 2.5% glutaraldehyde in phosphate buffer.
3. Rinse specimen in three changes of phosphate buffer, 5 min each.
4. Post-fix for 1 to 3 h at room temperature in 1% osmium tetroxide.
5. Rinse three times in distilled water, 5 min each.
6. Follow **steps 6–8** in **Subheading 3.1.1.**

3.4. Storage of Specimens Before the Completion of Drying

One would normally process the specimen to complete dryness and store the specimen in a desiccator. However, it is possible to store the specimen for long periods of time at several points in the process. After fixation with glutaraldehyde, specimens may be stored in buffer for several weeks or months, if refrigerated. Storage in 75% ethanol (of the dehydration series) is possible for several days to weeks.

3.5. Drying the Specimen

Before specimens can be viewed in the conventional SEM, they must be completely dried because the high vacuum conditions in the SEM chamber will cause hydrated specimens to boil, thereby destroying the integrity of the specimen surface. Specimens may be dried in a variety of ways, depending on the nature of the specimen (e.g., intact whole organisms, cell suspensions, excised portions). If there is any doubt as to the proper procedure to follow, especially with an important specimen, critical point drying is the safest method to use.

3.5.1. Air Drying

Air-drying may be used with some unfixed, hardy specimens such as insects or botanical specimens such as seeds or pollen (*see* **Fig. 5**). Some bacteria, such as the thick-walled Gram-positive organisms, may also be air-dried following chemical fixation and dehydration in ethanol. Air-drying can be achieved several ways:

- Put unfixed specimens in a drying oven at 30 to 40°C for several days to weeks. This works well with insects with exoskeletons, some botanical specimens, and possibly some bacteria and fungi.
- Air-dry certain chemically fixed and ethanol-dehydrated specimens using a solvent with high vapor pressure such as hexamethyldisilazane.

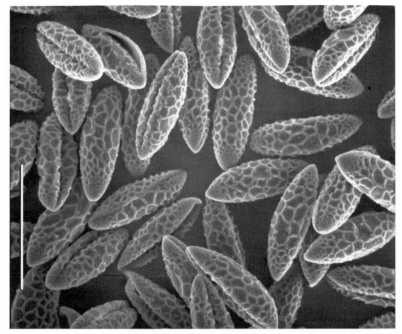

Fig. 5. Pollen from day lily. Pollen was shaken onto cover glass, air-dried and sputter coated as described in **Subheading 3.5.1.** Bar = 120 μm. (Courtesy of Steve Schmitt).

3.5.1.1. Basic Steps for SEM Preparation of Air-Dried Specimens

1. Clean specimen surfaces and fix in glutaraldehyde/osmium fixatives following the standard schedule presented earlier in this chapter (*see* **Subheading 3.3.1.**).
2. Dehydrate in ethanol series up to absolute ethanol.
3. Transfer specimen into hexamethyldisilazane for 5 to 10 min.
4. Air dry specimen in dust free environment at room temperature or in a drying oven.
5. Mount specimen on SEM stubs, coat with heavy metal, and examine in SEM.
6. Store specimens in dry, dust free environment.

3.5.2. Critical Point Drying (CPD)

CPD is the method most commonly used to complete the drying of chemically fixed and dehydrated specimens (*see* **Note 12**).

1. Transfer small specimens into holding devices (*see* **Fig. 6**) to prevent loss during CPD process. This step may be accomplished at any point after fixation in osmium tetroxide.
2. After the final change of absolute ethanol, place 10 to 15 mL of absolute ethanol in the prechilled chamber of the critical point dryer and quickly transfer specimens into the chamber.

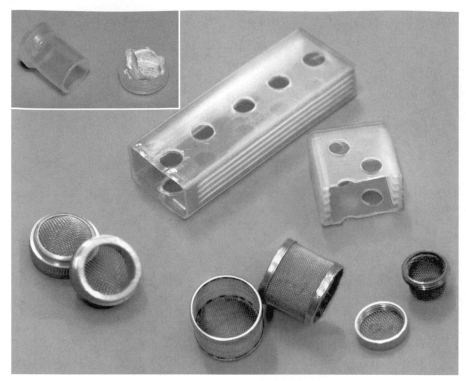

Fig. 6. Small holders used to protect small specimens during the critical point drying procedure. Inset, top left corner, shows holder fashioned from polypropylene embedding mold used in transmission electron microscopy. Large, rectangular container is a microscope slide mailer that has been modified to hold slides and cover slips. The three sets of stainless-steel mesh holders are commercially available.

3. Seal the chamber and fill with liquid CO_2, and maintain the recommended cold temperature, usually 0°C.
4. After several changes of liquid CO_2, to completely displace ethanol, raise the temperature of the CPD device to the critical point. For liquid CO_2, the critical point is 31.1°C at 1073 PSI.
5. After the liquid CO_2 is converted to gas, release the pressure while maintaining the elevated temperature to prevent re-condensation of the liquid CO_2.
6. Remove the fragile specimens, mount on specimen stubs, coat with heavy metal and view in the SEM.
7. After viewing, store specimens in dry, dust free environment.

3.5.3. Freeze-Drying

Freeze-drying is occasionally used on specimens that would be damaged by the CPD procedure. Although chemically fixed specimens are often used, this

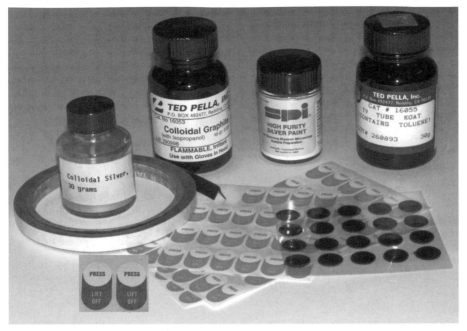

Fig. 7. Adhesives used to hold specimens on scanning electron microscopy stubs. These include conductive paints containing carbon or silver, conductive tapes containing carbon, sticky tabs (shown in lower left) that transfer a dab of adhesive material onto the stub, and carbon, double-stick disks.

is normally not needed since rapid freezing preserves the ultrastructure to a depth of 5 to 10 µm below the surface.

1. Rapidly freeze the specimen by plunging into isopentane or absolute ethanol that is chilled in liquid nitrogen to approx −85°C.
2. Transfer the specimen into liquid nitrogen (for storage) and then onto the −75°C cold stage of the freeze drying apparatus and activate the vacuum.
3. After several hours or days, depending on the mass of the specimen, warm the specimen stage to room temperature.
4. Remove specimens, mount on stubs, coat with heavy metal and view in the SEM. Store specimens in a dry, dust free environment.

3.6. Mounting Dried Specimen on SEM Stub

After specimens have been fixed and dried, they are mounted on an aluminum planchet, or specimen stub (*see* **Note 13**). Several adhesives or glues are available (*see* **Fig. 7**).

• Liquid adhesives may be used to hold solid specimens such as glass or plastic substrates containing cells, large insects, or botanical specimens such as seeds or hard stems or leaves. Examples of these adhesives include cyanoacrylate, china cement,

silver- or carbon-doped conductive adhesives available through electron microscopy supply houses, quick-setting epoxy cement, and even hot glue from a glue gun.
  • Tacky tapes, such as double-sided sticky tapes or special transfer tabs (from electron microscopy suppliers) are useful with specimens that might wick up liquid adhesive. Some double-sided tapes are electrically conductive since they contain carbon. The tapes and transfer tabs are used with small or porous specimens such as pollen grains, insects, and pieces of excised tissue.

3.7. Coating Specimens With Heavy Metal

Mounted specimens are normally coated with a thin layer of heavy metal which serves as a source of secondary electrons and electrically grounds the specimen to prevent the build up of a high voltage static charge from the electron beam. The deposition of heavy metal may be accomplished in several ways.

3.7.1. Sputter Coating

Sputter coating is the most commonly used method because it is fast, reliable, and the apparatus is relatively affordable.

1. Transfer several stubs of mounted specimens into the chamber of the sputter coater. Most sputter coater chambers will accommodate 5 to 10 standard-sized specimen stubs. Do not crowd the chamber, because it will interfere with proper coating.
2. Close the chamber and activate the rotary pump to remove atmospheric gases. This may take 5 to 10 min if the coater has not been used in several days.
3. After the system has reached the recommended vacuum level, open the valve to flow dry argon gas into the chamber.
4. Adjust the argon level based on the reading from the vacuum gauge, activate the high voltage to the recommended setting and coat the specimen with the proper thickness of heavy metal such as palladium/gold or gold. Some coaters have thickness monitors whereas others estimate the thickness based on argon levels, high voltage, and time (*see* **Note 14**).
5. Store the coated specimen in a dry, dust-free environment.

3.7.2. Vacuum Evaporation Coating

Vacuum evaporation is sometimes used when a sputter coater is not available or if higher resolution coatings are needed.

1. Transfer several specimen stubs into the chamber of the vacuum evaporator. The chamber should be equipped with an electrode capable of carrying a current of 30 to 50 ampere (*see* **Note 15**).
2. Evacuate the vacuum evaporator to the recommended vacuum level, typically 10^{-4} Pa for most heavy metals. This step may take 15 to 30 min, depending on the efficiency and design of the system.
3. Slowly apply current to the electrode to melt the metal (platinum, palladium/gold, palladium) on the electrode. After the metal has formed a molten ball, raise the current rapidly to completely evaporate the metal.

4. Allow the electrode to cool several min and then admit air to the specimen chamber.
5. Store specimen stubs in a dry, dust-free environment.

3.8. Nonmechanical Coating of Specimens

Nonmechanical coating techniques do not require equipment such as sputter coaters or vacuum evaporators because osmium is chemically reduced on the specimen surface *(6)*. Although more time consuming, the quality of coating is comparable with those obtained using mechanical devices.

1. Follow any of the protocols presented in this chapter to fix the specimens in glutaraldehyde and osmium tetroxide.
2. Rinse the specimen three times in distilled water, 10 min each.
3. Transfer specimen into freshly prepared (buffered or aqueous) solution of 8% glutaraldehyde and 2% tannic acid for 12 h. Change the solution three times during the 12-h period.
4. Rinse 3 times in distilled water, 10 min each.
5. Immerse in 2% aqueous solution of osmium tetroxide for 2 h.
6. Rinse three times in distilled water, 10 min each.
7. Repeat **steps 3** and **6**.
8. Continue normal processing of specimen: dehydrate, CPD, and mount on stub.
9. Store specimen under dry, dust-free conditions.

3.9. Storing SEM Specimen Stubs

Dried and coated or uncoated specimens must be stored in a clean, dry environment. Special plastic containers that firmly hold specimen stubs are available from microscopy supply houses (*see* **Fig. 8**). It is also possible to modify small cardboard or plastic boxes to safely hold stubs. These containers should hold stubs securely so that they do not spill out if the container is dropped or tipped. Stubs should be labeled on the bottom side using a permanent marker. Specimens must be stored desiccated, either inside of a sealed chamber with desiccant or in a drying oven set to 40 to 45°C. A sealed chamber can be readily fashioned from a large mayonnaise or other glass jar with a tight fitting lid. Desiccants, such as silica gel or calcium sulfate, should be placed inside the glass jar and changed when exhausted. If a large glass jar is not available, then specimens can be stored temporarily in a zip-lock plastic bag containing desiccant. Avoid contacting the specimen with the desiccant since this will transfer particulate dust onto the specimen surface.

4. Notes

1. Bacterial and fungal cells are processed in a similar manner as mammalian cultures with the exception that longer fixation times are used (several hours to overnight in each fixative).

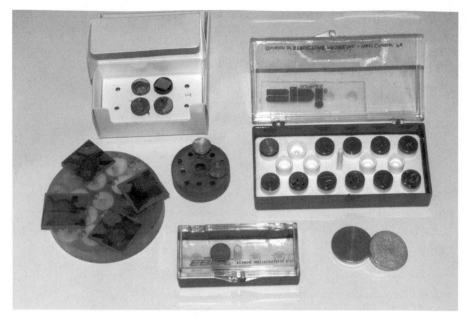

Fig. 8. Specimen stubs are best stored in rigid holders, shown. After securing the lids, the entire holders are placed in a dry, dust-free environment for storage.

2. To protect the cells from drying out during this procedure construct a humid chamber consisting of a Petri dish, or Tupperware® container, lined with paper towels soaked with water. Place several lengths of glass rod or applicator sticks on top of the paper to support the microscope slide. Cover the container during the incubation process.

3. The majority of cells are lost during the first dehydration step since the mixing of the ethanol with water generates vigorous swirling and heat. This can be lessened by using a very gradual dehydration, starting with 10% ethanol and slowly flowing the next alcohol (20%, etc.) in the series across the slide.

4. We have not been as successful with this procedure versus the method that uses a poly-L-lysine coated slide. Several causes of failure: clogging of filter because of an excessive number of cells, poor preservation of ultrastructure due to poor exchange of processing chemicals (especially in the CPD), curling or disintegration of the micropore filter. We have been most successful with the sturdier 13-mm silver micropore membrane, with 0.45-μm pores, available from Structure Probe, Inc.

5. The micropore membrane is fragile unless handled properly. Fluids must flow only toward the membrane. Never pull back on the syringe plunger with the filter attached since the back flow will rip the unsupported filter. The syringe must be loaded with liquid and then attached to the filter by means of a screw on lock rather than a friction fitting since the friction fittings do not hold the filter securely and it will fly off under pressure.

6. First, snip away the sides, then cut the bottom to size taking small cuts to avoid cracking the dish. A Dremel® high-speed rotary tool with a cutter wheel accomplishes this task quickly and neatly.

7. Be sure to maintain the orientation of the agar with the cells uppermost. Remove as much underlying agar as possible and handle the specimen very gently because the dried agar crumbles like a cigarette ash.

8. Hard organisms may not need to be fixed but simply air-dried. This includes specimens such as insects that have a sturdy exoskeleton, botanical specimens such as seeds or pollen, scales or hair from certain animals.

9. Because the surface is what one normally examines in the SEM, it is critical that all debris be removed. With some specimens, this step can be accomplished using light puffs of air, perhaps accompanied by sweeping with a soft brush. With wet specimens, buffer solution can be gently flowed across the surface. Sturdier specimens can sometimes be cleaned using a dental irrigation device set to dispense a low-pressure stream of buffer. In some cases, it may be necessary to use mild detergents, surfactants or enzymes to dislodge stubborn debris.

10. If the specimen is large, several cuts should be made in the specimen to allow fixative to diffuse into the interior or it should be cut into smaller, manageable pieces. The application of a low vacuum (from a faucet aspirator) may be applied to enhance penetration of fixative, especially in botanical specimens. With some botanical specimens, the inclusion of several drops of a detergent such as Tween-20® or PhotoFlo® will help break the surface tension and hydrophobic nature of the cell surface.

11. When specimens are so large that fixative cannot penetrate quickly, smaller pieces must be excised. The pieces should be no larger than 2 to 3 mm on one side. Place the specimen in the buffer solution and cut the desired pieces from the submerged specimen. Transfer specimens into fixative solution. To facilitate penetration, detergents may be added to the fixative and vacuum may be applied (*see* **Note 10**).

12. Most soft, biological specimens are damaged when they are allowed to dry by evaporation of water. The damage is caused by the passage of the air/water meniscus or interface through the specimen. This interface imposes surface tension forces of 2000 PSI and collapses most biological structure. The CPD substitutes ethanol, then liquid CO_2, for water. At the critical point, liquid CO_2 is converted to a gas that is then slowly released, thereby eliminating the air/water interface.

13. Because the specimens are quite fragile, special devices are used to transfer them onto the stub. Such devices include camel's hair brush, wooden applicator sticks, dissecting needles, jeweler's forceps, micropipettes, and eyelash probes consisting of a single eyelash glued to a wooden applicator. Electron microscope suppliers market a device called a vacuum needle that consists of a hollow, needle probe connected to a vacuum source generated by an aquarium pump. The operator controls the application of pressure by means of a hole in the handle. The specimen is picked up by touching the probe to the specimen, moving it onto the stub, and removing the finger pressure to release the specimen.

14. The thickness of the coating is very important: not enough and the specimen gives a poor signal and builds up an electrostatic charge; too much and fine details are buried under the metal coat. Some sputter coaters are equipped with a thickness monitor, a vibrating quartz crystal that accurately displays the thickness deposited on the specimen. Most manufacturers of sputter coaters provide a scale that correlates voltage and time to thickness of coating. It is also useful to construct a series of colormetric guides based on the deposition of metal on white paper. The colors change as the thickness increases, so that once one has a satisfactory coating (based on observation in the SEM), it is possible to approximate this coating for subsequent coating sessions.

15. In the thermal evaporation process, heavy metals are heated to their melting point in a high vacuum. Under these conditions, the metal is vaporized to a monoatomic state and travels in straight lines to condense onto the specimen stubs. The source of the evaporated metal is usually a fine gauge wire of pure metal such as palladium/gold, platinum, or gold. Typically, 1 inch of pure metal wire is wound around an electrode composed of a heavy gauge tungsten wire. The electrode is slowly heated under high vacuum, vaporizing the pure metal wire and depositing it onto the specimen. If the heat is applied too quickly, the metal will not melt properly and metal chunks will be ejected from the electrode, spoiling the specimen. Gold metal is easiest to evaporate, followed by palladium/gold alloy and platinum. The finest coatings are obtained with platinum, the coarsest with gold.

References

1. Bozzola, J. J. and Russell, L. D. (1999) *Electron Microscopy Principles and Techniques for Biologists.* Jones and Bartlett Publishers, Sudbury, MA.
2. Dykstra, M. J. (1992) *Biological Electron Microscopy Theory, Techniques, and Troubleshooting.* Plenum Press, New York and London.
3. Goldstein, J. I., Newbury, D. E., Echlin, P., Joy, D. C., Fiori, C., and Lifshin, E. (1992) *Scanning Electron Microscopy and X-ray Microanalysis.* Plenum Press, New York.
4. Clayton, D. F. and Alvarez-Buylla, A. (1989) In situ hydridization using PEG-embedded tissue and riboprobes: increased cellular detail coupled with high sensitivity. *J. Histochem. Cytochem.* **37,** 389–393.
5. Heckman, C. A., Oravecz, K. I., Schwab, D., and Pontén, J. (1993) Ruffling and locomotion: role in cell resistance to growth factor-induced proliferation. *J. Cell Phys.* **154,** 554–565.
6. Takahashi, G. (1979) Conductive staining method. *Cell* **11,** 114–123.

23

Variable Pressure and Environmental Scanning Electron Microscopy

Imaging of Biological Samples

Brendan J. Griffin

Summary

The use of elevated gas pressures in the sample chamber of a scanning electron micro-scope (i.e., variable pressure SEM, or VPSEM) together with specialized electron detec-tors create imaging conditions that allow biological samples to be examined without any preparation. Specific operating conditions of elevated pressures combined with sample cooling (usually restricted to the environmental SEM range) can allow hydrated samples to be maintained in a pristine state for long periods of time. Dynamic processes also can be easily observed. A wider range of detector options and imaging parameters introduce greater complexity to the VPSEM operation than is present in routine SEM. The current instrumentation with field emission electron sources has nanometer-scale beam resolution (approx 1 nm) and low-voltage beam capability (0.1 kV). However, under the more ex-treme variable pressure conditions, useful biological sample information can be achieved by skilled operators at image resolutions to 2 to 4 nm and with primary electron beam volt-ages down to 1.0 kV. Imaging relating to electron charge behavior in some biological sam-ples, generally referred to as charge contrast imaging, provides information unique to this VPSEM and environmental SEM that closely relates to luminescence imaged by con-focal microscopy.

Key Words: VPSEM; ESEM; uncoated samples; hydrated samples; charge contrast imaging.

1. Introduction

Conventional scanning electron microscopy (SEM) of biological samples has been hampered by the need for samples to have an electrically conductive outer surface, be fully dehydrated, and completely stable in the SEM chamber environment. The achievement of this sample state has employed many electron microscopists in ongoing sample preparation technique development since

From: *Methods in Molecular Biology, vol. 369*
Electron Microscopy: Methods and Protocols, Second Edition
Edited by: J. Kuo © Humana Press Inc., Totowa, NJ

SEM became established. Almost equal effort has been spent in artifact recognition and avoidance. As the SEM technology continues to improve and the scale of required information reduces, the sample preparation becomes more challenging. The most critical parameter has become the conductive coating, and it is very difficult to deposit featureless and sufficiently continuous sub-nanometer-thick metal films. Any technique that reduces or eliminates sample preparation is increasingly attractive and, consequently, new optical microscopy and variable pressure SEM (VPSEM) techniques have become very important research tools.

The use of a gas in the sample chamber to minimize sample charging is well established. The early techniques were aimed at rapid mineral sample imaging and microanalysis, using Ar gas at pressures to 15 Pa (0.1 torr) in conjunction with back-scattered electron detectors (BSEDs *[1]*). Extension of operation to elevated pressures (to 3000 Pa) was achieved in the late 1970s with a differentially pumped electron column and multiple pressure-limiting apertures *(2)*. Operation at these higher pressure levels allowed the stabilization of hydrated and even liquid samples, when used in conjunction with a cooling stage and with water vapor as the chamber gas *(3)*. The final barrier of imaging using the low energy secondary electron signal was overcome through the development of the "environmental secondary detector," which rapidly evolved to the modern "gaseous secondary electron detector" (GSED *[4,5]*). Recent developments in detector technology have allowed the realization of image quality (resolution and signal-to-noise) comparable with conventional SEM *(6,7)*.

The aim of this chapter is to provide the first practical guide on the application of these techniques to the study of biological samples. The literature is now quite extensive, but it is disseminated through many discipline areas. The terminology used also is confusing because of this cross-disciplinary activity. Therefore, the first section of the chapter aims to provide a clear and concise summary of the instrumentation available, the signal detection options, and the information content carried by the various detectable signals. Application examples are then discussed for the two discrete modes of operation, the "variable pressure" mode, in which hydrated samples are unstable, and the "environmental" mode, in which hydrated samples may be maintained in their pristine state.

VPSEM and environmental SEM (ESEM) are not simple extensions of conventional SEM and must be approached separately. The present of the gas influences both the image S:N and the information content of the images. The gaseous secondary electron detector operation controls not only image content but also the charge cancellation on the sample. One must gain an intuitive understanding of the processes of VPSEM and ESEM to maximize their benefits.

Postcollection image enhancement, through Adobe Photoshop® or a similar program, is also a major tool when using VPSEM and ESEM because optimal image contrast and brightness may not be achievable without introducing charging or other artifacts at the point of image capture.

2. Instrumentation and Operation

2.1. Terminology

A range of SEM that can operate with elevated chamber gas pressures is commercially available. Modern marketing has dictated that each manufacturer should have a unique descriptor for their products and, unfortunately, this has led to considerable confusion in the scientific literature.

The instruments are basically similar in that they all operate with a differentially pumped sample chamber relative to the electron source. There may be a series of pressure regimes in the electron column separated by pressure limiting apertures, typically in those instruments that can operate at relatively high chamber gas pressures. Typical column vacuum profiles are shown in **Fig. 1**.

The nomenclature describing the different instruments is not consistent in the literature because of the capability to image fully hydrated samples being dominated by one instrument for a long period. The original instrument was the ElectroScan Environmental Scanning Electron Microscope (ESEM®); it was released in 1988. The ESEM® model descriptor was subsequently acquired and retained by Philips Electron Optics, and it is still in use by the current manufacturers, FEI Pty Ltd. The ESEM was the sole instrument of its kind until late 2003, when Zeiss released their extended pressure range EVO® series and in 2004, when Hitachi released the extended range version of their 'Natural SEM®'. Consequently, the scientific community came to see ESEM as a *modus operandi* rather than a specific model of variable pressure SEM and hence the general confusion.

The distinction between an instrument that can operate with low sample chamber gas pressures as distinct to those that operate at pressures high enough to permit the unrestricted observation of hydrated samples is important. The most appropriate terminology seems to be to refer to all instruments of this general capability as VPSEM and then to use the term ESEM for the subset of instruments capable of imaging hydrated samples. Practically there is a real distinction as the basic VPSEMs will not operate at pressures greater than 133 to 270 Pa (1–2 torr) whereas the ESEM variants operate up to 1330 Pa (10 torr), and in some cases well beyond.

2.2. Operation and Imaging

2.2.1. The Chamber Gas

VPSEM and ESEM operate with elevated chamber gas pressures. A range of gases can be used, and the attributes of the different gases have been extensively

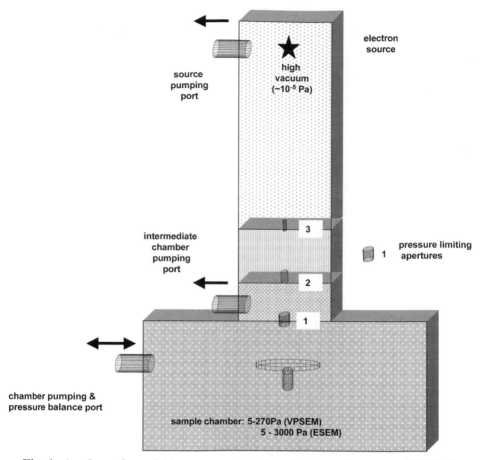

Fig. 1. A schematic variable pressure scanning electron microscope (VPSEM) or environmental SEM (ESEM) column illustrating the role of the pressure-limiting apertures (PLAs 1-3) and the differential pumping of the column. Some VPSEM may only have 1 or 2 PLAs.

investigated *(8)*. For imaging of biological samples two gases, air and water vapor, dominate current usage. Water vapor is without question the most effective gas, and the commercial ESEM range has used it since the original development. Water vapor has ideal ionization characteristics and so it behaves well as the imaging gas. Second, a major benefit of VPSEM and particularly ESEM is the ability to image hydrated samples and this requires water vapor as the chamber gas. In VPSEM, the rate of dehydration of biological samples (*see* **Section 3**) is obviously reduced significantly if water vapor is used.

Water vapor can be used in any VPSEM by the following simple modification, BUT only with support of the appropriately trained and suitably responsible service personnel. A vacuum flask, with protective wrap, can be attached to the air inlet used for supplying the chamber gas. The flask should be partially filled with DI water. Never fill the flask to more than 40% capacity because the space above the liquid is the vacuum reservoir. Overfilling will resulting in "boiling" of the water as the result of rapid pressure reduction over the liquid surface. The compatibility of the pumping system MUST also be checked prior to this conversion.

2.2.2. Gas–Electron Interactions

The basic electron-gas interactions that occur in VPSEM are reasonably well understood, and the physics is well defined *(9)*. The main practical effect is a reduction in the image signal-to-noise (SNR) occurs because of the loss of scattered electrons out of the primary beam probe as the chamber gas pressure is increased (*see* **Fig. 2**). These scattered primary electrons generate a signal from the sample that is a noise component as it lacks spatial coherence. The scattered electrons are a more significant factor when performing X-ray microanalysis as the X-rays generated by them contribute spurious data from outside of the primary analysis area.

1. Gas path length = distance from final aperture to sample surface (this is often NOT the same as working distance because in many instruments the final aperture projects into the sample chamber below the plane of the objective lens).
2. Chamber gas pressure.
3. Chamber gas characteristics.
4. Primary beam energy (accelerating voltage).

2.2.3. Secondary Electron Imaging

Conventional Everhard-Thornley and the newer in-lens SE detectors do not operate under elevated gas pressure conditions because a high-voltage discharge will occur. A range of secondary electron detectors have been developed based on the attraction of the secondary electrons to a biased element. In all cases, the emitted electrons interact with the chamber gas and a cascade amplification of the electron signal occurs toward the biased element. The emitted electrons accelerate to a point where they ionize a gas molecule, with the resultant electrons then repeating the process. The resultant electron signal is termed the gaseous secondary electron (GSE) signal. The most common GSE detector measures the signal flux in the biased element *(2)*; a second variant measures the specimen current that is in fact the ion flux, arising from the electron-gas amplification, flooding the sample surface *(5)*. The third common GSE detector variant measures the luminescence from the gas as it recombines *(10)*. The basic electron gas cascade amplification process and the various GSED detector arrangements are shown in **Fig. 3**.

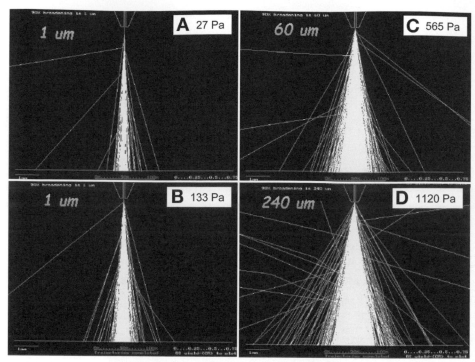

Fig. 2. Modeled effects of scattering a 10-kV primary electron beam by water vapor with a 6-mm working distance (= gas path length) and for chamber gas pressures of (**A**) 27 Pa (0.2 torr), (**B**) 133 Pa (1.0 torr), (**C**) 565 Pa (4.0 torr), and (**D**) 1120 Pa (8 torr). The values in each pangel represent the diameter containing 90% of primary beam electrons on the sample surface. Bar = 1 mm.

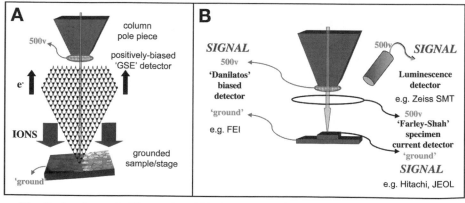

Fig. 3. A schematic illustrating (**A**) the gas amplification process and (**B**) the gaseous secondary electron (GSE) detector variants available for variable pressure scanning electron microscopy and environmental scanning electron microscopy.

Primary beam and backscattered electrons contribute to this detected GSED signal. The level of contribution is a function of the scattering parameters (*see* **Subheading 2.2.2.**), and among those parameters, scattering has a strong inverse relationship to the primary beam energy, i.e., at low beam energies very high degrees of scattering are encountered. Some detector variants collect in parallel a backscattered electron signal and subtract it from the GSE signal to provide a more pure SE type signal. Comprehensive fundamental accounts of the gas amplification processes and contrast mechanisms have been presented *(11)*.

All GSED depend upon the bias voltage on the positive detector element to generate the necessary gas amplification process. The bias voltage is commonly labeled "contrast" on the GSED controls. Practically, it is useful to note the value when good imaging conditions are present to allow consistency and reproducibility of imaging. The upper level of the GSED bias is normally limited by the value where an electrical discharge between the detector and the sample, or arcing, occurs rather than the operational range of the electronics. The specific voltage depends on gas pressure, type, detector design, and sample/stage configurations, i.e., the general operating conditions. The usual practice is to maximize the GSED bias to just below the discharge condition to provide the best image SNR ratio.

2.2.4. Sample Charge Cancellation

Positive ions are generated by the various electron–gas interactions in the sample chamber. The earliest and simplest VPSEM operated with low Ar gas pressures (typically 10–15 Pa) introduced in the chamber just above the sample. This arrangement produced sufficient positive ions to cancel charge and so to allow the BSE imaging (BSEI) and X-ray microanalysis of the uncoated and non-conductive mineral samples under study *(1)*.

The gas amplification present with GSED operation produces a positive ion flood that far exceeds the primary electron beam current. The ions move slowly relative to the electrons, which can be observed practically as a brief instability of the SE image when moving to a new area on the sample, depending on sample and conditions. Where ions dominate the signal, a smearing of the image is present at faster beam scan rates. Most importantly, the ions can recombine and thereby reduce the GSE signal. A second detector element, grounded or slightly negatively biased, introduced between sample and biased GSE detector element will reduce this effect and can markedly improve image quality under some conditions *(12,13)*.

2.2.5. BSE and Cathodoluminescence (CL) Imaging

The original imaging mode in VPSEM was based on both solid state and scintillator based BSED and excellent results remain obtainable on biological

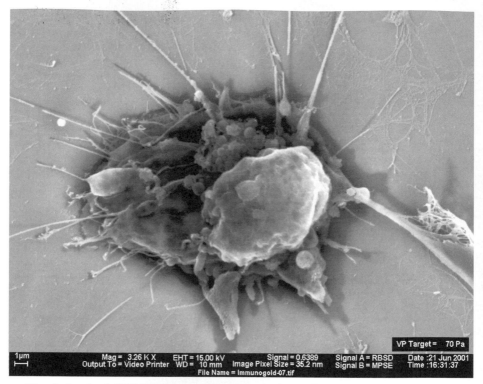

Fig. 4. A mixed GSE-BSE image of a biological sample (cancer cell grown on glass) at low magnification in the VPSEM compared with the same sample coated and imaged by conventional SE under high vacuum conditions (Courtesy of Zeiss SMT Pty Ltd.)

samples (*see* **Fig. 4**). The practical limitation for both VPSEM and ESEM is in the BSED assembly thickness as these detectors are normally positioned directly above the sample (*see* **Fig. 5**). A thick BSED assembly will severely compromise the minimum achievable working distance and should be avoided. Practically, the consequent long gas path length will limit the achievable image quality at moderate chamber gas pressures and/or low accelerating voltages.

Panchromatic CL imaging is routinely achievable with conventional CL detectors. The only compromise is that the emissions from gas luminescence and charge contrast imaging (*see* **Subheading 3.4.**) can interfere with the CL signal. Practically the contributions to CL from these other processes can be easily determined by imaging with the GSED bias on and then to identify the various signal contributions. Fiberoptic-based spectral CL is also effective in VPSEM *(14)*. There has only been limited application of CL in the study of biological samples to date but the technique is relevant to investigating bioluminescence

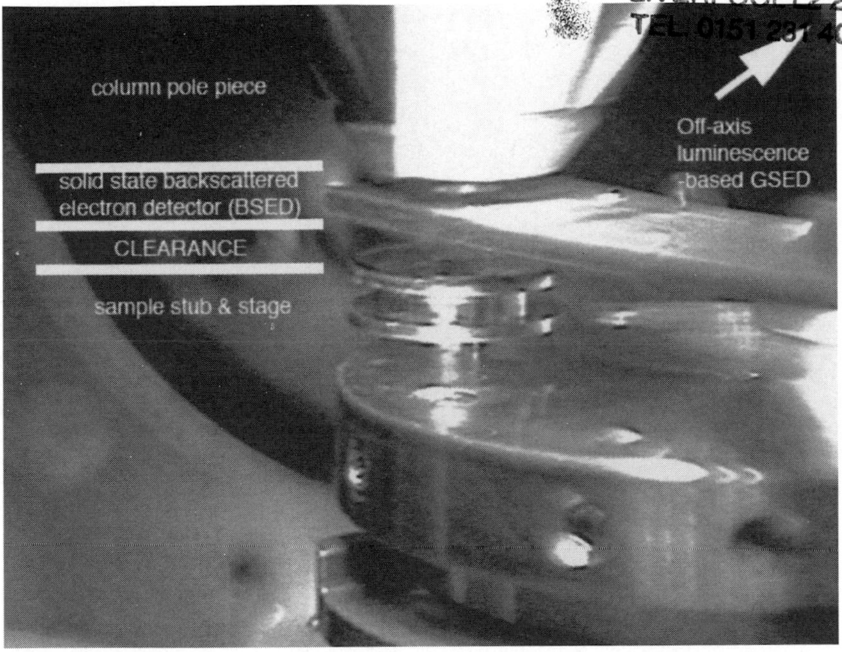

Fig. 5. A CCD camera view of a solid state backscattered electron detector (BSED) in place above the sample with an off-axis luminescence-based gsed. The narrow clearance between sample and detector and the increased gas path length due to the presence of the BSED are evident. GSED, gaseous secondary electron detector.

and similar processes. The new identification of a charge contrast effect in coral and other tissues (*see* **Subheading 3.4.**) will increase interest in this area.

2.3. Imaging of Hydrated Samples

Imaging fully hydrated samples is a well-established and mature technique *(2,3,15)*. It requires two components, an ESEM variant of VPSEM (to operate at the higher chamber gas pressures), and a specimen cooling stage, usually Peltier-cooling based (*see* **Fig. 6**). A recent detector development *(13)* that can operate at high chamber gas pressures has made the latter unnecessary but only with one of the commercial ESEM variants and also only at higher primary beam energies (20 kV).

The imaging procedure is simple in plan; operate under conditions at or close to 100% relative humidity. These conditions are achieved by using water vapor as the imaging gas and by operating at pressure and temperature conditions that lie anywhere on the water saturation curve (*see* **Fig. 7**). The usual practice is to operate with the sample close to, but always above, the freezing point. This allows the minimum gas pressure to be used and thus to minimize

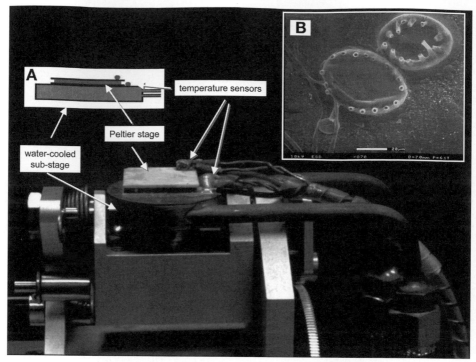

Fig. 6. An *in situ* photograph and operational schematic, insert (**A**), of a simple home-made Peltier cooling stage used for imaging of fully-hydrated samples in an ESEM. The stage has a 25 × 25-mm surface area and allows large samples to be examined. The example image, insert (**B**), shows fully hydrated diatoms with intact processes *in situ* on the surface of an algal strand of a modern stromatolite from Hamelin Bay, Western Australia.

the primary beam scatter. The limiting pressure is the equilibrium vapor pressure over ice at 0°C is 600 Pa (4.5 torr). Typically water vapor pressures around 700 to 800 Pa (~5–6 torr) and specimen temperatures around 3°C are used.

2.3.1. Practical Procedure

1. Precool the sample on the stage to usually approx 10 to 12°C (remember that cooling is from the base and thick samples will have some degree of thermal inertia).
2. Place several droplets of water around the sample (these are sacrificial and if conditions are above equilibrium then evaporation from these droplets will cool the sample down to equilibrium).
3. Evacuate the sample chamber ("flush" or inject the sample chamber with water vapor as soon as allowed by the vacuum system—this will exchange the air in the sample chamber for water vapor before the sample dehydrates. Remember the equilibrium conditions relate to the water vapour pressure only).

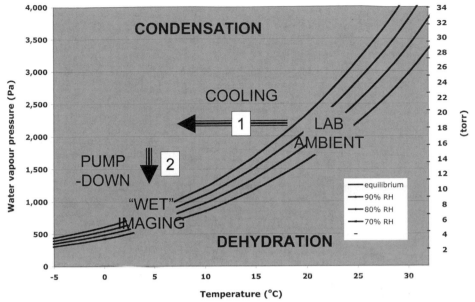

Fig. 7. The water saturation curve and relative humidity P-T regimes (source data from http://hyperphysics.phy-astr.gsu.edu/hbase/kinetic/watvap.html#c1). The sample hydration management from laboratory ambient through to ESEM imaging conditions is superimposed on the water saturation curve.

4. Use the *gas pressure* control to fine tune the conditions (pressure changes are rapid whereas the temperature control is governed by the cooling rate of the stage and sample thickness and thermal characteristics).

The ideal path of changing conditions from laboratory ambient to stabilized 'wet' imaging conditions is shown in **Fig. 7**.

2.3.2. Optimizing Image Quality Use

1. The smallest final aperture size (this restricts the field of view but it protects the column vacuum and minimizes beam scatter in the second pressure region behind this aperture).
2. The shortest possible working distance. (A simple rule of thumb is that if a gap is visible between your sample and the GSED, via a ccd-based chamberscope or viewing port, then you are too far away!)
3. A primary beam current that gives an acceptable image (usually this will be significantly greater than "normal" for your microscope because a high proportion of the primary beam is scattered under these conditions and, therefore, to achieve conventional SNR levels, these higher primary beam currents are required. The beam resolution will degrade in conventional SEM columns but this should be offset by the use of the short working distance).

4. An accelerating voltage approx 5 to 10 kV. (The lower end is preferable to maximize surface detail in the GSE images. At 5 kV, the primary beam scattering is very severe and operating at the shortest possible working distance and high beam current are essential.)

All ESEM with cooling stages will provide high-quality images of "wet" samples once a minimal level of operator skill is achieved. With care, samples can be introduced under a thin film of water than can then be evaporated away to expose the surface. This technique is useful for extremely sensitive samples but risks freezing the sample because of the cooling effect of the evaporation. The rule here is to proceed very slowly, that is, reduce the pressure by only 10 to 20 Pa from equilibrium and then be patient. Never adjust the temperature via the cooling stage unless you are certain that it is incorrectly set as the response is slow and surface damage (dehydration) can occur very quickly.

2.4. Novel Contrasts in Biological Samples

In VPSEM GSEI of various poorly or nonconductive samples, a novel contrast relating to charge behavior in the sample, termed charge contrast imaging (CCI), has been observed *(16)*. The contrasts relate to lattice-scale properties of the sample that affects charge transfer and electron emission in the near-surface region of the sample. The CCI images have been found to relate closely to CL imaging in many material samples. In some instances, the sample data evident by CCI is unique *(17)*. Additional novel GSEI contrasts have been observed. One example is the patterning seen on a termite leg that evidences either a thin secretion or bacterial presence (*see* **Fig. 8**). The contrast in this example appears similar to those seen and discussed for fluid mixtures, for example, oil-water emulsions *(18)*. These novel contrasts are seen in GSEI under specific operating conditions that relate to the sample and instrument. CCI is dynamic and they are visible only within a 'window' of operating conditions. Outside of these conditions the GSEI contain conventional SE contrast data *(16)*.

Recently, it has been discovered that CCI is observed in biological tissues that also exhibit autofluorescence and that this phenomena is particularly strong in some corals *(19)*. This exciting development also emphasizes the relatively undeveloped state of VPSEM and ESEM imaging of biological samples. Always keep "odd" images because they will almost certainly have some useful information—you just haven't realized it yet.

3. Variable Pressure (VPSEM) Applications: Robust Samples

Any biological sample that has a relatively robust outer surface can be immediately imaged by VPSEM. Suitable samples include most insects, pollen and, surprisingly, most plant samples. The reason most samples survive is that the VPSEM sample chamber is at pressures approx 133 Pa whereas the SEM sample

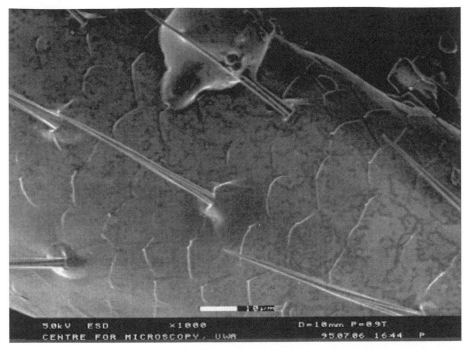

Fig. 8. A gaseous secondary electron image of an "as-collected" termite leg. The dark patterning is thought to be a coating or a secretion. The contrast revealing its presence is a result by variable charge behavior on the surface (charge contrast).

chamber is at approx 10^{-3} Pa, some five orders of magnitude difference. The simple rule is to test a "sacrificial" sample. Most of these samples will dehydrate but, commonly, this will occur over a time scale long enough to permit imaging. Similarly, any samples that have been air dried or otherwise prepared are immediately suitable.

One area of great VPSEM application is of imaging biological samples with complex three-dimensional structures. It is almost impossible to apply a continuous conductive coating to this type of sample and charging during SEM is generally inevitable. This type of sample can be immediately imaged using GSE and often are imaged most usefully using a mixed signal of GSE with BSE. Methylmethacrylate casts of vascular systems and of rare and delicate open ocean zooplankton samples are two excellent examples of this application (*see* **Fig. 9**).

The imaging of immunogold-labelled biomedical samples is also an application where VPSEM excels (*see* **Fig. 10**). The normal procedure is for samples to be labeled normally with a colloidal-gold conjugate and then exposed to a silver stain and finally air or critical-point dried. The silver attaches and enhances the nanometre-sized colloidal gold to form grains of a size readily visible by

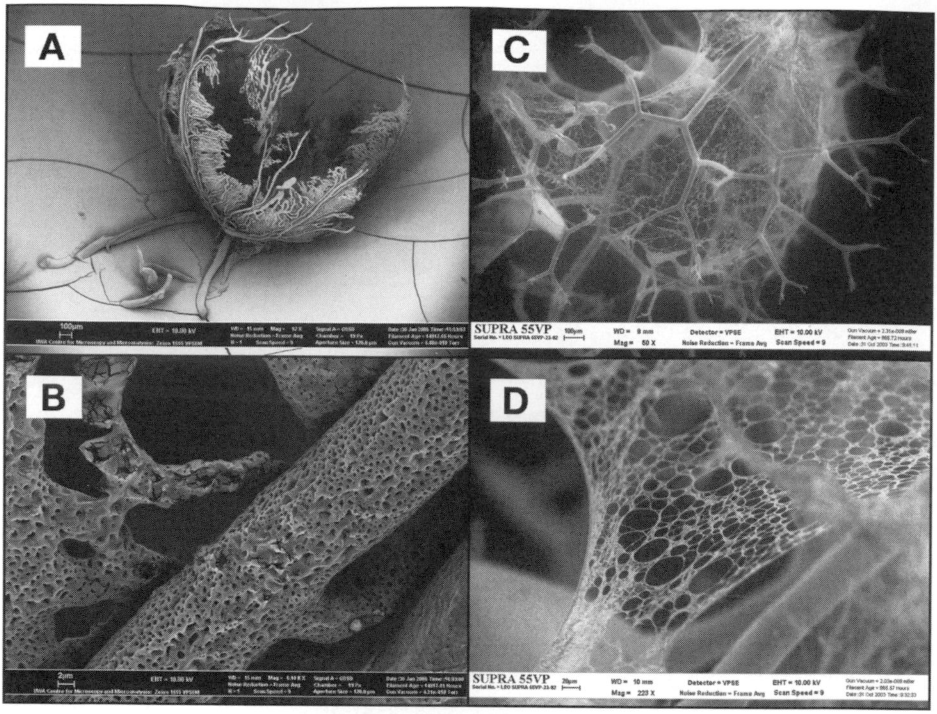

Fig. 9. Variable pressure scanning electron microscopy images of complexly structured samples. Images **(A)** and **(B)** are back-scattered electron images of a methylmethacrylate choroidal vascular cast of a vascular endothelial growth factor transgenic mouse. The sample has been gold coated, but this step failed to provide image stability in normal SEM. (Sample courtesy of Dr. L. Tee and Professor S. Dunlop). Images **(C)** and **(D)** are gaseous secondary electron images of an air-dried, uncoated zooplankton sample. (Sample courtesy of H. Paterson).

BSEI in VPSEM *(20)*. The preparation is very simple and, importantly, is attractive because there is no solution-based processing required after the colloidal gold labeling. Expressions of the labeling can also be investigated over several microns depth in the sample by imaging with moderate to high (15–30 kV) primary beam energies, as required. The new high resolution field emission-sourced VPSEM have the resolution potential to directly image the nanometre-scale primary gold labeling and this is being investigated at present as it would avoid

---

Fig. 10. *(Opposite page)* Variable pressure scanning electron microscopy back-scattered electron images at low **(A)** and high **(B)** magnification illustrating the distribution of macrophages labeled with the colloidal-gold conjugate ED1 monoclonal antibody and Ag-enhanced. The sample is the developing lens and tunica vasculo lentis transient vascular network from a 10-day postnatal Wistar rat. (Courtesy of P. McMenamin).

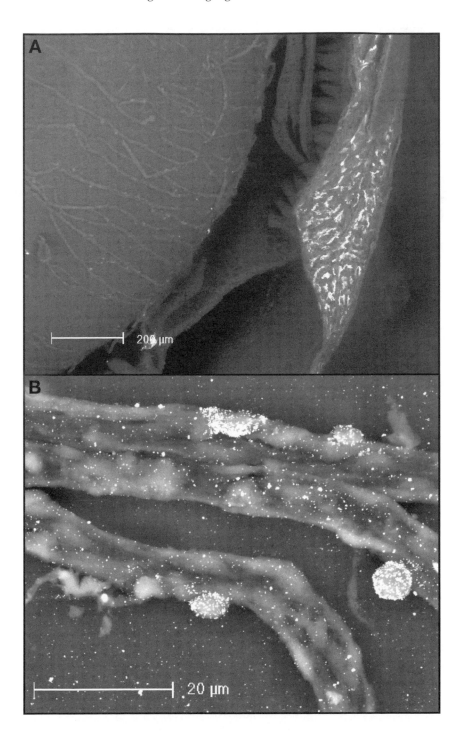

the background contamination that is seen to occur during the Ag-enhancement treatment.

3.1. Low-Magnification Imaging

The lower pressure operation of VPSEM mean that there is little or no restriction on the final aperture and large fields of view can be achieved, perhaps contrary to expectation. As in normal SEM, the field of view is sensitive to both working distance and accelerating voltage, the latter affecting the degree of deflection the scan coils can achieve. By way of example, at long working distance and low accelerating voltage, a 9-mm wide field of view can be obtained in the ElectroScan E-3 ESEM in VPSEM mode.

The GSED used for VPSEM and large field of view imaging may be positioned off-axis, as is the case with the FEI Large Field Detector and the Zeiss VPSE Detector. This off-axis positioning will provide a degree of directional sensitivity or shadowing that can enhance image detail.

At long working distance, both lower gas pressure and lower GSED bias can be used with a similar degree of charge cancellation. A practical summary of optimal imaging conditions can expressed as the product (V. Robinson, personal communication):

$$Pressure \times Working\ Distance \times GSED\ bias = an\ imaging\ constant$$

This simple "equation" is a guide to varying the different parameters and retaining similar signal quality.

A further unappreciated benefit of VPSEM is that the dynamic pressure regime allows the imaging of samples that outgas significantly. The chamber pressure sensors simply measure the total chamber pressure and any out-gassing from the sample is included within this measurement. This allows large samples, for instance, rodent skulls or human jawbones, to be examined intact. The only limitation is effectively if the chamber door will close.

3.2. High-Resolution and Low-Voltage Imaging

The greatest challenge to imaging is as the primary beam energy is reduced the probability of primary electron scattering increases significantly. A simple interactive modelling program is available to graphically illustrate this problem *(21)* and the difficulty of operating under low primary beam energy conditions (that suitable imaging of biological samples) is clearly illustrated in **Fig. 11** *(21)*.

The current practical limit for low voltage VPSEM operation is 1 kV and such images have been obtained from a number of instruments using the shortest possible gas path length/working distance, high beam current and very slow scan rates (because of very low SNR). Sub-1 kV images have been obtained from various experimental instruments down to 0.5 kV under VPSEM condi-

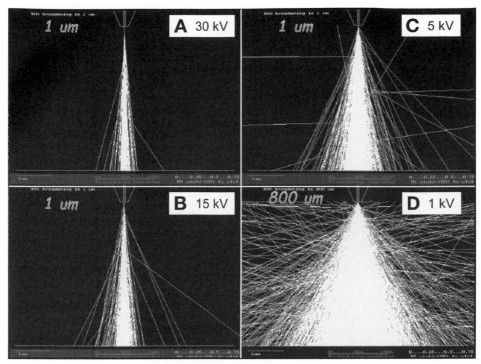

Fig. 11. Monte Carlo modeling *(21)* showing the scattering of the primary electron beam as a function of accelerating voltages from 30 kV to 1 kV with 270-Pa gas chamber pressure and 10-mm gas path length (working distance). The model illustrates the severe scattering effects encountered less than 5 kV.

tions (*see* **Fig. 12**). Routine operation around 3-5 kV provides an acceptable balance between image SNR and depth of information in the images of biological samples.

It is important to note that the primary beam current will drop with beam energy under constant spot size (condenser lens) conditions in conventional SEM (VPSEM and ESEM) columns. For example, at 5 kV, the primary electron beam current is generally only 20% of that at 20 kV. At low beam energies, the spot size must usually be increased to maintain the acceptable SNR in the image. A Faraday cage or even the sample current can be used to monitor this variation. The exceptions to this behavior are the (VP)SEM columns, where the primary electron beam is extracted at a constant energy and then decelerated to the required value at or near the pole piece of the column.

The best image resolution obtainable in VPSEM is typical a factor of two to three times worse than that achievable in SEM. The reasons are primary beam scattering and lower efficiencies of GSED relative to the current generation in-lens

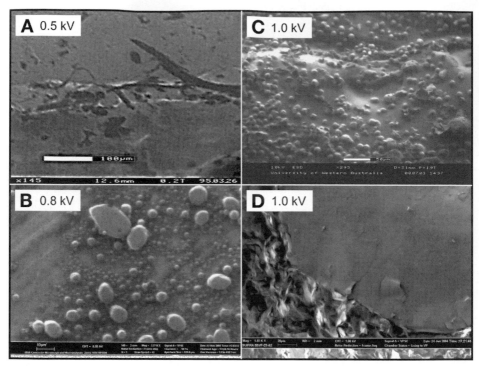

Fig. 12. Examples of very low accelerating voltage images (0.5–1.0 kV) obtained from variable pressure scanning electron microscopy and environmental scanning electron microscopy. Image (**A**) at 0.5 kV is of the edge of carbon tape on an aluminum stub. The true working distance was 2.0 mm with approx 30 Pa (H$_2$O), the instrument data bar was incorrect under these conditions. Image (**B**) at 0.8 kV is of tin spheroids at 10 Pa (air) and 2.0 mm; image (**C**) is of iron microspheres on carbon tape at 250 Pa (H$_2$O) and 3.1 mm; image (**D**) is of a mica (mineral) flake in a siltstone at 5 Pa (air) and 2.0 mm.

SE detectors. Actual image resolution is more a function of the sample and the primary beam energy. On suitable materials samples, resolutions of several nm at 5 kV have been achieved *(2,22)*.

In practice, a diverse range of biological samples have been successfully imaged. A selection of biological applications is shown in the composite **Fig. 13** by way of example. These images, and those from the other figures in this chapter, are a guide to what is achievable.

3.3. Sample Presentation

There are almost no limitations as to how biological samples can be presented for VPSEM examination. A common practice is to mount samples on SEM stubs using a double-sided conductive tape. Some users prefer to have a conductive 'drain' near to the area to be imaged on large, poorly conductive samples, for example, rodent skulls. The regular tapes will still "collapse" under the VPSEM conditions giving a pitted background and the carbon tabs should be used where a smooth background is required. It is worth noting that the use of SEM stubs is just a traditional practice and it was based on a need for a conductive (metal) mount. Today it is simply convenient as storage units are based on the SEM stub.

Loose or irregular samples can be presented with some form of simple containment, for example, a Petri dish. Larger samples can be sat directly on the stage or substage (*see* **Fig. 13A**). There is a risk that small and/or light samples could be lost in the initial turbulence associated with the evacuation of the chamber. A second risk is present when a very short working distance is used as the vacuum effect of the lower pressure electron column may siphon the sample up through the column or vacuum system, usually with unfortunate results. It is distressing to watch a researcher realize that a precious sample has been consumed by the VPSEM.

Surface cleaning of samples should also be considered in the context of the research and to any image database these VPSEM images may be compared with. The sample preparation techniques used for conventional SEM usually includes exposure to hydrocarbons. The hydrocarbon strips oils and waxes from the sample and, in effect, cleans the sample surface. The consequence is that many samples examined "as collected" by VPSEM do appear significantly different to their conventionally prepared counterparts. A typical example of this problem is with flower anthers and pollen. In "fresh" samples a liquid nectar component is usually present. In VPSEM of such pollen the surfaces appear smooth due to the nectar coating. The nectar has a high sugar content and the consequently low vapour pressure means that the nectar is relatively stable in a vacuum, even to the point of surviving sputter coating. An uncoated–coated pollen comparison is shown in **Fig. 14**. In this figure the residual nectar can be seen to obscure and "smooth" the surface detail of the pollen, even in the coated example. This can be a serious problem if VPSEM images are to be compared to an existing database collected by conventional SEM. The surface preparation aspect needs to be considered on a case-by-case basis. Samples for VPSEM can be processed as per routine SEM protocols prior to imaging for consistency of results.

Many insects are remarkably resilient and can tolerate the VPSEM conditions for significant periods of time. This represents a preparation challenge and our experience is that five minutes in liquid nitrogen is the most effective process to humanely slow them down without damage.

3.4. Optimizing Image Conditions

The project research aims dictate the selection of operating conditions and signal selection. General wisdom suggests low beam energy conditions should

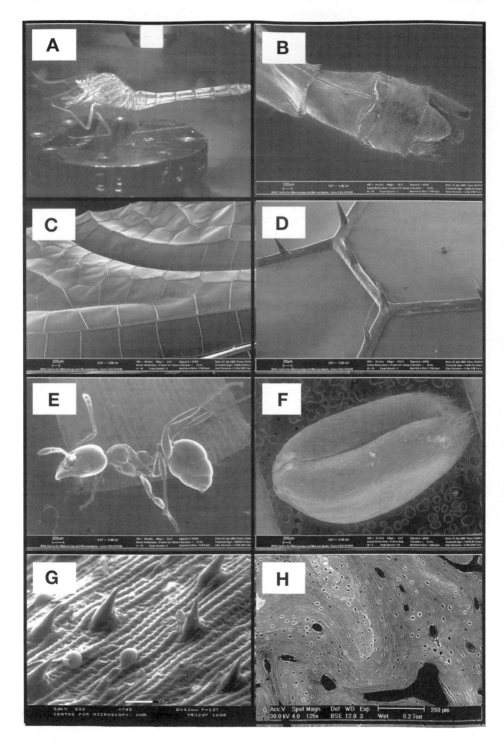

be used to capture surface detail from biological (low atomic number element-rich) samples BUT this requirement is in reality a function of the scale of imaging required. At low magnification (approx ×400 on-screen) the interaction volume of higher beam energies (15 kV) is within the pixel size (approx 1 μm) and so this protocol becomes less significant (*see* **Fig. 15**) although the edge-highlighting effects present in SEI may wish to be avoided. Under these conditions, BSEI may be more useful to image these types of samples (*see* **Fig. 9A**). At higher magnifications low primary beam energies must be used for biological sample imaging (*see* **Figs. 13C,D**) and as noted earlier the current practical lower limit for VPSEM is 3 to 5 kV.

In practice and, surprisingly, sample charging is actually the most common problem in VPSEM when it comes to recording images. The problem arises as at the scan rates slow enough to provide an acceptable image quality (high SNR) the ion drift is not fast enough to cancel the increased electron charge per unit area of the sample resulting from the slower scan rate. The detailed basis for this charging is discussed in detail elsewhere *(11)*. Increasing gas pressure may help but will reduce contrast and may introduce other artifacts. The best solution to minimize this problem is to use a faster scan rate (to the level where charging is not evident) combined with the digital integration or averaging of a number of scans. A simple rule of thumb is to use a combination of scan rate and number of scans that in total gives a usual image capture time, for example, 60 s. The actual value is very dependent on signal level, and it is important to stress that image SNR should be the controlling factor. Where charge cancellation of the whole sample is ineffective then the image will drift leading to a smearing of the integrated image. In this case chamber gas pressure should be adjusted, usually upwards, carefully and in minimal (approx 5 Pa) increments.

Fig. 13. *(Opposite page)* Variable pressure scanning electron microscopy examples. Images **(A)** to **(D)** are of a large (48-mm body length and 73-mm wingspan) dragonfly. The specimen was air dried and attached to the stage using a strip of double-sided adhesive tape **(A)**. The image of the tail of the insect **(B)** shows no charging artifacts despite its extreme projection into free space. Images **(C)** and **(D)** show the fine surface detail of the wing. The images were collected at 5 kV, 34 mm working distance and 10-15 Pa (air). Images were integrated over 91 scans with a 0.3-s scan rate to avoid charging. The ant **(E)** and wheat grain **(F)** were imaged at longer working distances (42–50 mm), by removing the sample stage assembly, to allow a larger field of view to be obtained. Image **(G)** is of silica-rich spikes on the upper surface of a bamboo leaf at short working distance (4.2 mm) and higher gas pressure (200 Pa H_2O). Image **(H)** shows the detailed structure of a cow bone sample. The image was collected by using a back-scattered electron detector at 30 kV and moderate working distance (12.9 mm) using 28 Pa H_2O.

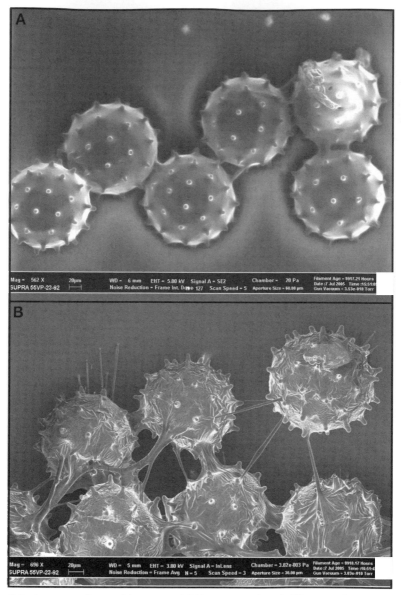

Fig. 14. A comparison of images of **(A)** uncoated and **(B)** coated nectar-covered pollen collected by variable pressure scanning electron microscopy and scanning electron microscopy, respectively. The variable pressure scanning electron microscopy gaseous secondary electron imaging at 5kV with 20 Pa (air) shows no surface structure due to the surface nectar layer which can be seen joining the grains. The in-lens SEI at 3 kV in normal vacuum again shows the presence of the nectar together with a crinkled surface texture on the pollen grains. This texture is not a dehydration effect of the grain but of the nectar layer and is an artifact of the preparation.

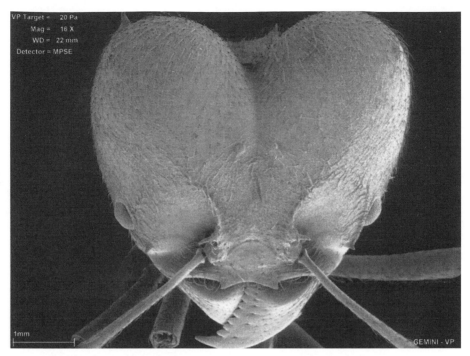

Fig. 15. A low-magnification mixed gaseous secondary electron image/back-scattered electron image of a large ant. The horizontal field of view is 7.3 mm. The image was collected at 20 kV, 20 Pa (air) and using a working distance of 22 mm.

Figure 16 is a compilation of images exhibiting charge-related 'drift' and charging effects and the optimized images.

A final point to emphasize is that VPSEM GSE and even BSE images typically lack the usual "strong" contrasts present in conventional SEM images. Primary electron beam scatter, sample charging, poor detector efficiencies and the lack of a metal (e.g. Au, Au/Pd) surface coating contribute to this problem. It is usual practice to use image manipulation software to optimize image contrast and brightness after collection. Noise reduction and edge-enhancing "unsharpen mask" filters also are very useful as image enhancement tools. *Remember* that unedited (raw) images should always be archived.

4. Environmental or Extended Pressure (ESEM) Applications

The procedures for imaging fully hydrated samples by ESEM have been described in **Subheading 2.3.** In practice a diverse range of biological samples have been successfully imaged, from live diatoms (*see* **Fig. 6B**) to cyanobacteria (*see* **Fig. 17**). A selection of further examples is presented in the composite **Fig. 18**.

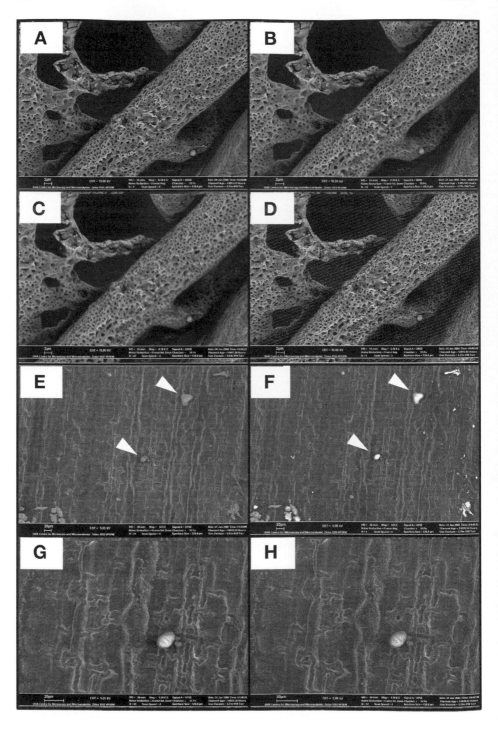

In practice, to record the best possible images under "wet" conditions, the pressure can be reduced down to a level in which the relative humidity is approx 70% to 80% *(23)*. The reduced scatter of the primary electron beam will improve the SNR and contrast in the image. All other aspects of the image setup, focus etc., should be performed under the stable "wet" conditions because the changing pressure has no influence on the electron beam optics. The higher pressure detectors provide good SNR *(24)* but have restricted (instrument specific) availability.

5. Summary

VPSEM and ESEM are important techniques for imaging biological samples. The techniques provide access to samples in their pristine state and with minimal risk of alteration. Operation requires a higher level of user skill/experimentation to develop optimum imaging protocols than with conventional SEM. The protocols are robust and reproducible once established. The variation in GSED detector designs results in a range of efficiencies but the VPSEM and ESEM operating protocols are universal. The expectation of imaging should always be tempered by awareness of a need for experimentation on new samples to establish optimum conditions.

Fig. 16. *(Opposite page)* A comparison of images collected using different scan rates and numbers of integrated scans. Series **(A)** to **(D)** are back-scattered electron images collected at 10 kV, 15 mm working distance, 19 Pa (air) and an on-screen magnification of 6100X. Images were collected for (A) 1 scan at 20.2 s, (B) 17 scans at 1.3 s = 22.1 s, (C) 62 scans at 0.33 s = 20.5 s, and (D) 1 scan at 0.33 s. The whole sample was charging and the resultant drift is evident in a defocusing of images **(b)** and **(c)**. The poor signal-to-noise ratio at the 0.33-s scan rate (D) is not completely eliminated even by integrating 62 scans (C). Image (A) is optimum in this case. Series **(E)** to **(H)** are lower-magnification images of a bamboo leaf surface. Small pollen grains are present. In this case the integrated image at collected at an intermediate scan rate (**E**, 21 scans at 1.3 s) avoids the charging artifacts present in the single 20.2-s slow scan (**F**). However, at increased magnification, the charging artifacts reappear even at a slightly faster scan rate (**G**, 61 scans at 0.67 s). The charging can be minimized by reducing the number of scans (**H**, 32 scans at 0.67 s).

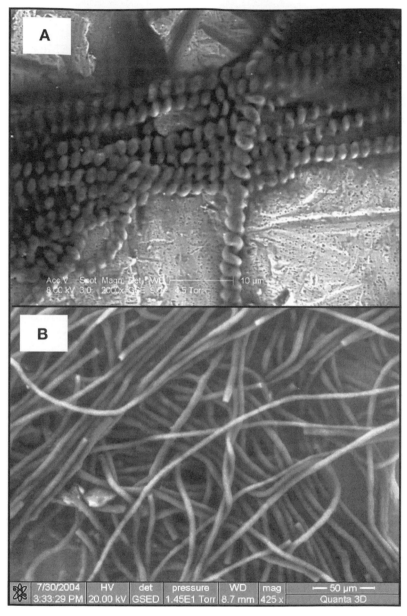

Fig. 17. ESEM images showing live cyanobacteria **(A)** *Spirulina* partially surrounded by
buffer solution. Water vapor pressure = 600 Pa (4.5 torr), temperature = 3°C. Although these
represent conditions at less than 100% relative humidity (relative humidity approx 80%), they
are adequate for maintaining the specimen in a hydrated state. **(B)** is the cyanobacteria *Leo* on
agar gel under relatively ambient conditions using a high pressure detector. Water vapor pressure
= 1920 Pa (14.5 torr), temperature = 24°C, giving a relative humidity of approx 75%. (Images
courtesy D. J. Stokes).

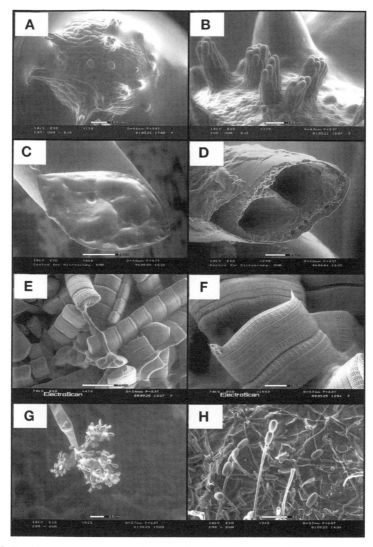

Fig. 18. Environmental scanning electron microscopy images of a range of biological samples. Images (**A**) and (**B**) are of orchid seedlings grown in a nutrient-rich agar gel. The agar was sliced into blocks and placed directly on the Peltier stage shown earlier. Water surrounds the seedling in image (A). Images collected at 10 kV, 4.6 mm working distance, and 720–780 Pa H₂O at 5°C. (Samples courtesy of K. Dixon). Images (**C**) and (**D**) are of a water-filled and empty brachial filament of a Western Australian marron, respectively. Imaging conditions as stated previously. The sample was sectioned under water on the laboratory bench and introduced in the environmental scanning electron microscope under a water layer (**c**) that was then evaporated to revel the structure (D). (Samples courtesy of Ms. P. Lindhjem). Images (**E**) and (**F**) are of diatoms imaged at 7 kV. The soft-tissue processes are clearly visible in (**e**), and the finer detail of the shells is evident in (F). Images (**G**) and (**H**) are of the grape mold *Botrytis nobilus* on a grape vine leaf. (Samples courtesy of J. Considine).

References

1. Robinson, V. N. E. and Robinson, B. W. (1978) Materials characterisation in a scanning electron microscope environmental cell. *Scanning Electron Microsc.* **1,** 595–602.
2. Danilatos, G. C. (1990) Theory of gaseous detector device in the environmental scanning electron microscope. *Adv. Electronics Electron Phys.* **71,** 1–102.
3. Peters, K. R. (1992) Principles of low vacuum scanning electron microscopy. *Proc. 50th Ann. Meeting Electron Microsc. Soc. Am.* (Bailey, G.W., Bentley, J., and Small, J.A., ed.), pp. 1304–1305.
4. Danilatos, G. D. and Robinson, V. N. E. (1979) Principles of scanning electron microscopy at high pressures. *Scanning* **2,** 72–82.
5. Farley, A. N. and Shah, J. S. (1988) A new technique for high pressure SEM. *Inst. Phys. Conf. Ser.* **93,** 241–242.
6. Knowles, W. R., Thiel, B. L., Toth, M., et al. (2004) Design of a two-stage gas amplification secondary electron detector for imaging insulating samples at the sub-1 nm scale. *Microsc. Microanal.* **10,** 1060–1061.
7. Thiel, B. T., Toth, M., and Knowles, W. R. (2005) Ultra-high resolution and metrology with Low Vacuum SEM. *Microsc. Microanal.* **11,** 384CD.
8. Fletcher A., Thiel, B. L., and Donald, A. M. (1997) Amplification measurements of alternative imaging gases in ESEM. *J. Phys. D: Appl. Phys.* **30,** 2249–2257.
9. Thiel, B. L., Bache, I. C., Fletcher, A. L., Meredith, P., and Donald, A. M. (1997) An improved model for gaseous amplification in the environmental SEM. *J. Microsc. (Oxford)* **187,** 143–157.
10. Phillips, M. R. and Morgan, S. W. (2005) Direct comparison of various gaseous secondary electron detectors in the variable pressure scanning electron microscope. *Microsc. Microanal.* **11,** 398–399.
11. Thiel, B. L. and Toth, M. (2005) Secondary electron contrasts in low-vacuum/environmental scanning electron microscopy of dielectrics. *J. Appl. Physics* **97,** 1–18.
12. Craven, J. P., Baker, F. S., Thiel, B. L., and Donald, A. M. (2002) Consequences of positive ions upon imaging in low vacuum scanning electron microscopy. *J. Microsc. (Oxford)* **205,** 96–105.
13. Stokes, D. J., Baker, F. S., and Toth, M. (2004) Raising the pressure: realizing room temperature/high humidity applications in ESEM. *Microsc. Microanal,* **10,** 1074–1075.
14. Griffin, B. J. and Browne, J. R. (2000) Fibre-optic based spectral cathodoluminescence: simple and economic option for use in convectional and environmental scanning electron microscopy. *Microsc. Microanal.* **6,** 42–48.
15. Griffin, B. J., van Riessen, A., and Egerton-Warburton, L. (1995) A review of detection strategies and imaging of hydrated biological specimens in the environmental SEM. *Scanning* **17,** 58–59.
16. Griffin, B. J. (2000) Charge contrast imaging of material growth and defects in Environmental Scanning Electron Microscopy—linking electron emission and cathodoluminescence. *Scanning* **22,** 234–242.

17. Roach, G. I. D., Cornell, J. B., and Griffin, B. J. (1998) Gibbsite growth history—revelations of a new scanning electron microscope technique, in *Light Metals 1998* (Welch, B., ed.), TMS Publications, pp. 153–158.
18. Stokes, D. J., Thiel, B. L., and Donald, A. M. (1998) Direct observation of water-oil emulsion systems in the liquid state by environmental SEM. *Langmuir* **14,** 4402–4408.
19. Clode, P. L. and Griffin, B. J. (2004) Charge contrast imaging of the soft tissue and mineralized skeletal phases in a scleractinian coral. *Microsc. Microanal.* **10,** 1078–1079.
20. McMenamin, P. G., Djano, J., Wealthall, R., and Griffin, B. J. (2002) Characterization of the macrophages associated with the tunica vasculosa lentis of the rat eye. *Invest. Ophthalmol. Visual Sci.* **43,** 2076–2082.
21. Joy, D. C. (1995) *Monte Carlo Modeling for Microscopy and Microanalysis.* Oxford University Press, New York.
22. Griffin, B. J. (2005) Challenges and progress towards low voltage imaging in VPSEM. *Microsc. Microanal.* **11,** 396–397.
23. Stokes, D. J. (2003) Recent advances in electron imaging, image interpretation and applications: environmental scanning electron microscopy. *Phil. Trans. R. Soc. Lond. A.,* **361,** 2771–2787.
24. Stokes, D. J., Baker, F. S., and Toth, M. (2004) Raising the pressure: realizing room temperature/high humidity applications in ESEM. *Microsc. Microanal.* **10,** 1074–1075.

24

Cryoplaning Technique for Visualizing the Distribution of Water in Woody Tissues by Cryoscanning Electron Microscopy

Yasuhiro Utsumi and Yuzou Sano

Summary

The protocol of cryoplaning techniques that to examine the distribution of water in living tree stems by cryoscanning electron microscopy have been developed and described. In brief, the procedures are as follows: First, a portion of transpiring stem is frozen in the standing state with liquid nitrogen to stabilize the water that is present in the conducting tissue. After filling with liquid nitrogen, discs are then collected from the frozen portion of the stem and stored in liquid nitrogen. The surface of disc is cleanly cut using a sliding microtome in a low temperature room at −20°C. Finally, the frozen sample is examined in a cryoscanning electron microscope after freeze-etching and metal coating.

Key Words: Cryoscanning electron microscopy; water transport; water distribution; cryoplaning; visualization.

1. Introduction

Cryoscanning electron microscopy (cryo-SEM) has been applied to examine the frozen hydrated plant tissues *(1–4)*. It is possible to observe the state of cytoplasm the cellular level using this method without chemical fixation and dehydration. This technique also enables the visualization of water distribution in the plant tissues. The presence or absence of water in the lumina of dead cells is clearly distinguished in each cell (*see* **Fig. 1**).

To observe the dynamic state of a plant tissue by cryo-SEM, samples usually are prepared by cryo-planing (*see* **Fig. 2A**) or freeze-fracturing (*see* **Fig. 2B** and also *[1,2,5–7]*). Freeze-fracturing is simple and allows the cellular ultra-structure of various plant tissues to be revealed *(8–12)*. However, freeze-fracturing is not adequate for examining the lumina of axial elements such as vessels

From: *Methods in Molecular Biology, vol. 369*
Electron Microscopy: Methods and Protocols, Second Edition
Edited by: J. Kuo © Humana Press Inc., Totowa, NJ

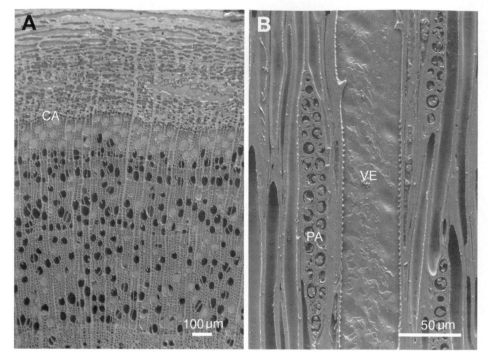

Fig. 1. Cryoscanning electron micrographs showing a cross-section of secondary xylem and phloem of *Cornus sericea* (**A**) and a tangential-section of secondary xylem of *Acer mono* var. *marmoratum* f. *dissectum* (**B**). CA, cambial zone; VE, vessel; PA, parenchyma.

and tracheids because it is difficult to cleanly expose a cross section of secondary xylem in a tree sample (*see* **Fig. 2B**). Although a longitudinal section can be cleanly exposed by freeze-fracture, the frequency of exposure of lumina of axial elements is low because the fracture tends to occur between cell walls of the adjacent cells. In addition, ragged surfaces often cause a substantial margin of error in quantitative morphological analyses. A series of suitable surfaces from the same specimen is usually impossible to obtain *(7,13)*.

In contrast, the cryoplaning method is useful for examining the state of lumina of axial elements *(13)*. With this method, 2- to 3-mm thick sections are cut away from the fractural surface of specimens with a steel blade on a sliding microtome under frozen conditions. It is possible to cleanly expose a wider range of surface of sample blocks (*see* **Fig. 2A**) and allows a desired smooth face of hard woody tissue to be cleanly exposed (*see* **Fig. 1** and also *[6,14–16]*).

This chapter describes a cryoplaning method that we devised. The method enables reproducible observations of the distribution and state of water in the

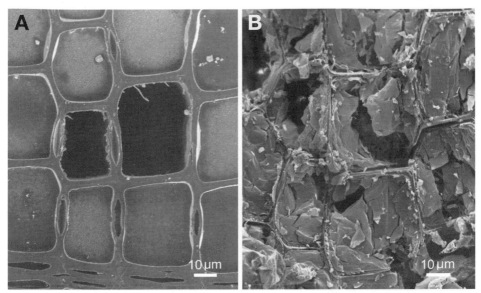

Fig. 2. Cryoscanning electron micrographs of cross section of secondary xylem in *Abies sachalinensis* exposed by cryoplaning (**A**) and freeze-fracture (**B**).

living xylem of woody plant stems *(2–4,6)*. In the text that follows, we will explain not only the sample preparation, but also the whole process from the sampling procedure in the transpiring woody stems to cryo-SEM observation.

2. Materials

2.1. Freeze Stabilization and Collection of Samples

1. Polyethylene funnel.
2. Steel blade.
3. Sticky (vinyl or duct) tape.
4. Fat clay.
5. Small Dewar flask.
6. Large liquid nitrogen (LN$_2$) Dewar flask.
7. Handsaw.
8. Cryo storage system.

2.2. Cryoplaning

1. Low-temperature room, maintained at −20°C.
2. Sliding microtome.
3. Steel blade.
4. Disposable blade for sliding microtome.
5. Glycerol.

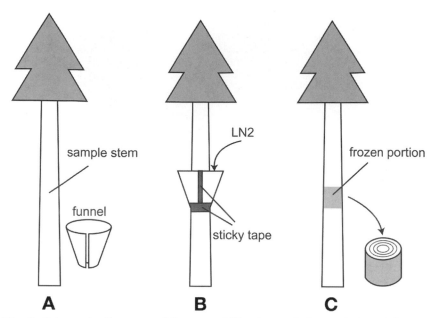

Fig. 3. Schematic diagrams of freeze stabilization techniques designed for woody plant stems.

6. LN_2.
7. Sample holder.
8. Dewar flask.

2.3. Cryoscanning Electron Microscopy

1. Freeze-etching unit.
2. LN_2.
3. Sample holder.
4. Dewar flask.
5. Heat insulation box.
6. Platinum-carbon pellet.
7. Cryoscanning electron microscope (JSM-840A, JEOL [1]).

3. Methods

3.1. Freeze Stabilization and Collection of Samples

1. Cut a polyethylene funnel with a steel knife or a pair of scissors used for metal-working so that the diameter of the narrower-side hole of the cut funnel becomes slightly larger than that of a sample stem around which the cut funnel is fitted (*see* **Fig. 3A**).

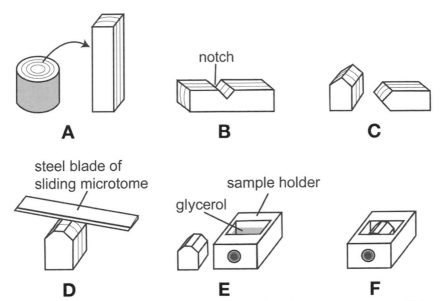

Fig. 4. Schematic diagrams of cryoplaning under a low temperature condition.

2. Wind the cut funnel around the sample stem and fix with waterproof sticky tape (*see* **Fig. 3B**). Tightly seal the gap between the funnel and the stem, and between the straight edges of the funnel to prevent leakage (*see* **Note 1**).
3. In addition, pack the gap between the funnel and stem with plastic material such as fat clay.
4. Run LN_2 from an LN_2 Dewar flask into a small Dewar flask.
5. Fill the collar made from the funnel with LN_2 (*see* **Fig. 3B**) and freeze the sample stem (*see* **Note 2**). Add LN_2 to the collar steadily to avoid leakage. The sample soaking time in LN_2 vary with the sample size (*see* **Note 3**).
6. Remove the collar after the freeze-stabilization treatment and cut out 2–3 cm thick discs with a hand saw from the frozen portion of the stem (*see* **Fig. 3C**).
7. Immerse the disc sample in the Dewar flask with LN_2 and then store in a cryo-storage system with LN_2, until the following treatment.

3.2. Cryoplaning in the Low-Temperature Room

1. Place the frozen sample in a low temperature room and to equilibrate at −20°C. The time of equilibration depends on the sample size (*see* **Notes 4–6**).
2. Cut the disc sample into rectangular parallelepipeds with the steel blade (*see* **Fig. 4A** and also **Note 7**).
3. Make notches in the sample block carefully (*see* **Fig. 4B**) to avoid compression and deformation of tissue by the steel blade.

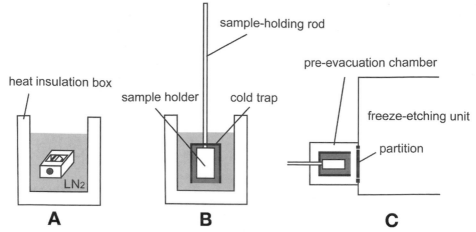

Fig. 5. Schematic diagrams of setting a sample in a freeze-etching unit.

4. Trim the sample into a small block ($5 \times 5 \times 5$ mm^3, *see* **Note 8**) with the steel blade (*see* **Fig. 4C**).
5. Plane the sample block surface cleanly with the steel blade of a sliding microtome to expose cell lumina (*see* **Fig. 4D**). The cutting strokes should be made slowly. Remove several 10- to 20-µm thick sections to clear away damaged tissue from the trimming process, and finally cut away sections of 3- to 5-µm thickness to expose a smooth surface on the specimen block.
6. Attach the specimen to a cubic hole of the sample holders with a drop of glycerol (*see* **Fig. 4E** and also **Note 8**). Keep the surface of the specimen horizontal and adjust to the upper face of the sample holder as close as possible (*see* **Fig. 4F**).
7. Move the sample holder with the specimen to the Dewar flask with LN$_2$ and transfer to a cryo-SEM system.

3.3. Cryoscanning Electron Microscopy

1. Keep a freeze-etching unit under a vacuum of approximately 1×10^{-4} Pa and equilibrate at –95°C with LN$_2$ and the heater in the cold stage (*see* **Note 9**).
2. Transfer the sample holder with the specimen from the Dewar flask to a heat insulation box with LN$_2$ (*see* **Fig. 5A**).
3. Screw a sample-holding rod into the sample holder and cover the holder with a cold trap (metal cylinder) in the heat insulation box (*see* **Fig. 5B**).
4. Transfer the sample holder in the state of being covered by the cold trap into the pre-evacuation chamber of the freeze-etching unit and evacuate (*see* **Fig. 5C**).
5. Open the partition between the chamber and the freeze-etching unit and set the sample holder onto the cold stage.
6. Freeze-etch the surface of specimens for 15 min to eliminate contamination by frost and to reveal cell outlines clearly (*see* **Note 10**).

7. Rotary-shadow the surface of the specimen with a platinum-carbon pellet.
8. Allow the cold stage of a cryo-SEM to cool (to approx −160°C in the apparatus that we used).
9. Transfer the specimen to the cold stage of the cryo-SEM from the freeze-etching unit.
10. Observe the secondary electron images and record at an accelerating voltage of 5 kV (*see* **Note 10**).

4. Notes

1. If the sample stem has thick bark, the outer bark should be removed with a steel blade before fitting the funnel.
2. Water in the vessels and tracheids of transpiring vascular plants has a potentiality of artificial redistribution because leaf transpiration lowers the water potential *(17)*. Water in the conducting tissue would be replaced by air if the conducting tissue is cut and exposed to air under conditions in which the water potential in the conducting tissue is below atmospheric pressure *(18)*. To avoid such artificial redistribution of water as much as possible, some researchers have frozen a portion of plant tissue before sample being cut *(3,19–21)* and has provided valuable information about water distribution in various types of tissues in many different plant species *(4,14,16,22,23)*. However, the possibility of artificial cavitations during freeze stabilization in low water potential conditions has been discussed *(24,25)*. When this technique is applied to samples at extremely low water potentials, attention should be paid to the interpretation of the results of cryo-SEM.
3. In cases of the tree stems, the sample of 8 cm in diameter was completely frozen within 15 min and the sample of 2 cm in diameter was frozen with in 5 min, respectively *(3,15)*.
4. The following procedures should be made to ensure that the sample surface is free of frost. The authors usually allow 15 min for equilibration in a 1-cm³ sample block with a cryomicrotome in low temperature condition.
5. In general, the low temperature room requires to be dehumidified ensure that machines are operated trouble-free. In the case of chamber of the box-type cryomicrotome, it is strongly recommended to maintain dehumidified condition. If the sample for cryo-SEM is prepared in an ordinary low temperature room or a conventional box-type cryomicrotome, water in the sample probably will be etched away quickly. In a low temperature room or box for the sample preparation, humidity should be maintained at a high level by such as placing ices surround the sample before the preparation.
6. Although a room temperature of −20°C is adequate to prepare clean sample surfaces, it is not low enough to prevent the recrystallization of ice *(26)*. Planing at −80°C by cryoultramicrotomy *(7,13)* would be more suitable for observations of very minute structures in sample tissues.
7. The use of a handsaw is not recommended, because if the sample is cut with a handsaw, fragments and debris of the ice and cell wall materials would plug some cavitated vessels to a depth of a few millimeters from the cutting surface.

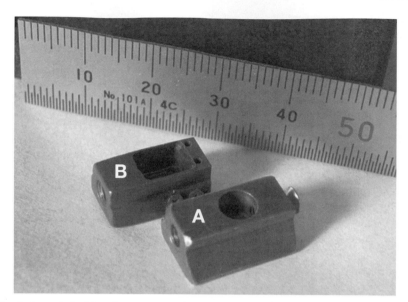

Fig. 6. Sample holder for cryoscanning electron microscopy observation. (**A**) Commercially sold holder with circular hole. (**B**) Custom-made holder with cubic hole.

8. A normal sample holder that is commercially sold has a circular hole to hold a plant specimen (*see* **Fig. 6A**). However, it requires more skill and is time-consuming to trim a specimen into a cylinder than into a cube. Thus, we made sample holders that have a cubic hole (*see* **Figs. 4, 6B**). This type of holder enables a wide range of observations at one time. There is a need to modify the holder to match the purpose and shape of the specimen *(7,13)*. The other types of sample holders are shown in Nijsse and van Aelst *(13)*.

9. Etching conditions for cryo-SEM observation depend on the performance of the freeze-etching unit. If the etching unit achieves higher vacuum conditions, lower temperatures are desirable for etching. A preliminary experiment to determine suitable etching conditions is important.

10. To control the precise etching amount, monitoring the degree of etching under uncoated conditions is useful *(7,22)*. First, the uncoated specimen surface is watched continuously in cryo-SEM at a low accelerating voltage such as 1 kV *(7)* whereas the sample stage is warmed and the specimen etched to an adequate condition. The specimen is then rapidly recooled and transferred back to the freeze-etching unit. Finally, the specimen is coated and returned to the sample stage of cryo-SEM for observation.

Acknowledgments

We wish to thank Dr. Seizo Fujikawa for explanations of the basics of cryo-SEM; Dr. Ryo Funada and the late Dr. Jun Ohtani for their helpful comments;

Dr. Takayuki Shiraiwa, Dr. Keita Arakawa, Dr. Daisuke Takezawa, and the late Dr. Takuya Fukuzawa for use of the low-temperature room of the Institute of Low Temperature Science of Hokkaido University; and Mr. Toshiaki Itoh, Mr. Kunio Shinbori, and Mr. Shigeo Kon'no for their technical support.

References

1. Fujikawa, S., Suzuki, T., Ishikawa, T., Sakurai, S., and Hasegawa, Y. (1988) Continuous observation of frozen biological materials with cryo-scanning electron microscope and freeze-replica by a new cryo-system. *J. Electron Microsc.* **37**, 315–322.
2. Sano, Y., Fujikawa, S., and Fukazawa, K. (1993) studies on mechanisms of frost crack formation in tree trunks. *Jap. J. Freezing Drying* **39**, 13–21.
3. Utsumi, Y., Sano, Y., Ohtani, J., and Fujikawa, S. (1996) Seasonal changes in the distribution of water in the outer growth rings of *Fraxinus mandshurica* var. *japonica*: a study by cryo-scanning electron microscopy. *IAWA J.* **17**, 113–124.
4. Utsumi, Y., Sano, Y., Funada, R., Ohtani, J., and Fujikawa, S. (2003) Seasonal and perennial changes in the distribution of water in the sapwood of conifers in a subfrigid zone. *Plant Physiol.* **131**, 1826–1833.
5. Ohtani, J. and Fujikawa, S. (1990) Cryo-SEM observation on vessel lumina of a living tree: *Ulmus davidiana* var. *japonica. IAWA Bull. n. s.* **11**, 183–194.
6. Sano, Y., Fujikawa, S., and Fukazawa, K. (1995) Detection and features of wetwood in *Quercus mongolica* var. *grosseserrata. Trees* **9**, 261–268.
7. Huang, C. X., Canny, M. J., Oates, K., and McCully, M. E. (1994) Planing frozen hydrated plant specimens for SEM observation and EDX microanalysis. *Microsc. Res. Tech.* **28**, 67–74.
8. Fujikawa, S., Suzuki, T., and Sakurai, S. (1990) Use of micromanipulator for continuous observation of frozen samples by cryoscanning electron microscopy and freeze replicas. *Scanning* **12**, 99–106.
9. Fujikawa, S., Kuroda, K., and Fukazawa, K. (1994) Ultrastructural study of deep supercooling of xylem ray parenchyma cells from *Styrax obassia. Micron* **25**, 241–252.
10. Fujikawa, S., Kuroda, K., and Ohtani, J. (1996) Seasonal changes in the low-temperature behavior of xylem ray parenchyma cells in Red Osier Dogwood (*Cornus sericea* L.) with respect to extracellular freezing and supercooling. *Micron* **27**, 181–191.
11. Kuroda, K., Ohtani, J., and Fujikawa, S. (1997) Supercooling of xylem ray parenchyma cells in tropical and subtropical hardwood species. *Trees* **12**, 97–106.
12. Kuroda, K., Kasuga, J., Arakawa, K., and Fujikawa, S. (2003) Xylem ray parenchyma cells in arboreal hardwood species respond to subfreezing temperatures by deep supercooling that is accompanied by incomplete desiccation. *Plant Physiol.* **131**, 736–744.
13. Nijsse, J. P. and van Aelst, A. (1999) Cryo-planing for cryo-scanning electron microscopy. *Scanning* **21**, 372–378.

14. Utsumi, Y., Sano, Y., Fujikawa, S., Funada, R., and Ohtani, J. (1998) Visualization of cavitated vessels in winter and refilled vessels in spring in diffuse-porous trees by cryo-scanning electron microscopy. *Plant Physiol.* **117,** 1463–1471.

15. Utsumi, Y., Sano, Y., Funada, R., Fujikawa, S., and Ohtani, J. (1999) The progression of cavitation in earlywood vessels of *Fraxinus mandshurica* var. *japonica* during freezing and thawing. *Plant Physiol.* **121,** 897–904.

16. Sakamoto, Y. and Sano, Y. (2000) Inhibition of water conductivity caused by watermark disease in S*alix sachalinensis. IAWA J.* **21,** 49–60.

17. Dixon, H. H. and Joly, J. (1894) On the ascent of sap. *Ann. Bot.* **8,** 468–470.

18. Zimmermann, M. H. and Brown, C. L. (1971) *Trees, Structure and Function.* Springer-Verlag, Berlin.

19. Canny, M. J. (1997) Vessel contents of leaves after excision—A test of Scholander's assumption. *Am. J. Bot.* **84,** 1217–1222.

20. Canny, M. J. (1997) Vessel contents during transpiration—embolisms and refilling. *Am. J. Bot.* **84,** 1223–1230.

21. McCully, M. E., Huang, C. X., and Ling, L. E. C. (1998) Daily embolism and refilling of xylem vessels in the roots of field-grown maize. *New Phytol.* **138,** 327–342.

22. Buchard, C., McCully, M., and Canny, M. (1999) Daily embolism and refilling of root xylem vessels in three dicotyledonous crop plants. *Agronomie* **19,** 97–106.

23. Sano, Y., Okamura, Y., and Utsumi, Y. (2005) Visualizing water-conduction pathways of living trees: selection of dyes and tissue preparation methods. *Tree Physiol.* **25,** 269–275.

24. Cochard, H., Bodet, C., Ameglio, T., and Cruiziat, P. (2000) Cryo-scanning electron microscopy observations of vessel content during transpiration in walnut petioles. Facts or artifacts? *Plant Physiol.* **124,** 1191–1202.

25. Canny, M. J., McCully, M. E., and Huang, C. X. (2001) Cryo-scanning electron microscopy observations of vessel content during transpiration in walnut petioles. Facts or artefacts? *Plant Physiol. Biochem.* **39,** 555–563.

26. Willson, J. H. M. and Rowe, A. J. (1980) *Replica, Shadowing and Freeze-etching Technique.* North-Holland Pub. Co., Amsterdam.

25

X-Ray Microanalysis
in the Scanning Electron Microscope

Godfried M. Roomans and Anca Dragomir

Summary

X-ray microanalysis conducted using the scanning electron microscope is a technique that allows the determination of chemical elements in bulk or semithick specimens. The lowest concentration of an element that can be detected is in the order of a few mmol/kg or a few hundred parts per million, and the smallest amount is in the order of 10^{-18} g. The spatial resolution of the analysis depends on the thickness of the specimen. For biological specimen analysis, care must be taken to prevent displacement/loss of the element of interest (usually ions). Protocols are presented for the processing of frozen-hydrated and freeze-dried specimens, as well as for the analysis of small volumes of fluid, cell cultures and other specimens. Aspects of qualitative and quantitative analysis are covered, including limitations of the technique.

Key Words: X-ray microanalysis; scanning electron microscopy; peak-over-background ratio; frozen-hydrated specimen; freeze-dried specimen; Sephadex beads; cell culture; airway surface liquid.

1. Introduction

X-ray microanalysis is a technique of elemental analysis that is conducted in an electron microscope. In this chapter, only X-ray microanalysis conducted in a scanning electron microscope will be discussed. The technique is based on the generation of characteristic X-rays in atoms of the specimen by the incident beam electrons (*see* **Fig. 1**). When an incident electron hits atoms in the specimen, it may cause an inner shell ionization, that is, the incident electron transfers so much energy to an electron in one of the inner shells of the atom that this electron leaves its shell and the atom. This results in a "gap" in the shell where this electron originally was present. This situation is an unstable one, which may be remedied by an electron transition, in which an electron

From: *Methods in Molecular Biology, vol. 369*
Electron Microscopy: Methods and Protocols, Second Edition
Edited by: J. Kuo © Humana Press Inc., Totowa, NJ

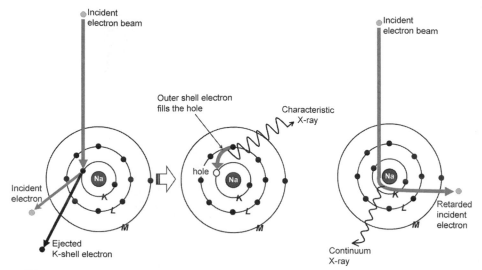

Fig. 1. Generation of characteristic and continuum X-rays in atoms of the specimen by incident electrons.

from a higher shell falls back to fill the gap. Because this electron has a higher energy than the electron originally removed, energy is liberated in this electron transition in the form of an X-ray. This X-ray has an energy that is equal to the energy difference between the shells between which the electron transition takes place. These X-rays are characteristic for the element from which they originate, and hence contain information on which elements are present in the specimen (and how much of each).

Because there are many electron shells and subshells, each with their partic-ular energy, there are, in principle, many potentially possible electron transi-tions. If the gap occurs in the K-shell, the resulting X-ray is called a K-line; if the gap occurs in the L-shell, the resulting X-ray is called an L-line, and so on. If the gap in the K-shell is filled by an electron coming from the L-shell, the resulting X-ray is called a K_α-line, if it is filled by an electron coming from the M-shell, the resulting X-ray is called a K_β-line. Not all electron transitions are equally probable; hence, the K_α-line of an element is about five times as intense as the K_β-line; some theoretically possible electron transitions do not occur at all. The K_β-line has a slightly higher energy than the K_α-line, since the M-shell has a slightly higher energy than the L-shell, and so E_M-E_K is slightly more than E_L-E_K. Therefore, the energy of the K_β-line of a particular element is higher than that of the K_α-line of the same element. The energy of an L-line from a par-ticular element is always lower than that of the K-lines from that element, but higher than the M-lines from the same element.

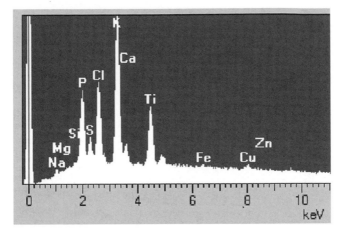

Fig. 2. Example of an X-ray spectrum with characteristic peaks and background.

In addition to characteristic X-rays, the beam electrons generate background or continuum X-rays, which are generated by the incident electron being retarded by the electric field of the nucleus of the atom. The energy loss that is suffered by the incident electron in this process is released in the form of a background or white X-ray (*see* **Fig. 1**). This white or background radiation does not contain any information about the chemical identity of the element from which this radiation originates, but the intensity of the radiation is related to the total mass of the analyzed volume in the specimen and is used in quantitative analysis *(1,2)*. A typical X-ray spectrum, showing characteristic peaks and background, is shown in **Fig. 2**. The lowest concentration of an element that can be detected is in the order of a few mmol/kg or a few hundred parts per million, and the smallest amount is in the order of 10^{-18} g. The spatial resolution of the analysis depends on the thickness of the specimen. The best spatial resolution is obtained in (thin) sections where as a rule of thumb the diameter of the analyzed volume is about half the section thickness. If bulk specimens are analyzed, as is done in the scanning electron microscope, the spatial resolution depends on the accelerating voltage and the composition of the specimen. A typical value for analysis of freeze-dried biological material at an accelerating voltage of 20 kV is approx 10 μm *(1,3)*.

The scanning electron microscope produces a narrow beam by demagnification in the condenser/final lens system of the beam leaving the filament. This beam can be directed to the feature of interest. However, as soon as the incident electrons hit the specimen surface, they collide with atoms in the specimen and are spread under a certain angle. Hence, an interaction volume is created from which X-rays may be generated. The size of this interaction volume depends

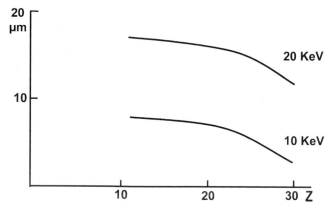

Fig. 3. Dependence of the interaction volume on the atomic number and the accelerating voltage.

on the density and composition of the specimen and on the energy of the incident electron (determined by the accelerating voltage) (*see* **Fig. 3**). X-rays may be generated throughout the interaction volume.

The scanning microscope also allows the visualization of the specimen to be analyzed. As will be discussed below, specimens for X-ray microanalysis are generally not coated with gold or impregnated with heavy metals, so the morphological information from the specimen is not optimal. However, at the level of resolution required, it is generally sufficient to coat the specimen with a conductive carbon layer, to avoid charging.

2. Materials

1. Scanning electron microscope equipped with an X-ray detection system and, optionally, with a cryo-chamber with cold stage and low temperature carbon-coating devices (*see* **Fig. 4** and **Note 1**).
2. Freezing device, liquid nitrogen, pressurized propane or ethane for rapid freezing the specimen (*see* **Note 2**).
3. Carbon evaporator and carbon rods (*see* **Note 3**).
4. Planchets of spectroscopically pure carbon for SEM.
5. Sephadex G-25 beads, diameter 20 to 40 µm (Pharmacia, Uppsala, Sweden).
6. Hydrophobic volatile silicone oil (Dow Corning 200/1cS, BDH, Poole, U.K.).
7. Nylon electron microscopy grids (Agar Scientific, Stansted, U.K.).
8. Sterile cellulose nitrate or cellulose nitrate/cellulose acetate filters, or another suitable substrate for cell culture.
9. Suitable software and standards for qualitative and quantitative analysis (*see* **Note 4**).
10. Double-sided tape (3M, Minneapolis, MN).
11. Filter paper (Whatman, Springfield Mills, U.K.)
12. Salt solutions, for example, NaCl or KCl, ranging from 50 to 250 mM.

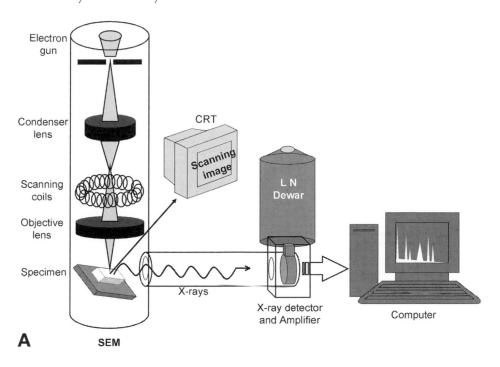

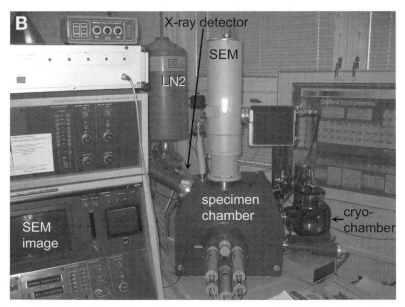

Fig. 4. (A) Schematic of the electron microscope with the X-ray detection system. **(B)** Photograph of the specimen chamber with X-ray detector.

13. Rinsing solutions, for example, 150 m*M* ammonium acetate or 300 m*M* mannitol in distilled water.

3. Methods

3.1. General Considerations

Most scanning electron microscopes operate at vacuum conditions; therefore, specimen preparation has to take this into account. In a conventional scanning electron microscope, specimens therefore have to be dried; alternatively, frozen-hydrated specimens can be used. It is, however, possible to conduct X-ray microanalysis in an environmental scanning electron microscope at atmospheric pressure *(4)*.

It should be realized that glutaraldehyde fixation will cause leakage of diffusible ions, and it has been shown that nearly all Na, K, Cl, Mg, and Ca are lost during such treatment. Hence, unless the interest is restricted to very firmly bound elements, cryotechniques have to be used in the preparation of specimens for X-ray microanalysis.

1. The first step of a cryotechnique is the rapid freezing of the specimen. Because analysis in the scanning electron microscope is not carried out at high resolution, requirements for freezing are somewhat less stringent than when X-ray microanalysis is carried out at high resolution in a transmission electron microscope. Similar freezing techniques can be used: freezing using a liquid coolant, or freezing against cold metal. As liquid coolants, liquid propane or ethane cooled by liquid nitrogen, or liquid nitrogen itself, can be used. Of these, liquid propane or ethane are to be preferred.

2. After the specimen has been frozen, several avenues can be chosen. It is possible to fracture the specimen and to analyze it in the frozen-hydrated state *(5)*; alternatively, after fracturing the specimen can be dried an analyzed in the freeze-dried state *(6)*. It is also possible to prepare thick cryosections of the specimen in a conventional cryostat; these cryosections can then be dried and analyzed *(3,7)*. A type of specimen that is suitable for analysis in the scanning electron microscope is cultured cells, that can be grown on a solid substrate, frozen and freeze-dried, and then analyzed *(8)*. Small amounts of fluid can be collected *in situ* by placing Sephadex beads into the fluid covering, for example, an epithelial surface *(9,10)*; these beads with the collected fluids are then dried and analyzed. This does not exhaust the types of specimen that can be analyzed in the scanning electron microscope: any type of specimen that can be viewed in a scanning electron microscope can be the subject of X-ray microanalysis, if the general rules given above are followed.

3.2. Cryofractured Frozen-Hydrated and Frozen-Dried Specimens

Analysis of frozen-hydrated bulk specimens has two advantages over analysis of frozen-dried specimens: (1) it allows analysis of large fluid-filled compartments and spaces, for instance, the lumina of glands and ducts, or fluid-filled spaces in the inner ear, and (2) it allows the elemental concentrations to be related to the wet weight of the specimen, which is physiologically more informative than the relation to the dry weight; by carrying out the analysis first in the frozen-hydrated state and then in the freeze-dried state, the ionic concentrations can be related to the water in the analyzed compartment, which physiologically speaking is the ultimately relevant parameter.

The procedure consists of the following steps:

1. Rapid freezing.
2. Transfer to the cold stage of the scanning electron microscope.
3. Fracturing to expose an internal surface.
4. Analysis at low temperature, typically close to −190°C.
5. **Step 4** may be followed by freeze-drying in the scanning electron microscope.
6. Analysis in the freeze-dried state (*see* **Fig. 5**).

As discussed previously, freezing should be conducted as rapidly as possible. Marshall *(5)* suggests freezing in melting nitrogen, but other methods, such as freezing in liquid propane or ethane, or against metal cooled by liquid nitrogen, should be adequate alternatives. After the freezing step, the specimen should be transferred to liquid nitrogen if it is not already in this medium. Fracturing at low temperatures (−170°C or less) provides, according to Marshall *(5)*, a rough-surfaced specimen with sufficient topographic detail. At higher temperatures, a smoother fracturing surface is obtained. Although this surface may have some advantages over a rough surface (a smooth surface will show less variation in X-ray intensity with topography), smooth surfaces provide less topographic information, and it is more difficult to localize features of interest on them. Frozen-hydrated specimens charge in the electron beam, but this problem can be reduced by coating the specimen with a conductive carbon layer. Most commercially available cold stages allow one to conduct coating with carbon at low temperature. Marshall *(5)* states that the analysis of these specimens should be performed with a rastered beam (preferably covering a size of 25 μm²), rather than with a stationary beam (*see* **Note 4**). An example of a frozen-hydrated specimen of rat trachea in cross section is illustrated in **Fig. 6**.

If one wishes to continue the analysis in the frozen-dried state, the specimen can be freeze-dried on the cold stage without breaking the vacuum, by raising the temperature. The specimen is then analyzed again in the freeze-dried state, and the water content of the specimens (the water fraction F_w) can be calculated from:

$$F_w = 1 - (R_h/R_d) \tag{1}$$

where R_h and R_d are the relative intensities (peak-to-background ratios) in the hydrated and the dried state, respectively, and, once the water fraction F_w is known, the elemental concentrations related to the cell water can be calculated from there *(2)*.

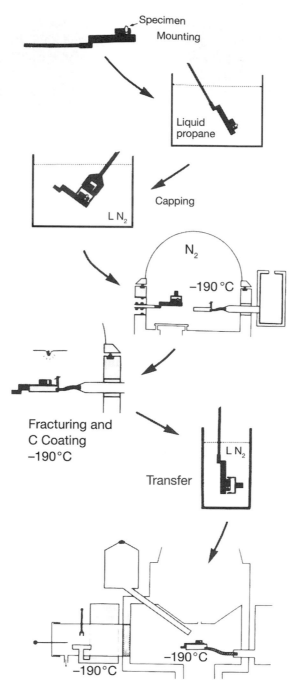

Fig. 5. Preparation steps for frozen-hydrated bulk specimens for X-ray microanaly-sis. After Marshall *(5)*.

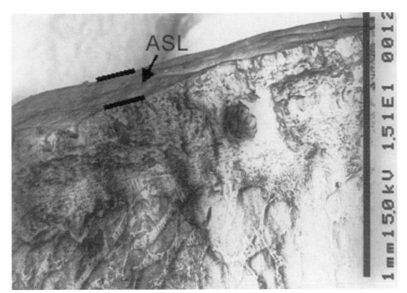

Fig. 6. Scanning electron micrograph of a thick cryo-section of pig trachea. The airway surface layer (ASL) is easily identifiable *(14)*.

3.3. Thick Freeze-Dried Cryosections

To simplify the quantitative procedure, the cryosections should be infinitely thick to the electron beam, i.e., the beam should not penetrate the section completely. Because a 20-kV beam penetrates freeze-dried biological soft tissue to approx 10 to 12 μm, sections of soft tissue to be analyzed at 20 kV should be at least 16-μm thick. Different values may apply if the analysis is conducted at a different accelerating voltage or if hard (calcified) tissue with a greater density is to be analyzed. The procedure is carried out as follows *(3,7,11)*:

1. The tissue is rapidly frozen and mounted on the specimen holder of a cryostat. The cryostat is set at a temperature of −20 to −30°C.
2. Cryosections (16-μm thick of soft tissue for analysis at 20 kV) are cut and mounted on a planchet of spectroscopically pure carbon (*see* **Note 5**).
3. Leave the carbon planchet for 48 h in the cryostat with a copper weight on top of the section. Then, remove the copper weight and store the planchet with the section in a desiccator over silica gel or other drying agent until analysis.
4. Coat the sample with carbon before analysis. Mount the carbon planchet onto the specimen stage of the scanning electron microscope and analyze at room temperature at 20 kV for 100 to 200 s.

An example of a scanning electron micrograph of a cryosection of freeze-dried biological tissue (human muscle) is shown in **Fig. 7.**

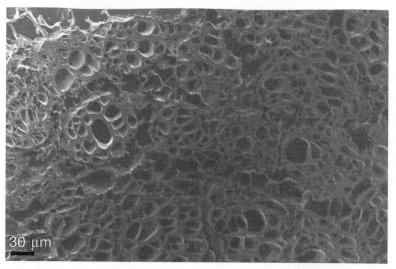

Fig. 7. Scanning electron micrograph of a thick cryosection of human muscle. The "holes" in the cells are due to ice crystals freezing.

3.4. Small Volumes of Fluid Collected on Sephadex Beads

A method that can be used to collect fluid from the surface of epithelia (e.g., uterine epithelium or airway epithelium in experimental animals or in humans) uses dextran (Sephadex G-25) beads (diameter 20–40 µm; *see* **Fig. 8**). The following procedure is used *(10,12–14)*:

1. The beads are equilibrated for 20 min with the fluid, for example, in the uterus or trachea by spreading the beads over the epithelial surface.
2. Then, the beads are collected, washed with hydrophobic volatile silicone oil to remove adhering fluid and debris.
3. Each bead is individually moved to a nylon electron microscopy grid, which had been submerged into the oil, until the grid contains at least 15 beads.
4. Alternatively, the Sephadex G-25 beads can be applied to double-sided tape attached to small pieces of filter paper (approx 1×1 mm^2).
5. The filter papers with the beads are placed onto the epithelium (beads facing downwards) for 20 min. Saturation of the beads with a salt solution is obtained after 5 min *(12)*.
6. Then, the filter paper with saturated beads is removed and washed with hydrophobic volatile silicone oil to remove adhering fluid and debris. Each bead is individually moved to a nylon electron microscopy grid as described above. The grid with beads is slowly lifted out of the oil bath and mounted onto an aluminum holder covered with carbon adhesive tape and left at room temperature for evaporation of the oil.

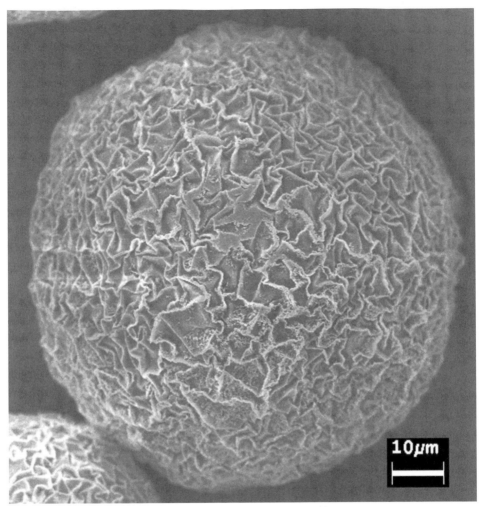

Fig. 8. Scanning electron micrograph of a Sephadex G-25 bead, used to collect fluid.

7. Nasal fluid from humans or laboratory animals can be collected on Sephadex beads as follows: The Sephadex G-25 beads are applied to double-sided tape attached to a filter paper (width 1-2 mm, length 5 mm). The filter paper with the beads is inserted carefully into one or both nostrils, with the beads facing the nasal septum, and kept there for 20 min. Then, the filter paper with saturated beads is removed from the nostril and washed in silicon oil as described above. With the help of a small needle, beads are individually transferred to nylon electron microscopy grids as described previously.

8. The grids with Sephadex beads are carbon-coated before analysis. X-ray micro-analysis of the beads is conducted at room temperature in the scanning electron

microscope, at 20 kV for 100 s with a beam size of 100 nm. Typically 12 to 14 beads are analyzed from each sample. For quantitative analysis, the data are compared with the results obtained on beads soaked in salt solutions of different concentrations (e.g., 50–250 mM).

3.5. Cultured Cells

A cell culture is a very useful type of specimen for X-ray microanalysis. Cell cultures have in general a number of advantages as an experimental system in biomedical research: they are relatively simple and avoid the problems associated with having to isolate tissue from an experimental animal or a human. Also, cell cultures can be exposed to a wider range of experimental conditions, compared with experimental animals and certainly compared with human tissue. Cultured cells can be analyzed in the scanning electron microscope, but also, if they are cultured on very thin substrates, such as ultrathin plastic films, in the transmission electron microscope *(7,15)*. Here, the discussion is limited to cultured cells that can be analyzed in the scanning electron microscope. These cells can be cultured on any substrate, but the simplest way is to culture them on a "thick" substrate such as a plastic Petri dish or plastic cover slips, or filters. Glass is not generally a suitable substrate, since glass contains several elements (e.g., Na) that may disturb the analysis. It has been shown that if X-ray microanalysis is used to analyze dynamic processes, e.g., chloride transport in cultured cells, the results are comparable with those obtained using dynamic techniques, such as fluorescent indicators for chloride *(16)*.

X-ray microanalysis of cells cultured on a solid substrate in the scanning electron microscope is conducted as follows *(7)*:

1. Cells are cultured in a medium appropriate for the particular cell type in plastic flasks in an incubator.
2. Harvest subconfluent cells with trypsin–ethylene diamine tetraacetic acid and seed the cells directly on sterile cellulose nitrate or cellulose nitrate/cellulose acetate filters, or another suitable substrate. Different types of solid substrates can be used. For the experimental approach it can be important whether the substrate is permeable or impermeable. With an impermeable substrate, a substance added to the experimental solution can only reach the apical membrane of the cells, whereas if the cells are cultured on a permeable substrate, also the basolateral membrane may be accessed. From a practical point of view, the substrate should be easy to handle, allow the growth of cells, should not show excessive charging in the electron microscope, and should not contain elements that disturb the analysis.
3. Place the filters in a Petri dish.
4. After the cells have attached to the filters, which takes approx 3 h, add 3 to 4 mL of complete culture medium and allow the cells to grow for 2 to 3 d before the experiment.

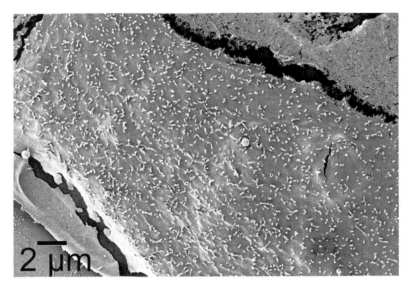

Fig. 9. Scanning electron micrograph of cultured human airway epithelial cells for X-ray microanalysis.

5. Check the cells using an inverted light microscope for growth and the possible presence of infection each time the medium is changed.
6. If the experiment consists of exposing the cells to physiological stimuli, this can be done in medium or in a physiological buffer (Ringer's solution), for the appropriate time.
7. Terminate the incubation by washing the filters respectively with one of the following washing fluids: (a) distilled water, (b) 150 mM ammonium acetate, or (c) 300 mM mannitol for 15 to 20 s only to remove the NaCl-rich experimental solution (*see* **Note 6**).
8. To wash the filters, the fluid is placed in a beaker on a magnetic stirrer, and the filter is dipped in the fluid, and held with forceps.
9. Take out the filter and remove excess fluid by blotting with a filter paper. Freeze the cells immediately in liquid nitrogen or liquid propane cooled by liquid nitrogen.
10. Freeze-dry the cells overnight.
11. Coat the dried filters with a conductive carbon layer to avoid charging in the electron microscope.
12. Store the filters in a vacuum container or over a drying agent if they are not analyzed immediately.
13. Mount the filters on a specimen holder (carbon is suitable) and analyze the cells on the filter in a scanning electron microscope. An accelerating voltage of 10 to 20 kV is appropriate. Acquisition times between 50 and 100 s are required. Only one spectrum should be acquired from each cell.

An example of a sample of cultured cells is shown in **Fig. 9**.

3.6. Other Specimens

In principle, any specimen that can be viewed in the scanning electron microscope can by analyzed. Examples are nail clippings *(17)* for the diagnosis of cystic fibrosis, and hairs *(18)*. The dried specimen is mounted, coated with a conductive carbon layer, and analyzed at an accelerating voltage that is appropriate with regard to its thickness, and the critical excitation energy of the elements of interest.

3.7. Qualitative Analysis

Qualitative analysis is restricted to answering the question, which elements are present in the specimen. Identification of a peak can be carried out with the software provided by the manufacturer of the analytical system and does generally not present special problems. It should be noted that before a peak can be identified as a minor peak for an element (e.g., a K_β-peak) the major peak for this element (i.e., the K_α-peak) should also be present in the spectrum. If a peak is identified as an L-line for a particular element, one should check whether the K-line for this element also can be seen (if the accelerating voltage is adequate to excite the K-line; this is, in practice the case if the accelerating voltage is 2 to 2.5 times the critical excitation energy). Since the X-ray detection process that underlies the spectrum is subject to statistical variations, there is always a certain amount of noise in the spectrum. This noise is governed by Poisson statistics, that is, for a peak to be an indicator of the presence of an element with 95% certainty, the peak should be at least two standard deviations above background, which in this case equals two times the square root of the background intensity.

3.8. Quantitative Analysis

From the point of view of quantitative analysis, the specimens analyzed in a SEM can be divided into two groups: (1) specimens on a substrate that are completely penetrated by the beam, and where the beam even excites the substrate ("semithick" specimens) and (2) specimens that are infinitely thick to the beam, that is, where the beam does not excite the underlying substrate ("thick specimens"). This latter type of specimens is simplest and will be discussed first.

3.8.1. Quantitative Analysis of Thick Specimens

When the electron beam hits the specimen surface, the electrons are spread in a teardrop-shaped volume within the specimen. X-rays may be generated throughout this volume. When incident electrons collide with atoms in the specimen, they change direction by elastic and inelastic collisions and lose energy by inelastic collisions, until they have lost (virtually all) their initial energy and come to a halt. The rate of energy loss and the penetration depth of the incident electron depend on the specimen composition and the energy of the incident electron. The property of the specimen with regard to slowing down of the incident electrons is called the stopping power of the specimen. Also, part of the electrons hitting the specimen will be backscattered from the specimen surface, and they will not penetrate the specimen and generate X-rays. Also this is dependent on the specimen composition and the energy of the incident electron. These two factors, the stopping power and the electron backscatter are mainly dependent on the (mean) atomic number of the specimen and, hence, they are called the atomic number (Z) factor.

Not all X-rays that are generated in the specimen are released from the specimen surface and can be detected by the detector. A portion of the generated X-ray is absorbed by atoms in the specimen. The amount of X-ray absorption is dependent on the specimen composition because different elements absorb X-ray from other elements to a different extent. This is called the absorption (A) factor.

When an X-ray is absorbed by an atom in the specimen, this may give rise to the generation of a secondary X-ray. Thus, an electron colliding with an element A in the specimen, may give rise to an X-ray for element A, that is then absorbed by element B, and gives rise to the generation of X-rays from element B, that are then detected. Hence, instead of X-rays from element A, the detector will see an X-ray from element B. This is called the secondary fluorescence (F) factor.

The correction that takes care of the atomic number, absorption and secondary fluorescence is called the ZAF correction. In principle, a comparison with a standard of known composition is carried out, and a correction is made for differences in atomic number factor, absorption factor and secondary fluorescence factor between specimen and standard. The ZAF correction factor has been developed well for metallurgical specimens, but is less well developed for biological specimens *(19,20)*. Hence, an alternative way for the quantitative analysis of biological specimens has been developed that is more closely related to the analysis of thin sections *(21)*. In this method, the P/B-method, one departs from the ration between the characteristic intensity (the peak intensity (P) and the background intensity (B) taken in the same energy range as the peak, the background under the peak. The reason for this is that of the three parts of the correction factor, the absorption correction A is the most important in the analysis of biological specimens. The atomic number correction Z can be minimized by an appropriate choice of standard, that is, the standard should resemble the specimen, and the secondary fluorescence correction is small in biological specimens. If the background is determined in the same energy region as the specimen, the background X-rays will be absorbed to about the same extent as the characteristic X-rays. There is a small difference, however, in that the background X-rays are generated at a slightly different depth than the characteristic

X-rays. Thus, they have a different way length through the specimen compared to the characteristic X-rays, and this may be noticeable in extreme cases, e.g., in specimens with a very uneven surface. However, this is the exception rather than the rule. For most specimens, the P/B-method provides an adequate simplification of the ZAF-correction and an adequate correction for the absorption effect.

The concentration for an element x in the specimen (sp) $C_{x,sp}$, is found from:

$$C_{x,sp} = C_{x,st} \frac{P/B_{sp} \, \overline{Z^2/A_{sp}}}{P/B_{st} \, \overline{Z^2/A_{st}}} \tag{2}$$

where $C_{x,st}$ is the concentration of element x in the standard (st), $(P/B)_{x,sp}$ is the P/B value of element x in the specimen, and $(P/B)_{x,st}$ is the P/B value of element x in the standard and $\overline{Z^2/A}$ is a correction factor calculated from the weighted values of Z^2/A for all elements present in the standard or specimen:

$$\overline{Z^2/A} = \Sigma C_x (Z^2/A)_x \tag{3}$$

where C_x is the concentration (as fraction) of element x and $(Z^2/A)_x$ is the value of the square of the atomic number divided by the atomic weight of element x.

In practice, the procedure is carried out as follows:

1. Analyze the specimen and determine for all elements of interest the P/B ratio (the ratio of peak to the background in the same energy range).
2. Analyze the standard and determine for all elements of interest the P/B ratio, or use stored values for the P/B ratios.
3. Calculate Z^2/A_{st} from the known composition of the standard.
4. Estimate a value for Z^2/A_{sp} and use this to calculate $C_{x,sp}$ using the known value of $C_{x,st}$. With the newly calculated value for $C_{x,sp}$ calculate a new estimate for Z^2/A_{sp} and use this to calculate a new $C_{x,sp}$. Repeat this procedure until the value of $C_{x,sp}$ in two subsequent calculations does not change more than, for example, 0.1% (**2**).

3.8.2. Semiquantitative Analysis of Semithick Specimens

If the electron beam excites the substrate, it is difficult and often impossible to calculate the elemental concentrations in the specimen. The best one can do often is to calculate relative elemental concentrations (relative to an internal standard). For the internal standard, an element occurring in a high concentration in the specimen should be chosen; often, the phosphorus (P) concentration can be used as an internal standard. The relative sensitivity factors for other elements can be determined as follows (**2**):

1. Make a solution of a salt with known molar elemental ratios, for instance, a solution of KH_2PO_4 will have a molar ratio of K to P of 1:1.

2. Place a small droplet (in the microliter range) of this salt solution on a carbon planchet and let it air-dry.
3. Analyze this droplet under the same conditions (accelerating voltage) as the specimen.
4. Determine the ratio of the (P/B)-values for the elements occurring in the salt (in this case P and K).
5. Express the relative concentration of (in this case) K to the "internal standard" P as a fraction or percentage.

Although methods theoretically can be devised to calculate the thickness of the semi-thick specimen, e.g., by having it on a substrate of an element that gives a measurable X-ray spectrum, such as an Al foil, and determining the thickness of the specimen by determining the quenching of the Al signal by the specimen, such methods are often inaccurate and do not give better results than semi-quantitative analysis.

3.9. Standards

As stated previously, it is preferable if the standard resembles the specimen in its physical and chemical properties. The reason for this is that the correction factors are not exact and parameters such as mass loss are difficult to determine. However, if specimen and standard resemble each other, corrections factors and mass loss will be similar in specimen and standard and errors will not have a great impact.

A suitable type of standard can be made up as follows *(2,22)*:

1. Dissolve the required amount of mineral salts in distilled water; for most elements, concentrations of 100 to 200 mM generally are adequate.
2. Add 20% gelatin to this solution. The gelatin should be free of salts; it may have to be dialyzed against distilled water to make sure that this is the case. Dissolve the gelatin by heating. The formation of ice crystals during freezing may be reduced by addition of a small amount (5%) of glycerol.
3. Let the gelatin solidify and cut out millimeter-size cubes and freeze these rapidly, for example, in melting nitrogen. Cut 16 μm-thick cryosections as described above and mount these on carbon planchets. Let the sections dry during 48 h in the cryostat. Coat with a conducting carbon film and analyze under the same conditions as the specimen.

3.10. Elemental Mapping

As a complement to the analysis made in individual point of interest, X-ray microanalysis instruments and software are capable of producing a map for one or more chemical elements of interest. This map is obtained by running the acquisition of X-ray spectra in the scanning mode and letting the software determine the concentration of the element of interest for each point. The map can be color-coded in order to indicate the absolute or relative concentration of the element of interest, giving thus a bidimensional image of the abundance and localization of the specific element. It is also possible to graphically overlap maps of two or more elements or to combine them with the scanning electron microscope image of the specimen in order to get information about the relative distribution of complementary or correlating elements.

It should be kept in mind that the spatial resolution of the elemental mapping is dependent both on the atomic number of the element of interest and on the accelerating voltage (*see* **Fig. 3**) and for biological bulk specimens one is limited to cellular level.

Instructions for acquiring the elemental maps and advanced features are software-specific and usually easy to follow. Elemental mapping is typically a time-consuming procedure (a map may take several hours to acquire) and this in combination with the mechanical instability of the specimen (specimen drift) has seriously limited the use of this procedure in biological X-ray microanalysis *(23)*.

4. Notes

1. The X-rays are detected by an X-ray detection system (*see* **Fig. 4**). The most common X-ray detection system in biological X-ray microanalysis is the lithium-drifted silicon semiconductor detector, in short the Si(Li)-detector. This detector consists of a silicon semiconductor, with lithium atoms interspersed at regular intervals between the silicon atoms. When an X-ray hits this detector, the energy of the X-ray is transferred to the material of the semiconductor, where it generates a small current (in the order of a few hundred to a few thousand electrons). The magnitude of this current is proportional to the energy of the X-ray impinging on the detector. The current is converted to a voltage, of which the magnitude depends on the magnitude of the current. The voltage is measured and sorted in a "bin" according to its magnitude; "bins" can have an energy width of 10 to 20 eV. On the monitor a spectrum is built up with time: each time an X-ray with a particular energy is detected, the "bin" corresponding to this energy is increased in size by one (*see* **Fig. 2**). Normally, it takes from one to several minutes to obtain a spectrum from a biological sample. A short time (in the order of microseconds) after the detector has been hit by an X-ray, it cannot register a second hit. This is called the detector "dead time." An automatic correction is applied for this dead time so that when the dead time is 10%, the counting time is automatically prolonged by 10% to obtain the correct "live time." Most detectors are separated from the vacuum of the microscope by a very thin beryllium window. This beryllium window has the disadvantage that it absorbs X-rays with a low energy. Therefore, the lightest element that can be detected by such a system is Na, and the sensitivity for lighter elements (Ca and lighter) is reduced. Although this is often acceptable, there are situations where this is a limitation, and therefore detectors with an ultrathin plas-

tic window or without a window at all have been developed. With these detectors, it is possible to detect also light elements, such as C, N, and O. The lithium-drifted silicon detector only functions well at low temperature, which is why this type of detector is usually cooled to liquid nitrogen temperature by a reservoir of liquid nitrogen. Alternative cooling methods, such as Peltier elements, are used in some circumstances. If the detector is not being used for some time, some manufacturers allow it to be warmed up to room temperature. The use of an older type of detector, the crystal spectrometer or wavelength-dispersive detector in biology is now rare, although this type of detector is still used in metallurgical or geological applications. The principle of this detector is that the X-rays emerging from the specimen are refracted by a crystal with parameters (lattice spacing, angle of incidence) such that only one wavelength (corresponding to a certain energy) is reflected by the crystal, whereas the X-rays with other wavelengths are extinguished. The X-rays that are reflected by the crystal are then counted by a proportional counter. This principle implies in practice, that only one element at a time can be detected by this technique, but elements can be detected sequentially by changing the settings of the detector, or by attaching multiple detectors to the microscope. The disadvantage of wavelength-dispersive detector systems for biological specimens is their low efficiency, so that the specimen has to be irradiated by an electron beam with a very high current. Organic specimens, such as biological specimens, do not withstand these high currents, whereas most inorganic specimens can be analyzed at high beam currents. The fact that wavelength-dispersive detectors are slightly more sensitive than semiconductor detectors has made that their use persists in the metallurgical and geological sciences, whereas the technique is rarely used with biological specimens *(24)*.

2. Liquid nitrogen is at its boiling point (−196°C) and when a "warm" specimen (at room or body temperature) is dropped in the liquid nitrogen, it will start to boil, thereby surrounding the specimen with an insulating gas layer ("Leidenfrost" phenomenon). If one wishes, this can be avoided by cooling the liquid nitrogen by adiabatic expansion (e.g., by placing a container with liquid nitrogen in vacuum, e.g., in a vacuum evaporator, until the nitrogen freezes, which happens at −214°C). If the specimens are small so that the liquid nitrogen is not warmed up to over its boiling point when the specimen is brought into the nitrogen, the insulating gas layer is not formed. Whether a liquid coolant or cooling against metal is used is a matter of personal preference. Liquid coolants make better contact with the specimen, which improves heat transfer, but metal is a better conductor of heat, and can therefore remove the specimen heat faster.

3. It should also be kept in mind that one should not chemically compromise the specimen by addition of elements. Hence, treatment with solutions of heavy metals (to decrease charging) or coating the specimen with a metal layer (Au, Pt, Cr) should be avoided. Coating the specimen with a thin C layer is permitted; in a conventional detector, C cannot be seen, and in a windowless detector, the amount of C added by the thin film is small in comparison to the C normally present in biological samples.

4. It should be pointed out that in analysis of frozen-hydrated samples, it is important to monitor the actual voltage of the electrons reaching the specimen surface (the "overvoltage"), since this determines the X-ray intensity; the relation between the X-ray intensity I and the overvoltage ratio U (the ratio between the effective accelerating voltage and the critical excitation energy for a particular element) is given by:

$$I = (U - 1)^{1.63} \qquad (4)$$

The actual voltage of the electrons hitting the specimen surface can be determined by measuring the intercept of the spectrum with the x-axis, that is, by determining above which energy no (background) X-rays occur in the spectrum. This is particularly important if quantitative analysis is to be conducted because then specimen and standard should be analyzed at identical overvoltage conditions.

5. This can be done by pressing the section with a piece of copper onto the carbon planchet; alternatively, the section can be "melted" carefully on the carbon planchet by slightly warming the planchet with the tip of a finger before the section is placed on the planchet *(25)*. During this procedure, it should be avoided that the section is completely melted; this can be ascertained from the color of the section after the freeze-drying step; if the section is white, only partial melting has taken place, but if the section is grey it has been melted through and through.

    Often sufficient topographic information can be obtained from the scanning electron microscopic image of the cryo-section (*see* **Fig. 7**). If this is not possible, a serial section can be mounted on a glass slide and stained for light microscopy *(15)*.

6. The washing fluid should have a temperature of approx 4°C. The rinsing procedure is possibly the most critical step in the preparation of whole cells for X-ray microanalysis. The procedure should be evaluated for each new cell type investigated. The rinsing procedure is necessary because salts in the culture medium or the experimental solution, if these were not completely removed, would after freeze-drying cover the surface of the cells and interfere with analysis. Distilled water has the advantage that it does not leave any remnants after freeze-drying. It may, however, cause osmotic shock, and induce efflux of ions either by general damage to the membrane or by specific ion transport mechanisms. Ammonium acetate is volatile, and most or all of this compound will disappear after freeze-drying, leaving little or no remnants. Although a 150 m*M* ammonium acetate solution is about iso-osmotic with the cytoplasm, ammonium acetate may cause pH changes in the cell and thereby induce ion transport. A 300 m*M* mannitol solution is isoosmotic and is not expected to cause changes in the ionic content of the cells. However, after freeze-drying the cells are covered with a thin layer of powdery mannitol. Some of this may be removed by careful brushing, but some will remain. This makes visualization of the cells more difficult and also adds a layer of organic substance to the specimen that "dilutes" the signal from the cell. This can be a problem in quantitative analysis. In our experience, some cell types withstand washing with ice-cold distilled water or ammonium acetate and in such a case these solutions are to be preferred. If the cells obviously lose ions after washing with water or

ammonium acetate, then mannitol or a substance with similar properties should be used. (We find sucrose, however, to be even more sticky than mannitol). The best way to make sure that the rinsing procedure does not introduce artifacts is to compare whole-mount specimens, rinsed in different washing fluids, with cryosections of the same cells, frozen without rinsing.

Acknowledgments

The technical assistance of Mr. Leif Ljung, Mr. Anders Ahlander, and Mrs. Marianne Ljungkvist is gratefully acknowledged, as is the contribution of our colleagues Dr. Romuald Wróblewski, and Mrs. Inna Kozlova, Mrs. Viengphet Vanthanouvong, and Mrs. Harriet Nilsson to various procedures summarized in this review. The studies reviewed in the paper were supported by the Swedish Research Council, the Cystic Fibrosis Research Trust, the Swedish Heart Lung Foundation, the Swedish Asthma and Allergy Association, and the Swedish Association for Cystic Fibrosis.

References

1. Roomans, G. M. (1988a) Introduction to X-ray microanalysis in biology. *J. Electron Microsc. Tech.* **9,** 3–18.
2. Roomans, G. M. (1988b) Quantitative X-ray microanalysis of biological specimens. *J. Electron Microsc. Techn.* **9,** 19–44.
3. Wróblewski, R., Roomans, G. M., Jansson, E., and Edström, L. (1978) Electron probe X-ray microanalysis of human muscle biopsies. *Histochemistry* **55,** 281–292.
4. Sigee, D. C. and Gilpin, C. (1994) X-ray microanalysis with the environmental scanning electron microscope: interpretation of data obtained under different atmospheric conditions. *Scanning Microsc.* Suppl **8,** 219–229.
5. Marshall, A. T. (1980) Quantitative X-ray microanalysis of frozen-hydrated bulk biological specimens. *Scanning Electron Microsc.* **1980II,** 335–348.
6. Zs.-Nagy, I. (1989) A review on the use of bulk specimen X-ray microanalysis in cancer research. *Scanning Microsc.* **3,** 473–482.
7. Wroblewski, J., Wróblewski, R., and Roomans, G. M. (1988) Low temperature techniques for microanalysis in pathology: alternatives to cryo-ultramicrotomy. *J. Electron Microsc. Techn.* **9,** 83–98.
8. Roomans, G. M. (2002) X-ray microanalysis of epithelial cells in culture, in *Methods in Molecular Biology, vol. 188, Epithelial Cell Culture Protocols* (Wise, C., ed.), Humana Press, Totowa, NJ, pp. 273–289.
9. Jin, Z. and Roomans, G. M. (1997) X-ray microanalysis of uterine secretion in the mouse. *J. Submicrosc. Cytol. Pathol.* **29,** 173–177.
10. Nilsson, H., Kozlova, I., Vanthanouvong, V., and Roomans, G. M. (2004) Collection and X-ray microanalysis of airway surface liquid in the mouse using ion exchange beads. *Micron* **35,** 701–705.

11. McMillan, E. B. and Roomans, G. M. (1990) Techniques for X-ray microanalysis of intestinal epithelium using bulk specimens. *Biomed. Res. (India)* **1,** 1–10.
12. Vanthanouvong, V. and Roomans, G. M. (2004) Methods for determining the composition of nasal fluid by X-ray microanalysis. *Microsc. Res. Tech.* **63,** 122–128.
13. Kozlova, I., Vanthanouvong, V., Almgren, B., Högman, M., and Roomans, G. M. (2005) Elemental composition of airway surface liquid in the pig determined by X-ray microanalysis. *Am. J. Resp. Cell Mol. Biol.* **32,** 59–64.
14. Vanthanouvong, V., Kozlova, I., Johannesson, M., Nääs, E., Nordvall, S. L., Dragomir, A., and Roomans, G. M. (2006) Composition of nasal airway surface liquid in cystic fibrosis and other airway diseases. *Microsc. Res. Tech.,* in press.
15. Hongpaisan, J., Mörk, A.-C., and Roomans, G. M. (1994) Use of in vitro systems for X-ray microanalysis. *Scanning Microsc.* Suppl. **8,** 109–116.
16. Andersson, C. and Roomans, G. M. (2002) Determination of chloride efflux by X-ray microanalysis versus MQAE fluorescence. *Microsc. Res. Tech.* **59,** 351–355.
17. Roomans, G. M., Afzelius, B. A., Kollberg, H., and Forslind, B. (1978) Electrolytes in nails analysed by X-ray microanalysis in electron microscopy. Considerations on a new method for the diagnosis of cystic fibrosis. *Acta Paediatr. Scand.* **67,** 89–94.
18. Roomans, G. M. and Forslind, B. (1980) Copper in green hair a quantitative investigation by electron probe X-ray microanalysis. *Ultrastruct. Pathol.* **1,** 301–307.
19. Boekestein, A., Stadhouders, A. M., Stols, A. L. H., and Roomans, G. M. (1983a) Quantitative biological X-ray microanalysis of bulk specimens: an analysis of inaccuracies involved in ZAF-correction. *Scanning Electron Microsc.* **1983/II,** 725–736.
20. Boekestein, A., Stadhouders, A. M., Stols, A. L. H., and Roomans, G. M. (1983b) A comparison of ZAF correction methods in quantitative X-ray microanalysis of light-element specimens. *Ultramicroscopy* **12,** 65–68.
21. Roomans, G. M. (1981) Quantitative electron probe X-ray microanalysis of biological bulk specimens. *Scanning Electron Microsc.* **1981II,** 344–356.
22. Roomans, G. M. and Sevéus, L. A. (1977) Preparation of thin cryo-sectioned standards for quantitative microprobe analysis. *J. Submicrosc. Cytol.* **9,** 31–35.
23. LeFurgey, A., Davilla, S. D., Kopf, D. A., Sommer, J. R., and Ingram, P. (1992) Real-time quantitative elemental analysis and mapping: microchemical imaging in cell physiology. *J. Microsc. (Oxford)* **165,** 191–223.
24. Pogorelov, A., Pogorelova, V., and Repin, N. V. (1994) Quantitative biological electron probe microanalysis with a wavelength dispersive spectrometer. *Scanning Microsc.* Suppl **8,** 101–108.
25. Wróblewski, R., Roomans, G. M., Ruusa, J., and Hedberg, B. (1979) Elemental analysis of histochemically defined cells in the earthworm *Lumbricus terrestris*. *Histochemistry* **61,** 167–176.

26

Botanical X-Ray Microanalysis in Cryoscanning Electron Microscopy

Beat Frey

Summary

Modern microscopy in plant sciences has evolved in the direction of providing ultra-structural and analytical information simultaneously. Energy-dispersive X-ray microanalysis (EDX) is a powerful technique that allows the qualitative and quantitative measurement of many elements of physiological interest at the cellular and subcellular level. The most significant advance has been the development of freezing techniques to study cells in plant tissues by EDX in the cryoscanning electron microscopy. Cryofixation is fast enough to retain the original distributions of inorganic elements of tissue electrolytes sufficiently for microanalytical studies. This approach may have broad application for various types of localizations of relevance to plant physiology, environmental pollution, and plant–microbe interactions. In this chapter, the experimental procedure of analytical cryoscanning electron microscopy applied to botanical samples is outlined.

Key Words: Energy-dispersive X-ray microanalysis; low-temperature scanning electron microscopy; freeze-fractures; cryofixation; elemental analysis.

1. Introduction

The physical principle of X-ray microanalysis is based on two types of electron-atom interactions, both leading to X-ray emission: deceleration of the electron beam by specimen atoms causes emission of energy as X-ray photons in a broad energy range, the so-called continuum radiation. In addition, atom-specific X-rays are emitted as a consequence of ionization of inner shell electrons of the specimen atom, the so-called characteristic radiation. X-rays can be detected by either wavelength-dispersive or energy-dispersive spectrometers. Energy-dispersive detectors (EDX) are the most frequently used type of detector in botanical applications because the higher probe currents required to generate sufficient

From: *Methods in Molecular Biology, vol. 369*
Electron Microscopy: Methods and Protocols, Second Edition
Edited by: J. Kuo © Humana Press Inc., Totowa, NJ

X-rays for detection by wavelength-dispersive systems (WDS) are damaging to the majority of botanical tissues.

The basics of EDX in a scanning electron microscope (SEM) is that the electron beam scanned across the specimen area of interest excites atoms to emit X-ray photons, which are collected by a semiconductor crystal. The energy of the X-ray photon hitting the crystal is converted into a voltage pulse proportional to the energy of the X-ray photon. The voltage pulses are sorted by a multi-channel analyzer, thus forming the X-ray spectrum. For more detailed information on physical and technical details of EDX, the reader is referred to the literature *(1–4)*.

EDX technique detects and measures elements within cells or subcellular compartments of botanical samples; it cannot distinguish "bound" from "free" elemental pools, and it cannot differentiate between isotopic or redox states. The fact that a spectrum of interest acquired in a short time (30–150 s) allows for a rapid evaluation of the specimen constituents and is therefore a promising method in botanical research. The spatial resolution of EDX is unsurpassed by any other analytical technique combing simultaneous compositional and morphological observations. However, the sensitivity is several orders of magnitude poorer than, for example, "bulk" analytical techniques, such as atomic absorption spectrophotometry and induction-coupled plasma analysis.

Reliable EDX measurements of botanical specimens require special preparation techniques because of two features of biological cells and tissues: (1) biological cells and tissues consist mainly of aqueous solutions that are not compatible with the vacuum of the electron microscope; and (2) the structure and the elemental composition of intracellular and extracellular compartments changes depending on the functional state of the cell. Cryofixation is fast enough to retain the original distributions of inorganic elements of tissue electrolytes sufficiently for microanalytical studies *(5–7)*. Such a frozen botanical specimen has to be exposed by freeze-fracturing or cryosectioning to image and analyze the inner part *(8–13)*. However, specimens with voluminous air-spaces, which is typical of many mature plant cells, often prove difficult to cryosection. Preparation and analysis of ultrathin cryosections of botanical samples for analytical purposes are demanding and far from routine. Therefore, this method is not included here. In this chapter, the experimental procedure of EDX on freeze-fractures of botanical samples is outlined (*see* **Fig. 1**).

2. Materials

2.1. Equipment

1. Scanning electron microscope with a cold stage.
2. Secondary electron detector.

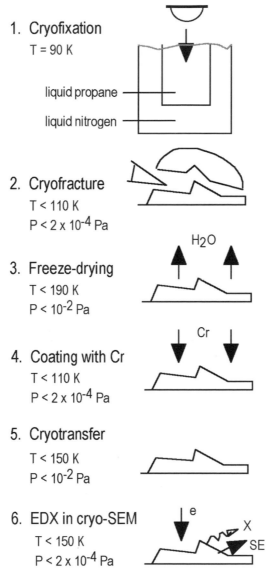

1. Cryofixation
   T = 90 K

   liquid propane
   liquid nitrogen

2. Cryofracture
   T < 110 K
   P < 2 x 10⁻⁴ Pa

3. Freeze-drying
   T < 190 K
   P < 10⁻² Pa

   H_2O

4. Coating with Cr
   T < 110 K
   P < 2 x 10⁻⁴ Pa

   Cr

5. Cryotransfer
   T < 150 K
   P < 10⁻² Pa

6. EDX in cryo-SEM
   T < 150 K
   P < 2 x 10⁻⁴ Pa

   e
   X
   SE

Fig. 1. Schematic flow sheet illustrating the steps for cryo-preparation of bulk specimens for EDX in the cryo-SEM. e: electron beam; SE: secondary electrons; X: X-rays.

3. Energy dispersive X-ray detector (*see* **Note 1**).
4. Software for the acquisition of spectra and spectra processing, and an image processing package for the display of images and digital X-ray maps.
5. Cryopreparation chamber attached to the SEM (dedicated system) or cryopreparation chamber isolated from the SEM (nondedicated system).

6. Transfer freezing device.
7. Sputter coating unit, fitted to the cryopreparation chamber.
8. Quarz thin film thickness monitor inside the cryopreparation chamber.
9. Vacuum pumps (rotary pump, turbomolecular pump) for the cryopreparation chamber.

2.2. Specimen Preparation

1. SEM specimen support holder.
2. Mounting stubs (*see* **Note 2**).
3. Double-sided adhesive tape.
4. Cryoglue (*see* **Note 3**).
5. Cryotubes (2 mL).
6. Large forceps for handling cooled cryotubes.
7. Fine tipped forceps.
8. Razor blades or scalpels.
9. Supply of primary cryogen (liquid nitrogen).
10. Secondary cryogen (e.g., propane).
11. High purity argon gas.
12. Coating metals.
13. Cryotank for LN_2 (5 L, 50 L).
14. LN_2 dewars for temporary storage or transfer of samples (1 L).

3. Methods

3.1. Cryofixation of Botanical Specimens

The cryopreparation of botanical specimen starts with cryofixation, which means solidification of the aqueous specimen by rapid cooling with minimal displacement of its components. Cryofixation requires rapid cooling of the aqueous specimen to a temperature less than 130K to prevent detrimental ice crystal growth. At temperatures less than this, the mobility of water molecules is too slow to form ice crystals. Cryofixation can be achieved by rapid contact of the specimen with a cold copper surface or by rapid plunging of the specimen into a cold liquid such as propane or ethane cooled by liquid nitrogen. However, even the closest contact with the cooling medium cannot balance the delay of heat flow within the specimen because of the low thermal conductivity of ice of approx 2 $W/(m \cdot K)$. The lower the cooling rate, the greater the rate of ice crystal that grows within the specimen. Growing ice crystals displace solutes and macromolecular components by phase separation.

1. Select a fresh leaf, root, etc. and cut the tissue in small pieces (<10 mm^3) using a scalpel and subsequently blot the excised sample with filter paper to remove adhering surface water.

2. Place the sample on a sample support. The samples are set vertically on stubs using a cryoglue to make freeze-fractures or mounted horizontally with double-sided adhesive tape for the analysis of fractural surfaces (*see* **Note 2**).
3. Using forceps, plunge-freeze the mounted specimen in a primary or secondary cryogen for at least 5 s (*see* **Note 3**). Then quickly transfer the specimen holder to the basket in the LN_2 and release it from the forceps.
4. Store the frozen specimens in a LN_2 storage dewar until required for observations (*see* **Note 4**).

3.2. Cryopreparation of Specimen for SEM-EDX Analysis

Modern cryo-SEM equipment consists of a high vacuum cryopreparation chamber directly attached to the SEM. In the cryopreparation chamber, which is an evacuated chamber containing a cold stage cooled by LN_2, the specimen can be kept at a controlled temperature. The specimen may be retained intact or fractured and kept either fully frozen hydrated, partially freeze-dried ("etched"), or fully freeze-dried. The samples may be coated for subsequent observations in the SEM. In the nondedicated system, the cryopreparation chamber stands separately from the SEM, whereas in the dedicated system the cryo-preparation chamber is connected to the SEM by a high vacuum valve. The transfer of the sample is easily performed by opening the gate valve between the cryopreparation chamber and the SEM. The sample is then transferred onto a temperature-controlled cryostage in the SEM, where it is examined at very low temperature (<150K). This cold stage is located in the electron microscope specimen chamber and is indirectly cooled (thermally insulated) from an external source.

1. Mount the stub with the frozen specimen on a precooled SEM specimen holder (*see* **Note 5**).
2. Transfer the specimen holder to the cryopreparation chamber set at <190K and partially freeze-dry in a high vacuum (p < 2×10^{-4} Pa) for 5 to 10 min. Sublimation of the specimen can be monitored visually in the cryopreparation chamber (*see* **Note 6**).
3. Fracture the sample with a microtome at <110K. After fracturing, etch the fracture plane (if necessary) by keeping the fractured specimen for 1 min at 190 K (*see* **Note 7**).
4. Purge the cryo-preparation chamber with argon gas raising the pressure to 2.2 Pa prior to coat and set the current at approx 60 mA for platinum or 150 mA for chromium, respectively. Allow coating to proceed until a programmed thin film thickness monitor terminates at a specific coating thickness (5–15 nm; *see* **Note 8**).
5. After coating, transfer the specimen with a manipulator through the sliding vacuum valve onto the SEM cold stage with the temperature set to less than 150 K (dedicated system) or transfer the specimen in a high vacuum cell onto the SEM cold stage (nondedicated system).

6. Observe the frozen botanical samples (surface or fracture plane) by secondary emission (SE) electron mode of the SEM at accelerating voltages between 12 and 18 kV.

3.3. Cryo-SEM Operating Conditions for EDX Analysis

The number of X-ray counts generated from the specimen is dependent on the electron microscope settings used for the analysis and exposition of the specimen to the detector. The steps in optimal parameter settings for EDX analysis are the following.

3.3.1. Choice of Accelerating Voltage

Higher accelerating voltages allow increased beam currents with increased X-ray production, and a voltage greater than 12 kV usually is needed. However, one should be aware of the fact that the X-rays collected are generated in a pear-shaped excitation volume below the specimen surface because of the scattering of the electron beam in the bulk specimen. In bulk specimens, the use of higher accelerating voltages results in greater penetration of the electron beam into the specimen and an increase in both depth and width of the volume from X-rays are produced. The excitation volume of a 12-kV beam in botanical specimens of low atomic number extends to several micrometers and is almost independent of the beam size. The voltage chosen for analysis is, therefore, a compromise between the requirement to increase X-ray production and the need for a good spatial resolution. Voltages between 12 and 18 kV usually are selected for X-ray microanalysis of bulk specimens.

3.3.2. Choice of Probe Current

The primary factor that determines the choice of probe current for analysis is the count rate, that is, the number of X-rays emitted from the specimen per unit time. Excessive current may result in specimen drift (particularly important in prolonged X-ray mapping). Damage to the specimen should be minimized by using the lowest probe current that is compatible with obtaining a statistically significant peak for the element of interest in a reasonable amount of time (60–100 s). If the aim of the analysis is to identify elements present in a dense particulate inclusion, it is unlikely that low count rates will be a problem, and low probe currents can be used. The probe current should not be so high that the emitted X-rays saturate the detector and so cause very high (greater than 50 %) values for the dead-time.

3.3.3. Duration of Analysis

The analysis should be carried out for a sufficiently long period of time to allow statistically significant peaks of the element of interest to be obtained. Typical analysis times are from 50 to 150 s live-time. Longer times will be required if the peak for the element of interest are overlapped by major peaks from other elements.

3.3.4. Choice of Magnification

The magnification and the probe diameter should be kept as low as possible within the constraints imposed by the spatial requirements of the study. Avoid unnecessary pre-exposure of the area to be analysed in order to minimize beam damage; "search" the specimen at low magnification and at low temperature.

3.3.5. Detector Specimen Geometry

The detector specimen geometry must be optimized. This involves moving the detector close to the specimen so that solid angle of X-ray collection is maximized, adjusting the tilt angle of the specimen, and adjusting the specimen height.

3.4. Steps of EDX Analysis of Bulk Specimens in the Cryo-SEM

1. Calibrate the EDX detector (*see* **Note 9**).
2. Set the microscope to SEM imaging.
3. Survey the specimen and select areas of interest. Choose a spot of the specimen or a small area of selected cell compartments and tissues for analysis (from 50 to 100 nm).
4. Perform the analysis with the selected and predetermined conditions. Collect and store the spectra on disks.
5. Establish the optimal conditions of accelerating voltage, probe current, duration of analysis, magnification and sample geometry (*see* **Subheading 3.3.**).
6. Make the analytical settings of the spectrometer (pulse processor; spectra acquisition time) and select the X-ray energy range for the collection of spectra (*see* **Note 10**).
7. Adjust the probe current to keep the count rate at approx 2000 counts per second (*see* **Note 11**).
8. Adjust the dead time to 20–30% and check the shape of the spectrum. Reduce excessive extraneous continuum x-ray signals and stray signals (*see* **Note 12**).
9. Acquire EDX spectra and record the areas analysed.
10. Store the spectra and image on disk.
11. Once the spectrum has been collected and stored, follow the manufacturer's software instructions for processing of the peak (*see* **Subheading 3.5.** and also **Note 13**).
12. Determine the energy of the characteristic peaks, number of counts in the peak and the counts in the background (B) under the peak (P), and calculate the P/B ratio (by the software package; *see* **Subheading 3.5.** and also **Note 14**).

3.5. Spectrum Processing and Methods for Quantification

3.5.1. Qualitative Identification of Peaks Present in the Spectrum

The first stage in the analysis of an unknown sample is the identification of the elements present. Qualitative information of the specimen, are obtained by

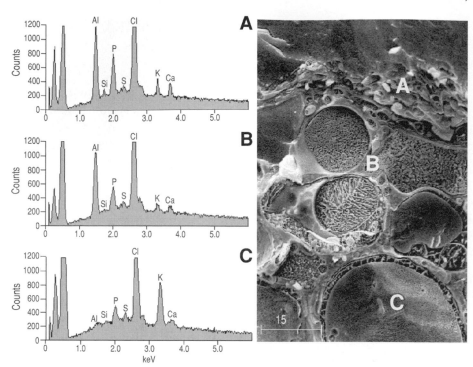

Fig. 2. Scanning electron microscopy image of a freeze-fracture of a mycorrhizal Norway spruce roots. The tree seedlings were grown in a normal nutrient solution supplemented with Al as chloride (10 ppm). The presence of elements in the root has been detected with EDX. Three spectra obtained from point analyses were recorded. (**A**) Cell wall of root cortical cells. (**B**) Cell wall of the fungal Hartig net. (**C**) Vacuole of a root cortical cell. The spectra show besides the characteristic peaks of the detected elements a high background under the peak. Bar = 15 μm.

a thorough examination of the spectrum, by determining the energy of the characteristic peaks from their positions in the spectrum (*see* **Fig. 2**). The energies of the X-rays from all elements in the periodic table are documented, and peak identification routines are usually included in the software for microanalysis. Until some years ago, only elements of an atomic number $Z \geq 11$ (Na) could be measured; recently introduced windowless EDX-detectors allow the measurement of all elements of $Z \geq 5$ (B) simultaneously. Elemental mapping is an alternative to spectrum collection. In this mode, short time X-ray spectra are collected from each pixel scanned by the electron beam. Then the number of X-ray counts is collected in specific energy windows and are represented by a pseudocolour scale. Qualitative maps can be achieved within less than one hour providing an overview on inhomogeneous element distributions (*see* **Fig. 3**).

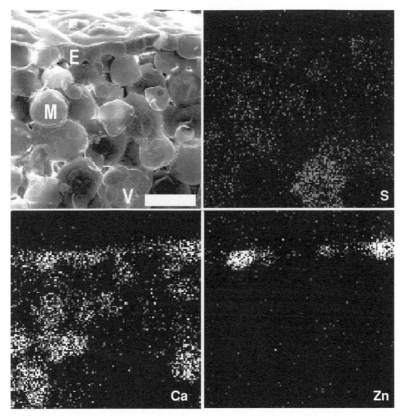

Fig. 3. X-ray mapping of a freeze-fracture of *Thlaspi caerulescens* leaf section. A scanning electron microscopy image and the corresponding elemental maps for S, Ca, and Zn of tissue section. *Thlaspi caerulescens* is a heavy-metal hyper accumulating plant. Here, zinc accumulated predominantly in the epidermal cells. Calcium was abundant in the epidermal (E) and mesophyll (M) cells, but the distribution was not homogeneous. The presence of sulfur was greater in the vein (V) compared with the epidermal and mesophyll cells. Bar = 50 μm (reproduced from Frey et al. *[10]*).

3.5.2. Background Subtraction and Estimation of Net Peak Integrals

Both qualitative and quantitative analyses require the ability to detect the presence of a characteristic peak (P) above the total background (B) under that peak. The number of counts under a peak is termed the net peak integral or the net peak intensity. An increase in peak-to-background (P/B) ratios leads to an increased sensitivity of analysis. Both subtraction of background and determination of the net peak counts for the constituent elements in a spectrum can be done by computer programs supplied with the analysis systems used. For more detailed information on qualitative and quantitative analysis of EDX the reader is referred to the literature *(4,14–17)*.

3.6. Quantification Using Standards

Quantitative analysis by the P/B ratios method requires the determination of calibration constants for the elements of interest of the analysis of a standard. Direct comparison of net peak intensities requires that standards are analysed at the same time as the specimen. Consequently, the preparation of standards is an important part of quantitative analysis, and the operator has to decide which type of standard is appropriate for a particular application, and to know about the possible problems which may be encountered with different types of standards. The different types of standards used for biological X-ray microanalysis have been reviewed by *(4)*. The standards available can be divided into two main types: mineral salts or matching standards with a matrix composition similar to that of the specimen, which are sometimes called ideal standards. They include protein-based standards and resin standards.

4. Notes

1. Principal suppliers of EDX and electron microscope manufacturers, materials and equipment for EM specimen preparation and EM accessories can be found on the Internet at http://www.microscopy-analysis.com/cgi-bin/item.cgi?id=238.
2. Standard stubs provided with cryo-SEM systems have a flat surface. These stubs can be readily modified to suit particular specimens. Mounting holes are useful for cylindrical objects and mounting slits are suited for transverse (or longitudinal) cryofractures of the specimen. A wide range of custom-made SEM stubs are now available, with holes or slots of various dimensions which are designed to hold a small, excised piece of tissue. SEM stubs are commonly made of carbon or aluminium. The choice of material from which the stubs are made is important in EDX analysis and depends on the elements of interest and their X-ray energies (e.g., overlapping peaks).
3. If the water content of the samples should not be altered beause of an aqueous cryo-glue (e.g., methyl cellulose or Tissue-Tek), then cryofixation without prior mounting in a cryoglue is essential. Cryofixed specimens are then mounted onto the stubs. Mounting is either conducted mechanically or with a cryoglue.
4. When using any cryopreparation equipment, all safety precautions should be observed. Always use precooled specimen holder and forceps when touching the frozen specimen. Beware of cold gas layers above coolants, which can freeze the specimen prematurely. Once the sample is cryofixed, care must be taken to assure that rewarming does not occur during subsequent manipulations. It should also not be forgotten that low temperature procedures carry a risk of cold burns, and care must be taken when handling liquid nitrogen and any cooled metal objects.
5. If fresh tissue samples are plunge-frozen immediately after harvest without prior mounting on a stub, they should be kept at low temperature (<150 K) until examinations in the SEM. Frozen samples can then be mounted under low temperature in a transfer freezing device. The frozen sample must be securely attached to the

stub so that it will not fall off during manipulations inside the specimen chamber. Furthermore, the stub must be firmly attached to the specimen holder in order to guarantee a high thermal conductivity and maximize the heat exchange rate at the cryogenic surface. The transfer of the specimen to the cryopreparation chamber should be as fast as possible in order to prevent contamination due to frost.

6. Freezing rapidly transforms freezable cellular and extracellular water into its solid state (ice) and the specimen is considered to be fully frozen-hydrated. Etching will remove superficial water droplets and films derived from environmental sources either naturally or as contaminants. If the sample is etched, the specimen may be then considered as partially freeze-dried and features of the etched surface can be exposed that would otherwise be obscured. If the objective is to measure element concentrations in matrix-free extracellular spaces and intracellular vacuoles, then the specimens should be analysed in the frozen-hydrated state.

7. The fracture plane is difficult to control. A small mound of cryoadhesive in the hole or slit will freeze around the specimen to form a rigid sleeve, and the specimen often breaks close to the top of this, at or above the weakest point. Unless the fracture plane of interest is exposed, repeat fracturing by transferring specimens again to the cold stage of the preparation chamber until the fracture plane of interest is exposed. In some cases, it is advisable to fracture the specimen (very thick samples or very delicate samples) outside the preparation chamber using a cold knife or scalpel. Repeated fractures of the specimen can be made outside and should be done with the aid of a dissecting microscope. If the fracture knife is not sufficiently precooled before fracturing or has insufficient clearance angle then smearing of the fracture face may occur.

8. After manipulation, fracturing, and sublimation, the specimen that is to be used for imaging must be coated to render its surface electrically conductive. For EDX analysis, leave the samples uncoated or to coat with chromium or carbon (5 nm) as these coating material normally do not interfere with the X-ray energies of elements of interest.

9. Detector calibration is subroutine work, called "gain calibration" within their programs. The manufacturer's instructions should be checked. Insert a specimen suitable for calibration. It must contain two peaks with well separated X-ray energies; for example a nickel or copper grid. Check the position of the characteristic peaks in the spectrum, for copper (CuLα 0.93 keV and CuKα 8.04 keV). If the peaks are not in their expected positions, adjust the zero and gain of the spectrum according to the manufacturer's instructions.

10. The energy range 0 to 20 keV is suitable for most applications because the K lines from elements of low atomic number and the L and M lines from heavy elements are in this range. Always remember to include all lines from a given X-ray shell when specifying the energy range of the peak. For example, for potassium, the range must include both Kα and K$_\beta$ peaks.

11. If a particular area of the surface gives a low counting rate compared with apparently similar neighbouring areas, this indicates that some of the X-rays emitted from the area are probably being absorbed. This is a common problem encountered in

the analysis of bulk specimens with surface roughness. Ideally a bulk specimen to be analysed should have a flat surface, but this is not always possible. It often happens that the line of sight between the point of analysis and the X-ray detector may be interrupted by a prominent part of the specimen (which may not be apparent to the observer) causing emitted X-rays to be absorbed. The user has to see whether it is possible to collect a spectrum by moving to another area of the specimen or to remount a new specimen.

12. X-rays from extraneous sources can cause large contributions to the total background continuum and continuum production may vary between analyses. Extraneous X-rays are generated when electrons scattered by the specimen, or X-rays, strike the specimen holder or any of the surfaces in the specimen chamber. Analysing specimens in the regions close to the holder are that may generate additional characteristic peaks by electrons, and X-rays scattered by the holder and absorption of some X-rays emitted by the specimen may occur. The shape of the spectrum should be observed during analysis as this will give an indication that absorption of low energy X-rays is occurring. Light elements such as C and O always show prominent peaks in frozen-hydrated or partially freeze-dried botanical samples.

13. Many detector systems include a "search-and-identify" routine with the software, which rapidly scans the spectrum and assigns a likely element to each peak. The biologists, however, should view the results from such a scan with some caution, since these programs are usually written for materials scientists, and mis-identification of elemental peaks in biological specimens often occurs.

14. The following guidelines should be considered when analyzing the spectrum: (1) only peaks which are significant, where the counts in the peak (P) are more than three times the standard deviation of the counts in the background (B) under the peak should be identified. (2) If the spectrum consists of only a few high peaks, examine the spectrum to see whether minor peaks could be the result of the occurrence of sum peaks which are in multiples of the energy of the main peak, or escape peak which appear at 1.74 keV less than the parent peak. (3) When assigning a peak, look for accompanying lines in the series for example for elements of atomic number greater than 17 (chlorine) both $K\alpha$ and K_β peaks should be present and (iv) there could be overlapped of some peaks, for example, K_β and $Ca\alpha$.

References

1. Chandler, J. A. (1977) *X-ray Microanalysis in the Electron Microscope*. North-Holland, Amsterdam.
2. Goldstein, J. I., Newbury, D. E., Echlin, P., et al. (2003) *Scanning Electron Microscopy and X-ray Microanalysis*. Kluwer Academic and Plenum, New York.
3. Sigee, D. C., Morgan, A. J., Sumner, A. T., and Warley, A. (1993) *X-ray Microanalysis in Biology: Experimental Techniques and Applications*. Cambridge University Press, Cambridge.
4. Warley, A. (1997) X-ray microanalysis for biologists. *Practical Methods in Electron Microscopy*, vol. 16 (Glauert, A. M., ed.), Portland Press, London and Miami.

5. Echlin, P. (1992) *Low-Temperature Microscopy and Analysis*. Plenum Press, New York.

6. Jeffree, C. E. and Read, N. D. (1991) Ambient- and low-temperature scanning electron microscopy, in *Electron Microscopy of Plant Cells*, (Hall, J. L. and Hawes, C., eds.), Academic Press, London, pp. 313–413.

7. Zierold, K. and Steinbrecht, R. A. (1987) Cryofixation of diffusible elements in cells and tissues for electron probe microanalysis, in *Cryotechniques in Biological Electron Microscopy* (Steinbrecht, R. A. and Zierold, K., eds.), Springer, Berlin, Heidelberg, New York, pp. 272–282.

8. Brunner, I. and Frey, B. (2000) Detection and localization of aluminium and heavy metals in ectomycorrhizal Norway spruce seedlings. *Environ. Poll.* **108,** 121–128.

9. Frey, B., Brunner, I., Walther, P., Scheidegger, C., and Zierold, K. (1997) Element localization in ultrathin cryosections of high-pressure frozen ectomycorrhizal spruce roots. *Plant Cell Environ.* **20,** 929–937.

10. Frey, B., Keller, C., Zierold, K., and Schulin, R. (2000) Distribution of Zn in functionally different leaf epidermal cells of the hyperaccumulator *Thlaspi caerulescens*. *Plant Cell Environ.* **23,** 675-687.

11. Frey, B., Zierold, K., and Brunner, I. (2000) Extracellular complexation of Cd in the Hartig net and cytosolic Zn sequestration in the fungal mantle of *Picea abies– Hebeloma crustuliniforme* ectomycorrhizas. *Plant Cell Environ.* **23,** 1257–1265.

12. Stelzer, R. and Lehmann, H. (1993) Recent developments in electron microscopical techniques for studying ion localization in plant cells. *Plant Soil* **155/156,** 33–43.

13. Zierold, K. (1988) X-ray microanalysis of freeze-dried and frozen-hydrated cryosections. *J. Electron Microsc. Tech.* **9,** 65–82.

14. Marshall, A. T. (1998) X-ray microanalysis of frozen-hydrated biological bulk samples. *Mikrochim. Acta Suppl.* **15,** 273–282.

15. Roomans, G. M. (1988) Quantitative X-ray microanalysis in biology. *J. Electron Microsc. Tech.* **9,** 19–43.

16. Van Steveninck, R. F. M. and Van Steveninck, M. E. (1991) Microanalysis, in *Electron Microscopy of Plant Cells*, (Hall, J. L. and Hawes, C., eds.), Academic Press, London, pp. 415–455.

17. Zierold, K. (2002) Limitations and prospects of biological electron probe X-ray microanalysis. *J. Trace Microprobe Tech.* **20,** 181–196.

27

Static Secondary Ion Mass Spectrometry for Biological and Biomedical Research

Nicholas P. Lockyer

Summary

Static secondary ion mass spectrometry (SSIMS) is a capable of providing detailed atomic and molecular characterization of the surface chemistry of biological and biomedical materials. The technique is particularly suited to the detection and imaging of small molecules such as membrane lipids, metabolites, and drugs. A limit of detection in the ppm range and spatial resolution <1 μm can be obtained. Recent progress in instrumental developments, notably cluster ion beams, and the application of multivariate data analysis protocols, promise further advances. This chapter presents a brief overview of the theory and instrumentation of static secondary ion mass spectrometry followed by examples of a range of biological and biomedical applications. Because of the ultrahigh vacuum requirements and extreme surface sensitivity of the technique, appropriate sample preparation and handling is essential. These protocols, and the analysis methodology required to ensure high-quality, reliable data are described.

Key Words: Static SIMS; TOF-SIMS; imaging; mass spectrometry; cellular analysis; tissue imaging; biomaterials.

1. Introduction

Secondary ion mass spectrometry (SIMS) is a technique in which material desorbed from a surface by energetic particle (usually keV ion) bombardment is analyzed by mass spectrometry *(1)*. Originally developed in the 1950s and 1960s by Herzog et al. and Honig et al. to analyze metals and oxides, the technique has matured into a powerful and versatile analytical tool. Various operational modes have arisen from the development of the instrumentation and improved understanding of the underlying physics. Static SIMS for true surface analysis was first demonstrated by Benninghoven in 1970 *(2)*. Applications now include a wide variety from the life sciences, including molecular biology

From: *Methods in Molecular Biology, vol. 369*
Electron Microscopy: Methods and Protocols, Second Edition
Edited by: J. Kuo © Humana Press Inc., Totowa, NJ

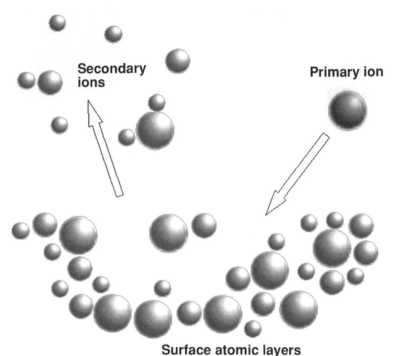

Fig. 1. Schematic representation of the ion beam induced sputtering of secondary ions from the surface of a solid specimen.

and medicine. This chapter briefly describes the basic theory and provides some examples of applications in biology and biomedicine before outlining the experimental protocols.

1.1. Ion Formation and the Static Limit

The impinging "primary" ions deposit their energy in the surface and subsurface of the sample, typically through collision cascades in sample atoms and molecules. Where these cascades return to the surface, "secondary" particles (electrons, neutrals and ions) can escape surface binding energies and be ejected, or sputtered (*see* **Fig. 1**). Only species in the top few atomic/molecular layers (approx 1–10 nm) of the sample are sputtered and, hence, their subsequent mass analysis and detection leads to a surface mass spectrum. The secondary ion flux sputtered from a molecular sample consists of atomic, molecular and fragment ions. The number of secondary particles sputtered per incident primary ion is termed the sputter yield. For organic materials the ionization probability of molecular secondary species is typically 10^{-3} to 10^{-7}. Together these parameters lead to the secondary ion yield, which is key feature in determining the sensitivity of the technique.

Secondary ion yields are influenced by the local chemical environment of the sputtered material. For atomic ions and inorganic samples, these matrix effects can result in secondary ion yield variations of several orders of magnitude for the same atom in different chemical environments. In organic systems the effects are generally less severe and with careful experimentation relative concentrations of similar analytes in similar matrices can be measured. With a good set of calibration data, quantitative information can be obtained from TOF-SIMS spectra using multivariate methods.

The nature of the sputtering process means that the SIMS technique is inherently destructive. The progressive removal of material and exposure of the subsurface to the primary beam leads to the possibility of depth profiling, the analysis of successively deeper regions of the sample. This approach is known as dynamic SIMS, and has become a powerful tool in the interfacial analysis of heterogeneous inorganic materials such as semiconductor devices. To ensure that the secondary ions reflect the undamaged surface each primary ion must impinge on a new area of the sample. This leads to static SIMS (SSIMS) analysis. Statistically this means that less than 1% of the sample can be impacted by primary ions to ensure static conditions. The surface density of atoms is typically 10^{15} atoms cm^{-2}; therefore, the so-called static limit equates to a primary ion dose density (PIDD) of 10^{13} ions cm^{-2}. The effective static limit will in practice be determined by the details of the sputtering process, which in turn depends on the nature of the sample and the primary ions.

1.2. Primary Ion Beams

A variety of primary ion sources are used in SIMS. Until recently mono-and diatomic ions were almost exclusively used, including Ar^+, Cs^+, Ga^+, and O_2^+. Sources based on the liquid metal ion gun (LMIG) design, such as Ga^+ are capable of being focused to less than 50-nm regions of the surface. In recent years, a number of cluster and polyatomic ion sources have been developed for SIMS. These include SF_5^+, Au_3^+, Bi_3^+, and C_{60}^+. These sources provide X10 to 100 increase in molecular secondary ion yield making them particularly suitable for SSIMS studies of (bio)molecular samples. Intriguingly, C_{60}^+ offers not only very high secondary ion yields but also much reduced subsurface damage for some materials. Using these unique properties, molecular depth profiling has been demonstrated. Cluster and polyatomic ion sources, and their sputtering characteristics are currently the subject of much research activity. For reviews, see Castner (*3*) and Winograd (*4*).

Sources based on the LMIG design, including Au_n^+ and Bi_n^+, deliver submicron spot sizes, whereas gas phase sources, including Ar^+ and SF_5^+, typically deliver spot sizes of the order 10 µm. Very recently, our laboratory has commenced testing of a C_{60}^+ beam with a minimum spot size approx 200 nm.

Digitally scanning a microfocused primary ion beam over the sample during the analysis results in a mass spectrum at each pixel element, that is, a chemical image. An alternative approach, using unfocussed primary beams, is to retain the spatial information related to the point of emission throughout the mass analysis and detection process. These two modes of imaging SIMS are termed microprobe and microscope imaging, respectively. In SSIMS, the microprobe imaging method is most commonly encountered because of the choice of mass analyzer.

1.3. Mass Spectrometry

Magnetic sector, quadrupole and time-of-flight (TOF) mass analyzers have all been used in SIMS. Each spectrometer has specific advantages and disadvantages. In SSIMS the TOF mass spectrometer (MS) has become dominant, mostly because of its very high transmission (>50%) and parallel detection of all masses. These features mean that TOF-SIMS instruments exhibit very high sensitivities (ppm-ppb detection limits), which are necessary to achieve analysis under static conditions of very low ion dose. In TOF-MS analyte ions are given the same kinetic energy by accelerating them to a given potential V. The flight time t of ions over a fixed distance L is related to the mass-to-charge ratio of the ion m/z by a simple formula.

$$t = L \, (m/2zV)^{1/2}$$

Generally, TOF-MS requires that the source of analyte ions be temporally pulsed. Mass resolution is limited by the precision with which the extraction time of the analyte ions is defined. In TOF-SIMS, the primary ion beam is pulsed by electrostatic deflectors, producing approx 10-ns primary pulses. These can subsequently be "bunched" to less than 1 ns, increasing mass resolution without loss in sensitivity. In TOF-SIMS mass resolution is also dependent on the inherent energy spread (10–100 eV) of secondary ions caused by the sputtering process. This energy distribution can be compensated for by an energy analyzer in the flight tube. The most common device is an ion mirror or *reflectron* through which ions of the same m/z but different kinetic energy experience different path lengths, bringing them back into temporal focus at the detector (*5*). Imaging TOF-SIMS instruments incorporating reflectrons use the microprobe technique (*see* **Fig. 2**). Microscope imaging TOF-SIMS instruments are much less common and employ an alternative design of electrostatic energy analyzer (*6*).

The SSIMS analysis of biological sample represents a number of specific challenges in terms of sample handling, sensitivity, and data interpretation. The methodology of specimen preparation and handling has been adopted from the cryoelectron microscopy community. Plunge-freezing and freeze-drying methods

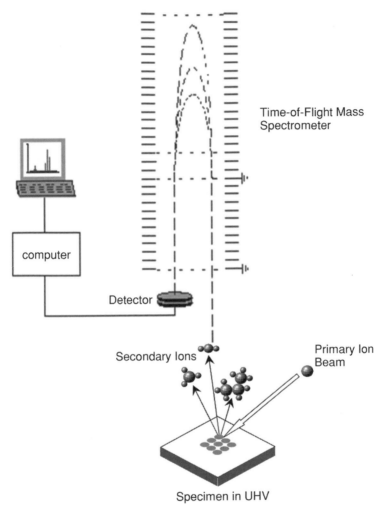

Time-of-Flight Mass
Spectrometer

computer

Detector

Secondary Ions

Primary Ion
Beam

Specimen in UHV

Fig. 2. Schematic representation of ion microprobe imaging time-of-flight secondary ion mass spectrometry with a reflectron-based mass analyzer.

make hydrated/volatile biological material compatible with the ultrahigh vacuum ($<10^{-8}$ mbar) of the SIMS instrument. Cryogenic specimen preparation was first applied in the SIMS field for dynamic SIMS ion microscopy of elemental species such as isotopic tags *(7)*. The enhanced sensitivity afforded by cluster/polyatomic primary ions and the latest TOF-MS technology means that relatively high mass (1–5 kDa) or low concentration biomolecules can be detected in single cells or in tissue. At least one instrument, the "BioToF-SIMS," has been specifically designed to meet the challenge of biological and bio-medical analysis (*see* **Fig. 3** *[8]*). It includes a cryogenic fast entry system and an *in-situ*

Fig. 3. The BioToF-SIMS instrument (reproduced with permission from C. Corlett, Kore Technology Ltd.).

freeze-fracture device to avoid the problems of surface ice contamination, which would otherwise render a surface-specific SSIMS analysis meaningless *(9,10)*. The analysis stage can be cooled to 100 K to maintain the chemical and physical integrity of the sample during prolonged analysis. The extraction gap is unusually large (10 mm) to facilitate topographically rough samples without loss in mass resolution, and to enable complementary analysis by laser postionization. Our system is currently configured with a Au_n^+ LMIG *(11)* and microfocused C_{60}^+ gun *(12)*. Finally, a 20-kV postacceleration detector ensures high detection efficiency (~100%) for secondary ions up to 4 kDa *(13)*.

The resulting data is typically extremely complex and interpreting the mass spectrum is a significant task. Multivariate data analysis methods *(14)* are increasingly being applied to extract information from SSIMS data, particularly in biological applications.

1.4. Applications in Biology and Medicine

Static SIMS, unlike many techniques for biological imaging/analysis, does not require derivitization/labeling of the compounds to be detected. Other advantages of the technique are its high sensitivity and specificity, particularly to small molecules such as drugs and metabolites and its high lateral resolution. SSIMS shows considerable promise in many areas of biological and biomedical research, including single cell and tissue imaging and the direct, quantitative analysis of biological fluids. Applications in these areas are rapidly developing as the instrumental and data-handling capabilities advance. This section provides an illustration of the breadth of applications in this field. For a more comprehensive discussion of the application of SSIMS in these and related areas the reader is directed towards a number of review articles covering applications to biological systems *(15–19)*.

1.4.1. Biomaterials

The performance of biomaterials in vivo depends largely on the extent to which the chemistry of their surface and underlying substrate can be controlled. Biocompatibility and specific functionality can be conferred by cellular and biomolecular adhesion, conformation, and interaction with the biomaterial surface. The surface and bulk chemistry of biomaterials are invariably different, and to understand the interactions that occur at the interface of biomaterial and the biological environment requires detailed *surface* analysis. TOF-SIMS and other surface analytical techniques, such as X-ray photoelectron spectroscopy (XPS), have been applied to reveal the detailed surface chemical composition, orientation and spatial arrangement of tailored biomaterials.

Organic polymers frequently are used as biomaterials and are well-suited to TOF-SIMS analysis, which yields a fingerprint spectrum of the monomer unit and information regarding the molecular weight distribution. In an assessment of biocompatibility, Chen and Gardella used TOF-SIMS to study the in vitro kinetics and mechanisms of the hydrolytic degradation of biodegradable polymers including poly(lactic acid) (PLA) *(20)*. There are numerous examples in the literature where TOF-SIMS, often in combination with X-ray photoelectron spectroscopy, has been used to monitor (specific) protein and cell attachment onto polymers and other potential biomaterials *(21–24)*. Castner et al. have used multivariate data analysis methods such as principal component analysis (PCA) *(25,26)* and linear discriminant analys *(27)* to identify specific adsorbed proteins based on the characteristic signals from their amino acids. Although, unlike other mass spectrometry techniques based on matrix-assisted laser desorption ionization (MALDI) or electrospray ionization (ESI), TOF-SIMS cannot generally detect specific diagnostic signals from proteins, these results show that it is sensitive to the adsorbed protein orientation and/or conformation due to its small sampling depth. Using sophisticated data analysis, these subtleties in the resulting mass spectra can be resolved. This work has been extended to the quantification of binary protein films on various substrates using partial least squares regression analysis *(28)*. Also in the area of biomaterials, Arlinghaus et al. have studied peptide nucleic acid biosensor chips with TOF-SIMS, identifying and investigating the immobilization and hybridization of deoxyribonucleic acid fragments *(29)*.

In drug-delivery applications, TOF-SIMS has been used to characterize the distribution of drugs in polymer delivery systems. John and co-workers *(30)* imaged the distribution and bioavailability of the orally inactive drug leuprolide in a freeze-fractured cross section of a medical patch made from hydroxypropyl cellulose. The distribution of the drug depended on the formulation of the drug/polymer system. TOF-SIMS has revealed directly the distribution of the drugs on cellulose beads *(31)* in drug pellet cross sections *(32)* and in various polymer coatings for drug-delivery beads *(33)*. Mahoney et al. *(34)* have used an SF_5^+ cluster beam to monitor the distribution of drugs in biodegradable polymer films. Little degradation of molecular ion yields was seen up to a primary ion dose of ~5×10^{15} ions cm^{-2}, far exceeding the accepted static limit for atomic primary ions. Such detailed chemical and spatial information aids the understanding of drug migration, diffusion and release in drug delivery systems.

1.4.2. Biological Cells and Membranes

Liposomes and planar lipid membranes have been used as model membranes with which to demonstrate the potential of SSIMS for studying cellular events such as cellular diffusion, fission and fusion. Leufgen et al. studied phase segregation in self-assembled Langmuir-Blodgett films of dipalmitoylphosphatidylcholine (DPPC) and nitrobenzoxadiazol-phosphatidylcholine using TOF-SIMS *(35)*. Bourdos et al. *(36)* used TOF-SIMS to image liquid-condensed and liquid-expanded domain formation from DPPC monolayers.

Winograd and co-workers have pioneered the adoption of fast-freezing and *in situ* freeze-fracture methodology in SSIMS analysis of *frozen hydrated* specimens using a variety of liposome and cellular systems *(9,37–39)*. Using characteristic phospholipid fragments, they observed fusion and mixing events between DPPC and dipalmitoylphosphatidyl-dimethylethanolamine cholesterol-containing liposomes using imaging TOF-SIMS *(9)*. The same freeze-fracture methodology was applied to successfully image components of the outer membrane and organelles of frozen hydrated red blood cells *(9,38)* and to image the distribution of cocaine doped into the single-celled organism *Paramecium (39)* TOF-SIMS imaging has been combined with bright-field and fluorescence microscopy to image frozen hydrated, freeze-fractured rat pheochromocytoma (PC12) cells *(40,41)*. This integrated approach aids identification of the fracture plane and subsequent interpretation of the TOF-SIMS data. Recently, the group used TOF-SIMS and PCA to demonstrate the heterogeneous reorganization of membrane lipids during the fusion of two *Tetrahymena* in the process of mating *(42)*.

In the area of microbiology Jungnickel et al. *(43)* have used multivariate TOF-SIMS analysis of cell surface chemistry to discriminate between yeasts at the species and strain level. The diagnostic power was attributed to differ-ences in the cell membrane lipid composition, which was in agreement with known biology. Thompson et al. *(44)* used a similar approach to discriminate between spores and vegetative cells of *Bacillus megaterium*. In a model drug–cell interaction experiment Cliff et al. *(45)* imaged frozen hydrated *Candida glabrata* yeast cells incubated in the presence of the antibiotic clofazimine. The drug was detected only on the outside of unfractured cells, implying that it does not penetrate the cell wall. This study suggests that TOF-SIMS analysis of pharmaceuticals on cell surfaces and within cells could provide a valuable new tool for pharmacological studies.

Gazi et al. *(46)* have imaged freeze-dried prostate cancer (PC-3) cell cultures, demonstrating chemical differences in the phospholipid composition of frac-tured and nonfractured cells. A TOF-SIMS study of different cell lines revealed that multivariate analysis can distinguish nonmalignant and malignant pros-tate cell lines derived from different metastatic sites, providing valuable infor-mation on the biochemical basis of metastasis (E. Gazi et al., unpublished data, 2005). Fartmann et al. *(47)* have imaged freeze-fractured, freeze-dried osteo-blasts and MeWo cancer cell cultures using TOF-SIMS and, to aid quantifica-tion of elemental ions, laser postionization of sputtered neutrals.

In a further demonstration of the enhanced analytical capabilities of poly-atomic (C_{60}^+) primary ions, Cheng and Winograd *(48)* have shown that molec-ular TOF-SIMS depth profiles of small peptides ($m/z < 500$) are possible with using a glassy trehalose matrix. Similar results have been obtained for histamine in an ice matrix *(49)*. These results suggest the exciting possibility of perform-ing three-dimensional molecular imaging of whole cells.

1.4.3. Biological Tissue

Pharmaceuticals and their metabolites may be present at ppm or sub-ppm levels in tissue. Several groups are investigating the use of TOF-SIMS to local-ize the distribution of these compounds in tissue sections. John and Odom *(50)* analyzed a number of representative pharmaceutical compounds in frozen tis-sues and tissue-like matrices, reporting detection limits of 100 to 1000 ppm for TOF-SIMS analysis with a Cs^+ primary ion beam. This study highlighted changes in relative ionization efficiencies in complex matrices and the problem of achiev-ing sufficient secondary ion yields to attain ppm level detection in small (μm^2) pixels. However, where concentrations are higher a number of successful appli-cations have been demonstrated. The distribution of epinephrine (approx 1 g/kg) in freeze-dried sections of rabbit adrenal tissue has been determined using a Cs^+ primary ion beam in TOF-SIMS microscopy, confirming its localization in the cortex versus the medulla region *(50)*. Using a Ga^+ microprobe, molecular species from a sun block cream were imaged on a freeze-dried artificial skin specimen *(50)*.

Todd and coworkers have used a quadrupole-based MS/MS ion microscope instrument to investigate the distribution of lipids in freeze-dried rodent brains *(17,51,52)*. The effect of stroke on gerbil brain was mimicked by injecting lipopolysaccharide. The phosphocholine distribution revealed evidence of brain damage by a demyelination process. The same group has demonstrated the imaging of acetocholine and methylphenylpyridinium iodide in chicken liver specimens *(53)*.

Methods for increasing sensitivity, in tissue imaging applications and in general, continue to be explored. The enhanced capabilities of cluster/polyatomic ion beams have already been discussed. Jones et al. (unpublished results, 2005) performed microprobe TOF-SIMS imaging of tissue specimens with a C_{60}^+ beam. The TOF-SIMS images in **Fig. 4**, which were obtained with a 20-keV C_{60}^+ beam, show the lipid distribution in freeze-dried cryosections of rodent brain. The corpus callosum is located at the inner part of the cerebrum and links left and right cerebral hemispheres. It is composed of myelin-coated nerve fibers and is relatively high in cholesterol and sphingomyelin concentration and low in phosphatidylcholine. The outer cerebrum contains a high concentration of phosphatidylcholine from the membranes of nerve cells and blood vessels. **Figure 4** shows that secondary ions diagnostic of cholesterol (m/z 369, $[M-H_2O +H]^+$) are localized in the corpus callosum whereas those diagnostic of phosphocholine (headgroup fragment m/z 184, $C_5H_{15}NPO_4^+$) are localized in the surrounding tissue signals in agreement with known physiology. Sjövall et al. *(54)* and Touboul et al. *(55)* have performed similar analyses using Au_3^+ beams. Various methods involving matrix addition have been reported to enhance the molecular sensitivity and spatial resolution in SSIMS. The enhanced cationization of (bio)organic species in combination with various metal salts or by deposition of a gold layer and has been reported by Delcorte et al. *(56,57)*. Wu and Odom *(58)* mixed matrix-assisted laser desorption ionization matrices such as 2,5,-dihydroxybenzoic acid with various peptides to enhance protonation and sensitivity, particularly at high mass. This methodology was adapted to retain imaging capability by Altelaar et al. *(59)*, who achieved micrometer resolution molecular images of cryosections of the cerebral ganglia of the freshwater snail *Lymnaea stagnalis* by depositing a thin layer of 2,5,-dihy-droxybenzoic acid on the nerve tissue. Sjövall and coworkers *(60)* have enhanced cationization by blotting freeze-dried cells onto a silver substrate or depositing a thin layer of silver onto cells *(61)* or tissue *(62)*. These instrumental and methodological developments promise real advances in the imaging of not only native lipids, but drugs and metabolites at trace levels, particularly in cells and tissue sections.

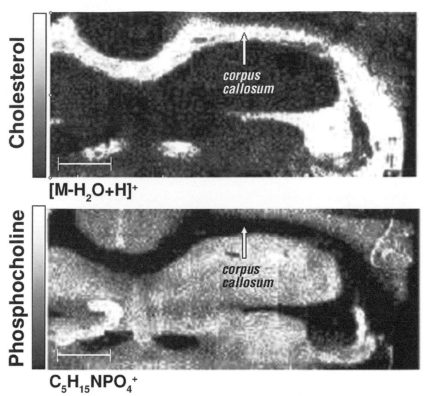

Fig. 4. Secondary ion images of a rodent brain cryosection showing localization of (**A**) cholesterol in the corpus callosum and (**B**) phosphocholine in the outer cerebrum. Time-of-flight secondary ion mass spectrometry analysis performed with 20 keV $C_{60}{}^{+}$, dose density $<1 \times 10^{10}$ ions/cm^2. Bar = 1 mm; pixel size 4.7 µm.

1.4.4. Biological Fluids

Applications of the quantitative analysis of biological fluids (blood and urine) by TOF-SIMS in screening for genetic disorders and in detecting drugs have been demonstrated. Zimmerman et al. *(63)* and Seedorf et al. *(64)* have used TOF-SIMS in screening for the genetic disorder Smith-Lemi-Opitz syndrome in infant blood samples *(63,64)*. Quantitative analysis of cholesterol and 7-dehydocholesterol revealed the abnormal levels of these metabolites in infants with Smith-Lemi-Opitz syndrome. Muddiman and colleagues have demonstrated the direct detection of cyclosporin A in blood *(65)* and multiple drugs, including cocaine, in urine *(66)* with limits of detection in the ng/mL range, using internal standards closely matched to analytes to aid quantification.

2. Materials

Items marked with (*) are needed for cryogenic sample preparation/analysis only.

2.1. Specimen Preparation and Handling

1. Substrate materials e.g., Si (100) wafer 99.999% (Advent Research Materials Ltd, Oxford, UK; *see* **Note 1**).
2. Ultrasonic bath.
3. HPLC-grade methylene chloride, hexane, acetone, methanol, and deionized water. (Sigma, St. Louis, MO). **CAUTION:** these are flammable/toxic solvents.
4. Polystyrene microspheres (5–50 µm; Duke Scientific, Palo Alto, CA).*
5. Propane gas cylinder. **CAUTION:** propane is highly flammable, with a risk of explosion.
6. Plunging-freezing machine, home-made or purchased.
7. Liquid nitrogen. **CAUTION:** nitrogen is a cryogenic liquid.
8. Cryostorage vials (Nalgene system 100, Nalge Co., Rochester, NY).*
9. Freeze-drying machine (e.g., Heto CT110, Thermo Electron Corp., www.thermo.com).*
10. Cryotransfer sample stub with thermocouple.*
11. Nitrogen gas clinder (5–10 bar) for flowing through TOF-SIMS cold stage heat exchangers. **CAUTION:** nitrogen gas is an asphyxiant.*

2.2. TOF-SIMS Acquisition

1. Faraday cup in specimen holder for measuring primary ion/electron dose (*see* **Note 2**).
2. Picoammeter or electrometer for measuring primary ion/electron dose (Model 485 Keithley Instruments Ltd, Theale, U.K.; www.keithley.com).
3. A suitable "test" sample consisting of various specimens for tuning and measuring instrumental parameters such as sensitivity, mass resolution and spatial resolution (*see* **Note 3**).
4. Provision for computer control of primary ion beam raster and position.
5. Optical microscope and/or CCD camera for viewing sample in analysis position.

2.3. Data Analysis

1. Visualization software.
2. Spectral data library (*see* **Note 4**).
3. Multivariate data analysis software (*see* **Note 5**).

3. Methods

This section describes the typical methodology required to achieve a successful TOF-SIMS analysis of a biological sample (*see* **Note 6**).

3.1. Specimen Preparation and Handling

The aim of a successful sample preparation for TOF-SIMS is to preserve the native chemical state of the sample in a form that is compatible with the ultra-high vacuum environment of the instrument. Where imaging is to be performed it is particularly important to minimize the migration of diffusible ions. Plunge-freezing and freeze-drying methods for various specimen types have been described elsewhere in this volume. Particular attention here is drawn to the measures necessary to facilitate a surface chemical analysis.

3.1.1. Plunge-Freezing

Cell suspensions, cells grown on silicon or steel substrates, and tissue sections mounted on silicon have been prepared for TOF-SIMS by plunge-freezing. The cryopreparation methodology has been described in detail elsewhere in this volume, and only a brief summary is presented here.

1. Prepare a dewar (approx 100 mL) of freezing medium (e.g., liquid propane or other cryogen) in a fume cupboard to ensure good ventilation and minimize condensation. **CAUTION:** for flammable liquids/gases, avoid sources of ignition.
2. Prepare a storage dewar of LN_2.
3. Deposit the cell suspension or fresh tissue section on a suitable substrate (*see* **Note 7**).
4. If the specimen is to be fractured, add polystyrene "spacer" microspheres (1:1 vol. mixture) to avoid crushing the specimen and place a second substrate on top, forming a "sandwich" structure.
5. Hold the specimen substrate(s) in an automated plunge-freezing device or in suitable tweezers taking precautions to avoid touching the specimen itself (*see* **Note 8**). **CAUTION:** when working with cryogenic liquids, avoid skin contact.
6. Plunge the specimen rapidly into the freezing medium. Transfer sample rapidly to LN_2.
7. While it is under the cryogenic liquid, using precooled tweezers, place the specimen substrate into a cryovial. If using a single-substrate, take care to avoid contaminating the specimen surface.
8. Place the cryovial under LN_2 in the storage dewar.
9. Store specimen under LN_2 until transfer to freeze-dryer or TOF-SIMS.

3.1.2. Freeze-Drying

Freeze-drying of plunge-frozen specimens has proven successful in the preparation for TOF-SIMS of cell cultures and tissue sections mounted on silicon or steel. The advantage of freeze-drying is that subsequent sample handling is simpler in that cryogenic temperatures do not need to be maintained (*see* **Note 9**).

1. Remove the frozen specimen from the LN_2 storage dewar.
2. If the sample is to be freeze-fractured this should be performed at this stage, on the bench under LN_2.

3. Load the frozen specimen into the freeze-dryer.
4. Freeze-dry according to manufacturers instructions (*see* **Note 10**).
5. Once freeze-drying is complete the specimen can be removed and stored in a desiccator for several days prior to analysis.

3.1.3. Specimen Mounting Procedures

Where specimens other than cell suspensions/growths or tissue sections are to be analyzed a variety of different mounting methods must be considered. As always, tools and substrates must be cleaned prior to use. For transfer to the TOF-SIMS, substrates can be mounted mechanically onto sample stubs, for example, using screws or a thin metal mask. Powdered specimens can be analyzed directly by pressing them into indium foil. Coverage should be kept to a sub-monolayer to avoid sample charging during analysis. Solid specimens may alternatively be analyzed as a thin (monolayer) film (*see* **Note 11**). This step generally results in the highest sensitivity. Liquid specimens can be deposited directly onto suitable substrates or spun cast (see above). Thin layers are less likely to present charging problems during analysis. Specimens should be air-dried before transfer to the TOF-SIMS. Individual fibers or hairs can be mounted on the usual substrate material using small strips of adhesive tape at each end (*see* **Note 12**). The specimen should be as flat to the substrate as possible to ensure secondary ions are extracted with the full kinetic energy. Fiber/hair mats can be mounted under a conducting mask that exposes a small region (few mm^2) to analysis. This minimizes contamination issues but requires more careful optimization of charge neutralization.

3.1.4. Transfer to TOF-SIMS

1. Isolate the fast entry vacuum lock from the remained of the TOF-SIMS instrument and vent it to dry nitrogen.
2. Insert the specimen, attached to the sample stub, into the fast entry lock and evacuate it to approx 10^{-3} mbar. If the sample is frozen then try to minimize the formation of frost on the sample stub or specimen in transit between LN_2 and the instrument. This can be achieved by working rapidly in a low-humidity environment and by using a transfer assembly with a hood over the sample stub.
3. Transfer the specimen to the preparation chamber and evacuate it to the base pressure of the analysis chamber (typically 10^{-9} mbar).

3.1.5. In Situ Freeze-Fracture

Assuming that a suitable device exists in the TOF-SIMS preparation chamber, the following procedure should be followed to successfully prepare a freeze-fractured, frozen-hydrated specimen.

1. Transfer the frozen sample to the pre-cooled (approx 100 K) freeze-fracture device.

2. Warm the specimen to the required fracture temperature (e.g., by using an on-stub heater filament) while maintaining the surrounding fracture stage elements (including the knife) as cold as possible (*see* **Note 13**).
3. Once the correct temperatures have been reached and the chamber pressure is at its minimum, use the knife to separate the two substrates of the specimen "sandwich," thereby fracturing the specimen.
4. Turn off the heater supply and re-cool the fractured specimen to approx 150 K (*see* **Note 14**).
5. Transfer the specimen to the precooled (approx 150 K) stage in the analysis chamber (*see* **Note 14**).

3.2. TOF-SIMS Acquisition

This section outlines the procedure for carrying out routine TOF-SIMS analysis of biological specimens. Additional information and tips for monitoring and optimizing instrument performance are given in the **Subheading 4., Notes**.

3.2.1. Choice of Primary Ion Beam

Many modern TOF-SIMS instruments are configured with more than one primary beam. Additionally some ion sources are capable of providing multiple ion species, for instance, Au_n^+ (n = 1–3). The choice of the most appropriate ion beam for a given analysis depends on the detailed requirements of the experiment and often the nature of the sample (*see* **Note 15**). The detailed tuning procedures for a given primary ion gun design vary widely and are documented by the manufacturers. The procedures for ensuring that the beam parameters are optimized for a given analysis are outlined to follow.

1. Obtain a primary beam on the sample surface (take care not to direct the beam onto the sample at this point) as detailed by manufacturers instructions.
2. Using a Faraday cup measure the primary ion current in the d.c. primary ion beam (*see* **Note 2**).
3. Ensure that the beam blanking is efficient by checking the measured current in "blanked" mode (*see* **Note 16**).

3.2.2. Alignment of Sample

It is important that the sample is correctly aligned to achieve optimum performance of the instrument. In particular, the primary ion axis and the secondary column (the ion axis of the TOF-MS) must intersect at the sample, in the region of interest. This situation can be established using a "test" specimen (*see* **Note 3**).

1. Using the TOF-SIMS optical microscope or CCD camera focus as a guide, adjust the sample stage micrometers such that the specimen is at the correct height for analysis (*see* **Note 17**).

2. Choose the region of interest for analysis. Some instruments provide software that can be used to automatically select a region of interest from the optical view.
3. It is good practice to verify that the sample position is correct by imaging the 'optical gate' of the instrument (*see* **Note 18**).

3.2.3. Choice of Operating Conditions

TOF-SIMS analysis can be performed in various modes, even with a single instrument. The user must decide which operational mode is most appropriate to the given analysis and set operating parameters accordingly. For example, the maximum mass resolution will be attained with the minimum primary ion pulse width and a narrow energy acceptance window for the mass analyzer. These conditions will degrade the sensitivity and lateral resolution of the measurement. Generally, a compromise condition can found that satisfies the various requirements of the experiment.

1. Using selectable apertures in the primary ion column, select the appropriate beam current and spot size (*see* **Note 19**).
2. By adjusting the pulse timings of the blanking voltage set the appropriate primary ion pulse width (*see* **Note 20**).
3. Select an appropriate field of view and pixel number for image acquisition (*see* **Note 21**).
4. Set the detector gain at the optimum value (*see* **Note 22**).
5. It is good practice to perform a sensitivity test at regular intervals (weekly) and after each time the detector setting has been changed (*see* **Note 23**).
6. Select a number of pulses to integrate for TOF-SIMS spectral acquisition or frames to sum for TOF-SIMS imaging, bearing in mind sample damage (*see* **Note 24**).

3.2.4. Charge Compensation

Many biological specimens are electrically insulating and will tend to acquire a positive charge during positive ion bombardment (*see* **Note 25**). To alleviate this phenomenon charge compensation is required, usually through the use of a low-energy (30 eV) electron flood-gun. The electron beam and secondary ion extraction field must be pulsed such that the low energy electrons are not accelerated by the extraction field. Pulse synchronization is such that the electron pulse arrives at the specimen between the primary ion pulses, when the extraction field is pulsed off.

1. Using a sample biased at +30 V measure the current from the electron flood-gun in continuous mode (*see* **Note 2**).
2. While monitoring the secondary ion signal on an adjacent specimen area to the region of interest, adjust the flood-gun current until the SIMS signal is maximized and stable (*see* **Note 26**).

3. Knowing the flood-gun pulse width and spot size calculate the electron dose density per pulse (*see* **Note 27**).
4. Calculate the maximum number of flood-gun pulses that can be used to ensure the integrated dose density is less than the damage threshold of 10^{14} electrons cm^{-2}.
5. If suitable TOF-SIMS data cannot be acquired within this number of pulses then consider adjusting the flood-gun spot size to reduce the dose density while maintaining charge compensation. Alternatively, consider other sample mounting methods, such a using a conducting grid over the sample to alleviate localized charging.

3.3. Data Analysis

TOF-SIMS spectra or (bio)organic specimens typically are composed of hundreds of peaks that contain information on the detailed chemical structure/ orientation and conformation. Extracting the maximum information is a significant challenge, particularly in images, where the data set can be extremely large, for example, 256×256 mass spectra each with 10^5 time channels. There is no single standard method of analyzing and interpreting TOF-SIMS data, but several approaches have proven successful, including the assignment of peaks from first principles, based on the rules of conventional (electron impact) mass spectrometry *(67)*. This approach is likely to prove most successful for single component systems. Another approach is to compare data with standard reference spectra, either acquired in-house, or available commercially *(68–70)*. This method is aided by accurate mass measurement, afforded by operating under high mass resolution conditions. Peak searching algorithms can help significantly in identifying possible matches. Finally, chemometric pattern recognition techniques which group data according to their location in multivariate space can aid the identification and classification of complex multi-component specimens such as cells *(14)*.

4. Notes

1. Single crystal silicon wafers provide an excellent and cost-effective substrate for many specimens. Precut sections (5×5 mm) are ideal for many applications and are available from Agar Scientific Ltd, Stanstead U.K. (www.agarscientific.com). All substrates must be cleaned before use (*see* **Note 8**).
2. A Faraday cup is used to accurately measure the current in the primary beam(s) without including contributions from secondary electrons. It is convenient to make a simple Faraday cup in each sample stub, or in a single "test stub" for routine use. For a focused ion beam, a 1-mm diameter aperture (such as used in many primary ion guns) attached with colloidal silver ("silver dag") over a hole in the sample stub 2 mm × 10 mm deep serves the purpose. The beam current is measured with a sensitive picoammeter or electrometer connected to the sample stage. If no Faraday cup is available, or where the primary beam is unfocused (for example the electron

flood-gun), then the sample can be biased to +30 V to suppress secondary electrons created during the primary ion/electron impact.

3. A typical test: sample might consist of the following specimens (1) a piece of domestic aluminum foil for measuring sensitivity. (2) a section of bulk polymer (PET, PMMA, etc.) with which to test charge compensation parameters. (3) a 200-mesh TEM (Cu/Ni, etc.) grid mounted over a hole, and (for higher resolution work) a small fragment of a used microchannel plate for measuring spatial resolution. (4) an approx 1 cm diameter thin layer of phosphor (e.g., Lumilux B47, Riedel-deltaën) for visualizing the primary beam(s) and electron beam.

4. For example, see Vickerman et al. *(70)* or www.surfacespectra.com

5. For example PLS Toolbox v.2.0 (Eigenvector Research, Manson, WA) for MATLAB v. 7.0 (MathWorks Inc., Natick, MA).

6. The vast majority of SSIMS instruments for this type of research are reflectron-TOF-MS, so the discussion is based on this generic form of TOF-SIMS instrument. Detailed procedures and functionality may vary slightly from one design of instrument to another, but the same general principles apply.

7. Ideally, substrates for TOF-SIMS should be flat and electrically conducting. Traditionally, single crystal silicon wafers have been used. Small pieces of shim-steel have been found to offer better cell adhesion during freeze-fracture.

8. SSIMS is extremely surface sensitive. Great care should be taken to avoid surface contamination, which often arises from packing and storage materials or from protective gloves that deposit organic compounds on substrates and tools used in sample preparation. Many of these compounds have a tendency to creep over previously clean surfaces! Scrupulous cleanliness must be observed. Domestic aluminum kitchen foil is very useful for providing convenient and clean surfaces to work on and transfer specimens with. Substrates, and specimen handling tools should be ultrasonically cleaned in appropriate solvents. Typically, a 5-min ultrasound in HPLC-grade methylene chloride, hexane, acetone, then methanol, followed by a DI water rinse and air drying is sufficient. A ubiquitous contaminant observed in many SSIMS spectra is (poly)dimethlysiloxane. The SSIMS spectrum of this compound includes the following peaks: m/z 73$^+$, m/z 147$^+$, and m/z 221$^+$.

9. It may be advisable to cool the sample (approx 250 K) before introduction it to the TOF-SIMS and to hold the sample at a similar temperature during analysis to avoid the loss of any semi-volatile material that remains after freeze-drying.

10. In general a freeze-drying temperature of −80°C and pressure of 10^{-3} mbar has been used for the preparation of cell cultures and tissue sections for TOF-SIMS.

11. To prepare a monolayer dissolve the specimen in a suitable solvent to make a 10^{-3} M solution and spin (~2000 rpm), cast a few drops onto a clean substrate. The use of etched silver as a substrate promotes the cationization of analyte molecules and further increases the sensitivity.

12. A brand of double-sided adhesive that has proven relatively free of mobile contaminant is "Scotch" tape (3M, St. Paul, MN, www.3m.com). The quantity of such tape used should be minimized to avoid excessive out-gassing in vacuum.

13. Experience with our BioTOF-SIMS instrument indicates that the optimum temperature for the specimen during freeze-fracture is approx 170 K. At this temperature, there is no significant net loss or gain of water ice from the exposed specimen surface. The optimum condition will depend on the temperature of the surrounding stage components (including the fracture knife) and the partial pressure of water in the vacuum chamber. To prevent recondensation of specimen water evolved during the fracture event the knife and surrounding stage should be as cold as possible. A pressure of 10^{-9} mbar or lower should be achieved in the fracture (preparation) chamber before the fracture is performed. A similarly low pressure must be maintained in the analysis chamber to avoid condensation of water or contaminants during the TOF-SIMS experiment.

14. The optimum specimen temperature for cryoanalysis must be determined experimentally and will depend on the nature of the sample, the residual vacuum components and the analysis time. The aim is to minimize sublimation or migration of component of the specimen while preventing condensation of contaminants from the vacuum. The BioTOF-SIMS cold stage has the facility for differential cooling such that the surrounding stage can be kept colder than the specimen and helps achieve this situation.

15. Atomic beams generally offer the smallest spot size and highest beam currents and may be appropriate for the detection of atomic and very low mass molecular ions at high spatial resolution. Electropositive primary ions such as Cs^+ enhance the yield of negatively charged secondary ions whereas electronegative ions such as O_2^+ enhance the yield of positive ions. As described in the Introduction, cluster beams give higher secondary ion yield for (bio)molecular species but not all of these can be microfocused routinely.

16. Leakage current from the primary ion gun when "blanked" is a common source of noise in TOF-SIMS spectra. It is a serious problem as it not only reduces signal-to-noise but also will contributed significantly to specimen damage and charging. To minimize it increase the blanking voltage or adjust the alignment of the primary beam in the column such that the blanking action is more efficient.

17. To ensure that the primary beam impact intersects the axis of the secondary column (and that of the electron gun) it is essential to compensate for differences in specimen thickness or mounting technique by finely adjusting the distance of the sample stage from the mass spectrometer. The term "height" here refers to the position of the specimen in the direction of the axis of the mass analyzer. Typically, the TOF-MS is mounted vertically.

18. The "optical gate" is the area on the specimen from which secondary ions can be efficiently detected. Essentially, it is the active area of the detector, projected back through the optical axis of the analyzer onto the specimen. It is visualized by expanding the raster size of the primary ion impact to an area bigger than the field of view of the mass analyzer and acquiring a SIMS image. When the primary and secondary optical axes are correctly aligned and the sample at the correct height the optical gate will appear as a bright circular region of signal centered in the

square raster frame. If the sample is at the wrong height the area of signal will appear off-centre in the resultant SIMS image.

19. Highest data rates will be obtained with high primary ion currents and correspondingly large spot sizes. For high lateral resolution imaging, smaller apertures should be used, giving small spot sizes and low currents. The choice of spot size is determined by the dimensions of the analyzed area and the number of pixel elements used (*see* **Note 21**). In all cases the effect of ion beam-induced chemical damage to the specimen must be considered. For atomic ion beams molecular analysis should be carried out within the *static limit*.

20. Shorter pulses offer the best mass resolution but reduce data rates. To ensure good peak shape and reproducible height in mass spectral signals, the primary ion pulse should exceed the time resolution of the counting electronics by a factor greater than 5. For example if a 1-ns time-to-digital converter is used the primary ion pulses should exceed 5 ns.

21. To avoid excessive oversampling or undersampling, the primary ion spot size should match the pixel size, which in turn is determined by the number of pixels available and the field of view of the analyzed area. For example, to achieve 0.5-μm lateral resolution with a similarly sized ion beam and a maximum or 256 × 256 pixels, the field of view should be set at approx 130 × 130 μm. The acquisition software may allow automatic selection of the appropriate field of view.

22. Most TOF-SIMS instruments use a microchannel plate detector (MCP), either in a dual "chevron" assembly or in combination with a scintillator and photomultiplier. The *gain* of the MCP is controlled by the applied voltage. As the MCP ages, a greater voltage must be applied to achieve the same gain, up to the voltage limit for a given MCP design. The detector performance should be monitored regularly (weekly) and the applied voltage adjusted to maintain optimum performance. A reliable and straightforward method of setting the gain is to measure a "gain curve," a plot of signal intensity vs applied voltage and to set the operating voltage at 30 V greater than the value that gives 50% of the plateau efficiency. It is advisable to carry out a sensitivity specification at the new detector setting (*see* **Note 23**).

23. A straightforward check of instrument sensitivity is conveniently performed on domestic aluminum foil (for example on the test sample). A section of the foil etched clean with approx 10^{15} ion cm^{-2} should yield a signal at $^{27}Al^+$ of 10^8 counts per nanocoulomb of primary ion dose. Care must be taken in choice of primary beam current to avoid saturation effects on this high-yield specimen; the number of counts in a given time channel of the spectrum should not exceed 1% of the number of pulses in the acquisition. This number is reasonable independent of the primary ion.

24. Beam-induced specimen damage is apparent by the loss of relative intensity of high-mass molecular ions caused by excessive fragmentation. The static limit (10^{13} ion cm^{-2} for atomic primary ions) has been discussed in the Introduction. The damage effects caused by cluster/polyatomic beams are subject to ongoing research. There is evidence that some such beams have a much reduced damage cross section, permitting molecular analysis beyond the "static limit."

25. Specimen charging arises principally through the loss of secondary electrons. It will distort the kinetic energy profile of secondary ions, shifting and broadening the peaks in the mass spectrum. In severe cases, the secondary ion kinetic energy will be moved outside the energy acceptance window of the TOF-MS and all signal will be lost. Differential charging will also affect lateral resolution. Conducting grids placed over the specimen have also been used to help reduce the effects of localized charge build-up, but the electron flood-gun technique has become the standard.

26. To avoid damaging the region of interest, this optimization should be performed on an adjacent region of the specimen with the same electrical properties. High-mass molecular signals are the best indicator of charge compensation performance because they are the first to be lost from the spectrum as the result of charge build-up. The minimum sufficient flood-gun current should be used to avoid additional damage to the specimen.

27. The flood-gun pulse width will typically be 10 to 100 μs. The spot size can be measured on a phosphor specimen floated to approx +200 V to ensure a high photon yield.

Acknowledgment

The author would like to thank Prof. J. C. Vickerman and members of his research group for their contribution. The financial support of the Engineering and Physical Sciences Research Council (EPSRC), Biotechnology and Biological Sciences Research Council (BBSRC), Association of International Cancer Research (AICR) is gratefully acknowledged.

References

1. Vickerman, J. C. and Briggs, D. (eds.) (2001) *ToF-SIMS: Surface Analysis by Mass Spectrometry.* Surface Spectra, Manchester and IM Publications, Chichester.
2. Benninghoven, A. (1973) Surface investigation of solids by the statical method of secondary ion mass spectroscopy (SIMS). *Surf. Sci.* **35,** 427–437.
3. Castner, D. G. (2003) View from the edge. *Nature (London)* **422,** 129–130.
4. Winograd, N. (2005) The magic of cluster SIMS. *Anal.Chem.* **77,** 143A–149A.
5. Niehuis, E., Heller, T., Feld, H., and Benninghoven, A. (1987) Design and performance of a reflectron based ToF-SIMS with electrodynamic primary ion mass separation. *J. Vac. Sci. Technol.* **5,** 1243–1246.
6. Schueler, B. (1992) Microscopic imaging by ToF-SIMS. *Microsc. Microanal. Microstruct.* **3,** 119–139.
7. Chandra, S., Smith, D. R., and Morrison, G. H. (1986) Imaging intracellular elemental distribution and ion fluxes in cultured cells with ion microscopy. *J. Microsc. (Oxford)* **144,** 15–37.
8. Braun, R. M., Blenkinsopp, P., Mullock, S. J., et al. (1998) Performance characteristics of a chemical imaging time-of-flight mass spectrometer. *Rapid. Commun. Mass Spectrom.* **12,** 1246–1252.

9. Cannon, D. M., Pacholski, M. L., Winograd, N., and Ewing, A. G. (2000) Molecule specific imaging of freeze-fractured, frozen-hydrated model membrane systems using mass spectrometry. *J. Am. Soc. Mass Spectrom.* **122,** 603–610.

10. Cliff, B., Lockyer, N. P., Corlett, C., and Vickerman, J. C. (2003) Development of instrumentation for routine ToF-SIMS imaging analysis of biological material. *Appl. Surf. Sci.* **203–204,** 730–733.

11. Davies, N., Weibel, D. E., Lockyer, N. P., Blenkinsopp, P., Hill, R., and Vickerman, J. C. (2003) Development and experimental application of a gold liquid metal ion source. *Appl. Surf. Sci.* **203–204,** 223–227.

12. Weibel, D., Wong, S., Lockyer, N., Blenkinsopp, P., Hill, R., and Vickerman, J. C. (2003) A C60 primary ion beam system for ToF-SIMS: its development and secondary ion yield characteristics. *Anal. Chem.* **75,** 1754–1764.

13. Gilmore, I. S. and Seah, M. P. (2000) Ion detection efficiency in SIMS: dependencies on energy. Mass and composition for microchannel plates used in mass spectrometry. *Int. J. Mass Spectrom.* **202,** 217–229.

14. Mellinger, M. (1987) Multivariate data analysis: Its methods. *Chemom. Intell. Lab. Syst.* **2,** 29–36.

15. Berry, J. I., Ewing, A. G., and Winograd, N. (2001) Biological systems, in *ToF-SIMS: Surface Analysis by Mass Spectrometry* (Vickerman, J. C., Briggs, D., eds.), Surface Spectra , Manchester and IM Publications, Chichester, pp. 595–626.

16. Belu, A. M., Graham, D. J., and Castner, D. G. (2003) ToF-SIMS: techniques and applications for the characterisation of biomaterial surfaces. *Biomaterials* **24,** 3635–3653.

17. Todd, P. J., Schaaff, T. G., Chaurand, P., and Caprioli, R. M. (2001) Organic ion imaging of biological tissue with secondary ion mass spectrometry and matrix-assisted laser desorption/ionisation. *J. Mass Spectrom.* **36,** 355–369.

18. Chilkoti, A. (2001) Biomolecules on surfaces, in *ToF-SIMS: Surface Analysis by Mass Spectrometry* (Vickerman, J.C, Briggs, D., eds.), Surface Spectra Ltd., Manchester and IM Publications, Chichester, pp. 627–650.

19. Lockyer, N. P. and Vickerman, J. C. (2004) Progress in Cellular Analysis using ToF-SIMS. *Appl. Surf. Sci.* **231–232,** 377–384.

20. Chen, J. X. and Gardella, J. A. (1999) Time-of-flight secondary ion mass spectrometry studies of in vitro hydrolytic degradation of biodegradable polymers. *Macromolecules* **32,** 7380–7388.

21. Ruiz, L., Fine, E., Voros, J., et al. (1999) Phosphorylcholine-containing polyurethanes for the control of protein adsorption and cell attachment via photoimmobilized laminin oligopeptides. *J. Biomat. Sci.-Pol. Ed.* **10,** 931–955.

22. Neff, J. A., Tresco, P. A., and Caldwell, K. D. (1999) Surface modification for controlled studies of cell-ligand interactions. *Biomaterials* **20,** 2377–2393.

23. Shi, H. Q. and Ratner, D. B. (2000) Template recognition of protein-imprinted polymer surfaces. *J. Biomed. Mat. Res.* **49,** 1–11.

24. Vargo, T. G., Thompson, P. M., and Gerenser, L. J. (1992) High lateral resolution imaging by SIMS of photopatterned self-assembled monolayers containing aryl azide. *Langmuir* **8,** 130–134.

25. Lhoest, J.-B., Wagner, M. S., Tidwell, C. D., and Castner, D. G. (2001) Characterisation of adsorbed protein films by ToF-SIMS. *J. Biomed. Mater. Res.* **57,** 432–440.
26. Wagner, M. S. and Castner, D. G. (2001) Characterisation of adsorbed protein films by ToF-SIMS with principal component analysis (PCA). *Langmuir* **17,** 4649–4660.
27. Wagner, M. S., Tyler, B. J., and Castner, D. G. (2002) Interpretation of static time-of-flight secondary ion mass spectra of adsorbed protein films by multivariate pattern recognition. *Anal. Chem.* **74,** 1824–1835.
28. Wagner, M. S., Shen, M., Horbett, T. A., and Castner, D. G. (2003) Quantitative analysis of binary absorbed protein films by ToF-SIMS. *J. Biomed. Mater. Res.* **64A,** 1–11.
29. Arlinghaus, H. F., Ostrop, M., Friedrichs, O., Feldner, J., Gunst, U., and Lipinsky, D. (2002) DNA sequencing with ToF-SIMS. *Surf. Interface Anal.* **33,** 35–39.
30. John, C. M., Odom, R. W., Salvati, L., Annapragada, A., and Lu, M. Y. F. (1995) XPS and ToF-SIMS microanalysis of a peptide/polymer drug delivery device. *Anal. Chem.* **67,** 3871–3878.
31. Davies, M. C., Brown, A., Newton, J. M., and Chapman, S. R. (1988) SSIMS and SIMS imaging analysis of a drug delivery system. *Surf. Interface Anal.* **11,** 591–595.
32. Belu, A. and Tarcha, P. (2001) Drug distribution by ToF-SIMS in phosphoryl choline-containing polymers used for drug delivery stent coatings. *Surfaces in Biomaterials Symposium 2001*, Scottsdale AZ.
33. Belu, A. M., Davies, M. C., Newton, J. M., and Patel, N. (2000) ToF-SIMS characterisation and imaging of controlled-release drug delivery systems. *Anal. Chem.* **72,** 5625–5638.
34. Mahoney, C. M., Roberson, S. V., and Gillen, G. (2004) Depth profiling of 4-acetamindophenol-doped poly (lactic acid) films using cluster secondary ion mass spectrometry. *Anal. Chem.* **76,** 3199–3207.
35. Leufgen, K. M., Rulle, H., Benninghoven, A., Sieber, M., and Galla, M. J. (1996) Imaging time-of-flight secondary ion mass spectrometry allows visualization and analysis of coexisting phases in Langmuir-Blodgett films. *Langmuir* **12,** 1708–1711.
36. Bourdos, N., Kollmer, F., Benninghoven, A., Sieber, M., and Galla, M. J. (2000) Imaging of domain structures in a one-component lipid monolayer by time-of-flight secondary ion mass spectrometry. *Langmuir* **16,** 1481–1484.
37. Pacholski, M. L., Cannon, D. M., Ewing, A. G., and Winograd, N. (1998) Static ToF-SIMS imaging of freeze-fractured, frozen-hydrated biological membranes. *Rapid Commun. Mass Spectrom.* **12,** 1232–1235.
38. Pacholski, M. L., Cannon, D. M., Ewing, A. G., and Winograd, N. (1999) Imaging of exposed headgroups and tailgroups of phospholipids membranes by mass spectrometry. *J. Am. Soc. Mass Spectrom.* **121,** 4716–4717.
39. Colliver, T. L., Brummel, C. L., Pacholski, M. L., Swanek, F. D., Ewing, A. G., and Winograd, N. (1997) Atomic and molecular imaging at the single-cell level with ToF-SIMS. *Anal. Chem.* **69,** 2225–2231.

40. Roddy, T. P., Cannon, D. M. Jr., Meserole, C. A., Winograd, N., and Ewing, A. G. (2002) Imaging of freeze-fractured cells with in-situ fluorescence and ToF-SIMS. *Anal. Chem.* **74,** 4011–4019.
41. Roddy, T. P., Cannon, D. M. Jr., Ostrowski, S. G., Winograd, N., and Ewing, A. G. (2002) Identification of cellular sections with imaging mass spectrometry following freeze fracture. *Anal. Chem.* **74,** 4020–4026.
42. Ostrowski, S. G., van Bell, C. T., Winograd, N., and Ewing, A. G. (2004) Mass spectrometric imaging of highly curved membranes during *Tetrahymena* mating. *Science* **305,** 71–72.
43. Jungnickel, H., Jones, E. A., Lockyer, N. P., Oliver, S. G., Stephens, G. M., and Vickerman, J. C. (2005) Application of ToF-SIMS with chemometrics to discriminate between four different yeast strains from the species *Candida glabrata* and *Saccharomyces cerevisiae. Anal. Chem.* **77,** 1740–1745.
44. Thompson, C. E., Jungnickel, H., Lockyer, N. P., Stephens, G. M., and Vickerman, J. C. (2004) ToF-SIMS studies as a tool to discriminate between spores and vegetative cells of bacteria. *Appl. Surf. Sci.* **231–232,** 420–423.
45. Cliff, B., Lockyer, N., Jungnickel, H., Stephens, G., and Vickerman, J. C. (2003) Probing cell chemistry with ToF-SIMS: development and exploitation of instrumentation for studies of frozen-hydrated biological material. *Rapid Commun. Mass Spectrom.* **17,** 2163–2167.
46. Gazi, E., Dwyer, J., Lockyer, N., et al. (2003) The combined application of FTIR microspectroscopy and ToF-SIMS imaging in the study of prostate cancer. *Faraday Discuss.* **126,** 41–59.
47. Fartmann, M., Dambach, S., Kriegeskotte, C., et al. (2002) Characterisation of cell cultures with ToF-SIMS and laser-SNMS. *Surf. Interface Anal.* **34,** 63–66.
48. Cheng, J. and Winograd, N. (2005) Depth profiling of peptide films with ToF-SIMS and a C60+ probe. *Anal. Chem.* **77,** 3651–3659.
49. Wucher, A., Sun, S., Szakal, C., and Winograd, N. (2004) Molecular depth profiling in ice matrices using C60+ projectiles. *Appl. Surf. Sci.* **231–232,** 68–71.
50. John, C. M. and Odom, R. W. (1997) SSIMS of biological compounds in tissue and tissue-like matrices. *Int. J. Mass Spectrom.* **161,** 47–67.
51. Todd, P. J., McMahon, J. M., Short, R. T., and McCandlish, C. A. (1997) Organic SIMS of biological tissue. *Anal. Chem.* **69,** 529A–535A.
52. Todd, P. J., McMahon, J. M., and McCandlish, C. A. Jr. (2004) Secondary ion images of the developing rat brain. *J. Am. Soc. Mass Spectrom.* **15,** 1116–1122.
53. Todd, P. J., McMahon, J. M., and Short, R. T. (1995) Secondary ion emission and images from a biological matrix. *Int. J. Mass Spectrom. Ion Proc.* **143,** 131–145.
54. Sjövall, P., Lausmaa, J., and Johansson, B. (2004) Mass spectrometric imaging of lipids in brain tissue. *Anal. Chem.* **76,** 4271–4278.
55. Touboul, D., Halgand, F., Brunelle, A., et al. (2004) Tissue molecular ion imaging by gold cluster ion bombardment. *Anal. Chem.* **76,** 1550–1559.
56. Delcorte, A. and Bertrand, P. (2005) Metal salts for molecular ion yield enhancement in organic secondary ion mass spectrometry: a critical assessment. *Anal. Chem.* **77,** 2107–2115.

57. Delcote, A., Médard, N., and Bertrand, P. (2002) Organic secondary ion mass spectrometry: sensitivity enhancement by gold deposition. *Anal. Chem.* **74,** 4955–4968.
58. Wu, K. J. and Odom, R. W. (1996) Matrix-enhanced secondary ion mass spectrometry: a method for molecular analysis of solid surfaces. *Anal. Chem.* **68,** 873–882.
59. Altelaar, A. F., van Minnen, J., Jiménez, C. R., Heeren, R .M. A., and Piersma, S. R. (2005) Direct molecular imaging of *Lymnaea stagnalis* nervous tissue at subcellular spatial resolution by mass spectrometry. *Anal. Chem.* **77,** 735–741.
60. Nygren, H., Eriksson, C., Malmberg, P., et al. (2003) A cell preparation method allowing subcellular localization of cholesterol and phosphocholine with imaging ToF-SIMS. *Coll. Surf. B:Biointerfaces* **30,** 87–92.
61. Nygren, H. and Malmberg, P. (2004) Silver deposition on freeze-dried cells allows subcellular localization of cholesterol with imaging TOF-SIMS. *J. Microsc. (Oxford)* **215,** 156–161.
62. Nygren, H., Malmberg, P., Kriegeskotte, C., and Arlinghaus, H. F. (2004) Bioimaging TOF-SIMS: localization of cholesterol in rat kidney sections. *FEBS Lett.* **566,** 291–293.
63. Zimmerman, P. A., Hercules, D. M., and Naylor, E. W. (1997) Direct analysis of filter paper blood specimens for identification of Smith-Lemli-Opitz syndrome using time-of-flight secondary ion mass spectrometry. *Am. J. Med. Gen.* **68,** 300–304.
64. Seedorf, U., Fobker, M., Voss, R., et al. (1995) Smith-Lemli-Opitz syndrome diagnosed by using time-of-flight secondary-ion mass-spectrometry. *Clin. Chem.* **41,** 548–552.
65. Muddimann, D. C., Gusev, A. I., Proctor, A., Hurcules, D. M., Venkataramanan, R., and Diven, W. (1994) Quantitative measurement of cyclosporin A in blood by time-of-flight mass spectrometry. *Anal. Chem.* **66,** 2362–2368.
66. Muddimann, D. C., Gusev, A. I., Martin, L. B., and Hurcules, D. M. (1996) Direct quantification of cocaine in urine by time-of-flight mass spectrometry. *Fres. J. Anal. Chem.* **354,** 103–110.
67. McLafferty, F. W. and Turecek, F. (eds.) (1993) *Interpretation of Mass Spectra, 4th ed.* University Science Books, Sausalito, CA.
68. Briggs, D., Brown, A., and Vickerman, J. C. (eds.) (1989) *Handbook of Static Secondary Ion Mass Spectrometry (SIMS).* Wiley, Chichester, U.K.
69. Newman, J. G., Carlson, B. A., Michael, R. S., Moulder, J. F., and Teresa, A. H. (eds.) (1991) *Static SIMS Handbook of Polymer Analysis.* Physical Electronics, Eden Prairie, MN.
70. Vickerman, J. C., Briggs, D., and Henderson, A. (2002) *The Static SIMS Library Ver. 3.* SurfaceSpectra Ltd., Manchester, UK.

Using SIMS and MIMS in Biological Materials

Application to Higher Plants

Nicole Grignon

Summary

Analytical imaging by secondary ion mass spectrometry allows the precise cartography of elements and isotopes at subcellular level in biological samples. Multielemental coacquisition from the same sputtered zone gives new prospects in the study of biological matrix, where many elements coexist and move according to cellular metabolism. For plant studies, localizing and quantifying compartmentations of mineral nutrients is important to understanding their metabolism. The use of stable isotopes as analogous markers is an efficient strategy of labeling for transport and metabolic studies. As for all microanalytical techniques, the stabilization of biological material is essential for the use of secondary ion mass spectrometry/multi-isotope imaging mass spectrometry microprobes. The techniques chosen are explained: decontamination, cryoprocessing, sections in embedded material (*Arabidopsis thaliana,* Brassicae) and analysis. In the Notes section, we will comment on the need to adapt the protocol described to the variety and complexity of the biological materials. Other proven protocols established in other laboratories for other type of organisms or biological questions will also be introduced.

Key Words: Multielemental mass spectrometry imaging; biological sample cryoprocessing; plant inorganic subcellular microlocalizations; nitrogen isotope labeling; high mass resolution; HMR; high spatial resolution.

1. Introduction

Analytical imaging using secondary ion mass spectrometry (SIMS) provides images of the chemical elements and isotopes and quantitative data from biological specimens. SIMS allows the use of stable isotopes as markers to localize microscopically physiological or molecular events in biological samples. One specificity of SIMS is the excellent results obtained for light elements,

From: *Methods in Molecular Biology, vol. 369*
Electron Microscopy: Methods and Protocols, Second Edition
Edited by: J. Kuo © Humana Press Inc., Totowa, NJ

which are the main components of organisms. The instrument used for SIMS analyses is 4F Cameca microprobe.

The new multi-isotope imaging mass spectrometry (MIMS) technology is a further development from SIMS. The Nanosims 50 Microprobe (CAMECA, Courbevoie, France), used for MIMS, is particularly well adapted to biological needs with: (1) a high lateral resolution and (2) a parallel detection of five different elements from the same microvolume, with a high collection efficiency. Nanosims 50 performances make it highly suitable for cellular and intracellular analyses. They offer the possibility of simultaneous studies of different soluble ions at the same place. Moreover, the simultaneous acquisition of signals allows precise isotopic ratio measurements with a perfect localization of labelings in metabolic studies. The sensitivity of the microprobe offers a spatial resolution similar to that of an electron microscope (50 nm) and a high analytical sensitivity (parts per million to parts per billion) in minute volumes of material. Thus, the preparation of biological samples has to combine the techniques used for transmission electron microscopy (TEM) microanalysis and mass spectrometry.

The problem of sample preparation for biologists is crucial: the material is highly fragile and heterogeneous in face of the demands of the apparatus. The aim of sample preparations is to respect both the native structures and the native location of soluble and insoluble elements. Avoiding contamination during sample preparation is of paramount importance to detect trace elements. The general conditions for decontamination are those recommended for mass spectrometric studies.

The method used for plant sample preparation is common to SIMS and MIMS. The chosen techniques are cryoprocessing, resin embedding, dry sectionning, and the depositing of sections on metal plates for direct analysis. Handling sections on water surfaces is excluded, because it leads to major displacement of elements. Images and quantifications are performed differently with 4F and Nanosims 50 instruments. Relative measurements of elements and isotope ratios of the same element are obtained with references to standards. The universally used mineral standards are replaced by biological standards, which may take into account isotopic discrimination and matrix effects specific to living cell context. Isotopes are used as markers in biological systems to trace element flows and to reveal dynamic processes which result in a given spatial distribution. Stable isotopes are of special interest for SIMS and MIMS analyses because they are safe and compatible with the relatively long duration of sample preparation.

Finally, it is essential to bear in mind that these techniques are not simple recipes. They must be adapted to each biological material and to the various types of equipment. General principles, such as the avoidance of trace contamination, the preservation of native element distribution, and protection against artifacts, must be kept in mind at all time during the procedures (*see* **Note 1**).

2. Materials

2.1. Equipment

2.1.1. Heavy Equipment

1. Laboratory: a room easy to clean equipped with a fume hood.
2. Freezer (−120°C). A home-made device is conceived for high speed cooling just after sampling: Dewar fitted with grids capable of holding an inox 70-mL becher and a homemade sample projector. The injector is fixed to the top center of an insulating plexiglass lid of the liquid nitrogen Dewar *(1)*. Rapid congelation is performed by plunging the samples with the spring-driven mechanical injector, through an hole managed across the lid.
3. Three gases: (1) liquid industrial propane in tanker equipped with safety manometer; (2) liquid nitrogen in a tanker equipped with a manometer; (3) argon container equipped with a manometer.
4. Ultramicrotome Reichert-Jung Ultracut E (Vienna, Austria) equipped with a 35° diamond knife (DIATOME Ltd., Bienne, Switzerland) and an antistatic device Static Line Ionizer (*[2]* DIATOME Ltd., Bienne, Switzerland).
5. Light microscope, equipped with a camera.
6. TEM.

2.1.2. Small Equipment

1. Magnetic stirrer, oven at 60 to 65°C, hot plate.
2. Polystyrene storage units, new, unpowdered gloves, aluminum foil.
3. Forceps, wooden needles, filter paper, hair dryer, polystyrene boxes, Melinex film (TAAB Lab. Equipment LTB, UK), embedding molds.

2.1.3. Reagents

1. Nonionic detergent.
2. Decontamination solvents: pure ethanol, pure acetone.
3. Preparation solvents: ultrapure ethanol: 3 L; pure acetone: 3 L; ultrapure dehydrated acetone (Merck, HPLC grade) 2 L.
4. Freshly prepared ultrapure water (MilliQ), 2-d stockage maximum.
5. Spurr's resin (TAAB Lab. Equipment LTB, UK), total volume 1 L (*see* **Table 1**).

2.2. Plant Sample Preparation Requirement

2.2.1. Small Equipment for Cryofixation and Embedding, Decontaminated in Advance

1. Argon dried by circulating through a tube filled with silica-gels, placed at the bottom of the freezer.
2. Thermometer (−150°C), two long 20-mL glass pipets.
3. Large metal boxes containing 50-mL covered pots half filled with decontaminated glass pellets, cryo-tubes (Nunc for example) 1.8 mL or 3 mL externally threaded,

Table 1
Spurr's Resin Embedding Baths

Bath No.	ERL	DER[a]	NSA	S1[b]	Acetone	Temperature
1	1	No	No	No	Yes	−95°C
2	1	0.6	No	No	Yes	−80°C
3	1	0.6	2.6	No	Yes	−40°C
4	1	0.6	2.6	No	No	+4°C
5	1	0.6	2.6	0.04	No	Room temperature

ERL, ERL 4206 (vinyl cyclo hexene-dioxide); DER, DER 736 (diglycidyl-ether of propylen glycol); NSA, nonenyl succinyl-anhydre; S1, dimethylamino-ethanol-alkyl alkanol–amine.

The indicated amounts are relative (w/w). When indicated, pure acetone is added. The volume of acetone is equal to that of the sum of the other components. The mixtures are liquid at the indicated working temperatures.

[a]Flexibilizer.

[b]Accelerator.

punched all over at the top and on the sides with minute holes to eliminate bubbles after immersion, ordinary woolen gloves covered with thin rubber gloves for the manipulations inside the cold freezer.

2.2.2. Small Equipment for Sections and Dry Sections

1. Acupuncture needles or teflon needles, Petri glass dishes, polyethylen tubes to store the sections.
2. Microprobe sample-holders: cylinders or plates made of pure metals: tantalum (99.9%, Goodfellow, England), aluminum, copper.
3. Polisher equipped with polishing disks (ESCIL, Chassieu, France): alumina, colloidal silica, or diamond powders of 100-µm, 75-µm, 10-µm, 3-µm, 0.3-µm, and 0.1-µm grade.

2.2.3. Histology

1. Stain: 0.02% Azure A–0.13% methylen blue in 50% water, 10% methanol, 10% glycerol, 30% phosphate buffer, 0.1 *M*, pH 6.9.
2. For TEM: uncoated copper grids, 75/300 mesh, contrasting metals: saturated uranyl acetate in 100% methanol, lead citrate, 10% aqueous osmium vapors (w/v).

2.3. Ionic Microscopy

2.3.1. SIMS 4F Microprobe
(CAMECA S.A., Courbevoie, France) Main Characteristics

1. Primary sources: Cesium+ (Cs^+) and Oxygen+ (O^+), produced by a duoplasmatron. The angle between primary column and secondary column is ca. 30 degrees.
2. Optical system composed of electrostatic lenses; analytical system: mass spectrometer, with optical collection of the secondary ions.

3. Electron gun for charge neutralization.
4. The analytical image of each selected element is visualised with a multichannel plate, a fluorescent screen and a camera linked to an image processing system.
5. Imaging and counting are performed using the microprobe mode (scanning mode).

2.3.2. Nanosims 50 Microprobe: Main Characteristics

For more information, please see Slodzian et al. *(3)* and Hillion et al. *(4)*.

1. Primary sources: Cs^+ and O^- (produced by a duoplasmatron), with normal incidence to the specimen surface ensuring optimal focalisation of the primary beam.
2. The primary beam and the secondary ion beam collected coexist in a coaxial column and are focused. This configuration calls for primary and secondary ions of opposite polarity and equal energy. Cs^+ bombardment sputters atoms with a high electron affinity; the secondary column optically collects the negative secondary ions. O^- bombardment enhances yield of species with a low ionization potential; secondary positive ions are collected. Movable detectors allow parallel detection of 5 different species from the same sputtered volume.

3. Methods

The sample preparation is common for SIMS and MIMS analyses. It is impossible to forsee the emission of secondary ions obtained from a compound included in a complex matrix like biological samples. There have been so far no theoretical studies on this point. Therefore, for analyses of new compounds, various protocols for preparations and analyses must be tested. The results can be compared to other techniques, when possible. Moreover, a thorough knowledge of artifacts produced by the different modes of preparations or analyses, and by the microprobe itself is helpful.

3.1. Preliminary Operations

3.1.1. Decontamination: Progressive Steps (see **Note 1**)

1. The room and the largest containers are cleaned with detergents or acids, the manipulations being done with non-powdered plastic gloves only. The containers are rinsed with distilled water, then with ultrapure water (10 successive baths) and dried. The flasks and tubes which will be directly in contact with the samples are washed in an ultrasonic bath with diluted detergents, then rinsed in 10 successive baths of tap water and 10 successive baths of ultrapure water, dried in an oven and stored in aluminum foil or in closed containers. All the instruments are decontaminated in the ultrasonic bath, using the same protocol.
2. Metal plates and holders: the biological sections are deposited on metal plates, first perfectly polished by the sliding scale polishing disks under water and decontaminated. A mirror finish is essential for good adherence and it is particularly important for the quantifications. To be cleaned, each plate is slightly rubbed between two fingers covered with gloves. All the plates are placed vertically in a home

made device (a plastic holder for pipet cones), then put into an ultrasonic bath
for a sequence of various cleaning solvents. For each manipulation the plates
are maintained under liquid to avoid air contaminations. They undergo 20 baths
altogether alternating boiling ultrapure water and boiling high-purity ethanol or
acetone. Between each bath, the plates are not dried, but excess liquid is removed
on a filter paper. For the final bath the plates are rinsed with ultrapure acetone in
quartz tubes. Finally, after drying in an oven, the holders are stored in polystyrene
Petri dishes.

3.1.2. Preparation of Plant Material

1. Plants are hydroponically cultivated.
2. An horizontal freezer turned off (at room temperature) is filled with all the materials which will be used for the dehydration and cold embedding.
3. Two-liter ultrapure dehydrated acetone *(5)*, 20-mL glass pipets, the −150°C thermometer.
4. Four metal boxes, open, containing 50-mL glass pots half-filled with decontaminated glass pellets.
5. An open vacuum (Pyrex) evaporator to contain the final standard medium. At the
   end of the embedding, this container will be filled with of argon, closed and taken
   out of the freezer (*see* **Subheading 3.2.2.**). The five mixtures of Spurr's resin (four
   intermediary baths plus final standard medium *[6]*) are prepared shortly before the
   experiment, at most a few days, and stored at 10°C, avoiding moisture. Handling of
   the resins must be performed under a fume hood. The different liquids (*see* **Table 1**)
   are added one by one and thoroughly mixed with a magnetic stirrer before the next
   constituant is added, then degased. The four intermediary media (baths 1 to 4, **Table
   1**) are distributed among the 50-mL pots at the bottom of the freezer at room temperature, in the metal boxes. The pots containing the fifth lot of resin (complete
   standard medium) are placed in the uncovered vacuum evaporator with its lid
   ready to close it inside the freezer.
6. The freezer is switched on only when all the necessary equipment has been placed
   inside. The freezer tank will be filled with argon every time it is open to transport
   tubes from one bath to the other.

3.2. Experiments

3.2.1. Excision and Cryofixation

All first manipulations must be particularly rapid to obtain the best possible
results respecting the native characteristics of the cells (biological structures
and true locations of elements, soluble and insoluble).

1. Excision of the samples is done in a few seconds or minute (1 mm^3) to limit the
   excision stress (*see* **Note 2** and also Glass *[7]*) and to avoid the use of buffer, fixative or water (*see* **Note 3**). Each sample, without any cryoprotectant *(1)*, is mounted
   at the end of fine forceps which are quickly clipped onto the vertical injector of
   the plunge device and projected at 3 m/s speed into the liquid propane (*see* **Note 4**

and also Steinbrecht and Zierold *[8]*). After each cryofixation the forceps are dried with a hair dryer. The ideal state for samples should be water freezed in a vitreous state in each part of the tissues (*see* **Note 5**).

2. After 5 min in propane, samples are stored in liquid nitrogen in the perforated Nunc tubes. At this stage, the frozen samples are brittle ; they can be easily cut up inside liquid nitrogen to select the parts of interest. All operations should be completed within 3 to 4 h.

3. Samples in the Nunc tubes can then be set aside and stored in liquid nitrogen for later use. Otherwise, the Nunc tubes are deposited in 50-mL glass pots full of acetone at −95°C avoiding gas bubbles (*see* **Note 6**).

3.2.2. Dehydration and Embedding

1. Samples are dehydrated using three successive baths (3000-fold the sample volume minimum) of ultrapure dehydrated acetone over the course of 3 wk, at −95°C, without any fixative or molecular sieve added. To prevent hydration by water condensation, the tank is filled up with dry argon before opening. Acetone is renewed by sucking with 20-mL pipets in each pot and new acetone is poured directly from the acetone bottle into the pots at the bottom of the freezer. While the freezer is open, the temperature is watched at sample level. It must remain less than −90°C (*see* **Note 7**). For impregnation, the Nunc tubes containing the plant samples are transferred from acetone pots to resin pots (*see* **Notes 8** and **9**).

2. Samples are progressively infiltrated in 4 successive 24-hr baths of intermediate resin mixtures (*see* **Note 10**). Between each bath, the freezer temperature is progressively raised to the new temperature: −80°C, −40°C, +4°C The samples are rapidly transferred to a new bath of complete resin (standard medium) at room temperature and then mounted in drops of fresh resin between Melinex sheets. These are deposited at 65°C. The small volume of the sample makes the solidification rapid, but polymerization is not completed before 6 hr minimum. The Melinex slides must be stored in dry conditions.

3.2.3. Sections

1. At cutting stage, the upper Melinex sheets are removed and resin blocks are glued onto each sample with superglue or with liquid resin which is repolymeryzed. The blocks are dry stored in vacuum flasks in the dark.

2. Dry sections SIMS being a surface analyzing technique, the biological section must be as flat as possible (*see* **Note 11**).

3. Embedded blocks are trimmed to obtain one mm-large maximum surface. For SIMS, 0.5-μm thick sections (500 nm) are dry-cut at room temperature with a 35° diamond knife and handled with acupuncture or Teflon needles and the antistatic device. The latter is also used in the next step to ensure the deposition onto the metal plates by reversing the current (*see* **Note 12**).

4. Sections are deposited on polished metal plates and glued by heating them on a hot plate (10 min at 60°C). The plates are heated in glass Petri dishes for 6 to 14 h at 135°C.

5. The sections are then individually stored in polystyrene tubes in a dry atmosphere at room temperature. These sections must be analyzed within 2 mo because of soluble element diffusion risk. The sections used for SIMS (0.5 μm) are cut alternating with 2-μm sections deposited on glass slides for an optical microscopy.

3.2.4. Histological Controls

For light microscopy, the slides are stained with a drop of Azur A/ Methylen blue, 5 min on a hot plate at 60°C, then rinsed with tap water, and finally rinsed with freshly distilled water to differentiate cellulosic and woody structures. Histological observations on photographs of optical sections parallel to the SIMS sections will help determine the areas to be analyzed when the samples are in the machines.

The use of TEM allows one to observe the quality of preservation of complex, compartmentalized cell structures. 60- to 80-nm sections on copper grids are contrasted by classical uranyl-acetate and lead citrate or 10% aqueous OsO_4 vapors.

3.3. Ionic Microscopy

Basic knowledge of main ionic-microscopy principles is a good assistance for the biologists (*9,10*). Instrumental considerations like bombardment choice, energy, sputtering, collection of secondary ions, chemical considerations like environment or matrix are helpful for managing experiments. Practical considerations like insulation of samples, sensitivity, mass interferences, are essential to avoid miss-interpretations of some SIMS results.

3.3.1. SIMS 4F Analysis

3.3.1.1. CHOICE OF OPERATING MODE (PRIMARY ION BOMBARDMENT)

Choose the more convenient detection mode from ionization potential and electron affinity. In scanning mode, the spatial resolution is defined by the size of the probe. The number of atomic layers sputtered per second depends on the density of the primary ion beam and the surface density. Part of the sputtered matter is composed of positive or negative single or cluster ions. These secondary ions are characteristic of the atomic composition of the sputtered azrea. They are energy and mass filtered for analysis.

With O^+ bombardment, the main soluble cations in the cells are clearly localized (K^+, Na^+, Ca^+, Mg^+; *see* **Note 13**). Cs^+ as a primary bombardment sputters negative secondary ions (elements with a high electron affinty). The major ions N^-, S^- map the membranes separating the different intracellular compartments. Some elements emit positive as well as negative secondary ions. For example, Ca^+ is usually analyzed with O^+ bombardment, but may be identified with a Cs^+ bombardment when abundant. It is possible to analyze a same zone with O^+ and Cs^+ successively (*11*).

3.3.1.2. General Spectra of Each Biological Sample

The qualitative mass spectra are obtained in the low mass range (1 to ca 120 dalton), with the two-source bombardments. They give complex characteristic profile, even when cells are disorganized. Furthermore the various elements appear as complex structures, with light elements being the most abundant. At higher sensitivity it is possible to determine the elements of interest.

Two examples of operating conditions in scanning mode are given (*see* **Note 14**): (1) Using O^+ primary beam: primary beam energy 12.5 KeV; primary ion current 10 nA; analyzed image field 150 μm × 150 μm. (2) Using Cs^+ primary beam: primary beam energy: 10 KeV, primary ion current: 10–50 pA, analyzed image field: 60 × 60 to 150 × 150 μm, negative secondary ion beam: –4.5 KeV or –7 KeV.

Detection mode: electron multiplier, typical mass resolution: 2000 (for nitrogen: 4500 or 6000). Counted spots: 8 μm in diameter, acquisition time: 1 s for ^{14}N and 60 to 100 s for ^{15}N.

3.3.1.3. High Mass Resolution (HMR) Spectra

At each mass, the simple element appears as the first peak on the spectrum, then polyatomic ions of other species which have a higher mass (*see* examples in **Note 15**). Morrison table *(12)* gives the list of possible polyatomic ions. At masses of interest, HMR is a prerequisite in biology (minimum m/Δm = 800). Peaks with dissymmetrical shape indicate superimposition of two or more element peaks, which will be separated with HMR.

3.3.1.4. Imaging Protocol

An ionic image is defined by the spatial resolution (which depends on the probe diameter). Image quality depends on the transfert-function adjustment, the contrast shows the analytical sensitivity of the apparatus. The mass spectrometer has focal properties with an (1) electrostatic sector and (2) a magnetic sector. Analysis specificity depends on mass resolving power (R) which is the ratio between the mass (m) of the chemical species of interest and the smallest mass difference with other masses (Δm) – polyatomic ions, for example: $R = m/\Delta m$.

1. Before observing the biological specimen, fit the microprobe to obtain grid-test image (at mass 63 for Cu grid) as perfect as possible, without aberrations.
2. For the first image of a biological sample, choose a mass giving a high emission of secondary ions, such as mass 12 or mass 26 for Cs^+ source, or mass 39 or mass 40 for O^+ source. This image will appear only after a few minutes sputtering. At one given mass, the emission intensity (in fact the ionization efficiency) depends on numerous factors: primary ion implantation, chemical state of the sample surface (matrix and primary ions of the structures), element concentration and physical characteristics of the ionized element, characteristics of each section (adherence,

conductivity) and various source changes. The diameter of the primary beam spot is the determining factor of the spatial resolution. Other factors limit the spatial resolution, but with lesser effect. They are a too low intensity of the secondary beam, the electronic noise and chemical interferences (*see* **Note 15**).

3. Analytical images of various masses are acquired alternately at each passage in scanning mode, separated by a few seconds' relaxation time. Raw mass images of 256×256 pixels coded in integer values are integrated for a few seconds for each mass (*see* **Fig. 1**).The acquisition time for each element depends on the emissivity or the chemical concentration of the element. This process can be repeated 50 to 100 times. The best lateral resolution power is about 200 nm. To increase the mass resolving power of the mass spectrometer, without transfert-optic lenses, contrast and aperture diaphragms must be reduced. The collection efficiency and the field are reduced, thus the sensitivity is reduced. Mass resolution is determined by the width of the filter slit. The limit of sensitivity depends on the equilibrium between the reduction of the entrance slit and the use of transfert optics. The current intensity of the primary ion beam cannot be known precisely, both the diaphragm aperture and position of the beam must be kept constant.

4. To compare two images, it is necessary to refer to a set of species with uniform distributions in the section, for example, carbon, oxygen, and nitrogen contained the resin matrix. The sensitivity of SIMS depends on the purity of the sample. In biology, mass interferences due to a high number of elements componing biological systems hamper analysis performances of the apparatus.

5. When choosing between different isotopes of one same element (*see* **Note 16**), generally, the most abundant isotope is used for the analysis. Images of the different isotopes are virtually identical, except when one of the isotopes has been introduced as a tracer and has not reached the equilibrium distribution. In some cases, it may be advisable to work at mass of an isotope other than the main one if polyatomic ions interfere with the latter. Comparison between mass spectra and images obtained at the different isotopic masses enable to select the mass leading to the "purest" image.

3.3.1.5. CHECKING OF SAMPLE QUALITY

Comparison between the locations of several ions can be used to check the quality of the sample preparation (*see* **Note 17**). With O^+ primary beam, the most abundant cations (K^+, Na^+, Ca^{2+}, Mg^{2+}) are all located in various cellular compartments, with different concentrations, specific of living state (*see* **Fig. 1**). Indications of artifacts are: (1) each cation is no longer restricted to habitual compartments; (2) all the cations are colocalized *(13)*.

3.3.1.6. QUANTIFICATION PROTOCOL (USING NITROGEN AS AN EXAMPLE)

For plant preparation *see* **Note 18**.

1. Cs beam, *see* **Subheading 3.3.1.** The grid image must be as symmetrical as possible. Images in scanning mode are performed to select tissues or cells of interest.

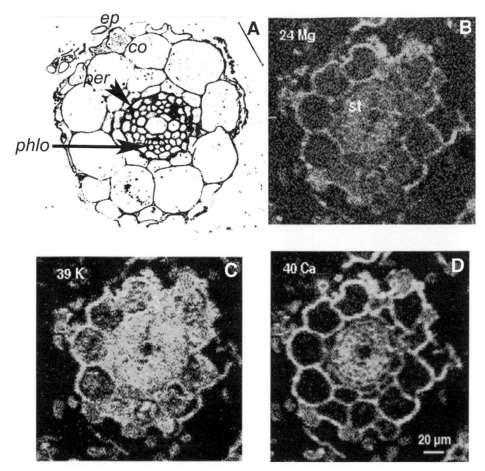

Fig. 1. Optical (A) and SIMS 4F images (B, C, D) of *Arabidopsis thaliana* root. (A) Transversal section shows different layers of the root tissues: ep, epidermis ; co,: cortical parenchyma ; per, pericycle ; phlo, phloem. Natural distribution of major soluble cations inside the cells. ^{24}Mg (**B**), ^{39}K (**C**), and ^{40}Ca (**D**) elements are naturally localized in the cell wall surrounding all the cells because they are present in the nutrient medium. Mg^{2+} and K^+, known as the main cytoplasmic cations, are detected inside the cells; Ca^{2+} is not visible inside the cells (Ca is known to be strongly excluded from cytoplasm by various transport systems). Bar = 20 μm.

2. Spectra at mass 27 (first peak is ^{15}N^{12}C$^-$) and mass 26 (^{14}N^{12}C$^-$), in the resin or unlabeled tissues, with HMR (m/Δm 4800 minimum is required; *see* **Note 19**). At each mass, respectively, three (mass 27) or two (mass 26) separate peaks must be obtained. High mass resolution is obtained by reducing the size of the crossover in the entrance stop with the transfert optics and by filtering the beam in the energy slit to reduce chromatic aberrations. The natural (terrestrial) ^{15}N/^{14}N ratio is used

as reference to match the spectrometer. It is also used as reference standard in un-labelled standards *(14)*.

3. Conditions for determining isotope ratio differ from those used for imaging. To quantify local enrichments of ^{15}N nitrogen, the $^{15}N/^{14}N$ ratio is calculated as R_2/R_1, where R_1 stands for $^{15}N^{12}C^-/^{12}C^-$ and R_2 for $^{14}N^{12}C^-/^{12}C^-$. In fact, the $^{15}N/^{14}N$ ratio may be directly measured when the beam intensity is constant along the two measurements. In our cryoprocessed plant samples, $^{15}N/^{14}N$ ratio is determined with numbers of repetitions ranging from three to five plants and to several hundreds of cells in various tissues. In parallel, measurements of reference elements known as constant in all samples are made. $^{12}C^-$ and $^{15}N/^{14}N$ natural ratio are periodically measured in the resin surrounding the sample or in intercellular spaces, one of 5 to 8 spots (*see* **Note 20**). With SIMS 4F apparatus, the natural $^{15}N/^{14}N$ ratio must be frequently verified because of possible variations of the magnetic field: a pro-gressive shift of ratio (growth of 27 as compared with 26) indicates a lower mag-netic field. If internal standard measure is below the natural value, indicating a change in signal intensity, the previous 8 measurements are discarded.

4. (Semi) quantification (*see* **Fig. 2**). The measured $^{15}N/^{14}N$ values are compared with a calibration curve of biological standards made in the same conditions as the experiments (*see* **Note 21**). These standards are values measured by SIMS or conventional mass spectrometry in plants expected to be at isotopic equilibrium with nutritive solutions with different $^{15}N/^{14}N$ ratios, after long period of growth. The localized quantifications of isotope enrichments in labeled samples are easier and more accurate when referenced to this curve (*see* **Note 22**).

3.3.2. Nanosims Analyses

Many characteristics of technical basic knowledge are the same as for SIMS; in the text they will be noted "as is for SIMS." Basic principles and handling will be described here ; however, it is impossible to manage the apparatus without specialist advices and several weeks of training with technicians. Optimal per-formances of the microprobe depend on the quality of the samples and a per-fect management of the apparatus.

3.3.2.1. PRIMARY ION BOMBARDMENT

The primary ion bombardment is characterized by normal incidence of the primary ion beam: Cs^+ or O^-. (The first steps, loading and pumping, are sensi-tive steps for a beginner.) The vacuum in the analysis chamber is 1.10^{-9} torr and 2×1.10^{-9} torr for Cs^+ and O^- respectively. Immersion lens ensures the focusing of the primary beam anf the extraction of the secondary ions. With Nanosims 50, the short working distance between the sample surface and the extraction system allows good spatial resolution but require a perfect surface of section. The double focusing mass spectrometer is characterized by good transfer opti-cal system, entrance and aperture slits with 5 apertures.

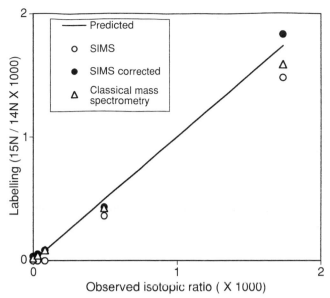

Fig. 2. Calibration curves made from biological standards. Measurements by secondary ion mass spectrometry (SIMS) and classical (gaseous) mass spectrometry. Shown are SIMS (open circles) isotopic ratios in the tissues of plants fed for 21 d with medium containing KNO_3 enriched with 0%, 0.3%, 1%, 5%, and 17% ^{15}N (tissues at isotopic equilibrium with the medium). SIMS results (solid circles) were corrected for the N content of the resin. Isotopic ratios (triangles) were determined using conventional mass spectrometer measurements on extracts from the same plants.

The standard analysis conditions for the SIMS curve were generated by isotopic ratios measured in several sections with 50 μm× 50-μm sections or by several 2500 μm² from repeated analyses. The value of the isotopic ratio in each standard was considered as the sum of the natural isotopic ratio (0.366%) plus the corresponding ^{15}N enrichment. The mean isotopic ratios measurements obtained with SIMS are linear with the average ratios defined by the culture conditions. Note the excellent correlation between the conventional mass spectrometry determination of $^{15}N/^{14}N$ in the biological standards and the corresponding determination of $^{15}N^{12}C^-/^{14}N^{12}C^-$ emission ratio by SIMS. This justifies the choice of the two secondary ion masses for ^{15}N and ^{14}N measurements.

3.3.2.2. General Spectra

One should use general spectra in the mass range of 1 to 120 Da (as is for SIMS) and HMR spectra of masses of interest (as is for SIMS).

3.3.2.3. Imaging Protocol

Use the grid-test image (as is for SIMS). Surface cleaning is the same as for SIMS. Sample position is spotted with an optical microscope, and the scanning position of the beam is controlled with the computer. Parallel imaging of several atomic mass images with a high mass separation allow fine locations inside the cellular compartments. Simultaneous images are acquired.

Cross-section analyses can be obtained at each mass. The major ions, N (CN^-), P^-, S^-, and C^-, are visible in all parts of the cells but in different concentrations, appearing as different grey-level intensities (*see* **Fig. 3**). They image different organelles with localizations consistent with those known through biochemical approaches. The lateral resolution is in the range of 60 to 80 nm with Cs^+ primary source, 150 nm with O^- primary source; this allows the possibility to study variations of ion repartitions at the level of intracellular organelles.

Because quantification of different elements is obtained from simultaneous measurements, the use of internal reference is not necessary.

3.3.2.4. ISOTOPICAL RATIO

An example is shown for ^{15}N nitrogen isotope as a tracer *(15)*. Cs^+ primary ion beam sputtering the sample at 8 to 15 keV energy. Probe diameter: 50 nm to 100 nm, carrying 1.5 pA to 2 pA. Sputtered areas: 5×5 to 150×150 µm areas. The images are recorded as a matrix of 256×256 pixels.

Separation of $^{12}C^{15}N^-$ and $^{13}C^{14}N^-$ at mass 27 (*see* **Fig. 4**): sensitivity at high mass resolution (m/Δm = 6000) is 60%, first peak used. The $^{15}N/^{14}N$ isotope ratios are obtained by direct measurement of the secondary ion current. The natural ratio is used as reference to match the spectrometer.

Quantification of local ^{15}N enrichment: by nitrogen isotope ratios measurements. The multielemental acquisition makes automatic reference to the another ion unnecessary. The ratios are obtained by direct measurement of the secondary ion currents. The precision depends on the secondary ion current intensity. To

---

Fig. 3. *(Opposite page)* MIMS images: distributions of carbon **(A)**, nitrogen **(B)**, phosphorus **(C)**, sulfur **(D)**, and transmission electron microscopy image of cryoprocessed leaf of *Arabidopsis thaliana*. Three types of cells are present: parenchyma (Pr), vascular tissue (phloem, Ph), and bundle sheath (Bs). The cells are enclosed by a cellwall (*w*) and contain a large vacuole (*vac*) surrounded by a 2- to 3-µm thick layer of cytoplasm (*cy*) containing organelles: mitochondriae (*m*), chloroplasts (*ch*), and nucleus (*n*). Bar = 3.9 µm. **(A)** At mass 12: $^{12}C^-$ images regions sugar rich, that is, cellulosic cell walls (*w*) or starch rich: starch grains inside chloroplasts (*arrow head, sch*). **(B)** At mass 26, $^{14}N^{12}C^-$ images mainly protein distributions in cellular membranes (*m*) (from proteins inserted in the phospholipidic bilayer) and organelles (*ch*: chloroplast), and also soluble N compounds inside the vacuole (*vac*). **(C)** At mass 31, $^{31}P^-$ (less emissive than the other ions) images (1) the nucleic acids of the nucleus appearing in the parenchyma cell (*arrow n*), (2) the ribosome part (ARN) inside the cytoplasm, and (3) the cytoplasmic membranes constituted by phospholipids. **(D)** At mass 32: $^{32}S^-$ images the

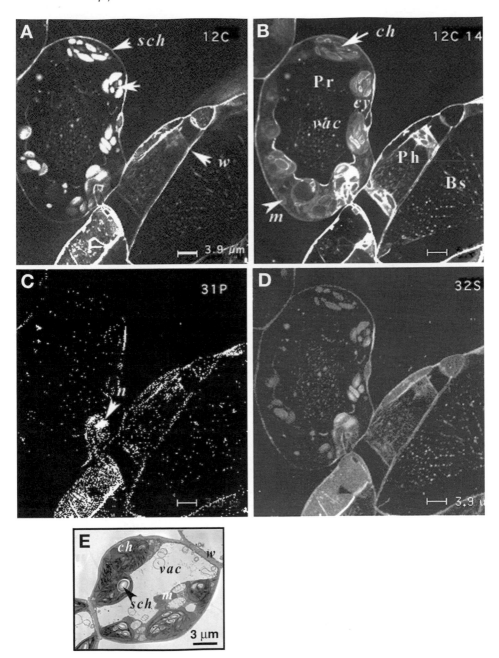

Fig. 3. *(Continued)* membranes containing high-sulfur proteins, particularly in the chloroplasts (which are sites of SO_4^{2-} metabolic assimilation). **(E)** Transmission electron microscopy image of a parenchyma cell showing cell walls *(w)*, chloroplasts *(ch)* containing starch grains *(sch)*, vacuole *(vac)*. Bar = 3 μm.

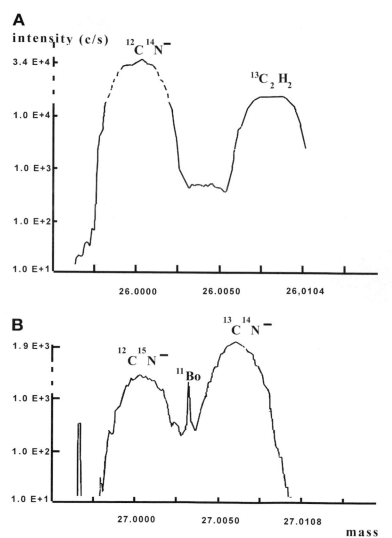

Fig. 4. High-resolution mass spectra at masses 26 (**A**) and 27 (**B**) ; m/Δm = 6 000, 10% valley definition. (**A**) At mass 26, the major peak is caused by $^{12}C^{14}N^-$ (m = 26.00307), and the minor one by $^{12}C_2H_2$ (m = 26.01565). (**B**) At mass 27, three peaks appear: $^{12}C^{15}N^-$ (m = 27.0001), $^{11}Bo^-$ (m = 27.00422), and $^{13}C^{14}N^-$ (m = 27.00643). Boron is a possible contaminant brought from polishing media or Pyrex glass.

obtain the $^{15}N/^{14}N$ ratios, one divides the counts obtained at mass 27 by the counts obtained at mass 26 simultaneously recorded.

The protocol for measurements in plant samples is as follows: in parallel, measurements of elements, constant in all samples: $^{12}C^-$ emission and $^{15}N/^{14}N$

isotopic ratio – (0.366% natural ratio $^{15}N/^{14}N$) measured in the Spurr's resin surrounding the sample).

In the *Arabidopsis* root, ^{15}N labelings are clearly identified by the isotopic ratios and localized. ^{15}N enrichments are directly measured in situ (*see* **Fig. 5**).

Quantification as for SIMS: comparison with a calibration curve of biological standards made in the same conditions as the experiments.

4. Notes

1. The most common sources of contamination come from the environment or the products used. One must study the contaminants on product packaging. In biology, another source of contaminants is the operator himself/herself. Thus, contaminant-free experiments cannot be envisaged. Before preparing samples, the researcher must decide which element he/she wants to study to avoid introducing particular element through manipulations and products.

2. In plants, wounding or even slight mechanical stimulations are known to change the ionic repartition in the compartments *(7)*. For instance, gentle touching may result in shifts in ion distribution, either transient or irreversible (loss of water and solutes). Transport of plants may induce such effects and a 2- to 3-h delay is necessary to restore equilibrium after transport. Excising plant organs changes the electric potential of membranes and induces ion fluxes (K^+, Ca^{2+}, Cl^-, H^+) between cellular or extracellular compartments. These changes occur within 1 min of cutting and can spread up to several millimetres beyond the wounded region. Thus the time elapsed between sample excision and cryofixation must be as short as possible, to prevent ion decompartmentation in the observed tissues.

3. In all aqeous manipulations, soluble compouds are washed; even insoluble compounds may be displaced. Thus, chemical fixatives in water or buffers must be excluded. Classical TEM treatments (aldehydes and heavy metals, e.g., OsO_4) are entirely inappropriate, as in all analyzing techniques. Indeed SIMS images of samples prepared for TEM, all elements appear co-localized, revealing cell decompartmentalization. After OsO_4 fixation, we observed that K^+ seems absent. Note that many cryoprotectants extract water from the samples.

4. Liquid industrial propane was chosen because its freezing point is lower than that of pure propane.

5. The quality of freezing may vary within organs and tissues. The vitreous state of water is impossible to obtain in all parts. Some tissues are naturally cryoprotected by a low water content (seeds) and a high osmolarity (meristema, tissues transporting sap, phloem, storage tissues). In stems, leaves, and roots, the epidermal cells generally are well preserved despite their huge vacuoles, because of their immediate proximity to the freezing agent (liquid or plate). In the center of organs, vessels and storage tissues are well preserved because of their high osmolarity. Intermediate tissues (parenchyma) are less well preserved, with shrinking of cytoplasm and swelling of organelles. Thus, we consider that a sample can be analyzed if the most mobile ions are restricted to areas corresponding to "biological"

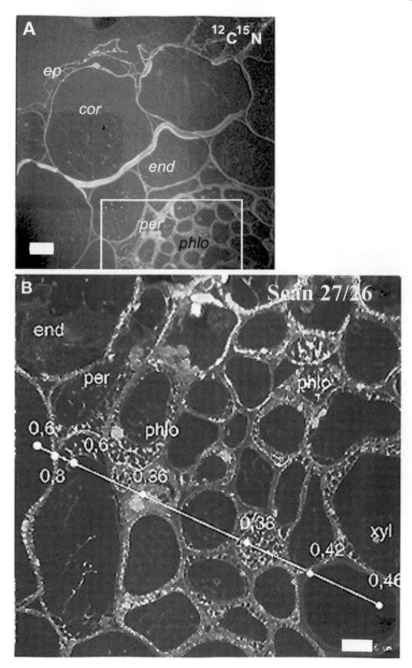

Fig. 5. Nanosims 50 images in *Arabidopsis thaliana* root section: $^{12}C^{15}N^-$ image **(A)** and intracellular isotopic ratios $^{15}N/^{14}N$ along a scan line in the inner part of the root **(B)**. **(A)** Some different tissues: (1) cortical part: epidermis (*ep*), cortical parenchyma (*cor*),

areas (compartments). If mobile ions (Na⁺ for example) appear in rings rather than being limited in organelles the samples are dubious and eliminated.

6. Major precaution: never break the cold chain until full dehydration and embedding; −88°C to −90°C is the superior limit to be accepted.

7. Alternative techniques for cryofixations: high-pressure freezing. After the rapid excision, the ideal protocol should be the most rapid freezing combined with the lowest temperature to obtain vitreous ice *(16)*. The rapid (3 ms^{-1}) sample projection in liquid propane we use, produces crystals in bulk samples, particularly inside the large central vacuole of higher plants. This compartment contains a dilute aqueous solution, with a freezing point higher than that of the contents of the other compartments. With some precautions, ice crystals can be limited to the vacuole and the solutes are redistributed around each crystal ("ghosts"). The corresponding MIMS maps show net-like patterns for elements. With high-pressure freezing techniques (*[17]*, *see* also Chapter 8) vitreous water is obtained in larger volumes of tissues. The hexadecene used to ensure a thermal conductivity has no significant osmotical properties. The results are excellent for unicellular systems and, in bulk samples, for cell structures and protein preservations. For rapid physiological events, the time between excision and immobilization by cold is too long. For the moment, we are unable to insert the sample in the different apparatus commercially available in less than 5 min, a too-long delay in regard to decompartmentalization phenomena. The sample quality is checked for the presence and size of the ghosts and general morphology with TEM before MIMS analyses.

8. Alternative techniques for dehydration include the following : (a) Acetone is efficient as dehydrating agent when it contains less than 1% water. For this reason, a big volume of dehydrated acetone must be used. (b) Other solvents can be chosen (*see* Harvey et al. *[5]*) concerning ether or acetone ion-extractions from tissues at different temperatures. Many solvents are contaminated by aluminum. Molecular sieve, largely used as dehydrating agent, contains this metal. (c) Another solvent,

---

Fig. 5. *(Continued)* endodermis (*end*) and (2) inner part "central cylinder": pericycle (*per*), phloem (*phlo*), xylem (*xyl*). m/Δm = 6 000. Bar = 3.1 µm. The added rectangle indicates the upper part of the area imaged in **B**. **(B)**N image of the inner part of the root (central cylinder). Along the line scan, spots indicate localized isotopic ratios ($^{15}N^{12}C^-/^{14}N^{12}C^-$) in different tissues. Isotopic ratios are directly measured in the cells with the multi-isotope mass spectrometer. m/Δm = 6 000. Bar = 1.6 µm. The plant has been fed with $^{15}NO_3^-$ before sampling. Natural ratio without labeling: $^{15}N/^{14}N = 0.366\%$. The referent natural isotopic ratio is measured in the resin surrounding the sample. The values indicated in different cells are $^{15}N/^{14}N$ in %: 0.6% and 0.8% in endoderm cells; 0.36% in phloem cells; 0.42% and 0.46% in parenchyma cells near xylem. The corresponding ^{15}N enrichments are: 0.2%, 0.4%, 0%, 0.06%, and 0.1% respectively. The highest enrichments are in the tissues involved in lateral transport (pericycle: per) or in storage and in the tissues involved in transports toward the aerial part of the plant (xylem system). These results are in keeping with the experimental protocol (unpublished data).

dimethoxypropane, has been proposed to shorten dehydration time *(11,18)*. However, it causes shrinking. Furthermore, it solidifies at -40°C. Therefore, it is not efficient at lower temperatures. (d) Freeze-drying at ultra low temperatures *(19)*. Freeze-drying is theoretically the best technique to avoid any loss. Excellent results were obtained in plant microanalyses by Bücking *(20)*. I used this technique only once, to localize soluble ions in big, highly vacuolated cells. The samples were freeze-dried for three weeks at −120°C. After drying, the embedment was high-pressure infiltrated with Spurr's resin. Unfortunately, the vacuolar content had been moved to the periphery of the vacuole, near the cytoplasm.

9. Resin embedment was chosen to try to standardize and limit matrix effects *(11,21)*. Mass discrimination strongly depends on the chemical composition of the sample. Resin gives a homogeneous matrix made of carbon, hydrogen, oxygen and, in some cases, nitrogen, giving ions which softens the interferences between other elements. LR White (or gold; TAAB Lab. Equipment LTB, UK), and Spurr's resin were chosen because of their low viscosity, which is suitable for plant tissue impregnation. Spurr's resin is contaminated by chloride. It is also particularly sensitive to hydration. LR White/Gold are used in experiments combining SIMS analyses with immunolocalization or *in situ* hybridation. These metacrylic resins do not contain nitrogen. Polymerization occurs in 8 h at −10°C for LR White and −30°C for LR Gold, in nitrogen atmosphere or vacuum, and under ultraviolet light (mercury). In pigmenteded zones (such as chloroplasts), ultraviolet absorption is high, and localized heating can modify the polymerisation or produce bubbles.

10. Embedding is the protocol chosen for bulk samples to obtain flat sections and to reduce matrix effects. But it is not an exclusive protocol: for example cells can be directly observed after cultured on silicon ships and freeze-dried *(22)*.

11. Samples must be solid, conductive and planar because the quality of analysis depends on surface collisions. Adherence and planarity ensure homogeneity of the spatial charge inside the plasma, which is the origin of ionic images. If the surface is not electrically equipotential, the image is distorted.

12. Alternative technique to sections: because sections for SIMS and MIMS are relatively thick, a histological microtome (LEICA, RM 2561) and glass knives (made by a Knife-Maker, LKB II) could be used. Note that sections 800-to 500 nm-thick used for MIMS analyses are easier to obtain than the 50- to 60-nm sections necessary for TEM or 30 nm for electron energy-loss spectroscopy analyses. This is a basic practical advantage. In all cases, dry sections provide better images than sections glued with water but they require for manual skill. A minute drop of 99% pure polyethylene-glycol (Merck) plus 1% ethanol can be used to glue the sections onto the metal plate. Polyethylene-glycol is a dehydratant and limits the diffusion of soluble elements through resins. However, it never must overflow the section. I think that thicker (800 nm to 500 nm) sections dry cut give better analyses than 300-nm sections cut on water using the TEM technique.

13. SIMS preliminary implantation of O^+ enhances the emission of negative ions obtained by Cs^+ bombardment. A secondary analysis with an oxygen bombardment becomes difficult on Cs^+ implanted sections.

14. "Cleaning" the surface: in all organic compounds, interferences are produced by C_mH_n ion series and different combinations of major species (like carbon, hydrogen, oxygen, and nitrogen). Gas adsorption, particularly hydrogen, contaminates the surface of each sample giving polyatomic ions at each atomic mass. Thus, at each chosen mass, some sputtering of external layers is necessary before the first image appears. How to test the influence of contaminants: the emission of the sample varies in proportion to the primary beam density; in contrast, the emission of contaminants vary inversely to the primary ion density.

15. Example of interferences between monoatomic and polyatomic ions are ^{27}Al and $^{27}C_2H_3$, ^{56}Fe and ^{56}CaO, $^{127}I^-$ and $^{127}PO_6$, and $^{14}N^{13}C$ and $^{15}N^{12}C$ *(23)*.

16. Choice between isotopes from one same element: example of Selenium. Se has 6 isotopes. ^{80}Se represents 49.6% and ^{78}Se 23.52% of total Se. The logical choice should be to localize Se using the emission at mass 80. In fact, at mass 80, with a Cs^+ source, two polyatomic sulfur ions hamper Se location. Sulfur is much more abundant than Se in organisms. The study of the mass spectra and the comparison of images at masses 32, 80, and 78 shows precisely the regions where S and Se are superimposed.

17. The repartition of some major soluble elements (K^+, Mg^{2+}, Na^+, Ca^{2+}) inside cells may reveal the moment of solute decompartmentation during the preparation. (a) In plants, K^+/Ca^{2+} ratios much higher inside cells than outside indicate that no major solute displacement has occurred. (b) In the tissue native state, K^+ is abundant in numerous vacuoles of mesophyll cells. Stress results in a rapid decompartmentation of K^+ out of the vacuole. (c) When the cells are not in good physiological conditions, Ca^{2+} abundance augment inside the cells, first in the vacuole, second, in the cytoplasm. (d) When the cells are dying, the permeability barriers disappear and all the ions appear colocalized. The artifacts appear different for losses or displacements of elements during the excision, surface contaminations, or diffusion of elements after sectioning, in the surrounding resin or cells. In our dry sections, stocked in a dry atmosphere and analyzed within 2 mo after cutting, we did not observe diffusion of solutes around zones of abundance.

18. Nitrogen analyses. For our experiments of nitrate absorption, we fed the roots with $K^{15}NO_3$. Nitrate is highly mobile in water. To check possible experimental artefacts such as losses of ^{15}N solutes during the sample preparation, we used conventional gaseous mass spectrometry. In the total acetone bath used for 3 weeks, only uncountable traces of N were detected.

19. Quantitative SIMS is mainly used with Cs^+ primary bombardment. CN^- used for SIMS analyses in place of N^-. Elemental N is not emitted as a negative ion because of low electron affinity. CN^- ions has a high electron affinity (3.82 eV) and is produced with a high yield from biological samples bombarded by Cs^+ beam. CN^- polyatomic ions accurately reflect nitrogen.

20. The embedment consists in replacing water with acetone, then with resin. In the Spurr's resin, the carbon proportion is nearly the same as in the biological material and the concentration of nitrogen is low (0.059% N w/w). The resin $^{15}N/^{14}N$ natural ratio is easily detected, with SIMS in the resin surrounding the samples

and is used as an internal reference within each section analyzed. The Spurr's resin embedded material is systematically enriched in natural nitrogen and ^{15}N labeling is diluted. The resin volume corresponds to the volume of water content. For this reason, the standard curve is corrected on the basis of 90% mean water content in the tissues. With this correction, the SIMS curve is identical to the curve obtained by gaseous mass spectrometry.

21. SIMS is considered as perfectly quantitative for isotope analysis, but the experiments have to be corrected using standards of adequate composition to eliminate discrimination effects. We prefer to use biological standards rather than mineral, as they can take into account differences in the absorption and transport of isotopes, and local isotopic fractionation, known to exist in some biological reactions.

22. The bombarded surface is eroded at variable speed, depending on the beam intensity. When the element is not abundant or when the yield is weak, some analyses need a very long time. We observed, using scanning electron microscopy on some SIMS preparations, differential erosions even in embedded material (vacuoles, cytoplasm, cell-wall). For this reason, depth profiles could not be studied in these samples.

Acknowledgments

I thank Dr. Croisy, Dr. Fragu, Dr. Guerquin-Kern, and Dr. Hillion for having allowed me to use the SIMS 4F and Nanosims 50 apparatus in the research division of the Gustave-Roussy Institute (Paris), the Curie Institute (Paris) and at CAMECA Inc. (Paris), respectively. I thank Dr. Corbiere for helpful corrections.

References

1. Echlin, P. (1992) *Low Temperature Microscopy and Analysis,* Plenum Press, New York.
2. Michel, M., Gnäggi, H., and Müller, M. (1992) Diamonds are a cryosectioner's best friends. *J. Microsc.* **166,** 43–56.
3. Slodzian, G., Daigne, B., Girard, F., Boust, F., and Hillion, F. (1992) Scanning secondary analytical microscopy with parallel detection. *Biol. Cell* **74,** 43–50.
4. Hillion, F., Daigne, B., Slodzian, G., and Schumacher, M. (1993) A new high performance SIMS instrument: the Cameca "Nanosims 50". *Proceed. SIMS IX,* Yokohama, Japan, p. 254.
5. Harvey, D. M. R., Hall, J. L., and Flowers, T. J. (1976) The use of freeze-subtitution in the preparation of plant tissue for ion localization studies. *J. Microsc. (Oxford)* **107,** 189–198.
6. Spurr, A. R. (1969) A low viscosity epoxy resin embedding medium for electron microscopy. *J. Ultrastruct. Res.* **26,** 31–43.
7. Glass, A. D. M. (1978) Influence of excision and aging upon K^+ influx into barley roots. *Plant Physiol.* **61,** 481–483.
8. Steinbrecht, R. A. and Zierold, K. (eds.) (1987) *Cryotechniques in Biological Electron Microscopy,* Springer-Verlag, Berlin.

9. Benninghoven A., Rudenauer, F. G., and Werner, H. W. (1987) *Secondary ion mass spectrometry*, in *Basic Concepts, Instrumental Aspects, Applications and Trends* (Elving J. P. and Winefordner J. D., eds.), John Wiley and Sons, New York.
10. Blaise, G. (1978) Fundamental aspects of ion microanalysis, in *Material Characterization using Ion Beams*, vol. 28 (Thomas, J. P. and Cachard, A., eds.), Plenum Pub. Co., New York, pp. 143–281.
11. Grignon, N., Halpern, S., Jeusset, J., and Fragu, P. (1996) SIMS microscopy of plant tissues. *J. Microsc. Soc. Am.* **2**, 53–63.
12. Burdo, R. A. and Morrison, G. H. (1971) Table of Atomic and Molecular Lines for Spark Source Mass Spectrometry of Complex Sample-Graphite Mixes . Report #1670. Marerials Science Center, Cornell University, Ithaca, New York.
13. Grignon, N., Halpern, S., Jeusset, J., Briançon, C., and Fragu, P. (1997) Localization of chemical elements and isotopes in the leaf of soybean (*Glycine max*) by SIMS microscopy: critical choice of sample preparation procedure. *J. Microsc. (Oxford)* **186**, 51–66.
14. Grignon, N., Jeusset, J., Lebeau, E., Moro, C., Gojon, A., and Fragu, P. (1999) SIMS localization of nitrogen in the leaf of soybeans: basis of quantitative procedures by localized measurements of isotopic ratio. *J. Trace Microprobe Technol.* **17**, 477–490.
15. Peteranderl, R. and Lechene, C. (2004) Measure of carbon and nitrogen stable isotope ratios in cultured cells. *Amer. Soc. Mass Spectrom.* **15**, 478–485.
16. Gilkey, J. C. and Staehelin, L. A. (1986) Advances in ultrarapid freezing for the preservation of cellular ultrastructures. *J. Electron Microsc. Tech.* **3**, 117–210.
17. Studer, D., Graber, W., Al-Amoudi, A., and Eggli, P. (2001). A new approach for cryofixation by high-pressure freezing, *J. Microsc. (Oxford)* **203**, 285–294.
18. Kaezer, W. (1989) Freeze substitution of plant tissues with a new medium containing dimethoxypropane. *J. Microsc. (Oxford)* **154**, 273–281.
19. Edelmann, L. (2002). Freeze-dried and resin-embedded biological material is well suited for ultrastructure research. *J. Microsc. (Oxford)* **207**, 5–26.
20. Bücking, H. and Heyser, W. (1997) Intracellular compartmentation of phosphorus in roots of *Pinus sylvestris* L. and the implications for transfer processes in ectomycorrhiza, in *Trees. Contribution to Modern Tree Physiology* (Rennenberg, H., Eschrich, W., Ziegler, H. eds.), SPB Academic Publ., The Hague, pp. 377–391.
21. Burns, M. S. (1986) Observations concerning the existence of matrix effects in SIMS analysis of biological specimens, in *Secondary Ion Mass Spectrometry, SIMS 5* (Springer Services Chemical Physics, 44), Springer-Verlag, New York, pp. 426–428.
22. Lechêne, C. and Harris, R. C. (1987) Electron probe analysis of cultured renal cells, in *Contemporary Issues in Nephrology Modern Techniques of Ion Transport* (Brenner, B. M. and Stein, J. H., eds.), Churchill Livingstone, New York, pp. 173–198.
23. Briançon, C. (1991) Effet de la surcharge iodée sur la régulation du métabolisme thyroidien de l'iode. *Thése Doct. Sc.* 165 pp., Paris-Sud-Orsay, France.

Index